68 Springer Series in Solid-State Sciences

Edited by Manuel Cardona

Springer Series in Solid-State Sciences

Editors: M. Cardona P. Fulde K. von Klitzing H.-J. Queisser

40 **Semiconductor Physics** An Introduction
3rd Edition By K. Seeger

41 **The LMTO Method**
Muffin-Tin Orbitals and Electronic Structure
By H.L. Skriver

42 **Crystal Optics with Spatial Dispersion, and Excitons**
By V.M. Agranovich and V.L. Ginzburg

43 **Resonant Nonlinear Interactions of Light with Matter**
By V.S. Butylkin, A.E. Kaplan, Yu. G. Khronopulo, and E.I. Yakubovich

44 **Elastic Media with Microstructure II**
Three-Dimensional Models
By I.A. Kunin

45 **Electronic Properties of Doped Semiconductors**
By B.I. Shklovskii and A.L. Efros

46 **Topological Disorder in Condensed Matter**
Editors: F. Yonezawa and T. Ninomiya

47 **Statics and Dynamics of Nonlinear Systems**
Editors: G. Benedek, H. Bilz, and R. Zeyher

48 **Magnetic Phase Transitions**
Editors: M. Ausloos and R.J. Elliott

49 **Organic Molecular Aggregates,** Electronic Excitation and Interaction Processes
Editors: P. Reineker, H. Haken, and H.C. Wolf

50 **Multiple Diffraction of X-Rays in Crystals**
By Shih-Lin Chang

51 **Phonon Scattering in Condensed Matter**
Editors: W. Eisenmenger, K. Laßmann, and S. Döttinger

52 **Superconductivity in Magnetic and Exotic Materials**
Editors: T. Matsubara and A. Kotani

53 **Two-Dimensional Systems, Heterostructures, and Superlattices**
Editors: G. Bauer, F. Kuchar, and H. Heinrich

54 **Magnetic Excitations and Fluctuations**
Editors: S. Lovesey, U. Balucani, F. Borsa, and V. Tognetti

55 **The Theory of Magnetism II**
Thermodynamics and Statistical Mechanics
By D.C. Mattis

56 **Spin Fluctuations in Itinerant Electron Magnetism** By T. Moriya

57 **Polycrystalline Semiconductors,**
Physical Properties and Applications
Editor: G. Harbeke

58 **The Recursion Method and Its Applications**
Editors: D. Pettifor and D. Weaire

59 **Dynamical Processes and Ordering on Solid Surfaces**
Editors: A. Yoshimori and M. Tsukada

60 **Excitonic Processes in Solids**
By M. Ueta, H. Kanzaki, K. Kobayashi, Y. Toyozawa, and E. Hanamura

61 **Localization, Interaction, and Transport Phenomena**
Editors: B. Kramer, G. Bergmann, and Y. Bruynseraede

62 **Theory of Heavy Fermions and Valence Fluctuations**
Editors: T. Kasuya and T. Saso

63 **Electronic Properties of Polymers and Related Compounds**
Editors: H. Kuzmany, M. Mehring, and S. Roth

64 **Symmetries in Physics, Group Theory Applied to Physical Problems**
By W. Ludwig

65 **Phonons: Theory and Experiments II**
Experiments and Interpretation of Experimental Results By P. Brüesch

66 **Phonons: Theory and Experiments III**
Phenomena Related with Phonons
By P. Brüesch

67 **Two-Dimensional Systems: Physics and New Devices**
Editors: G. Bauer, F. Kuchar, and H. Heinrich

68 **Phonon Scattering in Condensed Matter V**
Editors: A.C. Anderson and J.P. Wolfe

Volumes 1–39 are listed on the back inside cover

Phonon Scattering in Condensed Matter V

Proceedings of the Fifth International Conference
Urbana, Illinois, June 2–6, 1986

Editors: A. C. Anderson and J. P. Wolfe

With 303 Figures

Springer-Verlag Berlin Heidelberg New York
London Paris Tokyo

Professor Ansel C. Anderson
Professor James P. Wolfe

Department of Physics, University of Illinois,
Urbana, IL 61801, USA

Series Editors:

Professor Dr., Dr. h. c. Manuel Cardona
Professor Dr., Dr. h. c. Peter Fulde
Professor Dr. Klaus von Klitzing
Professor Dr. Hans-Joachim Queisser

Max-Planck-Institut für Festkörperforschung, Heisenbergstrasse 1
D-7000 Stuttgart 80, Fed. Rep. of Germany

ISBN-13: 978-3-642-82914-7 e-ISBN-13: 978-3-642-82912-3
DOI: 10.1007/978-3-642-82912-3

Softcover reprint of the hardcover 1st edition 1986

2153/3150-543210

Preface

This volume contains the proceedings of the Fifth International Conference on Phonon Scattering in Condensed Matter held June 2–6, 1986 at the University of Illinois at Urbana-Champaign. The preceding conferences were held at St. Maxime and Paris in 1972, at the University of Nottingham in 1975, at Brown University in 1979, and at the University of Stuttgart in 1983. The Illinois conference dealt with both traditional and newly developing topics in the area of phonon scattering. Papers were presented on phonon scattering in glassy and crystalline dielectrics, semiconductors, metals (both normal and superconducting), and in the areas of phonon imaging, large wave vector phonons, optical techniques and new experimental methods. The 12 invited papers and 100 contributed papers were presented by the 125 scientists from 14 countries.

A citation was presented to Professor Paul Klemens of the University of Connecticut for his pioneering contributions to the physics of phonon scattering in solids.

Paul Gustov Klemens
Born – Vienna (1925)

B.Sc. – Sydney (1946)
D.Phil. – Oxford (1950)
National Standards Lab., Sydney (1950–1959)
Westinghouse Research Labs., Pittsburgh (1964–1969)
Univ. of Connecticut (1967–)

Fellow:
American Physical Society
British Institute of Physics & Physical Society

A long career dedicated to the understanding of thermal transport. Few papers are published on phonon thermal transport that do not reference his work.

The Conference was generously sponsored by the International Union of Pure and Applied Physics, the Argonne Universities Association Trust Fund, the U.S. National Science Foundation, and the Physics Department, Materials Research Laboratory, and College of Engineering at the University of Illinois.

The papers in these proceedings underwent a refereeing process which was mainly undertaken during the conference itself. We are grateful to L.M. Smith for his help in coordinating this effort. We also wish to thank Karen Freeman for her untiring help in coordinating the many details of an international conference. We are very grateful to our colleagues, H. Zabel and A.V. Granato, for their many contributions to the smooth running of the program. The staff of the Physics Department at the University of Illinois provided valuable assistance as did many of the graduate students.

We thank all members of the International Advisory Committee for their helpful suggestions and recommendations.

Urbana,
July 1986

A.C. Anderson
J.P. Wolfe

Contents

Part III **Phonons in Semiconductors**

Part V Quantum Matter and Kapitza Resistance

Part VI Phonon Scattering in Insulators

Part VII Phonon Imaging

Part VIII Large-Wavevector Phonons and Optical Techniques

Part IX **New Methods and Phenomena**

Introduction

True to the form of the previous four phonon-scattering conferences, the Illinois conference was characterized by lively discussions of traditional subjects, punctuated with new ideas and descriptions of new techniques. A newcomer might think that the subjects of thermal conduction in amorphous materials or the perennial Kapitza resistance problem were in their infancy, as judged by the diversity of viewpoints expressed. Indeed, much progress is being made in characterizing phonon scattering in wide varieties of condensed matter, but descriptions of the microscopic scattering processes are remarkably elusive.

Phonon scattering and thermal conductance in disordered and amorphous materials were some of the major topics of interest in this conference. The thermal conductivity of an amorphous solid, and of certain disordered crystals, varies as T^2 at low temperatures, exhibits a temperture-independent plateau between roughly 1 and 10 K, then increases again at higher temperatures. The phonon mean free path l as a function of wavelength λ (measured either by thermal conductivity or ultrasonic attentuation) is remarkably similar for all these materials. At temperatures below $\approx$1 K the phonons are scattered by "two-level states" (TLS) which are believed to arise from quantum tunneling of atoms or molecules. But there is no microscopic description of the TLS, nor an explanation of the universal dependence of l on λ. In the plateau, l appears to be dominated by Rayleigh scattering. But the soure of this scattering has not been identified. At temperatures above $\approx$10 K, even a description of the thermal excitation responsible for thermal transport has not been identified.

The Conference presentations addressed all of these questions. There have been studies of low-density sol-gel glasses in an attempt to identify localized phonons having fractal character, and a study of the influence of free volume on the number density of TLS. Evidence was presented indicating that TLS are produced in quartz crystals, not only by neutron irradiation, but also by electron irradiation where the damage produced is much less extensive. An intriguing observation was the apparent change in the number density of TLS for a glassy metal between the normal and superconducting states. Clearly, glassy solids will remain the object of active research.

Phonon scattering in crystalline systems constituted another major area of interest. Traditional thermal conductivity and acoustic measurements as well as various heat-pulse techniques are used to characterize the phonon scattering from defects and impurities. Scattering from point defects, dislocations, grain boundaries and magnons (spin waves) were touched upon. Phonon scattering from impurities in semiconductors is being studied by phonon spectroscopy techniques. These include the use of superconducting tunnel junctions and optical luminescence techniques for isolating phonons of a given frequency. A promising new technique both for the characterization of impurity levels in semiconductors and the development of new frequency-sensitive detectors is phonon-induced electrical conductivity, termed phonoconductivity in analogy with the optical photoconductivity effect. Phonons of variable frequency incident on a donor or acceptor can ionize the impurities if the phonon energy is sufficient. The resulting free electrons give rise to a detectable current. Optical excitation of the crystal is found to be helpful in changing the ionization state of the donors or acceptors and thus the frequency threshold for ionization by the incident phonon. In the future one can hope that by appropriate ion implantation of the semiconductor surface and optical excitation, the experimentalist can tailor the frequency selectivity of a phonon detector.

Also in the semiconductor area, the interaction of phonons with superlattices is gaining interest. In the now-common GaAs/GaAlAs system, minigaps have been observed by Raman scattering and phonon spectroscopy. A remarkable result reported at this conference is that high-frequency acoustic phonons (up to 900 GHz in Si) are transmitted through amorphous superlattices, in particular, seven 100 Å layers of SiO_2/Si. Well-defined phonon stopbands were observed using tunnel-junction phonon-spectroscopy techniques.

Even the ballistic propagation of phonons (i.e., propagation without scattering) in crystalline materials turns out to be a rich subject. At the Brown Conference in 1979, several experimental methods for observing the highly anisotropic propagation of heat pulses were unveiled. The technique is commonly known as phonon imaging and the phenomenon which gives rise to the large heat-flux anisotropies in any known crystalline system is called phonon focusing. Ballistic propagation is characterized by a spatial pattern of caustics, which depends on the elastic constants of the particular crystal. Systematic studies of cubic crystals and the effects of piezoelectricity were reported at this meeting. It is also found that the phonon images can be strongly modulated by the existence of pseudosurface waves at the generator or detector surfaces. Phonon images are now observable with frequency-selective detectors, and this opens up a new area of dispersive phonon imaging which bears on the theories of lattice dynamics. Images of phonon focusing at 30% of the zone boundary wave vector in InSb were reported. A new twist of the phonon imaging method is to examine defects and scattering centers within the bulk of a crystal. This is a potentially useful form of imaging in which sub-surface spatial structures can be characterized. Scattering from oxide precipitates in silicon was detected, and buried doping structures in Si were imaged with a few micrometer resolution.

The propagation and scattering of large wave vector phonons (i.e., those with wavelengths only a few times the lattice spacing) is an area of continuing interest and progress. This is a complicated regime marked by frequency dispersion, frequency down-conversion and elastic scattering. Inelastic neutron scattering still plays an important role in characterizing the phonon dispersion curves; however, the lifetime and scattering of large-k phonons is being elucidated by a variety of new techniques. Nonequilibrium phonons of high frequency are produced by optical excitation in the visible, near infrared or far infrared. Luminescence techniques have been employed to study the frequency distribution of these optically generated phonons as they down-convert along the acoustic branches. Theoretical techniques are rapidly developing which include both dispersion and anisotropy of the phonons. Efforts are being made to determine the appropriate phonon density of states and predict the thermalization times within the acoustic branches. The theoretical prediction of the absolute decay rate of a high-frequency phonon also involves matrix elements which are not well known at present. At the Stuttgart meeting in 1979, a good deal of attention was paid to the idea of quasi diffusion, which seemed to explain some of the anomalous transport observed in heat-pulse experiments. In a regime where both elastic (e.g., isotope) scattering and inelastic (e.g., down-conversion) processes are operative, neither pure ballistic nor pure diffusive transport is appropriate. Monte Carlo simulations have been reported which examine this intermediate regime; however, no solid agreement yet exists between experiment and theory.

The evolution of thermal energy following intense optical excitation remains one of the important problems in this field. The problem is being attacked from both the low-frequency and high-frequency sides. On the picosecond time scale coherent Raman scattering techniques are being developed to measure the lifetime and dephasing time of optical phonons. This is a promising new technique which makes use of the rapidly developing field of picosecond and subpicosecond optical spectroscopy.

Phonon scattering at surfaces and interfaces is still an active and perplexing area of research. Several authors have studied the effects of surface preparation or gas overlayers on the specularity or efficiency of phonon reflection from crystal surfaces. Highly ideal surfaces were prepared, allowing a phonon to reflect $\approx$100 times before experiencing a diffusive or thermalizing event. Under such conditions the thermalization, and hence the apparent thermal conductance of a crystal, can be severely influenced, for example, by the placement of thermometers.

There has been additional work on the transmission of phonons across an interface, the Kapitza conductance. It has been well established that classical acoustic theory can account for the Kapitza conductance between two solids at low phonon frequencies (i.e., low temperatures). Arguments were presented at the Conference suggesting that diffusive scattering at the interface may decrease the conductance at $T \gtrsim 10\,K$. It was also suggested that TLS at the interface between a solid and liquid helium can account for the large increase in conductance observed at $T \gtrsim 1\,K$ for any interface which is less than perfect. Here, an example of a near-perfect surface is one that has been laser annealed within the cryostat.

The Illinois conference contained some discussions on the propagation of phonons in quantum liquids, mainly ^4He. A predicted one-dimensional mode of second-sound propagation in superfluid ^4He has been observed, having a velocity close to that of first sound. This is to be contrasted with the "ordinary" second-sound velocity, which is $1/\sqrt{3}$ that of the first, or acoustic, sound velocity.

In addition to those already mentioned, a number of new phonon techniques were described. Picosecond laser pulses were used to generate high-frequency phonons and measure the acoustic attenuation in thin films. Superconducting tunnel junctions were employed in conjunction with optical excitation to measure the vibrational spectra associated with monolayers of molecules adsorbed onto a crystal surface. Large-area tunnel-junction detectors were combined with laser scanning to produce a frequency-selective detector for phonon imaging, and this detector was used to measure the anisotropic emission of phonons from a two-dimensional electron gas in silicon. At the other extreme in the frequency spectrum, it was demonstrated that small movable hypersound beams could be used to examine variations in the acoustic phonon scattering associated with surface and sub-surface imperfections. A novel phonon-spectroscopy method in superfluid ^4He is being developed to measure the gap energy associated with the fractionally quantized Hall effect in semiconductor superlattices. Finally, a potential application of phonon techniques to high-energy physics was discussed: it is suggested that the focusing properties of ballistic phonons could be used as a spatially resolving detector for solar neutrinos.

Part I

Glassy Materials and Two-Level Systems

Phonon Scattering in Disordered Systems

A.M. de Goër

Département de Recherche Fondamentale, Service des Basses Températures, Laboratoire de Cryophysique, Centre d'Etudes Nucléaires de Grenoble, 85 X F-38041 Grenoble Cedex, France

1. Introduction

It is well known that glasses, which are completely disordered homogeneous solids, display acoustic and thermal properties at low temperature which deviate markedly from those of crystals [1]. This is especially (but not exclusively) true below 1 K, and these differences are explained by the presence of low-lying excitations, described as "Two-Level Systems" (TLS) with a constant energy density of states. The TLS are supposed to be caused by tunnelling motion of a group of atoms in a double-well asymmetric potential [2]. The acoustic experiments have shown that long wavelength phonons exist in a large range of frequency, and are strongly scattered by TLS via resonant and relaxation processes [1]. It has also been established that phonons are responsible for heat transport [3], the resonant scattering by TLS leading to the T^2 dependence of the thermal conductivity K(T) below 1 K. The tunneling model gives a satisfactory phenomenological description of the acoustic and thermal anomalies below 1 K, but the microscopic nature of the TLS is still unknown. On the other hand, there is no definitively accepted model to describe the anomalies above 1 K, especially the plateau of K(T) around 10 K. In view of these two unresolved problems, a great deal of work has been done during the last years to search for "glassy" properties either in "model" systems where the tunneling units are known.or in partly disordered systems. We have collected in Table 1 a list of specific bulk insulating systems in which "glassy" anomalies have been observed by phonon experiments. Part of these has been already discussed in several reviews [1][4][5][6]. In this paper, we present and discuss some recent phonon scattering studies on irradiated crystals, KBr/KCN and spin-glasses.

Table 1

Type of disorder	Systems	References
Orientational	Cyclohexanol	[7]
	KBr/KCN	[8][9][10][4][5]
Non stoechiometry	βAl_2O_3	[11][5]
	Y_2O_3 stabilized ZrO_2 and Bi_2O_3	[12][4]
Irradiated crystals	Quartz	[13][14][15][16]
	Ni^{3+} doped Al_2O_3	[17]
Two-phases	Zr-Nb	[18][4]
	Ferroelectrics	[19]
Magnetic	$Eu_xSr_{1-x}S$	[20][21]
	Mn alumino silicate	[22][23]
	Mn fluorophosphate	[24]

2. Irradiated Crystals

2.1 Quartz

It is known for many years that the thermal conductivity of heavily neutron irradiated quartz displays a plateau similar to that of glasses [13]. A great amount of work has been done following the detection of TLS in lightly neutron irradiated quartz from acoustic attenuation experiments [14]. We present in Figs. 1 and 2 a selection of thermal conductivity data we have obtained on natural quartz crystals irradiated by γ rays, electrons and neutrons respectively [15]. These results are in qualitative agreement with those obtained independently by other groups [16]. The most important qualitative results are the following :

(i) below 1 K, phonon scattering appears after electron irradiation as well as neutron irradiation ; the two curves shown in fig.1 correspond to doses such that the total number of defects is estimated to be similar. Therefore electrons are as efficient (or perhaps more) as neutrons to create TLS. This first evidence of the presence of TLS in electron irradiated quartz has been obtained in 1980, and is confirmed by very recent ultrasonic experiments [25]. The nature of the disorder created by electrons and neutrons is very different : neutron irradiation introduces small "amorphous" (or very defective)

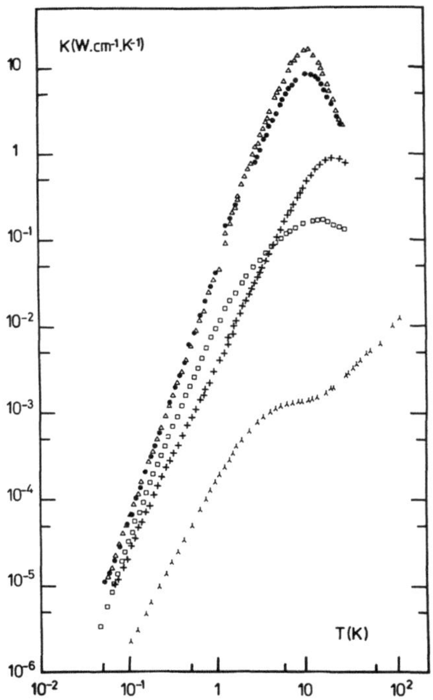

Fig.1 - Thermal conductivity of quartz crystals : virgin (Δ) ; irradiated with γ-rays (•), electrons (+), neutrons (□). a-SiO_2 (Suprasil W) : (λ).

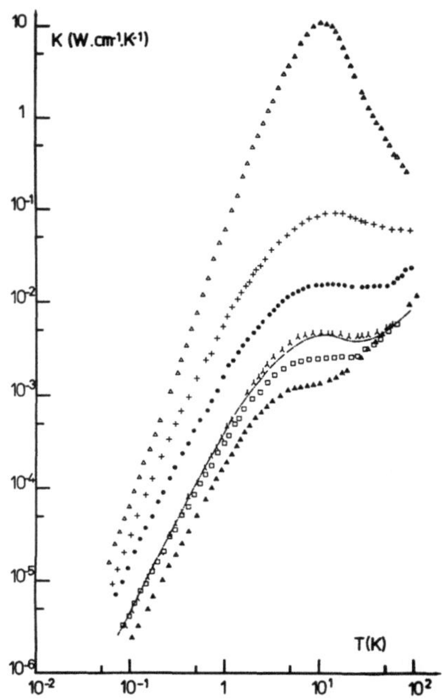

Fig.2 - Thermal conductivity of neutron-irradiated quartz with doses (n/cm^2) : 3 10^{18} (+), 1.8 10^{19} (•), 5.5 10^{19} (□), 2.2 10^{20} (λ). Virgin quartz (Δ) and Suprasil W (▲) are also shown. Curve is neutron-irradiated a-SiO_2 from [30].

regions even at low doses, as observed by SAXS [26] but such regions are absent in electron irradiated quartz [27], where it is supposed there are only point defects. Therefore these results demonstrate directly that the presence of TLS is not restricted to "amorphous" regions in irradiated crystals.
(ii) In the plateau region, the results of K(T) (figs.1 and 2) display two features. Firstly the plateau is not induced by electron irradiation but only neutron irradiation. Secondly, for neutron doses > 10^{20} n/cm^2, the value of K at the plateau increases with dose, whereas the low temperature scattering due to TLS is saturated (fig.2). Therefore the occurrence of the plateau is not systematically correlated to the presence of TLS. This leads to the elimination of some interpretations of the plateau based on phonon scattering by TLS, either by relaxation [28] on Raman [29]process.
(iii) Finally we note that for the highest neutron dose, the thermal conductivity of the quartz is nearly identical to that of heavily neutron irradiated amorphous silica in the whole temperature range. This result is in agreement with the fact that both crystalline quartz and a-SiO_2 are then transformed into the same highly disordered phase [31].

Quantitatively, we have fitted the K(T) curves using the simple Debye model. The low temperature parts are well described in all cases by a simple phonon relaxation rate including boundary scattering and resonant interaction with TLS [15] :

$$\tau^{-1} = V/L + (\pi\,\bar{P}\,\gamma^2/\rho V^2)\,\omega \qquad (1)$$

For electron irradiated samples, good fits have been obtained in the whole temperature range by including additional Rayleigh scattering ($\tau_D^{-1} = A\omega^4$) and phonon-phonon interactions determined for the virgin sample ; the value of A compares favorably with a crude evaluation from the estimated number of Frenkel defects [15]. In the case of neutron irradiation, different models have been tested for the plateau region and are discussed elsewhere [32]. Fits of overall similar quality have been achieved up to ≃ 30 K using three different models : (i) Rayleigh scattering as above, (ii) resonant scattering ($\tau_D^{-1} = D\omega^4/(\omega^2 - \omega_0{}^2)^2$) (iii) a cut-off frequency $\omega^* = kT^*/\hbar$ is introduced, separating low-energy propagating phonons ($\omega < \omega^*$) from localized excitations which do not contribute to the heat flow. This would be the case in the phonon/fracton model of Alexander and Orbach, who suggested it could be applicable to glasses [33]. As T^* depends weakly on dose, the corresponding wavelength of dominant phonons varies only from 40 to 60 Å. This result is against the fracton model, as there is no characteristic length related to the state of disorder. Therefore we think that resonant scattering of long wavelength phonons by few localized low-energy modes is the most probable explanation of the plateau. We note that a plateau is observed in crystalline polydiacethylene - which nevertheless displays a normal T^3 dependence of K(T) below 1 K - and explained by such a process of resonant scattering of acoustic modes on low-lying optic modes [34].

2.2 Al_2O_3 and Ni doped Al_2O_3

The effect of neutron irradiation on K(T) of a very pure Al_2O_3 single crystal is completely different from that observed in quartz, as illustrated in fig.3. The induced phonon scattering in Al_2O_3 is essentially of the Rayleigh type, and very much smaller than in quartz for the same doses. There is no strong low-temperature scattering due to TLS nor plateau. These results are not unexpected as Al_2O_3 cannot be amorphized in the bulk. These measurements were made as a reference for the study of Ni^{3+} doped Al_2O_3 crystal as a "model" system for TLS [17].

The energy level scheme of the strongly coupled Ni^{3+} ion in trigonal site is well known : the ground state ^2E is submitted to dynamic Jahn-Teller effect,

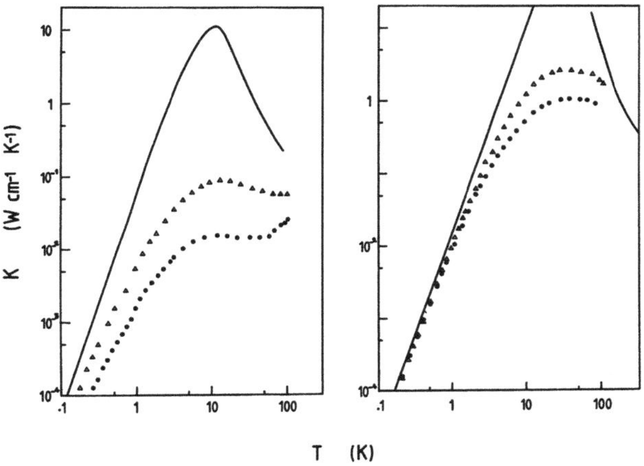

Fig.3 - Thermal conductivity of neutron-irradiated quartz (a) and Al_2O_3 (b) crystals :
(Δ) 3 10^{18} n/cm^2
(•) 1.8 10^{19} n/cm^2.
Solid lines correspond to virgin samples.

but the two lowest vibronic levels are well separated from the first excited state. Therefore we are dealing with an actual two-level system at low temperature, the splitting in zero stress being 0.47 cm^{-1}. Resonant phonon scattering between these levels asymmetrically broadened by residual strains has been observed and is apparent in fig.4. The interesting effect of neutron irradiation is the increase of K(T) below 1 K : the dip near 0.5 K disappears and is replaced by a monotonic variation as $T^{2.5}$ from 70 mK to 5 K. This low temperature behaviour is similar to that of irradiated quartz ; it is easily explained if the original half-lorentzian distribution of the energy splittings of the Ni^{3+} ions is replaced by a constant in a large energy range (fig.5). This extended distribution is due to the strain fields induced at the Ni^{3+} sites by the large number of created defects. The fit obtained using equ. (1) is quite satisfactory (fig.4). The value of $P\gamma^2$ is compared in Table 2 with those for a quartz irradiated at the same neutron dose (3.10^{18} n/cm^2) and for unirradiated Suprasil W. From the known coupling constants [17] we have calculated the density of states $\bar{P}$ as well as the total number n of TLS for E_{max} = 1 K. The interesting point is that $\bar{P}$ (and then n) is larger in the Ni

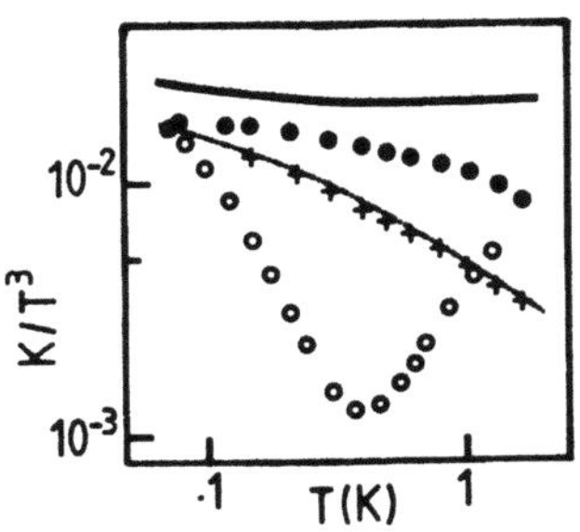

Fig.4 - Thermal conductivity divided by T^3 as a function of temperature. Pure Al_2O_3 : before irradiation (solid line), after neutron irradiation (•). Ni-doped Al_2O_3 : before irradiation (o), after neutron irradiation (+). The dashed line is calculated (see text).

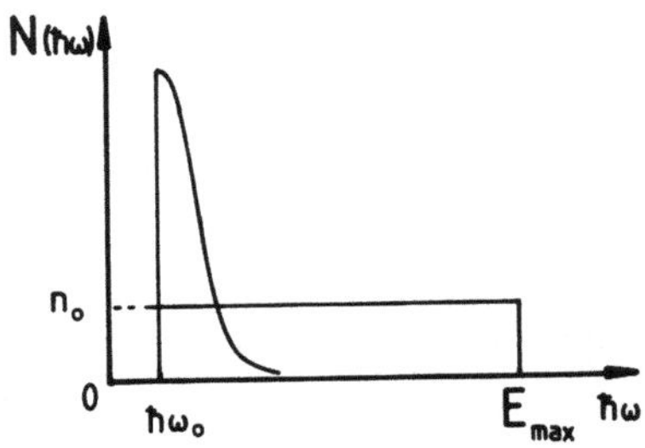

Fig. 5 - Energy density of states of Ni^{3+} ions in Al_2O_3 before and after neutron irradiation.

Table 2. TLS parameters

	$\bar{P}\gamma^2$ (erg cm^{-3})	γ (eV)	$\bar{P}$ (erg^{-1} cm^{-3})	$n(E_{max}$ = 1 K) (cm^{-3})
Ni-doped Al_2O_3	1.1 10^7	0.37	3.10^{31}	4.1 10^{15}
Quartz	4.3 10^6	1	1.7 10^{30}	2.3 10^{14}
Suprasil W	1.5 10^8	1	6.10^{31}	8.3 10^{15}

doped Al_2O_3 than in quartz, though the total concentration of Ni^{3+} ions is only 4 ppm. Therefore specific electronic defects of this type, that is "Jahn-Teller" defects with degenerate orbital ground state, could be a source of fast-relaxing TLS in irradiated quartz and glasses. The above results also suggest that the puzzling effects observed on K(T) of neutron irradiated MgO crystals [35] could be due to residual impurities, the strongly coupled Jahn-Teller valency states being induced by ionizing radiation.

3. $(KBr)_{1-x}(KCN)_x$ mixed crystals

This system is considered as giving an example of well known TLS as :
(i) for small x, reorientation of the isolated rod-like CN^- molecular-ions occurs via tunnelling at low temperatures, and the coupling to phonons is quite strong [36]
(ii) at intermediate concentrations (few % < x < 0.6) there is evidence of an "orientational glass" phase below some freezing temperature T_F [37], and thermal and acoustic "glassy" properties have been indeed observed [4][8][9][10]. Examples of results of thermal conductivity and acoustic attenuation are illustrated in Figs. 6 and 7 respectively. Moreover, time-dependent specific heat measurements have convincingly demonstrated the existence of a broad distribution of relaxation times in agreement with the tunnelling model of TLS [8][38]. A very small fraction of the CN^- still free to tunnel are thought to be responsible for the glassy properties, the CN^--CN^- elastic interaction leading to the broadening of the tunnelling states. Recently, a specific microscopic model of 180° reorientation has been proposed [38]. Actually the K(T) curves shown in fig. 6 are higher (below 1 K) than that of a crystal with x = 3.10^{-3} [8] so that the behaviour of K(T) is in a sense similar to that of the Ni^{3+}/Al_2O_3 system discussed above, though the origin of the broadening is completely different. On the other hand, many features of the orientational freezing are similar to that of spin-glasses (i.e. the frequency dependence of T_F as measured form the a.c. dielectric constant [37]) and the possibility that collective CN^- excitations would be at the origin of the phonon scattering below 1 K has been already suggested [8]. Finally we note that the "plateau" region presents in fact a true minimum of K which is shifted to higher temperatures from x = 0.25 to x = 0.5 (cf. fig. 6 ; the case x = 0.7 is specific, as it is slightly above the upper limit for the glassy phase). The ratio of these temperatures is about 1.7, very comparable to that of the temperatures of the minimum frequency of the TA phonon, for the same concentrations, as measured from Brillouin [39] and neutron [8][40] experiments. Therefore it is tempting to relate the pseudo "plateau" to the strong interaction of acoustic phonons with the CN^- rotational modes, which leads to coupled modes observed by neutron scattering in less doped samples [41].

4. Spin-glasses

Phonon scattering studies on insulating crystalline spin-glasses are limited to the system $Eu_xSr_{1-x}S$, and to thermal conductivity measurements [20][21]. Results extending well below the freezing temperature Tg are available only

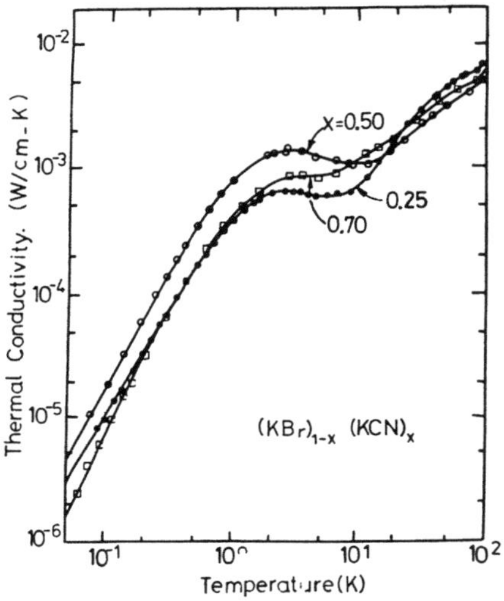

Fig. 6 - Thermal conductivity of $(KBr)_{1-x}(KCN)_x$ for x = 0.25, 0.5, and 0.7 (from [4])

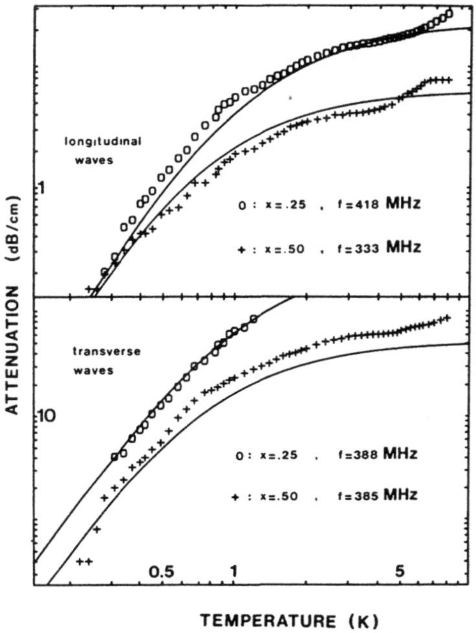

Fig. 7 - Temperature dependence of ultrasonic attenuation for LA and TA waves along the (001) axis in $(KBr)_{1-x}(KCN)_x$. Solid lines are calculated according to the tunneling model (from [9])

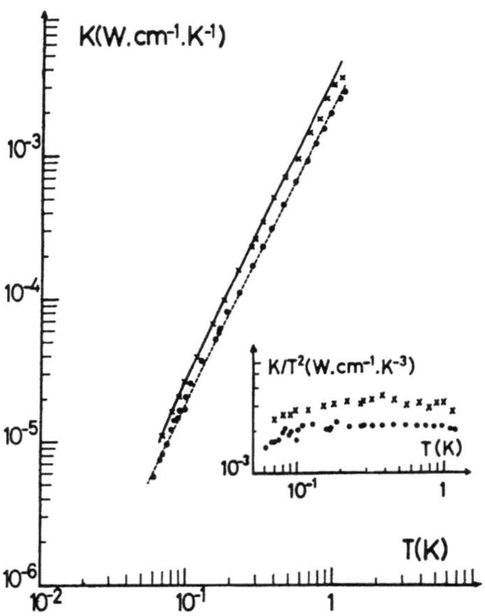

Fig. 8 - Thermal conductivity of $Eu_{0.44}Sr_{0.56}S$ crystal : (•) H = 0 ; (x) H = 7 T. The dashed and solid lines are calculated.

for x = 0.44 (Tg = 1.8 K), and are illustrated in fig. 8. K(T) varies nearly as T^2 and increases with magnetic field. We have interpreted these results as an evidence of the existence of "magnetic TLS" which resonantly scatter phonons in zero field. Quantitative fits of the curves (Fig. 8) give a value of $(\bar{P}\gamma^2)_{magn}$ = 5.8 10^6 erg cm^{-3}, which is more than one order of magnitude smaller than that found in typical structural glasses. The low-temperature specific heat has been measured on the same sample [42] and from the linear part in zero magnetic field, a large value of $\bar{P}$ = 1.2 10^{37} erg^{-1} cm^{-3} was obtained so that the coupling constant γ = 4.3 10^{-4} eV is quite small. It is now thought that this approach is not appropriate for this concentrated magnetic system : the physical phonon scattering process is probably reminiscent of the interaction phonon-magnon in ordered magnetic compounds rather than due to localized MTLS [42]. This point of view is supported by recent numerical calculations of the collective excitations spectrum of the spin system, which give a satisfactory description of all available specific heat data [43].

The other phonon studies on insulating spin-glasses concern systems which are also structural glasses, so that the analysis of the results is more difficult. An example is the Mn aluminosilicate with 13 at % Mn [22][23] (Tg ≃ 3 K [44]). The thermal conductivity, shown in fig. 9 is identical to that of a-SiO_2 below 1 K ; in the plateau region, it is very comparable to that of an alkaline-earth germanate glass (fig. 9) or a Mg aluminosilicate [45]. It is clear that magnetic phonon scattering, if present, is too small compared to structural scattering to be detected. A detailed study concerns the ultrasonic properties of the magnetic glass $(MnF_2)_{0.65}(BaF_2)_{0.2}(NaPO_3)_{0.15}$ (Tg = 3.4 K) [24]. Below Tg, the attenuation (and sound velocity change) is larger than in the non magnetic Zn based glass. Nevertheless the quantitative analysis developed by the authors leads to a spectral density $\bar{P}$ much smaller and coupling constants γ_L and γ_T much larger in the magnetic case. Clearly much more work is needed to understand the role of magnetic ions in structural glasses.

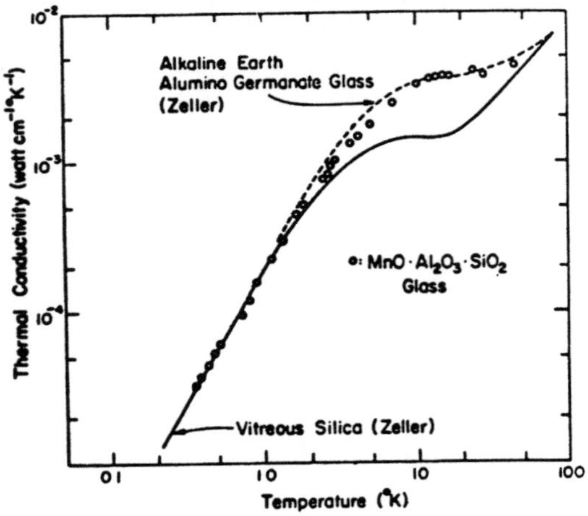

Fig. 9 - Thermal conductivity of MnO. Al_2O_3.SiO_2 glass (from [22])

Conclusions

The available experimental evidence of "glassy" properties (from the point of view of phonon scattering) in disordered systems leads to the following conclusions :

(i) Phonon scattering by TLS below 1 K is not systematically connected to

the occurrence of a plateau in K(T). Therefore interpretations of the plateau involving the TLS are unlikely
(ii) TLS are present in slightly disordered crystalline environment and not restricted to amorphous regions ; are there two types of TLS ?
(iii) Studies of "model" systems show that candidates for the microscopic nature of the TLS are localized defects with degenerate ground states (arising either from atomic position or electronic orbitals) and strong coupling to the lattice, that is, "Jahn-Teller" defects.
(iv) The most probable origin of the plateau is a strong interaction between acoustic phonons and low-energy modes ; this explanation has been in fact proposed a long time ago [46] and recent neutron work on a-SiO_2 shows that such modes could be somewhat localized intrinsic excitations [47]. Recent work on weak phonon localization could also be a powerful theoretical approach [48]. We think that the extreme situation of complete localization of the excitations above some cross-over frequency, as in the phonon-fracton model [33], is not realized in ordinary glasses, but perhaps in highly porous materials such as aerogels [49].

References

1. For recent reviews, see S. Hunklinger and A.K. Raychaudhury : Progr. Low Temp. Physics, ed. by D.F. Brewer, vol. 9 (Elsevier - North Holland, Amsterdan 1985) p. 265 ; R.O. Pohl, Phase Transitions 5, 239 (1985)
2. P.W. Anderson, B.I. Halperin and C.M. Varma : Philos. Mag. 25, 1 (1972) W.A. Phillips : J. Low. Temp. Phys. 7, 351 (1972)
3. R.O. Pohl, W.F. Love and R.B. Stephens : In Amorphous and Liquid Semi-Conductors, ed. by J. Stuke and W. Brenig (Taylor and Francis 1974) p.1121 M.P. Zaitlin and A.C. Anderson, Phys. Rev. B12, 4475 (1975)
4. R.O. Pohl, J.J. De Yoreo, M. Meissner and W. Knaak : In Physics and Chemistry of Disordered Systems ed. by D. Adler and M. Fritzsche (Plenum Press 1985)
5. A.C. Anderson : Phase Transition 5, 301 (1985)
6. C. Laermans : In Structure and Bonding in non Crystalline Solids ed. by G. Walrafen and A. Revesz (Plenum Press 1986), in press
7. R. Calemczuk, L. Jacqmin, G.P. Singh, K. Dransfeld, R. Vacher : Phys. Rev. B29, 3767 (1984)
8. J.J. De Yoreo, M. Meissner, R.O. Pohl, J.M. Rowe, J.J. Rush, S. Susman : Phys. Rev. Letters 51, 1050 (1983)
9. D. Moy, J.N. Dobbs, A.C. Anderson : Phys.Rev. B29, 2160 (1984)
10. J.F. Berret, P. Doussineau, A. Levelut, M. Meissner, X. Schön : Phys. Rev. Letters 55, 2013 (1985)
11. P.J. Anthony, A.C. Anderson : Phys. Rev. B14, 5198 (1976)
12. F.J. Walker, A.C. Anderson : Phys. Rev. B29, 5881 (1984) and references therein
13. R. Berman, Proc. Roy. Soc. London Ser. A208, 90 (1951)
14. C. Laermans, Phys. Rev. Letters 42, 250 (1979), see also [6]
15. C. Laermans, A.M. de Goër, M. Locatelli : Physics Letters 80A, 331(1980) ; A.M. de Goër, M. Locatelli, C. Laermans : J. de Physique, Colloque C6, C6-78 (1981)
16. J.W. Gardner, A.C. Anderson : Phys. Rev. B23, 474 (1981) ; M. Hofacker, H.v. Löhneysen : Z. Physik B42, 291 (1981)
17. A.M. de Goër, B. Salce : Europhysics Letters 1, 141 (1986)
18. L.F. Lou : Sol. State Comm. 19, 335 (1976)
19. J.J. de Yoreo, R.O. Pohl, G. Burns : Phys. Rev. B32, 5780 (1985)
20. C. Arzoumanian, A.M. de Goër, B. Salce, F. Holtzberg : J. Physique-Lettres 44, L-39 (1983) ; C. Arzoumanian, B. Salce, A.M. de Goër : In Phonon Scattering in Condensed Matter, ed. by W. Eisenmenger, K. Lassmann and S. Döttinger (Springer-Verlag 1984), p.460
21. G.V. Lecomte, H.v. Löhneysen, W. Zinn : J. Magn. Magn. Mat. 38, 235 (1983)

22. A.K. Raychaudhuri, R.O. Pohl : In Amorphous Magnetism, ed. by R.A. Levy and R. Hazegawa (Plenum Press 1977) p. 571
23. M.J. Lin, R.L. Thomas : In Phonon Scattering in Condensed Matter, ed. by H.J. Maris (Plenum Press 1980) p. 53
24. P. Doussineau, A. Levelut, M. Matecki, W. Schön, W.D. Wallace : J. Physique 46, 979 (1985)
25. C. Laermans, A. Vanelstraete : This Conference (1986)
26. D. Grasse, O. Kocar, J. Peisl, S.C. Moss : Rad. Effects 66, 61 (1982)
27. D. Grasse, D. Müller, H. Peisl, C. Laermans : J. Physique Colloque C9 C9-119 (1982)
28. M.P. Zaitlin, A.C. Anderson : Phys. Stat. Sol (b) 71, 323 (1975)
29. A.J. Leadbetter, A.P. Jeapes, C.G. Waterfield, R. Maynard : J. Physique 38, 95 (1977)
30. A.K. Raychaudhuri, R.O. Pohl : Sol. State Comm. 44, 711 (1982)
31. J.B. Bates, R.W. Hendricks, L.B. Shaffer : J. Chem. Phys. 61, 4163 (1974)
32. A.M. de Goër, N. Devismes, B. Salce, This Conference (1986) (and to be published)
33. S. Alexander, C. Laermans, R. Orbach, H.M. Rosenberg : Phys. Rev. B28, 4615 (1983)
34. M.N. Wybourne, B.J. Kiff, D.N. Batchelder, D. Greig, M. Sahota : J. Phys. C : Sol. State Phys. 18, 309 (1985)
35. D.S. Kupperman, G. Kurz, H. Weinstock : J. Low Temp. Phys. 10, 193 (1973); J.W. Gardner, A.C. Anderson : Phys. Rev. B23, 1988 (1981)
36. W.D. Seward, V. Narayanamurti : Phys. Rev. 148, 463 (1966)
37. A. Loidl, R. Feile, K. Knorr : Phys. Rev. Letters 48, 1263 (1982) and references therein
38. M. Meissner, W. Knaak, J.P. Sethna, K.S. Chow, J.J. de Yoreo, R.O. Pohl : Phys. Rev. B32, 6091 (1985)
39. S.K. Satija, C.H. Wang : Sol. State Somm. 28, 617 (1978)
40. J.M. Rowe, J.J. Rush, D.G. Hinks, S. Susman : Phys. Rev. Letters 43, 1158 (1979)
41. A. Loidl, R. Feile, K. Knorr, B. Renker, K. Daubert, D. Durand, J.B. Suck : Z. Phys. B38, 253 (1980)
42. C. Marcenat, A. Benoit, A. Briggs, C. Arzoumanian, A.M. de Goër, F. Holtzberg : J. Physique Lettres 46, L569 (1985)
43. J. Wosnitza, H.v. Löhneysen, W. Zinn, U. Krey : Phys. Rev. B 33, 3436 (1986) and references therein
44. R.W. Kline, A.M. de Graaf, L.E. Wenger, P.H. Keesom : In Magnetism and Magnetic Materials, ed. by J.J. Becker, G.H. Lander, J.J. Rhyne (A.I.P. New York, 1976) p.169
45. E. Bonjour, R. Calemczuk, A.M. de Goër, B. Salce : This conference (1986)
46. B. Dreyfus, N.C. Fernandez, R. Maynard : Phys. Letters 26A, 647 (1968)
47. U. Buchenau, N. Nücker, A.J. Dianoux : Phys. Rev. Letters 53, 2316 (1984)
48. E. Akkermans, R. Maynard : Phys. Rev. B32, 7850 (1985)
49. R. Calemczuk, A.M. de Goër, R. Maynard, B. Salce : This conference (1986)

The Thermal Conductivity of Amorphous Insulators

S. Alexander, O. Entin-Wohlman**, and R. Orbach*

Department of Physics, University of California
Los Angeles, CA 90024, USA

The temperature dependence of the thermal conductivity, κ, of amorphous materials exhibits a nearly universal behavior which is very different from that observed for crystalline materials. One finds that at low temperatures ($T \lesssim 5$ K), $\kappa \propto T^2$ for amorphous materials. This is followed by a plateau in the thermal conductivity ($5 \lesssim T \lesssim 10$ K) where κ changes very little with temperature. At higher temperatures ($10 \kappa \lesssim T$), κ rises slowly with increasing temperature.

The low temperature regime has been studied very carefully and confirms the predictions of Anderson, Varma, and Halperin[1] and of Phillips[2] based on phonon scattering from two-level-systems (TLS). There is no generally accepted explanation for the origin of the plateau, and there are virtually no theoretical models with which the higher temperature behavior of κ can be compared. There is also relatively little detailed experimental data available for the latter temperature regime, primarily because of the need for radiative corrections to the determination of κ.

The plateau in κ has been explained by resonant scattering from a large peak in the density of localized non-propagating vibrational (or torsional) states at a suitable energy,[3] and by the resummation of maximally crossed diagrams in two and three dimensions involving elastic scattering.[4] Both of these approaches derive a frequency ω_o at which the phonon diffusion constant vanishes. However, as we shall show below, any model which allows for high-frequency propagating phonons for $\omega > \omega_o$ will have difficulty generating the very slow rise in the thermal conductivity above the temperature of the plateau.

We have argued previously[4,5] that the thermal conductivity reflects a crossover behavior from (extended) phonon excitations at low energies ($\omega < \omega_c$) to (localized) fracton excitations at higher energies ($\omega > \omega_c$). Here, ω_c is termed the crossover energy. An amorphous material is homogeneous macroscopically, and is therefore homogeneous on relatively large length scales. The short-range properties reflect the disorder and large local inhomogeneities. The transition between the two regimes defines a crossover length which Alexander terms the Cauchy length ξ_c.[6] Such a length scale arises naturally in simple models of disordered materials, such as percolation and models for gelation and rubber elasticity.[6,7] The crossover frequency is related to the Cauchy length by the relationship $\omega_c \sim \xi_c^{-(D/\bar{\bar{d}})}$, where D is the fractal dimensionality[8] and $\bar{\bar{d}}$ is the fracton dimensionality.[9] The long wavelength, or low

---

* Permanent address: The Racah Institute, The Hebrew University, 91904 Jerusalem, Israel.

** Permanent address: School of Physics and Astronomy, Tel Aviv University, Ramat Aviv, 69978 Tel Aviv, Israel.

frequency vibrational modes, are phonons. The short length scale, or higher frequency harmonic modes, are strongly affected by the disorder, and are therefore qualitatively different. In particular, they exhibit a different frequency dependence of the vibrational density of states, different dispersion, and a strong tendancy for localizaion.[10,11] We have termed them fractons.[9]

We shall describe the behavior of the thermal conductivity of amorphous materials in the framework of a scaling model for the phonon and fracton excitations. Our purpose is to show that the fracton model predicts a linear temperature dependence of the thermal conductivity above the plateau temperature as a manifestation of fracton hopping, and that this prediction is quite insensitive to the detailed assumptions of the model. Conversely, observation of a linear temperature dependence for κ above the plateau temperature can be regarded as strong evidence for the unique applicability of the fracton model.

In what follows we shall first treat the thermal conductivity of the phonon excitations in an amorphous material. We shall show that, at the extreme limit for phonon scattering (when the phonon mean free path equals the phonon wavelength), the thermal conductivity increases as T^2. We use this to argue that <u>any</u> phonon scattering mechanism will generate κ which increases with increasing temperature at least as T^2. Hence, observation of a slower temperature dependence of κ signifies some other conduction mechanism. We then give a scaling argument to show that fracton hopping gives rise to a thermal conductivity which increases linearly with temperature. Observation of a temperature dependence for κ less than T^2 thereby can be used as evidence for the fracton model.

One can write quite generally for the thermal conductivity of the harmonic vibrational modes,

$$\kappa(T) = \int d\omega\ N(\omega)\ k_B C(\hbar\omega/k_B T)\ D(\omega,T) \quad , \tag{1}$$

where $N(\omega)$ is the vibrational density of states, $C(x) = x^2[\exp(x)][\exp(x)-1]^{-2}$ is the (single oscillator) specific heat, and $D(\omega,T)$ is the diffusion constant for the vibrational mode with frequency ω at temperature T.

We consider first the situation for phonons. One has,

$$D \propto v_s^{\,2}\tau \propto v_s\Lambda \quad , \tag{2}$$

where v_s is the sound velocity, τ is the phonon scattering time, and Λ is the phonon mean free path ($\Lambda = v_s\tau$). Consistency of the phonon description for a phonon of wave vector q requires $\Lambda q > 1$. Equivalently,

$$\tau > \omega^{-1}. \tag{3}$$

Substitution of Eqs. (2) and (3) into Eq. (1) [with $N(\omega) \propto \omega^2$] gives a lower bound to the temperature dependence of the thermal conductivity κ for propagating phonons:

$$\kappa_{min}(T) > \lfloor a^3 k_B/(2\pi)^3 v_s \rfloor \int d\omega\ \omega\ C(\hbar\omega/k_B T) \sim T^2 \quad , \tag{4}$$

as in the TLS scattering regime.[1,2] The implications of the bound, Eq. (4), should not be exaggerated. When ω is much larger than its bound, one can certainly devise frequency and temperature dependences of τ quite freely. It does remain true, nonetheless, that it is extremely difficult to construct a model which will exhibit a slower temperature dependence than T^2 over a considerable temperature range in which additional phonon states continue to be excited.

Consider now the contribution of localized vibrational modes (fractons) to the thermal conductivity. We suggest that the anharmonic vibrational interaction allows for hopping transitions between different fracton states localized at spatially distinct sites. The diffusion constant has the form,

$$D(\omega) \propto R^2(\omega)/\tau(\omega) \quad , \tag{5}$$

where R is a mean hopping distance and $\tau(\omega)$ a hopping time. Energy conservation for the localized states has strong geometrical constraints.[12] The dominant transfer process is therefore energy transfer between two fracton states localized at different sites, accompanied by the emission or absorption of an extended phonon.

There are no extended phonons above the crossover frequency, ω_c. A Mott[13] estimate of the hopping distance gives,

$$N_s(\omega)R^D(\omega)\omega_c \sim 1 \quad , \tag{6}$$

whence $[R(\omega)]^D \sim 1/[N_s(\omega)\omega_c]$. Thus, the hopping distance is a decreasing function of the density of states, and therefore of the frequency ω.

The hopping rate, $1/\tau(\omega)$, depends exponentially on the ratio $R(\omega)/\ell_\omega$, where ℓ_ω is the fracton localization length. Within the scaling fracton model[4–10] one has,

$$N_s(\omega)\ell_\omega^D \propto \omega^{-1} \quad , \tag{7}$$

so that [see Eqs. (6) and (7)] the ratio $R(\omega)/\ell(\omega)$ increases with a power of ω/ω_c. This is crucial in determining the fracton contribution to the thermal conductivity. The diffusion constant has a factor $\exp[-A(\omega/\omega_c)^x]$ by virtue of the fracton localization, where $x = d_\phi/D$, with d_ϕ a measure of the strength of the fracton localization {i.e. the fracton wave function is given by $\phi \propto \exp[-(R/\ell_\omega)^{d_\phi}]$. Typically, d_ϕ is expected to lie between d_{min} and unity, with d_{min} the exponent governing the relationship between the Pythagorian distance and the minimum distance along the network between two points.[14]}

The strong force of the exponential factor thereby leads to a fracton contribution to the thermal conductivity dominated by the lower bound of the integral [Eq. (1)], i.e. by the lowest frequency fractons near the crossover frequency ω_c. We have shown elsewhere[15] that one predicts a large excess density of states in the vicinity of ω_c. It is easy to see that the temperature dependence of the diffusion constant for these fractons is that of the phonon occupation number, and therefore linear in T when $k_BT/\hbar\omega_c > 1$.

The fracton model therefore generates a coherent picture for the thermal conductivity of amorphous materials. The thermal conductivity at low temperatures is due to propagating phonons. Because there are no phonons above ω_c, the phonon contribution saturates when $k_BT/\hbar\omega_c$ becomes of order unity, and subsequently decreases as τ decreases at higher temperature see Eq. (2). This is quite analogous to the behavior found for crystalline materials for temperatures above the conductivity maximum temperature]. At higher temperatures, the thermal conductivity of amorphous materials is dominated by the hopping of localized fractons. A linear temperature dependence for κ is found from quite general arguments. This prediction seems to have been confirmed by experiments[16] over a very wide temperature range.

A detailed calculation within the fracton model which allows a determination of the various power-law indices and numerical estimates will be published elsewhere.[17] Because the results for $\kappa(T)$ are not sensitive to the details, we do not reproduce them here.

In conclusion, we summarize the main points of our argument. We first showed that it is very difficult to reconcile in a consistent manner a model which has high-frequency propagating phonons with a thermal conductivity which increases more slowly than T^2. A high temperature thermal conductivity dominated by hopping of localized modes generates a κ linear in temperature. This occurs when the ratio of the hopping distance to the localization length increases with ω, so that the dominant contribution to κ arises from fractons lying near the crossover frequency. The linear temperature dependence arises from the domination of the hopping process by low-frequency phonons whose energies are less the k_BT. These are natural predictions of the fracton model.

This research has been supported by the U.S. National Science Foundation under grant DMR 84-12898, and by the Fund for Basic Research administered by the Israel Acadamy of Sciences and Humanities.

References

1. P. W. Anderson, B. I. Halperin, and C. M. Varma, Phil, Mag. 25, 1 (1972).
2. W. A. Phillips, J. Low Temp. Phys. 7, 351 (1972).
3. V. G. Karpov and D. A. Parshin, Pisma v Zh. Eksp. Theor. Fiz. (USSR) 38, 536 (1983) [Engl. trans.: JETP Lett. 36, 648 (1983)].
4. S. Alexander, C. Laermans, R. Orbach, and H. M. Rosenberg, Phys. Rev. B28, 4615 (1983).
5. R. Orbach and H. M. Rosenberg, LT-17 Proceedings, Ed. by U. Eckern, A. Schmid, W. Weber, and H. Wühl (Elsevier Science Publishers B. V., Amsterdam, 1984), p. 375.
6. S. Alexander, Physics of Finely Divided Matter, Ed. by M. Daoud, Springer Proc. in Phys. 5 (Springer-Verlag, Heidelberg, 1986), p. 162.
7. S. Alexander, Ann. Isr. Phys. Soc. 5, 149 (1983).
8. B. B. Mandelbrot, The Fractal Geometry of Nature (Freeman, New York, 1983); Ann. Isr. Phys. Soc. 5, 59 (1983); J. Stat. Phys. 34, 895 (1984).
9. S. Alexander and R. Orbach, J. Phys. (Paris) Lett. 43, L-625 (1982).
10. R. Rammal and G. Toulouse, J. Phys. (Paris) Lett. 44 L-13, (1983).
11. O. Entin-Wohlman, S. Alexander, R. Orbach, and Kin-Wah Yu, Phys. Rev. B29, 4588 (1984).
12. See e.g., S. Alexander, O. Entin-Wohlman, and R. Orbach, J. Phys. (Paris) Lett. 46, L-549 and L-555 (1985).
13. N. F. Mott, Philos. Mag. 19, 835 (1969).
14. K. M. Middlemiss, S. G. Whittington, and D. S. Gaunt, J. Phys. A: Math Gen. 13, 1835 (1980); R. Pike and H. E. Stanley, J. Phys. A: Math. Gen. 14, L169 (1981); D. C. Hong and H. E. Stanley, J. Phys. A: Math. Gen. 16, L475 (1983); D. C. Hong and H. E. Stanley, J. Phys. A: Math. Gen. 16, L525 (1983); H. J. Hermann, D. C. Hong, and H. E. Stanley, J. Phys. A: Math. Gen. 17 L261 (1984); J. Vannimenus, J. P. Nodal, and C. Martin, J. Phys. A: Math. Gen. 17, L351 (1984);

S. Havlin and R. Nossal, J. Phys. A: Math. Gen. 17, L427 (1984); S. Havlin, Z. V. Djordjevic, K. Majid, H. E. Stanldy, and G. H. Weiss, Phys. Rev. Lett. 53, 178 (1984).
15. A. Aharony, S. Alexandar, O. Entin-Wohlman, and R. Orbach, Phys. Rev. B31, 2565 (1985).
16. A. K. Raychaudhuri, Ph.D. Thesis, Cornell University (1980), unpublished; J. E. de Oliveira and H. M. Rosenberg, private communication (1986).
17. S. Alexander, O. Entin-Wohlman, and R. Orbach, Phys. Rev. B15, submitted for publication, 1986.

The Thermal Conductivity and Specific Heat of Glasses

C.C. Yu and J.J. Freeman

Department of Physics, University of Illinois at Urbana-Champaign
1110 W. Green St., Urbana, IL 61801, USA

It is well known that in a wide variety of amorphous materials an empirical correlation exists between the plateau in the thermal conductivity κ and the bump in C/T^3 where C is the specific heat [1,2]. Both occur at roughly the same temperature for a given material and this temperature lies between 3K and 10K. Recent theoretical efforts to explain the thermal conductivity have included fractons [3] and phonon localization [4,5]. It is not obvious, however, in the first case how one can map a glass onto a self-similar percolating network, or in the second case how one can explain the rise in κ above the plateau.

The thermal conductivity is given by the expression

$$\kappa(T) = (1/3)\int_0^{\omega_D} C(T,\omega)v\ell(T,\omega)d\omega \tag{1}$$

where $C(T,\omega)$ is the phonon specific heat, v is the velocity of sound, and $\ell(T,\omega)$ is the phonon mean free path. One possible explanation for the plateau is to view it as a crossover from a low-frequency region where $\ell \sim 150\lambda$ to a high-frequency region where $\ell \sim \lambda$ [6]. Here λ is the phonon wavelength. Suppose that the mean free path decreases at higher energies due to a sudden increase in the density of states of local excitations which scatter the phonons. This enhancement in the density of states n(E) would also account for the anomalous specific heat. However, if we use a step function for the density of states, the plateau in amorphous SiO_2 requires a rise in n(E) at 7K whereas the maximum in C/T^3 demands an increase at 43K. This discrepancy is eliminated if Rayleigh scattering is included at frequencies below the step. Since Rayleigh scattering breaks down when $ka \sim 1$, this implies spatial inhomogeneities slightly larger than the molecular unit (e.g. $a \sim 7$Å for SiO_2). We assume that the increased density of states is due to a flat distribution of local modes which we model as Einstein oscillators. Phonons can cause resonant transitions between the oscillator levels. For convenience we consider resonant scattering from the three lowest levels which is given by

$$\ell_{res}^{-1} = \frac{2\bar{P}\gamma^2}{\rho v^3}\,\omega\left[1 - \frac{3\exp(-\frac{\hbar\omega}{kT})}{1 + 2\cosh(\frac{\hbar\omega}{kT})}\right] \tag{2}$$

where ρ is the mass density, $\bar{P}$ and γ are the density of states and the strain field coupling of the standard two levels systems (TLS). We assume that to lowest order phonons do not change the level separation. Thus we will neglect the relaxation contribution to ℓ from the local modes. In addition we include at all frequencies phonons scattering via resonance and relaxation mechanisms from the conventional tunneling centers which have a flat distribution of asymmetries and barrier heights [7]. The

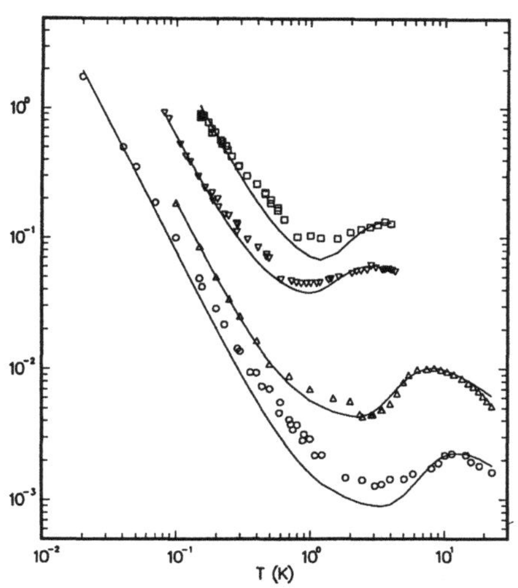

Fig. 1. Specific heat C divided by the cube of temperature; data are o-SiO_2 [9,10], △ - GeO_2 (×2) [10,11], ▽ - PS [12,13], and □ - PMMA (×4) [12,13]; curves are fits to the data as described in the text

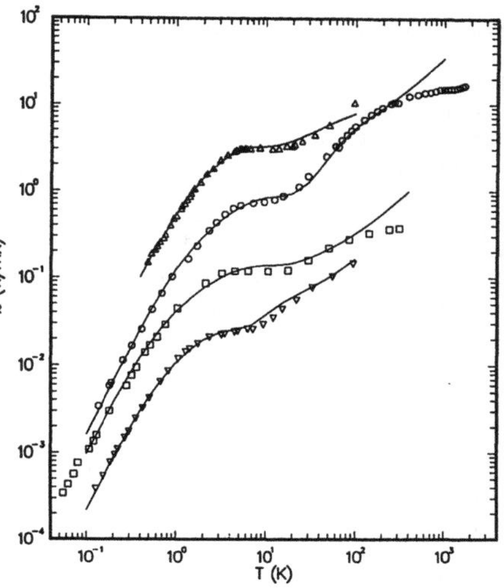

Fig. 2. Thermal conductivity κ plotted using the same symbols as in Fig. 1 for SiO_2 (×8) [14,15], GeO_2 (×16) [16], PS (×2) [17] and PMMA [18],; curves are those calculated using the parameters listed Table 1

TABLE I

	$\hbar\omega_c$(K)	S_κ	S_c	$B(10^{-45}\frac{s^4}{cm})$	γ(eV)	$\bar{P}\gamma^2(10^8\frac{erg}{cm^3})$
aSiO_2	43	75	415	16.5	0.3 [19]	3.2
aGeO_2	28	175	330	10.3	0.13	1.7
PS	10	45	65	619	0.09 [20]	0.56
PMMA	12	135	70	378	0.015	0.33

results of this model are shown in Figs. 1 and 2, and the parameters are given in Table 1. Our three-level approximation breaks down at high temperatures as can be seen in Fig. 2. There are four adjustable parameters-the frequency ω_c where the onset of the step occurs, the height of the step which we denote by S_κ for the thermal conductivity and by S_c for the specific heat, and the strength B of the Rayleigh scattering $\ell^{-1}_{ray}=B\omega^4$, which is that deduced from computer fits to the low temperature data [8].

In general $S_\kappa \bar{P} < S_c N_o$ because not all the modes which contribute to the specific heat will couple to phonons strongly enough to increase the thermal resistance. N_o is the density of states of TLS associated with the specific heat. We have neglected structural relaxation which involves thermal activation of atoms over a potential barrier. This mechanism involves frequencies which are orders of magnitude lower than those in which we are interested at a given temperature.

To our knowledge this is the first time that anyone has fit both the specific heat and thermal conductivity with a single set of parameters. We have used conventional tunneling centers, a rather sudden onset of local modes modelled as Einstein oscillators, and Rayleigh scattering at frequencies below this onset. This does not solve the problem. Rather it helps to define it. The microscopic nature of these features is unknown. Clearly, any microscopic theory must include more than just phonons moving in a random static potential. It must also include the interactions between phonons and local dynamic scatterers.

Acknowledgements

We would like to thank A. C. Anderson for many illuminating discussions and Y. Fu for helpful suggestions. This work was supported in part by National Science Foundation Grant No. DMR 83-16981 and by National Science Foundation-Low Temperature Physics-Grant DMR 83-03918.

References

1. D. P. Jones, J. Jackle and W. A. Phillips, in Phonon Scattering in Condensed Matter, ed. by H. J. Maris (Plenum, New York, 1980) p. 49, and papers cited therein.
2. D. A. Ackerman, D. Moy, R. C. Potter, A. C. Anderson and W. N. Lawless, Phys. Rev. B23, 3886 (1981) and papers cited therein.
3. R. Orbach, J. Stat. Phys. 36, 735 (1984).
4. J. E. Graebner, B. Golding and L. C. Allen, unpublished.
5. E. Akkermans and R. Maynard, Phys. Rev. B32, 7850 (1985).
6. J. J. Freeman and A. C. Anderson, submitted to Phys. Rev. B.
7. A general review may be found in Amorphous Solids, ed. by W. A.Phillips (Springer, Berlin, 1981).
8. J. J. Freeman and A. C. Anderson, this conference.
9. J. C. Lasjaunias, A. Ravex and M. Vandorpe, Solid State Comm. 17, 1045 (1975).
10. R. C. Zeller and R. O. Pohl, Phys. Rev. B4, (1971).
11. A. A. Antoriou and J. A. Morrison, J. Appl. Phys. 36, 1873 (1965).
12. R. B. Stephens, Phys. Rev. B8, 2896 (1973).
13. C. L. Choy, R. G. Hunt and G. L. Salinger, J. Chem. Phys. 52, 3629 (1970).
14. T. L. Smith, Ph.D. Thesis (University of Illinois, 1975), unpublished.
15. Y. S. Touloukian, R. W. Powell, C. Y. Ho and P. G. Klemens, Thermophysical Properties of Matter, Vol. 2 (Plenum, New York, 1970), p. 193.
16. K. Guckelsberger and J. C. Lasjaunias, Compt. Rend. 270, B1427 (1970).
17. J. J. Freeman, Ph.D. Thesis (University of Leeds, U.K., 1985), unpublished.
18. R. B. Stephens, G. S. Cieloszyk and G. L. Salinger, Phys. Lett. 38A, 215 (1972).
19. L. Piche, R. Maynard, S. Hunklinger and J. Jackle, Phys. Rev. Lett. 32, 1426 (1974).
20. J.-Y. Duquesne and G. Bellessa, J. Phys. (Paris) 40, L-193 (1979).

The Thermal Conductivity Plateau in Disordered Systems

B. Golding, J.E. Graebner, and L.C. Allen

AT & T Bell Laboratories, Murray Hill, NJ 07974, USA

1. Introduction

The thermal conductivity of disordered substances exhibits a region of weak temperature dependence known as the "plateau" [1]. It occurs in glasses in a temperature range typically between 1 and 20 K. At much lower temperatures, the thermal conductivity is governed by the resonant scattering of thermal phonons from tunneling systems and follows a T^2 power law. Above the plateau the temperature dependence is roughly linear in T. Explanations offered for the occurrence of a plateau include: phonon scattering by enhanced densities of tunneling systems [2], elastic scattering by density fluctuations [3], dimensional crossover of the vibrational density of states [4], and weak localization of phonons [5]. As yet there has been no convincing description of this phenomenon.

Part of the difficulty in resolving the issue stems from the lack of spectroscopic probes which yield direct information on the phonon lifetimes in this regime. Another critical shortcoming is the lack of detailed information on the structure of glass in the region below 100 Å. We shall therefore in this discussion limit ourselves to a critical evaluation of thermal conductivity data, most of which has been in the literature for many years. The picture which emerges from our analysis is one in which the Ioffe-Regel criterion, $kl=1$, is approached at the upper end of the thermal conductivity plateau. The short mean free path l is brought about by elastic scattering from static density fluctuations. We suggest that phonon localization occurs near the high temperature end of the plateau.

2. Heat Transport in Aggregates

Aggregates possess a number of significant advantages in attempting to quantify thermal transport processes. One can introduce characteristic lengths by choosing particle sizes judiciously and for relatively large particles, the structures may be imaged directly. Fractal structures may be produced, their dimensionality determined, and the porosity varied over a significant range. Elastic phonon scattering has maximal strength and, finally, the aggregate may be composed of crystalline or glassy particles.

The temperature of the plateau region varies with particle size. A scaling analysis of heat transport in amorphous silica and crystalline alumina aggregates has been carried out [6]. The characteristic phonon wavelength associated with the temperature at the upper end of the thermal conductivity plateau is plotted in Fig. 1 versus the characteristic particle size or correlation length of density defects of the medium. The solid line has slope unity and is consistent with all of the data. The conclusion is drawn that the upper end of the plateau occurs as a result of the matching of a characteristic length ξ of the medium by the phonon wavelength, $k\xi=1$. At temperatures higher than the plateau the conductivity should be representative of the homogenous medium.

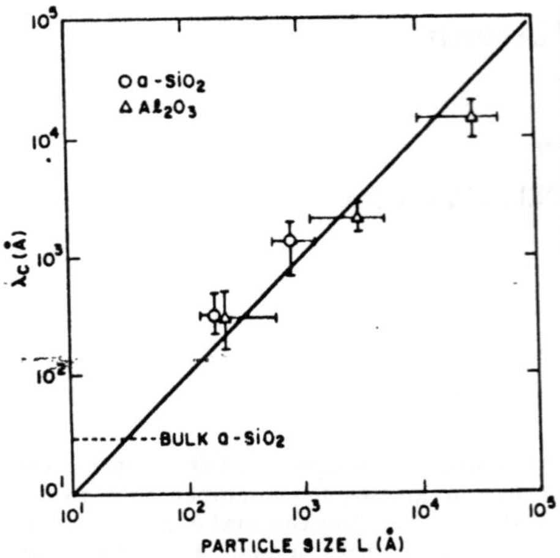

Fig. 1 Plot of thermal phonon wavelength λ at upper end of plateau vs. aggregate particle size. The data, for amorphous SiO_2 and crystalline (Al_2O_3), particles, are consistent with slope unity (solid line). This result suggests that matching of λ with the aggregate correlation length is responsible for a crossover in the thermal transport mechanism.

3. Heat Transport in Glasses

In glasses, the absence of any known characteristic scale requires a different approach. We shall ask whether the idea that heat is carried by propagating phonons is consistent with the following simple model. The thermal conductivity of glasses can be fit by incorporating into the kinetic model a temperature and frequency dependent phonon mean free path which is composed of two contributions: a term originating in resonant and relaxational scattering by tunneling systems and a term arising from Rayleigh scattering from static density fluctuations. The procedures have been described in detail in a recent publication [7] and only the results will be described here. It is found for glasses including silicas, chalcogenides, and organics that the condition $kl=1$ is universally satisfied at the upper end of the plateau. This is shown in Fig. 2 for silica glass. The existence of such short mean free paths for these high energy phonons means that they do not propagate. It may be assumed that excitations at these energies are localized vibrational modes which do not carry heat. Thus, the origin of the plateau may be thought of as thermal excitation of a band of modes with zero or low group velocity.

Attempts to explain the plateau by Rayleigh scattering have generally assumed scattering by microscopic density fluctuations, eg., due to bond length variations [3]. In the present analysis, we assume that the characteristic scale of the fluctuations is given by the correlation length L, typically 30-40 Å. Since the scattering strength varies as $(kL)^4$, a relatively small density fluctuation $O(10^{-2})$ is sufficient to explain the data.

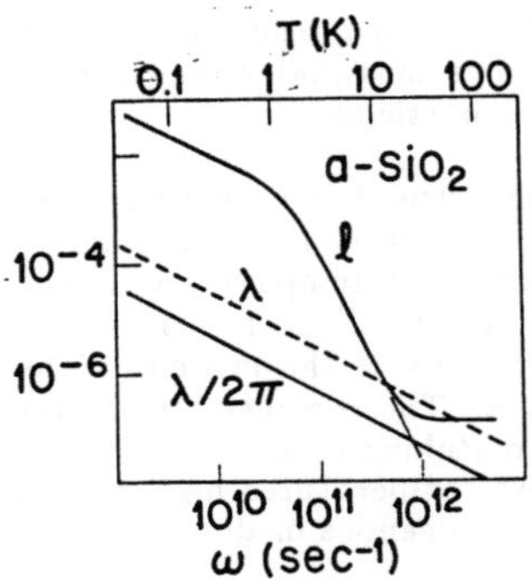

Fig. 2 Mean free path l and wavelength $\lambda=2\pi/k$ of thermal phonons in a-SiO_2. Note that the localization condition $kl=1$ is satisfied for thermal phonons excited at temperatures near the upper end of the plateau. Similar behavior occurs in all glasses analyzed including a-GeO_2, a-As_2S_3, a-As, a-Se, epoxy, PMMA, and polystyrene.

4. Comparison with Other Models

The fracton description of the thermal conductivity plateau postulates a crossover from propagating modes at low energies to localized fractons at high energies [4]. This is believed to occur when the fractal dimensionality of the medium becomes influential as the phonon wavelengths begin to sample the fractal character of the medium. Since there is no evidence that glasses are fractal on any scale, the most appropriate substances to test these ideas are aggregates.

It is worth noting that thermal conductivity experiments on epoxy resin [8], frequently cited as providing support for the fracton model, may be analyzed from a different perspective. It has been proposed that the shift of the plateau to higher temperatures as the hardener/resin ratio is increased results from shorter characteristic lengths in the epoxy due to crosslinking [9]. By fitting the data to a sum of tunneling system and elastic scattering contributions it is possible to show that the small changes in the thermal conductivity in the plateau region arise from significant changes in the tunneling system contribution alone [7]. The Rayleigh scattering coefficient is nearly independent of composition. It is therefore not necessary to invoke a fracton picture to explain these data.

To conclude, we have shown that models based on propagating phonons break down at the upper end of the thermal conductivity plateau and suggest that a phonon mobility-edge exists there. Recent theoretical ideas based on weak localization [5] have suggested that the problem of the plateau may be treated on a footing conceptually equivalent to electronic localization. This idea is particularly intriguing and we await predictions which will permit a quantitative test.

References

1. See for example articles in **Amorphous Solids: Low Temperature Properties**, ed. by W. A. Phillips (Springer, Berlin 1981).
2. M. P. Zaitlin and A. C. Anderson: Phys. Rev. B 12, 4475 (1975).
3. C. Kittel: Phys. Rev. **75**, 972 (1949); R. C. Zeller and R. O. Pohl: Phys. Rev. **B** 4, 2029 (1971); and D. P. Jones and W. A. Phillips: Phys. Rev **B** 27, 3891 (1983).
4. P. F. Tua, S. J. Putterman, and R. Orbach: Phys. Lett. **98A**, 357 (1983); and S. Alexander, D. Laermans, R. Orbach, and H. M. Rosenberg; Phys. Rev. **B** 28, 4615 (1983).
5. E. Akkermans and R. Maynard: Phys. Rev. **B** 32, 7850 (1985).
6. J. E. Graebner and B. Golding: Phys. Rev. **B** (to appear).
7. J. E. Graebner, B. Golding and L. C. Allen (to appear).
8. S. Kelham and H. M. Rosenberg: J. Phys. C **14**, 1737 (1981).
9. R. Orbach and H. M. Rosenberg: Proceedings of LT-17, ed. by U. Eckern, A. Schmid, W. Weber, and H. Wuhl (Elsevier Science Pub., 1984) p. 375.

Thermal Properties of Silica Gels

R. Calemczuk[1], *A.M. de Goër*[1], *B. Salce*[1], *and R. Maynard*[2]

[1]Département de Recherche Fondamentale, Service des Basses Températures, Laboratoire de Cryophysique, Centre d'Etudes Nucléaires de Grenoble, 85X F-38041 Grenoble Cedex, France

[2]C.N.R.S./C.R.T.B.T., 25, avenue des Martyrs, 166X F-38042 Grenoble Cedex, France

Silica aerogels are very exceptional materials in the huge panoply of solids : densities as low as 0.1 times that of amorphous silica can be obtained. In such systems, the distribution of matter is highly inhomogeneous, and this fact raises the problem of the nature of low-energy excitations (phonons, fractons, two-level systems) and their contributions to the low temperature properties. Fractons have been conjectured by Alexander and Orbach [1] to contribute in glasses and particularly in epoxy-resins [2]. As in ideal fractal system the density vanishes for large sample, it is opportune to look for such a contribution in gels where the density is very small. One more reason is the observation of the fractal nature of the surface of gels [3]. In this communication, we report the first results on the specific heat (down to 0.3 K) as well as thermal conductivity (down to 0.1 K).

Experiments

Samples for the different measurements have been cut from three blocks of aerogels [4] with densities given in table 1. The specific heat C_p has been measured in a thermocouple controled adiabatic configuration. One face of the samples was metallized and glued to the sapphire sample-holder using a little amount of silicon grease. Cooling from room temperature was performed without He gas to avoid parasitic contribution to C_p from adsorbed He. The thermal conductivity measurements have been performed using the classical stationary heat flow method.

The specific heat results are reported in Fig.1a in a C_p/T^3 plot as a function of T. It is remarkable that C_p of gels is higher than that of silica, up to two orders of magnitude for gel (c). However the most striking feature of the results is that C_p of gels is below the expected phonon contribution C_D in the low temperature Debye limit. The evaluation of C_D (table 1 and arrows in fig.1a) has been done using the measured samples densities and the sound velocities from the available room temperature data [5]. (In the case of gel (a), we have controled that only minor variations of the sound velocities - less than 5 % - occur during cooling down to 1.5 K). This situation is completely different from that in vitreous silica, where C_p is above the Debye limit, due to the contribution of TLS [6]. Therefore we believe that the specific heat of gels has nothing to do with two-level defects, but gives evidence of another regime of harmonic excitations, perhaps a fractal one. The specific heat obeys actually a power law T^α in the temperature range below 2 K. The values of α are given in table 1.

The thermal conductivity results are given in fig.1b. The K(T) values lie well below those of a-SiO_2 (also shown in the figure) in the whole temperature range. The general shape of K(T) evocates a sort of plateau between 0.5 - 1 K (gel (a)) and 0.1 - 0.5 K (gel (c)). The occurrence of a quasi-plateau in a temperature range where the specific heat has no bumps but displays a monotonic power law confirms the idea that aerogels have a quite different behaviour from glasses.

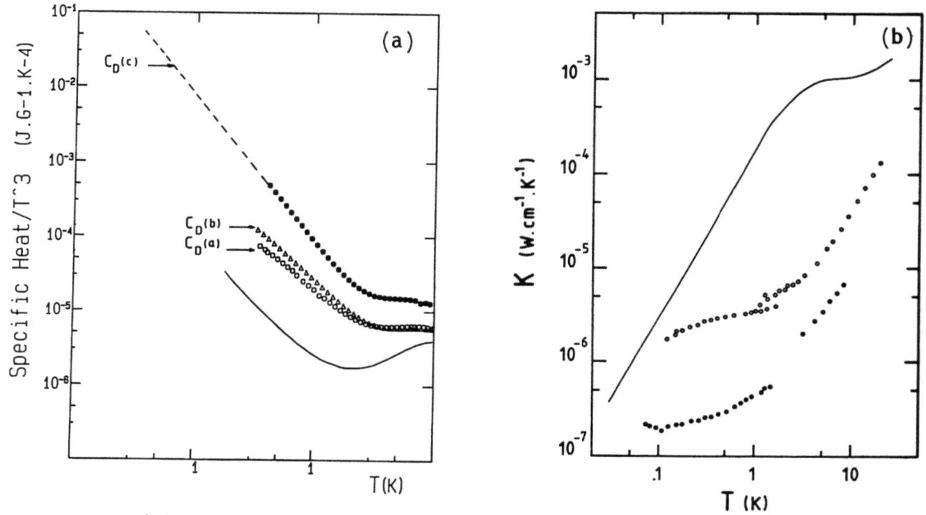

Figure 1 - (a) Specific heat and (b) Thermal conductivity of SiO_2 aerogels : o sample (a) ; Δ sample (b) ; • sample (c) ; solid lines, bulk a-SiO_2

Table 1.

Gel	ρ (g/cm^3)	V_L (cm/s)	V_T (cm/s)	C_D (erg/g K^4)	α	T^+_{co} (K)	T^*_{co} (K)	ℓ (μm)
a	0.87	1.85 10^5	1.18 10^5	640	1.75	0.4	0.3-0.5	3.5-9
b	0.72	1.5 10^5	9.6 10^4	1400	1.6	0.3	-	-
c	0.27	4.25 10^4	2.6 10^4	1.9 10^5	1.1	0.06	< 0.1	< 3.6

+ from specific heat   * from thermal conductivity

Analysis

For materials of such a high porosity, there are two main possibilities for the structure : either the pores have rather well defined size or the local structure is fractal up to a characteristic dimension ξ, as in the case of percolation above the threshold. In view of the lack of structural information on our aerogels, we have choosen to modelize them as fractal objects in order to compare our results with available theory. We suppose that the gel structure is built up by fractal blobs of size ξ with characteristic dimensions $\bar{d}$ (fractal dimension) and $\tilde{d}$ (spectral dimension). This is the simplest analysis, with only two parameters. This assumption allows us to associate the characteristic length ξ to the density ρ by the simple relation $\rho(\xi)/\rho(a) = (a/\xi)^{3-\bar{d}}$ where a is of the order of the atomic distance and $\rho(a)$ the density of the most compact gel : the silica. As the blobs are fractal for length scale up to ξ we assume as in [7] that the sound velocity varies as $v(\xi)/v(a) = (a/\xi)^{(\bar{d}-\tilde{d})/\tilde{d}}$. By comparison with the measured sound velocities as a function of density (exponent 1.25 [5]) we have found the simple relation $(\bar{d}-\tilde{d})/\tilde{d} = 1.25\ (3-\bar{d})$ (1). The continuity hypothesis for the dispersion relation produces then the cross-over frequency between phonons and fractons : $\omega_{co}(\xi) = v(\xi) \times 2\ \pi/\xi$. Above ω_{co} the density of modes tends to be fracton-like beyond the cross over frequency range. In the pure fracton regime the specific heat varies as $T^{\tilde{d}}$. We take the gel (c) with the lowest density as giving a pure fracton contribution to C_p below 2 K. Then we get $\tilde{d} = \alpha = 1.1$. Using the sound velocity relation (1) we deduce $\bar{d} = 2.1$. Both values of $\bar{d}$ and $\tilde{d}$ are plausible.

The values of α for gels (a) and (b) with larger densities are different from 1.1 (table 1). A possible explanation would be that the fractal structure of the different gels is not the same. However, for these gels, ξ is smaller, and we think that the pure fracton regime is not detectable. We can nevertheless estimate rough values of the temperature T_{co} corresponding to the cross-over frequency ($\omega_{co} = kT_{co}/\hbar$) from the intersection of the extrapolated C_p curve with the calculated Debye limit C_D (see fig.1). The values of T_{co} are given in table 1. Now, in the fracton model, the quasi-plateau in the thermal conductivity is explained within a simple hypothesis : only the phonon part of the spectrum contributes to the heat transport while the fractons are localized. Therefore the upper limit of the usual thermal conductivity integral must be taken as ω_{co} instead of the Debye frequency. Using a mean free path ℓ independent of frequency (as we have no experimental information on phonon scattering below 0.1 K), we have obtained calculated values of K in the plateau region in agreement with experiment, with the parameters T_{co} and ℓ given in table 1. The values of T_{co} are well compatible with those obtained from the specific heat and the values of ℓ are reasonable. We note that the fractal hypothesis is not necessary to obtain the coherence between the C_p and K results discussed above, but only the existence of a transition between phonons and localized modes.

References

1. S. Alexander, R. Orbach : J. Phys. Lett. (Paris) 43 L-625 (1982)
2. S. Alexander, C. Laermans, R. Orbach, H.M. Rosenberg : Phys. Rev. B28, 4615 (1983)
3. P. Pfeifer, D. Avnir, D. Farin : Surface Science 126, 569 (1983)
4. The gels have been prepared by M. Bourdineau (D.Ph.P.E. - C.E.A.) and are the same as used in the ultrasonic study [5]. We thank MM. Bourdineau and Zarembovitch for supplying these samples.
5. B. Nouailhas, F. Michard, R. Gohier, A. Zarembovitch : in 11e I.C.A. Paris p.179 (1983)
6. W.A. Phillips, Amorphous Solids, Low-Temperature Properties (Springer-Verlag, Berlin 1981)
7. A. Aharony, S. Alexander, O. Entin-Wohlman, R. Orbach : Phys. Rev. B31, 2565 (1985)

Phonon Scattering in a Sol-Gel Glass

J.M. Grace and A.C. Anderson

Department of Physics, University of Illinois at Urbana-Champaign, 1110 W. Green St., Urbana, IL 61801, USA

Oxide glasses can be produced by a sol-gel process involving the heat treatment of gels made by hydrolysis and gelation of metal alkoxides. The glasses thus obtained are virtually indistinguishable from those produced by conventional melt techniques [1]. Prior to final heating, however, the gels differ noticeably from the conventional glasses. For example, the gels are porous and, consequently, have densities and sound velocities that are reduced relative to the final product. A comparison of the low-temperature thermal properties of a sol-gel product (prior to compaction) and those of the final product is of interest because of the potential to provide information about the gel-to-glass conversion, as well as the nature of glasses in general.

The sample discussed in this paper is an SiO_2 gel made by the hydrolysis of $Si(OC_2H_5)_4$ and has a mass density 0.8 that of vitreous silica. Furthermore, the measured ultrasonic velocities, both longitudinal and transverse, are roughly one half those measured in vitreous silica. Below 4 K, the specific heat is an order of magnitude greater than that of vitreous silica [2]. In contrast to these differences, the sol-gel thermal conductivity is quite similar to that of vitreous silica [3], as can be seen from Fig. 1, even though the phonon mean free path in the sol-gel (below 1 K) is roughly one third that in vitreous silica. For both substances, the thermal conductivity can be accounted for empirically by phonons having frequency-dependent mean free paths given by

$$\ell^{-1} = A\nu^1 + B\nu^4 \quad (1)$$

The first term on the left of (1) gives rise to the well-known T^2 thermal conductivity observed below 1 K. The second term produces the weakly temperature-dependent region, or plateau, near 1-10 K that is also characteristic of amorphous solids.

The T^2 behavior of the thermal conductivity of amorphous solids has been attributed [4] to phonon scattering from two-level states (TLS). The tunneling model of TLS regards the localized low-energy excitations in glassy materials as resulting from units of molecular dimensions tunneling in double-well potentials. By consideration of the interaction of Debye phonons with TLS sites having broad distributions of energy splittings and equilibration times, the thermal conductivity κ is found to be of the form [4]

$$\kappa \propto \rho v T^2/(\gamma^2 P) \quad (2)$$

where v is an average phonon velocity, ρ is the mass density, γ is the TLS-phonon coupling constant, and P is the density of the TLS interacting most strongly with phonons. The strength of the phonon-TLS coupling is often given in the form of the dimensionless parameter $C_0 \equiv \gamma^2 P/\rho v^2$. From (2) and the first term of (1) it is found that C_0 is ≃ 70% larger for the sol-gel than for vitreous silica. On the other hand, the values of C_0 for both materials are similar in magnitude to those values obtained for other silicate glasses and fluoride glasses [5], borate glasses [6], amorphous polymers [7], and even metallic glasses [8]. Hence, despite the large density of TLS in the sol-gel (as indicated by the large specific heat below 1 K), the effective phonon-TLS coupling as expressed by C_0 is similar to what is observed in a variety of glasses.

Whereas the phenomenological tunneling-states model can account for the T^2 thermal conductivity of amorphous solids, there has been no well-established explanation for the plateau. The plateau is believed to arise from strongly frequency-dependent phonon scattering as in (1), but the origin of the scattering is unknown [4,9]. This same unknown mechanism may be responsibile for the plateau in the sol-gel thermal conductivity. On the other hand, for the porous silica glass Vycor, a much smaller conductivity in the plateau, as shown in Fig. 1, can be explained by the scattering of phonons from pores [10,11]. As the sol-gel also contains pores and has a conductivity near 10 K somewhat smaller than for vitreous silica, it is possible that phonon-pore scattering gives rise to the observed plateau. From the fitting parameter B of (1) it is found that pore diameters of 20-30 Å can account for the plateau [2]. The range of diameters reflects the uncertainty in how much residual ethanol and water is in the pores, as well as the uncertainty as to the geometry of the pores. The computed pore diameters of 20-30 Å are consistent with pore sizes observed in similar sol-gel glasses prior to compaction [12].

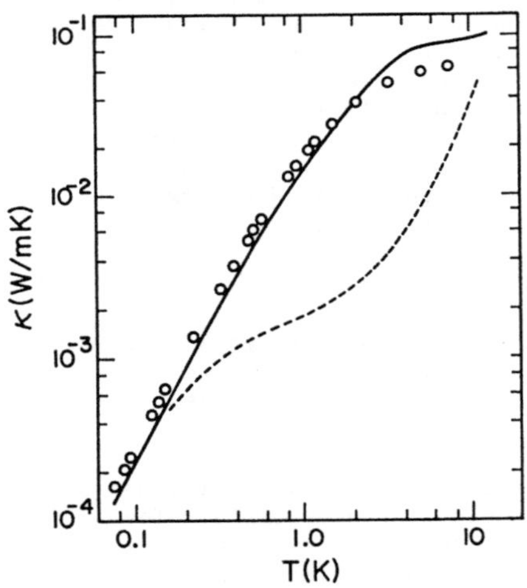

Fig. 1. Thermal conductivity κ versus temperature for a sol-gel glass (O) and, for comparison, vitreous silica (solid line) and porous Vycor (dashed line)

A third explanation for the plateau for sol-gel derives from the possible fractal nature of this low-density material [13]. In this case, the phonon mean free path is expected to become increasingly smaller with increasing ν until the nature of the excitation spectrum crosses over from phonons to localized fractons. The fractons can scatter phonons [14], and the phonon mean free path is expected to vary as ν^{-4} at frequencies below the cross-over. The parameter, B, of (1) is related to the correlation length of the fractal network [15]. Although the predicted frequency dependence is consistent with the thermal conductivity data of Fig. 1, the magnitude of ℓ has not been computed from the fracton theory and, therefore, a quantitative test is not possible.

In conclusion, measurements on a silica sol-gel glass indicate that it is acoustically soft and has an order of magnitude higher density of TLS than, for example, does vitreous silica. Yet, the phonon scattering parameter C_0 is roughly the same as for other glasses. The plateau in the thermal conductivity of the low density sol-gel cannot unmistakably be ascribed to a phonon-fracton crossover.

Acknowledgments

The authors thank S.D. Brown for providing the sol-gel sample and H.E. Jackson and B. Roughani for making Brillouin scattering measurements. This work was supported in part by the National Science Foundation - Low-Temperature Physics - under Grant DMR 83-03918. J.M.G. has been the recipient of a Kodak Fellowship.

References

1. C.J. Brinker, E.P. Roth, G.W. Scherer and D.R. Tallant: J. Non-Cryst. Solids 71, 171 (1982)
2. J.M. Grace and A.C. Anderson: Phys. Rev. B (to be published)
3. T.L. Smith, P. J. Anthony, and A.C. Anderson: Phys. Rev. B 17, 4997 (1978)
4. W.A. Phillips, ed.: Amorphous Solids (Springer, Berlin, 1981)
5. P. Doussineau, M. Matecki and W. Schon: J. Phys. 44, 101 (1983)
6. M. Devaud, J-Y. Prieur and W.D. Wallace: Solid State Ionics 9&10, 593 (1983)
7. M. Schmidt, R. Vacher, J. Pelous and S. Hunklinger: J. Phys. 43, C9-501 (1982)
8. W. Arnold, P. Doussineau, Ch. Frenois and A. Levelut: J. Phys. Lett. 42, L 289 (1981)
9. D. P. Jones, N. Thomas and W.A. Phillips: Philos. Mag. B38, 271 (1978)
10. R. H. Tait: Ph. D. Thesis, Cornell University, 1975 (unpublished)
11. T-C. Hsieh, W.M. MacDonald, A.C. Anderson: J. Non-Cryst. Solids 46, 437 (1981)
12. M. Nogami and Y. Moriya: J. Non-Cryst. Solids 37, 191 (1980)
13. R. Orbach: J. Stat. Phys. 36, 735 (1984)
14. P.F. Tua, S.J. Putterman and R. Orbach: Phys. Lett., 98A, 357 (1983)
15. O. Entin-Wohlman, S. Alexander, R. Orbach and K-W. Yu: Phys. Rev. B29, 4588 (1984)

Normalized Thermal Conductivity of Amorphous Solids

J.J. Freeman and A.C. Anderson

Department of Physics, University of Illinois at Urbana-Champaign, 1110 W. Green St., Urbana, IL 61801, USA

Thermal conduction in amorphous dielectrics is not understood. It is established that, for temperatures less than 10 K, thermal transport is provided by thermal phonons [1]. Therefore, in the Debye approximation, a scaled thermal conductivity may be defined as

$$\kappa/K = (T/\Theta)^3 \int_0^{\Theta/T} (k\Theta)(\tau/h) x^4 e^x (e^x - 1)^{-2} dx \tag{1}$$

where $K = 4\pi k^3 \Theta^2 / h^2 v$, $x = h\nu/kT$, k and h are the Boltzmann and Planck constants and Θ is the Debye temperature. For a crystal, Θ may be calculated using known atomic units and bulk properties, but for glasses, there is no well defined procedure for calculating Θ [2]. Nevertheless, the values of Θ needed to bring the scaled conductivities [3-10] into common register as in Fig. 1, agree to within 40% with those computed for a "simple" unit of the glass (e.g. SiO_2 for vitreous silica). Similar scaled conductivities may be computed for glassy metals and some disordered crystals [11]. The scaled conductivities in Fig. 1 differ significantly only in the range $10^{-2} \lesssim T/\Theta \lesssim 10^{-1}$, commonly referred to as the "plateau" region.

At $T/\Theta \lesssim 10^{-1}$ in Fig. 1, the scaled data are given by Eq. 1 with a phonon mean free path

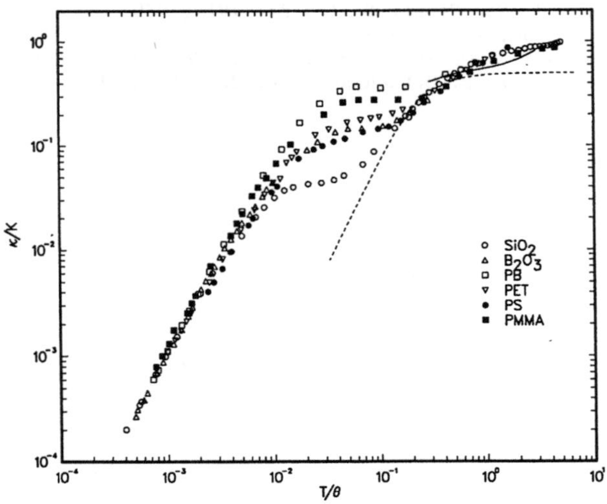

Fig. 1 Reduced thermal conductivity, versus reduced temperatures, for six glassy solids. The dashed line is from (1) with $\ell = \lambda$; the solid line indicates a possible contribution from low-frequency phonons

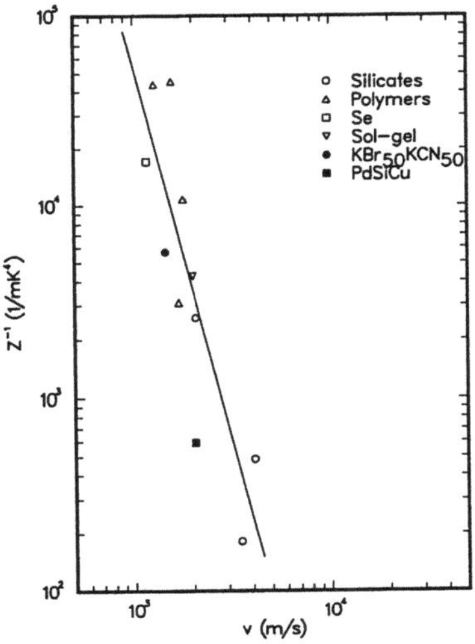

Fig. 2 The phonon scattering coefficient Z, responsible for producing the plateau in Fig. 1, versus phonon velocity for amorphous solids and a crystal exhibiting "glassy" behavior. The line has a slope of v^{-4}. (In computing Z, we used for ℓ_{TLS} that given by the tunneling model rather than $\ell_{TLS}/\lambda = 150$)

$$\ell^{-1} = (v\tau)^{-1} = \ell_{TLS}{}^{-1} + \ell_p{}^{-1}$$

with $\ell_{TLS}/\lambda \approx 150$ and $\ell_p = Z/\nu^4 \approx (4 \times 10^{-24} m^{-3})\lambda^4$ for all glassy materials for which the average phonon velocity v and other parameters are available. The term ℓ_{TLS}, which dominates κ/K below $T/\theta < 10^{-2}$, is related to phonon scattering from two-level states (TLS). The rather succesful tunneling model of TLS neither identifies the TLS sites, nor explains why $\ell_{TLS}/\lambda \approx 150$ for such a variety of glassy solids.

The fitted coefficient Z is plotted in Fig. 2 for several glassy solids, and reveals a reasonable fit to $\ell = Z'\lambda^4$ where $Z' = Z/v^4$. The λ^4 dependence is consistent with previous studies[12]. There has been much recent theoretical effort to explain the plateau [13,14,15] but no model has gained general acceptance. Nor does any model explain why Z should be so nearly the same for all glasses, a result consistent with that [16] derived from a different set of considerations.

The common registry at $T/\theta > 10^{-1}$ in Fig. 1 suggests that phonon-like excitations are responsible for thermal transport [17]. A suggested minimum thermal conductivity has the phonon mean free path ℓ equal to the wavelength λ [18]. This assumption, when used in Eq. 1, produces the dashed line in Fig. 1. The data track this line for $10^{-1} \lesssim T/\theta \lesssim 5 \times 10^{-1}$, but lie well above at higher temperatures. We suggest that since κ/K near $T/\theta \approx 1$ is very small, even a limited number of low-frequency phonons having large ℓ could be important. To demonstrate this possibility, we used in Eq. 1, the measured ultrasonic attenuation ℓ^{-1} for $\nu \lesssim 5 \times 10^{11}$ Hz [19] and $\ell/\lambda = 1$ at higher frequencies. The result is shown as the solid line in Fig. 1. This curve has a positive temperature dependence and a magnitude similar to that of the data, indicating that some fraction κ/K at $T/\theta > 10^{-1}$ may indeed be attributed to relatively low-frequency phonons.

In conclusion, the thermal conductivity of amorphous materials may be brought into common register, both at low and high temperatures, by a one-parameter scaling procedure. At high temperatures it is suggested that long wavelength phonons contribute to heat transport. The plateau appears to be caused by an explicitly wavelength-dependent process which is nearly independent of the nature of the glass.

Acknowledgements

This work was supported in part by National Science Foundation - Low Temperature Physics - Grant DMR83-03918.

References

1) M.P. Zaitlin and A.C. Anderson, Phys. Rev. B12, 4475 (1975)
2) O.L. Anderson, J. Phys. Chem. Solids 12, 41 (1959)
3) T.L. Smith, Ph.D. Thesis (University of Illinois, 1979) unpublished
4) Y.S. Touloukian, R.W. Powell, C.Y. Ho and P.G. Klemens, Thermophysical Properties of Matter, Vol. 2 (Plenum, New York, 1970) p. 193
5) R. B. Stephens, Phys. Rev. B13, 852 (1976)
6) D.S. Matsumoto and A.C. Anderson, J. Non-Cryst. Solids 44, 171 (1981)
7) D. Greig and M.S. Sahota, J. Phys. C16 L1051 (1983)
8) C.L. Choy and D. Greig, J. Phys. C8, 3121 (1975)
9) J.J. Freeman, Ph.D. Thesis (University of Leeds, U.K., 1985) unpublished
10) R.B. Stephens, G.S. Cieloszyk and G.L. Salinger, Phys. Lett. 38A, 215 (1972)
11) J.J. Freeman and A.C. Anderson, submitted to Phys. Rev. B.
12) W. Dietsche and H. Kinder, Phys. Rev. Lett. 43, 1413 (1979). The authors fit to $\ell \propto \nu^{-3}$, however a dependence of $\ell \propto \nu^{-4}$ would also fit the data.
13) V.G. Karpov and D.A. Parshin, JETP 61, 1308 (1985)
14) S. Alexander, O. Entin-Wohlman and R. Orbach (private communication)
15) D.P. Jones, J. Jackle and W.A. Phillips, in Phonon Scattering in Condensed Matter, ed. by H.J. Maris (Plenum, New York, 1980) p. 49 and papers cited therein
16) J.E. Graebner, B. Golding and L.C. Allen, unpublished.
17) Clare C. Yu and J.J. Freeman, (this conference)
18) M.C. Roufosse and P.G. Klemens, J. Geophys. Res. 79, 703 (1974)
19) S. Hunklinger and W. Arnold, in Physical Acousties, ed. by W.P. Mason and R.N. Thurston (Academic, New York, 1976), Vol. 12 p. 155.

Thermal Properties of Amorphous Cross-Linked Polybutadiene

H. v. Löhneysen[1 +], *E. Ratai*[1], *and U. Buchenau*[2]

[1]Zweites Physikalisches Institut der RWTH Aachen
D-5100 Aachen, Fed. Rep. of Germany
[2]Institut für Festkörperforschung der KFA Jülich
D-5170 Jülich, Fed. Rep. of Germany

In amorphous polymers, like in other glasses, phonons are scattered by localized two-level tunneling states (TLS) at temperatures below ≃ 1 K /1/. This leads to a T^2 dependence of the thermal conductivity κ, while the TLS themselves contribute with a linear term to the specific heat C. At higher temperatures (1 to 10 K), the origin of the excess specific heat (over the Debye contribution calculated from the sound velocity) and of the ubiquitous plateau in κ is unknown. The plateau must arise from a rather sharp drop of the phonon mean free path with increasing frequency. Various models have been offered to explain the κ plateau and the enhanced C /2-5/. It is therefore important to experimentally provide evidence which either supports or dismisses any of these models.

In particular, the phonon-fracton model /4/ predicts a vibrational density of states D (ω) which changes from phonon-like at low frequency ($D(\omega) \sim \omega^{d-1}$) to fracton-like ($\sim \omega^{p-1}$) where d is the Euklidean dimension and $p \leq d$ is (for infinite fractals) equal to the fracton dimension, with a pronounced feature occurring at the cross-over frequency ω_c. ω_c is inversely related to a characteristic length L_c of the system. It has been suggested that amorphous materials in general, and polymers in particular, would be a suitable realization of a fractal structure /4/. For polymers, it is tempting to associate L_c with the mean distance R between the cross-links of a polymer chain segment. Cross-linking can be achieved e.g. by γ irradiation. In this paper, we report specific-heat and thermal-conductivity results of amorphous cross-linked polybutadiene, $(-CH_2-CH=CH-CH_2-)_n$, where R was varied such as to span the whole range of "dominant phonon" wavelengths of the T range of our measurements.

Fig. 1 shows the specific heat of γ-irradiated polybutadiene (PB) plotted as C/T^3 vs. T on a log-log plot. The upturn at low T signals the linear contribution due to TLS. These data are in good agreement with earlier measurements below 1 K /6/. At higher T the data approach the Debye contribution, with only a small (if any) excess T^3 term. Finally, a shallow maximum occurs around 5 K. The eminent feature of Fig. 1 is the complete absence of a cross-linking effect, save a slight enhancement of the TLS contribution below 1 K for the highest dose.

Fig. 2 shows the thermal conductivity κ vs. T, also on a log-log plot. κ displays the features well-known from glasses. An overall systematic increase of κ with increasing cross-link density is observed. Below 1 K, κ varies as T^m with m increasing from 1.5 to 1.7 with increasing dose. At higher temper-

---
+ Present address: Physikalisches Institut der Universität Karlsruhe, D-7500 Karlsruhe, FRG

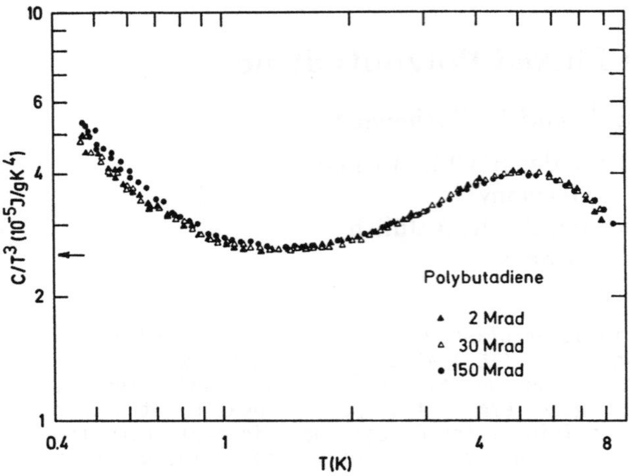

Fig. 1. Specific heat of polybutadiene irradiated with different γ doses plottes as C/T^3 vs. T. Arrow indicates the Debye contribution

atures, κ displays a shallow minimum around 10 K instead of the plateau. This feature has been occasionally found in glasses /7/. The largest κ increase (up to 30 %) occurs in this T region. This is much smaller than the 400 % increase reported previously for chemically cross-linked PB /6/. Apparently, in this early work some crystallization ocurred inadvertantly during chemical cross-linking /8/.

In this short paper, we discuss only the results for $T > 1$ K. The data below 1 K can be qualitatively accounted for in terms of the TLS model. Regardless of the nature of the elementary vibrational excitation, its frequency ω should increase monotonically with λ^{-1}, where λ is the characteristic length of the mode. For dominant phonons, the wavelength is $\lambda_d \simeq hv/4K_BT$ where v is an average over the measured /9/ sound velocities. λ_d varies between 20 Å at 10 K and 500 Å at 0.4 K in PB. The mean distance R between cross-links covers just this range of length-scales: For doses between 11.2 and 19.7 MRad, the number of monomers N between cross-links has been measured /10/. From $R = aN^{-1/2}$ with a = 6.9 Å /11/, R lies between 110 and 60 Å and is expected to cover a larger range in our samples. We would therefore expect a pro-

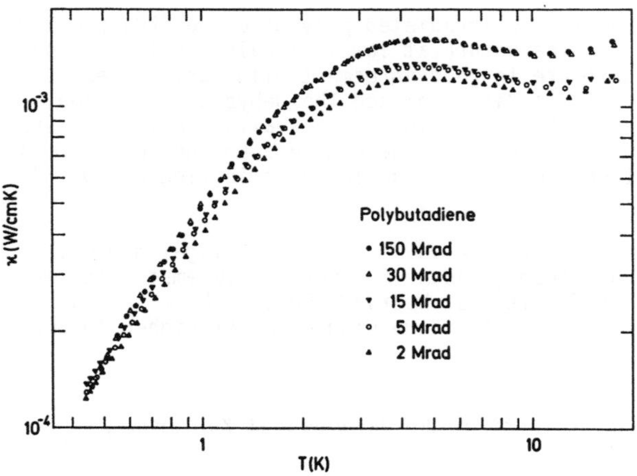

Fig. 2. Thermal conductivity κ vs. temperature T of polybutadiene irradiated with different γ doses.

nounced feature in the specific heat at a cross-over temperature T_c for which $\lambda_d/R \simeq 1$, i.e. at different temperatures T_c for our different samples, arising from the predicted distinct increase of the vibrational density of states at the phonon-fracton crossover frequency ω_c. This is not observed in our measurements.

According to the fracton model, the heat transport should change drastically at T_c when changing from propagating phonons to nonpropagating fractons. Again, no significant change of κ is seen when passing through the anticipated cross-over region although the increase of κ with cross-link density calls for further attention.

The failure to observe a fracton vibrational spectrum in the specific heat and related strong features in the thermal conductivity indicates that, apparently, amorphous polymers are not a good realization of a fractal as far as the vibrational properties are concerned. On the other hand a polymeric chain can be structurally described as a fractal. Hence it might be concluded that the polymer entanglement restores the Euclidean dimensionality for vibrations. By the same token, the neutron scattering data on epoxy resins /12/ provide only circumstantial evidence for the fracton model, as has been pointed out before /4/. A suitable model substance for a crucial test of this intriguing model remains yet to be found.

This work was carried out within the research program of Sonderforschungsbereich 125, supported by the Deutsche Forschungsgemeinschaft.

REFERENCES

1. Amorphous Solids: Low-Temperature Properties, ed. by W.A. Phillips, (Springer, Berlin 1981)
2. For a review of early models, see: D.P. Jones, N. Thomas, W.A. Phillips: Phil.Mag. B 38, 271 (1978)
3. M.I. Klinger: Solid State Commun. 51, 503 (1984)
4. A. Aharony, S. Alexander, O. Entin-Wohlmann, R. Orbach: Phys.Rev. B 31, 2565 (1985) and refs. therein
5. E. Akkermans, R. Maynard: Phys. Rev. B 32, 7850 (1985)
6. D.S. Matsumoto, A.C. Anderson: J. Non-Cryst. Solids 44, 171 (1981)
7. R.B. Stephens, Phys. Rev. B 8, 2896 (1973)
8. J.J. Freeman, G.X. Mack, A.C. Anderson: these Conference Proceedings
9. A. Bhattercharyya, T.L. Smith, A.C. Anderson: J. Non-Cryst. Solids 31, 395 (1979)
10. T.-K.Su, J.E. Mark; Macromolecules 10, 120 (1977)
11. F.S. Bates, G.D. Wignall, W.C. Koehler: Phys. Rev. Lett. 55, 2425 (1985)
12. H.M. Rosenberg: Phys. Rev. Lett. 54, 704 (1985)

Effect of Crosslinking on the Thermal Conductivity of cis-1,4 Polybutadiene

J.J. Freeman, J.X. Mack, and A.C. Anderson

Department of Physics, University of Illinois at Urbana-Champaign, 1110 W. Green St., Urbana, IL 61801, USA

The thermal conductivity κ of the polymer cis-1,4 polybutadiene (PB) had been shown [1] to increase by a factor 4 when the distance between crosslinks became as small as 50 chemical repeat units (RU). This phenomenon was not explained by the authors. Subsequent measurements on other polymers have shown similar types of behaviour [2,3] and suggest that the change in κ is due to the presence of crystallites. Although the authors of Ref. 1 had thought that cis-1,4 polybutadiene with a cis content of $\gtrsim$ 98% would not crystallize, we have investigated this possibility as a cause for the anomalous behavior.

The crystallization of polymeric materials is both strongly time and temperature dependent [4,5], since it takes a finite time for crystallites to form. Hence, as a test of the crystallization hypothesis, we have subjected the PB to various thermal treatments using differential scanning calorimetry (DSC) and by quenching the low-crosslink material to 77 K before measuring κ at low temperature.

New samples of PB crosslinked with dicumyl-peroxide were prepared as before [1], with crosslink densities of 7 ± 2 RU and 70 ± 15 RU. Data obtained for the new samples, together with those of Ref. 1, are shown in Fig. 1. The new data for samples cooled slowly in the cryostat (~ 1.5 K/min between 300 K and 180 K), clearly agree with the previous measurements.

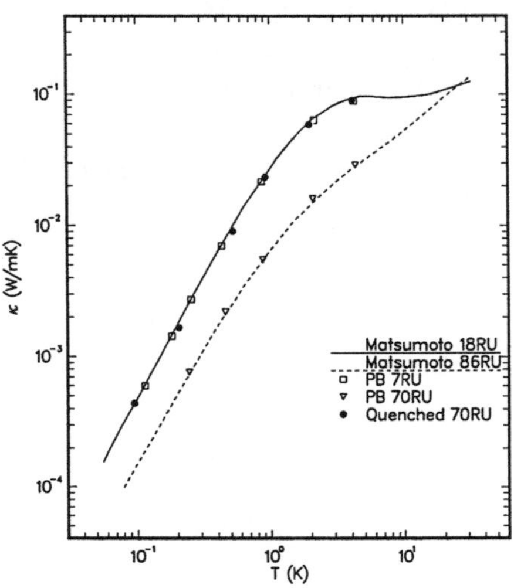

Fig. 1. Thermal conductivity κ for glassy and partially crystalline rubbers. Points are the present data, lines are from Ref. 1

However, the quenched sample of low-crosslink density now has the same thermal conductivity as the highly crosslinked material. That is, by rapidly cooling to 77 K, the low temperature thermal conductivity of the same sample increases by a factor of 4.

Since it is difficult to cool samples in the cryostat at intermediate rates (i.e., less than the quench rate of ~ 3000 K/min, but faster than 1.5 K/min) runs were also made using DSC techniques for which the cooling rate could be varied between ~ 1 K/min and 50 K/min. Data taken on the 70 RU sample [6] indicate that, for cooling rates $\lesssim$ 5 K/min, the volume fraction crystallinity attains a maximum ≈ 40%, and that for cooling rates faster than 20 K/min, no crystallites are formed. These results are in agreement with the thermal conductivity data in Fig. 1 and show that the 70 RU sample is ≈ 40% crystalline when cooled "normally" in the cryostat, but is amorphous when quenched. On the other hand, even for cooling rates $\lesssim$ 1 K/min, the 7 RU sample remained completely amorphous.

We now turn our attention to the magnitude of the change in κ. The reduction in κ due to the presence of crystallites has been considered before [7]. Below 1 K, phonons scattered by the crystallites can both reduce the phonon mean-free path and exclude heat flow from some of the sample volume. Scattering from the cyrstallites would be expected to produce a constant mean-free path and hence a T^3 dependence for κ. Since κ still depends quadratically on T as in Fig. 1, it appears that the mean-free path is not greatly effected by the crystallites, and that the dominant effect is merely to exclude the phonon flux from the crystalline portions of the sample. Models may be found in the literature [8,9] which predict the reduction in κ due to the presence of voids with various geometries. These predictions however, when compared to low-temperature PET data [2], are only found to work for crystalline fractions $\lesssim$ 30% [6]. Above this value, the measured reduction in κ becomes greater than that predicted. Experimentally however, the reduction in κ for ≈ 40% crystalline PET is of order 4 and is in good agreement with the reduction reported here for PB.

In summary, the anomalously large change in the low-temperature thermal conductivity of polybutadiene with chemical crosslinking, is due to the formation of crystallites in samples having low crosslink densities. The low-temperature thermal conductivity of rubber therefore, depends both on the crosslink density and on the cooling rate. It has been shown how the degree of crystallinity may be changed using appropriate thermal treatments.

References

1) D.S. Matsumoto and A.C. Anderson, J. Non-Cryst. Solids, 44, 171 (1981).
2) D. Greig and M.S. Sahota, J. Phys. C16, L1051 (1983).
3) J.J. Freeman and D. Greig, in Non-Metallic Materials and Composites at Low-Temperatures, ed. G. Hartwig and D. Evans (Plenum, New York, 1985) Vol. III.
4) L.A. Wood and N. Bekkedahl, J. Appl. Phys. 17, 362 (1946).
5) J.D. Hoffman, G.T. Davis and J.I. Lauritzen, in Treatise on Solid State Chemistry: Crystalline and Noncrystalline Solids, ed. N.B. Hannay (Plenum, New York, 1976) Vol. 3, Ch. 7.
6) J.J. Freeman, J.X. Mack and A.C. Anderson, J. Non-Cryst. Solids, to be published.
7) E.P. Roth and A.C. Anderson, J. Appl. Phys. 47, 3644 (1976).
8) R.E. Meredith and C.W. Tobias, J. Appl. Phys. 31, 1270 (1960).
9) M.P. Zaitlin and A.C. Anderson, Phys. Rev. B12, 4475 (1975).

Low-Temperature Thermal Diffusivity in Polyethylene

Y. Kogure, T. Mugishima, and Y. Hiki

Faculty of Science, Tokyo Institute of Technology
Oh-okayama, Meguro-ku, Tokyo 152, Japan

1. Introduction

The thermal diffusivity D (= $\kappa/\rho C$, where κ is the thermal conductivity, ρ the mass density, and C the specific heat) is an important quantity to study the phonon propagation in insulators, because D is directly proportional to the phonon mean free path when the simple gas kinetic theory is considered. In order to measure the thermal diffusivity directly, some kinds of dynamical experimental method should be used. We have developed a laser pulse method to measure the low-temperature thermal diffusivity, and polyethylene was adopted as the sample material.

2. Experimental Method

A pulse heating method was applied to measure the thermal diffusivity of the material at temperatures between 300 K and 4.2 K . The specimens used were disk-shaped, 10 mm in diameter and 1 mm in thickness. To absorb the laser pulse energy, a graphite carbon powder was sprayed on the front face of the specimen. A pulse ruby laser of usual type was used for the pulse heating. The pulse width was less than 1 msec , and the maximum power was 1 J/pulse . To detect the temperature variation $\Delta T(t)$ at the back face of the specimen after the pulse heating, a sputtered gold film bolometer and a painted carbon film bolometer were prepared on the face. The electrical resistance of the gold film decreases linearly with decreasing temperature in the temperature range of 300 K - 15 K , but the resistance change levels off at lower temperatures. On the contrary, the resistance of the carbon film increases rapidly with decreasing temperature at temperatures below 20 K . Both of the two bolometers are to be used for the diffusivity measurements in the wide temperature range of 4.2 K - 300 K .

The specimen was set to a sample holder by using polystyrene sheet washers or thin nylon strings. A chromel-constantan thermocouple and a germanium resistance thermometer were attached to the sample holder to calibrate the bolometers. This assembly was set in the specimen chamber of a cryostat. Low-pressure He exchange gas (10^{-2} - 10^{-3} torr) was contained in the chamber to attain thermal equilibrium between the specimen and the thermometers. The laser beam was introduced onto the specimen through fused-quartz optical windows of the cryostat.

When the laser pulse is irradiated on the front face of the specimen, the temperature of the back face changes with time. The temperature variation is converted to the electrical signal by the bolometers, being recorded in a sensitive digital recorder. The stored data are transfered to a micro-computer and are recorded in a floppy disk.

3. Analysis

Examples of the recorded data of the temperature variation ΔT are shown in Fig. 1. It is seen that the increase of ΔT becomes more rapid when the ambient temperature is lowered. After reaching a maximum, ΔT decreases gradually due to the heat loss from the specimen. Theoretical expressions for the temperature variation after the pulse heating have been derived by several authors. They treated, however, the effect of thermal radiation loss from the specimen, because the pulse method was usually used for the measurement at elevated temperatures. On the contrary, the loss through the heat conduction becomes noticeable when the temperature is lowered. We have derived theoretical formulas for the temperature variation with taking the conduction loss effect into account [1].

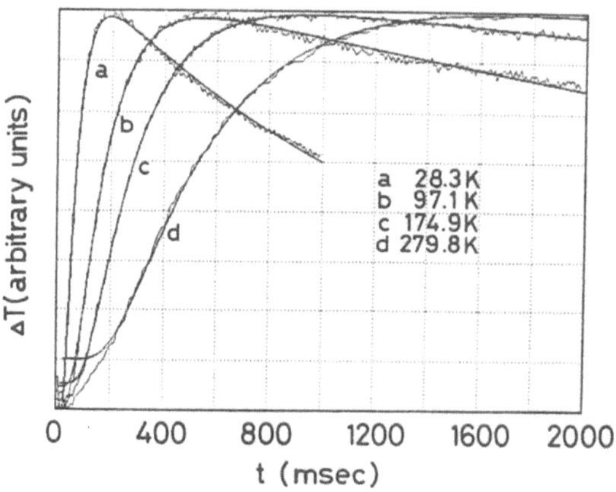

Fig. 1. Temperature variation ΔT(t) after the pulse-heating detected by a gold thin film bolometer, The solid curves show theoretically fitted ones. The part of the data immediately after the pulse-heating was excluded for the fitting because of the lack of stability of the data.

The thermal diffusion equation has been solved under such a boundary condition that the heat current density at the specimen surface was given by $\Delta T/R_s$, where ΔT was the temperature increase at the surface and R_s the surface thermal resistance per unit area. When values of R_s are uniform for all of the specimen faces, our formulas take the same forms that CAPE et al. have obtained for the radiation-loss case [2]. However, the loss parameter appearing in their forms should be changed to $Y_x = \dot{a}/\kappa R_s$, where a is the specimen thickness and κ the thermal conductivity of the specimen. We also examined the cases where the conduction loss occurred non-uniformly. Such a situation seems to be likely in the case of the present experiment, because the paths of the heat leak are rather complicated. After some numerical calculations, the effect of the non-uniformity on the finally determined value of the specimen diffusivity D was found to be very small, when the loss parameter was not large. Such a situation is always realized under usual experimental conditions [1].

The theoretical formula we obtained for the temperature variation after the pulse heating contains four unknown parameters. These are the thermal diffusivity D, the loss parameter Y_x, and the normarization constants related to the input power and the background temperature. These parameters can be determined independently by the least-squares fit of the experimental data to the theoretical formula. The fit was made for data at various ambient temperatures, and the temperature dependence of diffusivity was obtained.

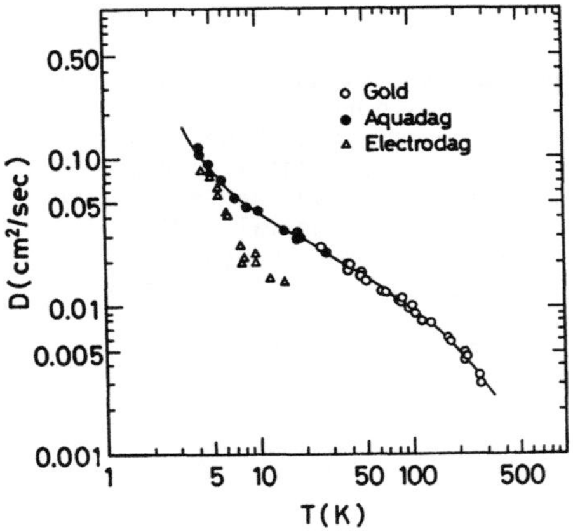

Fig. 2. Temperature dependence of thermal diffusivity D measured by a gold film and two kinds of carbon film bolometers.

4. Result and Discussion

The experimentally determined values of thermal diffusivity of polyethylene (60 % crystallinity, supplied by Idemitsu Kosan Co. Ltd.) are shown in Fig. 2. The data at higher temperatures ($T > 20$ K) were taken by using the gold film bolometer. For the bolometer at lower temperatures, two kinds of graphite carbon solution were tested: Aquadag and Electrodag 112 (Acheson). It is seen that the data by the use of Aquadag bolometer are smoothly connected to the high-temperature data by the gold bolometer. However, the measured values by the Electrodag bolometer deviate from the above data, probably due to the slow response of this carbon film.

The obtained values of the diffusivity are consistent with those derived from the results of static thermal measurements by other authors [3,4]. It is seen that, at high temperatures, the temperature dependence of the thermal diffusivity is nearly proportional to 1/T. The diffusivity values here are in agreement with those estimated by the phonon U-process mechanism [5]. At low temperatures, a noticeable deviation from the 1/T relation is seen. Similar temperature dependence has also been observed by ZIRKE et al. [6]. Polyethylene consists of crystal and amorphous parts (lamella structure), both composed of linear chains of CH_2. We are planning further experiments on samples with different crystallinity and on highly extruded samples to investigate the phonon propagation in such well characterized materials.

References

1. Y. Kogure, T. Mugishima and Y. Hiki: to be published
2. J. A. Cape and G. W. Lehman: J. Appl. Phys. 34, 1909 (1963)
3. S. Burgess and D. Greig: J. Phys. C 8, 1637 (1975)
4. T. A. Scott, J. de Bruin, M. M. Giles and C. Terry: J. Appl. Phys. 44, 1212 (1973)
5. Y. Hiki, T. Mugishima, Y. Kogure and K. Kawasaki: Proc. 2nd Intern. Conf. on Phonon Physics (World Scientific, Singapore, 1985) p. 771
6. J. Zirke and M. Meissner: Infrared Phys. 18, 871 (1978)

The Thermal Conductivity of Several Fluoride Glasses

K.A. McCarthy, H.H. Sample, and M.B. Koss

Physics Department, Tufts University, Medford, MA 02155, USA

1. Introduction

For all amorphous materials, irrespective of composition, the thermal conductivity, K, is nearly the same function of temperature. This fact is believed to be a consequence of the existence of a more or less universal, frequency-dependent phonon mean free path, leading to a conductivity which first increases as T^2, flattens out into a temperature-independent plateau region, and finally again increases slowly with temperature up to just below the glass transition temperature, T_g [1]. In this paper we present thermal conductivity results for a family of fluoride glass specimens, which we have measured in the 1.6 to 100 K temperature range: the materials include pure beryllium fluoride glass, a fluoroberyllate glass, three fluorozirconate glasses, and a fluoride glass with equal majority components of ZnF_2, YbF_3, and ThF_4 [2]. Some of these results have been reported previously [3]. We find that, below the plateau region, the thermal conductivities can be well correlated with specimen composition.

2. Results

Table 1 lists the compositions (in mol %) of the six specimens we studied. Note that the cation mass for the several fluorides increases downward, and

Table 1. Composition (mol percent) of fluoride glass specimens

	BFG (15.7) △△△	FBG1 (21.6) ▲▲▲	FZG1 (36.0) ○○○	FZG2 (34.7) ●●●	FZG3 (34.9) ×××	FG1 (52.6) ***
LiF			21	21		
BeF_2	100	48				
NaF					20	
AlF_3		10	3	3	4	
KF		27				
CaF_2		14				
ZnF_2						27
ZrF_4			50	50	56	
BaF_2			16	20	14	19
LaF_3			5	6	6	
NdF_3		1				
YbF_3						27
PbF_2			5			
ThF_4						27

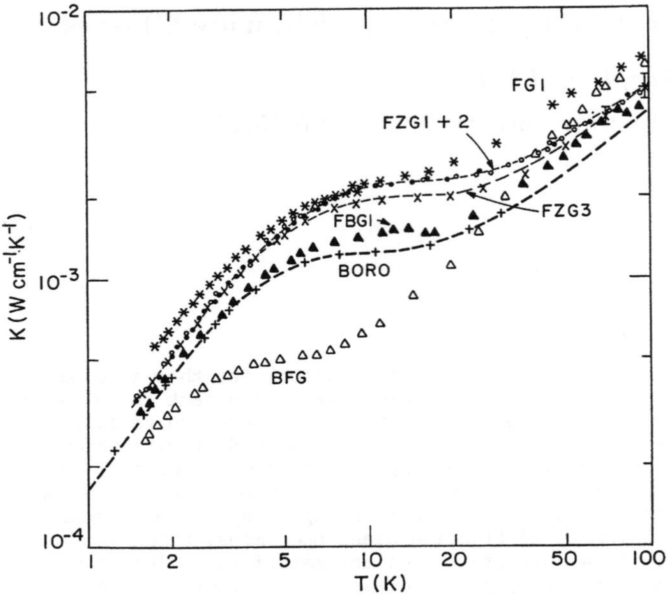

Fig. 1. Thermal conductivity versus temperature for six fluoride glass samples (see Table 1 for compositions and identification of plot symbols). The crosses are for a borosilicate glass reported by Zaitlin and Anderson (Ref. 4). The lines are fits to the data as described in the text

the samples are arranged so that the heavier glasses are on the right. The table also lists $\overline{M}_{at}$, the average mass per atom in amu, below each sample's designation.

Figure 1 shows the thermal conductivities of the six glasses. All exhibit the characteristic behavior of amorphous materials at high temperatures, in the plateau region, and at lower temperatures (although the data do not extend to low enough temperatures to attain the limiting T^2 dependence). Certain general features are immediately evident. First, the very similar composition fluorozirconate glasses (FZG1, 2, and 3) have identical conductivities at low temperatures, and are nearly the same at all temperatures, despite the fact that the three glasses come from two different sources. Second, the lighter glasses (BFG and FBG1) have lower conductivities at low temperatures and in the plateau region; to our knowledge, the BeF_2 (BFG) sample shown has the lowest plateau region conductivity of any glass reported to date. The heaviest glass (FG1) has the highest low-temperature conductivity. For comparison, we also show the results of Zaitlin and Anderson [4] for a borosilicate glass (BORO).

As described previously [3], these data can be fit to a high degree of precision using a tunneling states model [4]; see the lines through the fluorozirconate glass data in Fig. 1. The values of the parameters [3] used were: Θ_D=300 K, v=2.4x10^5 cm/s, P(1)=0.29-0.32 cm^{-1}K^{-3}, P(2)=(1.4-1.5)x10^3 cm^{-1}K^{-1}, P(3)=1.9-2.4 cm^{-1}K^{-4}, and l_c=(1.4-1.6)x10^{-8} cm. We also note that our BeF_2 results are consistent with the measurements of Leadbetter, et al. [5] and of Stephens [6] at lower temperatures.

In Fig. 2 we plot K(1.5 K), the conductivity of the fluoride glass samples extrapolated to 1.5 K, versus the average atomic mass, $\overline{M}_{at}$. The line illustrates that a linear relationship exists between these two parameters, in this family of similar glasses. (In fact, the relationship is perfect, within

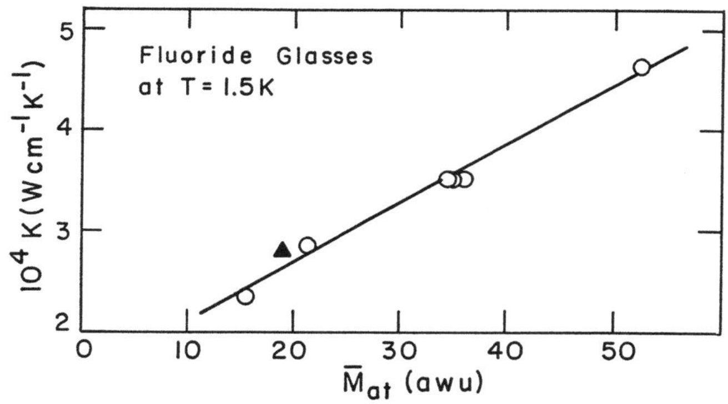

Fig. 2. Thermal conductivity at T=1.5 K (extrapolated) versus average atomic mass in amu. Circles: fluoride glasses, triangle: borosilicate glass (Ref 4)

the few percent reproducibility of our measuring apparatus.) The solid triangle is for the borosilicate glass shown in Fig. 1. A somewhat less perfect linear relationship also exists between K(1.5 K) and mass density, since the average atomic volumes of all these glasses are similar. We also attempted to correlate K(1.5 K) with inverse glass transition temperature, $1/T_g$, as suggested by the acoustic measurements on fluoride glasses of Doussineau, et al. [7], and by the specific heat measurements of Raychaudhari and Pohl [8]. We find no correlation: for example, the T_g's of the BFG and three FZG samples are the same to within 3%, and are some 20% lower than for the FG1 sample. This non-correlation of $1/T_g$ with low temperature thermal properties has also been reported for potassium borate glasses by MacDonald, et al. [9].

As seen in Fig. 1, in the plateau regions the sequence of conductivities is the same as at low temperatures, although the dependence of conductivity on $\overline{M}_{at}$ "saturates" at high mass. At still higher temperatures, several of the curves cross one another, and we have been unable to establish a regular dependence of conductivity on any physical parameter, including average atomic mass, average cation mass, average volume per atom, density, or glass transition temperature.

Acknowledgements

We thank M.G. Drexhage, D. Tran, and M. Weber for kindly supplying the samples used in this work. This work was partially supported by a grant from the Tufts Faculty Research Fund.

1. A.C. Anderson: in Amorphous Solids, Low Temperature Properties, ed. by W.A. Phillips (Springer-Verlag, New York 1981), pp. 65-80
2. The samples were supplied by M.G. Drexhage of RADC, D. Tran of NRL, and M. Weber of Lawrence Livermore
3. K.A. McCarthy, H.H. Sample and W.G.D. Dharmaratna, J. Non-Cryst. Solids 64, 445 (1984); K.A. McCarthy, H.H. Sample and M.B. Koss, Mat. Res. Soc. Symp. Proc. 61, 441 (1986)
4. P.W. Anderson, B.I. Halperin and C.M. Varma, Phil. Mag. 25, 1 (1972); W.A. Phillips, J. Low Temp. Phys. 7, 351 (1972; M.P. Zaitlin and A.C. Anderson, Phys. Rev. B12, 4475 (1975); Phys. Stat. Sol. (b) 71, 323 (1975)
5. A.J. Leadbetter, A.P. Leapes, C.G. Waterfield and R. Maynard, J. Physique 38, 95 (1977)
6. R.B. Stephens, Phys. Rev. B8, 2896 (1973)
7. P. Doussineau, M. Matecki and W. Schön, J. Physique 44, 101 (1983)
8. A.K. Raychaudhuri and R.O. Pohl, Phys. Rev. B25, 1310 (1982)
9. W.M. MacDonald, A.C. Anderson and J. Schroeder, Phys. Rev. B32, 1208 (1985)

Phonon Scattering by Low-Energy Excitations and Free Volume in Amorphous PdCuSi

H.W. Gronert[1], *D.M. Herlach*[2], *and G.V. Lecomte*[1]

[1]Labor für Tieftemperaturphysik, Universität/GH Duisburg
D-4100 Duisburg, Fed. Rep. of Germany
[2]Institut für Raumsimulation, DFVLR
D-5000 Köln 90, Fed. Rep. of Germany

It is well established that the low-temperature properties of amorphous solids are governed by structure-induced low-energy excitations (LEE). The LEE can be described by the two-level-system (TLS) tunneling model [1]. Cohen and Grest [2] correlate these tunneling states with the excess free volume v_f of glasses frozen in during the quenching process: the TLS are assumed to arise from tunneling of single atoms or small groups of atoms into voids, i.e. local agglomerations of free volume. In this model, one expects the density of TLS to increase with increasing v_f, or equivalently to decrease with increasing mass density ρ. Recently, we have demonstrated the similarity of the changes in v_f and in the phonon scattering from LEE upon altering the amorphous structure by variation of the quenching rate [3], by structural relaxation [4.5] or by plastic deforming [3] amorphous Pd-Cu-Si. These findings give direct evidence for the validity of the free volume model to explain the presence of LEE in metallic glasses.

Amorphous $Pd_{77.5}Cu_6Si_{16.5}$ samples are prepared by water quenching and splat cooling at various quenching rates (10^3K/s - 10^6K/s). Details of the heat treatments and the thermal conductivity $\kappa(T)$ measurement have been given elsewhere [3-6]. In nonmagnetic, normal conducting metallic glasses like Pd-Cu-Si, phonons and electrons act as heat carriers, $\kappa=\kappa^{ph}+\kappa^{el}$. The electronic contribution, κ^{el}, can be separated via the Wiedemann-Franz law. The phonon-LEE scattering coefficient $1/\alpha$ can be determined by plotting T/κ^{ph} versus $1/T$, because in Pd-Cu-Si for $T \lesssim 0.85$ K the phonon scattering on TLS is resonant, and

$$T/\kappa^{ph} = D_{el} + 1/(\alpha T) \qquad (1)$$

as shown earlier [6]. D_{el} is the phonon-electron scattering coefficient. $1/\alpha$ is given by the slope of the straight line, see Fig. 1. Within the TLS tunneling model $1/\alpha \sim n\gamma^2$ with n the density of TLS-states and γ the coupling parameter of phonons with TLS [1].

Recently, Cahn et al. [7] have shown that in Pd-Cu-Si the increase of ρ after annealing at 573 K for 1 h can be partially reversed by subsequent cold rolling. We applied the equivalent treatments to a water-quenched sample and measured the thermal conductivity in the various states. The results are shown in Fig. 1, where T/κ^{ph} is plotted vs. $1/T$. Below $T \lesssim 0.85$ K, a linear dependence is found for each state, confirming the validity of eq. 1. One can see that after structural relaxation the slope ($1/\alpha$=431 K^3cm/W) is reduced with respect to the as-quenched state ($1/\alpha$=740 K^3cm/W). After plastic deformation, the initial conductivity values are reached again ($1/\alpha$=794 K^3cm/W). In other words, an increase of the mass density by structural relaxation causes a corresponding decrease of $1/\alpha$, whereas a decrease in ρ by plastic deformation of the relaxed sample leads back to the original state. This observation is in qualitative conformity with the free volume model.

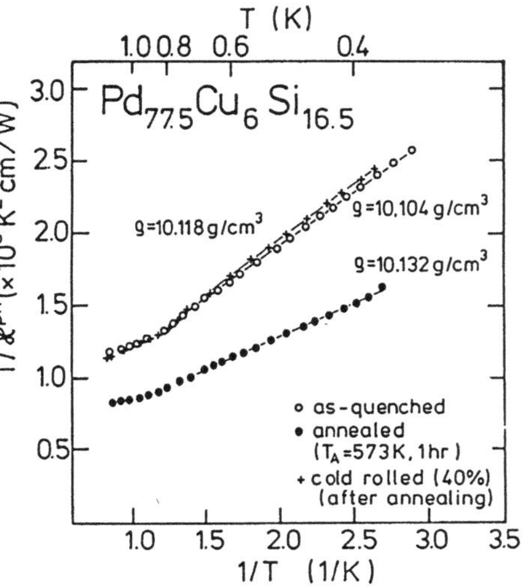

Fig.1: T/κ^{ph} vs. 1/T. The mass densities are taken from Ref. [7]

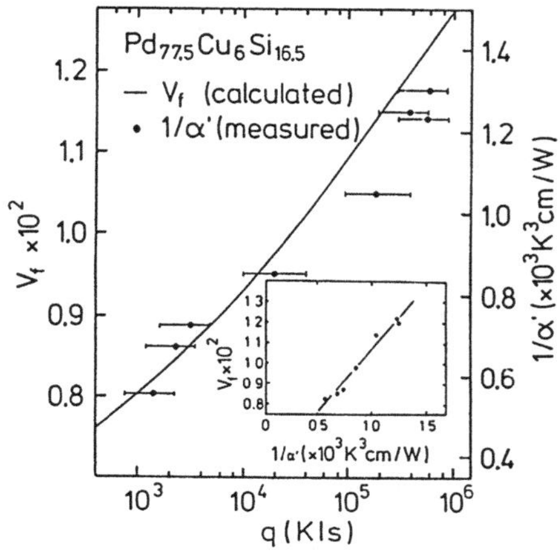

Fig. 2: v_f and $1/\alpha$ vs. the quenching rate q

We have also measured $\kappa(T)$ in a-$Pd_{77.5}Cu_6Si_{16.5}$ prepared with different quenching rates q. Fig. 2 shows a semi-logarithmic plot of $1/\alpha$ (right hand scale) as a function of q (closed dots). The free volume of the corresponding as-quenched states, calculated according to v.d.Beukel and Radelaar [10], is also given (left hand scale) vs. q (drawn line). As can be seen, both v_f and $1/\alpha$ increase in the same way with q. The insert shows v_f as a function of $1/\alpha \sim n\gamma^2$. We find the strong correlation $n\gamma^2 \sim v_f$ according to the model of Cohen and Grest, assuming that γ does not vary strongly.

As mentioned above, structural relaxation induces changes in the low temperature thermal conductivity of amorphous solids. Irreversible and reversible effects have been found [8,9,11]. Consequently, we have investigated systematically the changes of v_f and the accompanying changes of the phonon-LEE scattering coefficient $1/\alpha$ of amorphous Pd-Cu-Si by stepwise annealing differently quenched samples at various temperatures T_a below the glass temperature T_g=642K, see Fig.3. Closed circles represent measurements taken after fixed anneal durations (15min) and successively increasing anneal temperatures. They follow closely the behavior of the calculated free volume variation, in qualitative agreement with the Cohen and Grest model.

However, when the sample was annealed at a temperature of 463 K (open square), lower than the previous anneal temperature (513 K), it exhibited a markedly lower value of $1/\alpha$. Further anealing at 513 K lead back to a higher value of $1/\alpha$, lying again on the calculated v_f-curve. A reversible change in $1/\alpha$ had first been observed by Cotts et al. [9]. As shown in Fig.3 the free volume changes only can not satisfactorily explain the whole annealing behavior. This observation leads to the conclusion that, beside the irreversible reduction in $1/\alpha$ which strongly correlates to the v_f variation, there also exists a reversible mechanism that affects $1/\alpha$ while the free volume remains practically constant. The existence of reversible as well as of irreversible changes in $1/\alpha$ can possibly be related to differences between topological short range ordering (TSRO) and chemical short range ordering (CSRO), as proposed by Egami [13] for the magnetic properties. A phenomenological explanation would then be that gross changes in the alloy topology are reflected in the irreversible variations of the free volume. After initial relaxation, they correspond mostly to rearrangements of whole Si-metall groups which are predominantly covalently bond. The diffusion of metallic atoms between themselves is however easier and can occur faster. This leads to a chemical equilibrium of metal atoms at annealing temperatures T_a > 470 K. The degree of CSRO is a monotonic decreasing function of T_a [10], and can be reversibly altered. So, the irreversible decrease in $1/\alpha$ can be correlated to changes in TSRO, while the reversible changes can be attributed to variations in CSRO [5].

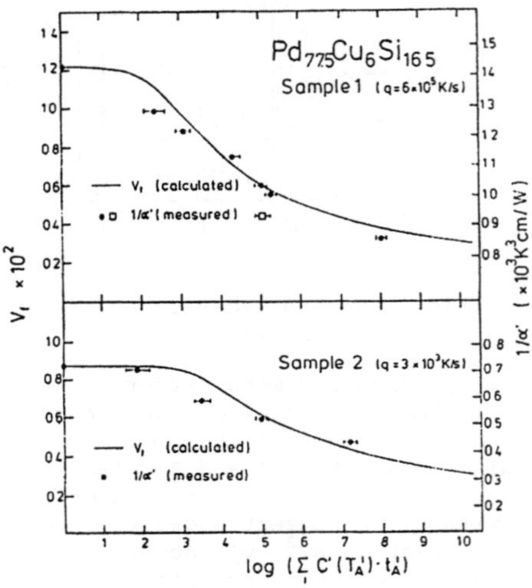

Fig. 3: v_f and $1/\alpha$ vs. the annealing conditions. Log. $\Sigma\ (C'(T_a^i)t_a^i)$ describes the step by step annealing procedure [4,5], with

$C'(T_a^i)=C\ \exp(-Q_a/k_B T_a^i)$ where $C=1.8\ 10^{18}s^{-1}$ and

Q_a=160KJ/mole is the activation energy for annihilation of the free volume [10,12].

1 For a review, see: Amorphous solids: Low-temperature properties, W.A. Phillips (Ed.), Springer verlag, Berlin, 1981
H.v. Loehneysen, Phys. Rep. 79, 161 (1981)

2 M.H.Cohen and G.S.Grest, Phys. Rev. Lett. 45, 1271 (1980); J.Non-Cryst. Solids 61&62, 749 (1984)

3 D.M.Herlach, H.W.Gronert and E.F.Wassermann, Europhys. Lett. 1, 23 (1986)

4 H.W.Gronert and D.M.Herlach, in: Proc. 2nd Int. Conf. on Phonon Physics (Budapest, 1985), p 89

5 D.M.Herlach, H.W.Gronert, E.F.Wassermann and W.Sander, to appear in Z. Physik B

6 R.Willnecker, D.M.Herlach and E.F.Wassermann, Phys. Rev. B 31, 6234 (1985)

7 R.W. Cahn, N.A.Pratten, M.G.Scott; H.R. Sinning and L.Leonardson: Materials Research Society, ed. by B.H.Kear and B.C.Giessen, Vol. 28 (North Holland, Amsterdam, 1984), p. 241

8 J.R.Matey and A.C.Anderson, Phys. Rev. B 16, 3406 (1977)

9 E.J.Cotts, A.C.Anderson ans S.J.Poon, Phys. Rev. B 22, 6127 (1983)

10 A.v.d.Beukel and S. Radelaar, Acta Met. 31, 419 (1983)

11 H.v. Loehneysen, H.Ruesing and W.Sander, Z. Physik B 60, 323 (1985)

12 A.J.Taub and F. Spaepen, Acta Met. 28, 1781 (1980)

13 T.Egami, Mat. Res. Bull. 13, 557 (1978)

Heat Flow in Glasses on a Picosecond Timescale

D.A. Young, C. Thomsen, H.T. Grahn, H.J. Maris, and J. Tauc

Department of Physics and Division of Engineering, Brown University, Providence, RI 02912, USA

The transient thermal properties of amorphous materials have been the subject of several recent investigations [1-4]. These experiments have mainly been designed to study the time-dependence of the specific heat at low temperature ($T \leq 2K$) on time scales $\geq 0.1 \mu sec$. In this paper we describe a new technique which enables measurements of thermal properties to be made on a time scale into the picosecond range, and we present preliminary results of measurements of this type.

In the experiment a metal film (thickness d~100A) is deposited onto the surface of the sample. This film is heated by light pulses of duration 0.2 ps, energy 0.1 nJ, and repetition rate 110 MHz from a CPM dye laser. The light is focussed onto a 10μ diameter spot on the film surface. The optical reflectivity of the film is temperature-dependent, and so a measurement of the reflectivity enables the film temperature to be found as a function of time t after the heating pulse. The reflectivity of the film is measured by means of a time-delayed probe light pulse also of duration 0.2 ps. In a typical experiment the transient temperature rise of the film is a few K, and the fractional change in reflectivity $\Delta R/R$ is of the order of 10^{-5}.

In Fig. 1 we show results we have obtained in this way. The thermometer is a gold film 160A thick, and the sample is an 1800A layer of SiO_2 which was electron-beam evaporated onto a sapphire substrate. On the time scale of the experiment the SiO_2 may be considered to be of infinite thickness. For times less than about 4 ps after the application of the heating pulse the thermometer does not read accurately. This is because the heat is absorbed initially by the electrons, thus raising their temperature above that of the phonons in the Au film [5]. After 4 ps, however, the electrons and phonons in the Au are strongly coupled; the high thermal conductivity of the Au ensures that the temperature is uniform throughout the thickness of the thermometer and ΔR is proportional to ΔT. The time resolution of the thermometer is of the order of 1 ps. Lateral heat flow in the thermometer is unimportant, and the heat flow is to a very good approximation only in the direction normal to the surface. The data shown in Fig. 1 are for an ambient temperature of 300K.

We assume first that there is no time-dependence to the specific heat C_S of the SiO_2, and that the thermal conductivity κ_S on these short length and time scales still has its macroscopic value. The temperature of the Au film $\Delta T(t)$ at time t after it was heated is then given by

$$\Delta T(t) = \Delta T(0)\, f(\beta t^{1/2}) \tag{1}$$

where f is a function which can be calculated, and $\beta = (\kappa_S C_S)^{1/2}/C_A d$, with C_A and d the specific heat and the thickness of the Au film respectively. To determine d accurately we use an interesting secondary effect present in the data. For t in the range 0 to 30 psecs ΔR contains an oscillatory part, shown on an expanded scale in Fig. 2. This occurs because the heating light pulse generates a strain wave (coherent acoustic phonon) which bounces back and forth in the Au film. Because ΔR depends on strain, as well as on temperature, this wave gives an oscillatory contribution to ΔR. From the period of the oscillation (9.9 ps) and the

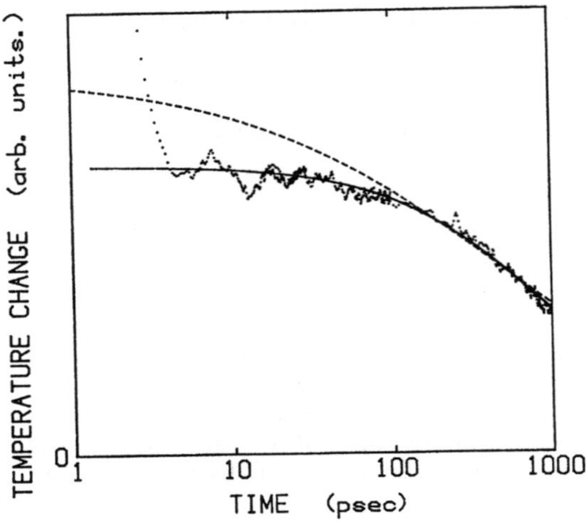

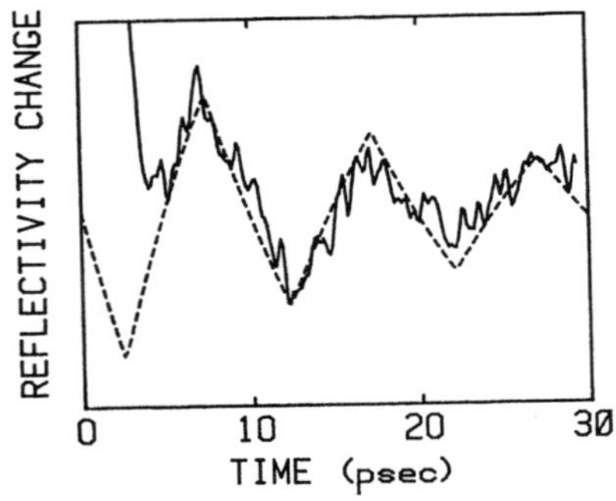

Fig. 1. Temperature change of the thermometer as a function of time after the heating pulse; dashed and solid curves are theoretical fits discussed in the text

Fig. 2. Reflectivity changes due to the excitation of acoustic phonons in the thermometer; the dashed line shows a fit discussed in the text

known sound velocity in Au we can calculate d accurately. The dashed line in Fig. 2 shows a fit to the oscillations based on the assumption that the strain wave is damped by transmission (according to the acoustic-mismatch theory) into the SiO_2 film. The agreement with the data implies that the bonding of the Au to the SiO_2 is good (but see comments below). Using the value of d and the known values of κ_S, C_S and C_A, we then obtain the fit to the data shown by the dashed curve in Fig. 1. The only adjustable parameter is the magnitude of the initial temperature rise, and clearly there is no value of this parameter that will produce a good fit for all t. Rothenfusser et al. [6] have shown that electron-beam evaporated SiO_2 may have properties quite different from bulk. Hence, we have tried to get a better fit by changing the values of κ_S or C_S, but this does not improve things significantly.

To try to get a better fit to the data we have considered the effect of a Kapitza resistance R_K at the interface between the Au and the SiO_2. An R_K of $1.45\times10^{-4}KW^{-1}cm^{-2}$ gives the fit shown by the solid line in Fig. 1. This agrees with the data very well, but the assumed R_K is somewhat large. We write

$$R_K = \frac{4}{C_A v_A \tau_A} \tag{2}$$

where v_A is an average velocity in the Au and τ_A is the average phonon transmission coefficient from the Au to the SiO_2. Then τ_A has the value ~ 0.05. This is considerably smaller than the acoustic-mismatch value which, as discussed above, appears to apply at least up to 100 GHz. However, it is certainly possible that τ_A decreases considerably as the frequency is increased from 100 GHz into the 2-3 THz range characteristic of thermal phonons in Au at room temperature.

This method makes possible the study of transient thermal properties in amorphous materials down to the picosecond time scale. We will report measurements over a broad temperature range in a subsequent paper. This work was supported in part by the National Science Foundation through the Materials Research Laboratory at Brown University.

1. W. M. Goubau and R. A. Tait: Phys. Rev. Lett. 34, 1220 (1975)
2. J. E. Lewis and J. C. Lasjaunias: In Phonon Scattering in Condensed Matter, edited by H. J. Maris (Plenum, New York, 1979), p. 33
3. W. Knaak and M. Meissner: In Phonon Scattering in Condensed Matter, edited by W. Eisenmenger, K. Lassmann, and S. Dottinger (Springer, Berlin, 1984), p. 416
4. M. T. Loponen, R. C. Dynes, V. Narayanamurti, and J. P. Garno: Phys. Rev. B25, 1161 (1982)
5. G. L. Eesley: Phys. Rev. B33, 2144 (1986)
6. M. Rothenfusser, W. Dietsche, and H. Kinder: In Phonon Scattering in Condensed Matter, edited by W. Eisenmenger, K. Lassmann, and S. Dottinger (Springer, Berlin, 1984), p. 419

Acoustic Attenuation in Amorphous $Pd_{30}Zr_{70}$ and $Cu_{30}Zr_{70}$ in the Superconducting and the Normal State

S. Hunklinger, P. Esquinazi, H.M. Ritter, H. Neckel, and G. Weiss*

Institut für Angewandte Physik II der Universität Heidelberg, Albert-Überle-Str. 3–5, D-6900 Heidelberg, Fed. Rep. of Germany

During the last years the low temperature properties of amorphous solids have been a puzzling experimental and theoretical problem /1/. In spite of intense efforts the "anomalies" can only be described on the basis of the phenomenological Tunneling-Model /2,3/. There it is assumed that small groups of atoms form Tunneling States (TS) with a wide distribution of their energy splitting. In metallic glasses these TS not only couple to strain fields but also to conduction electrons. This interaction has been described in analogy to the Korringa relaxation of nuclear spins /4/ and was later on also applied to superconducting metallic glasses /5/. Recently this "standard" description has been extended /6/ assuming that "bound" states are formed below a certain temperature if the coupling between TS and conduction electrons is strong enough.

To study the TS-electron interaction in more detail we have measured the ultrasonic absorption of a splat-cooled $Pd_{30}Zr_{70}$ sample and the low-frequency internal friction of $Pd_{30}Zr_{70}$ and $Cu_{30}Zr_{70}$ prepared by melt-spinning. In all cases not only the superconducting (SC) but also the normal conducting (NC) state was investigated.

At low temperatures a sound wave is attenuated by the TS via resonant and relaxation absorption /1/. The resonant absorption is observed at high frequencies and can be saturated at higher acoustic intensities. At low frequencies only the relaxation process contributes. In metallic glasses the relaxation time τ is determined by the interaction of TS with thermal phonons and conduction electrons as well. In general relaxation via electrons dominates, but above 2K phonons take over. Even if the the temperature is kept constant amorphous solids exhibit a characteristic distribution of the relaxation times with a minimal value τ_m/1,7/. This leads to a relaxation absorption with a characteristic temperature and frequency dependence. Its magnitude depends on the value of $\omega\tau_m$, where ω is the angular frequency. Two different regimes can be distinguished: at low temperatures ($\omega\tau_m >> 1$) absorption increase with temperature, whereas at high temperatures ($\omega\tau_m >> 1$) a "plateau" is expected. How fast absorption α rises with temperature is determined by the relaxation mechanism: if electrons or phonons are dominant, this rise should be proportional to T or T^3, respectively. The height of the plateau is expected to be independent of the relaxation mechanism and

---
*permanent address:
Centro Atomico, 8400 Bariloche, Rep. Argentina

should be given by $\alpha = C\pi\omega/2v$ or $Q^{-1} = C\pi/2$ where Q^{-1} is the internal friction and v the sound velocity. The coupling constant $C = \bar{P}\gamma^2/\rho v^2$ is determined by $\bar{P}$, the density of states of TS and γ the deformation potential reflecting the strength of the coupling of the TS to strain fields. ρ is the mass density.

The attenuation of PdZr in the SC state (see Fig.1) rises at low temperatures because of the increase of the relaxation rate τ^{-1}due to the rising number of phonons and quasiparticles. From the height of the plateau observed above 1.5 K we deduce a coupling constant $C = 6\cdot10^{-5}$. A completely different behaviour is seen when SC is suppressed. Then relaxation by free electrons dominates in the whole temperature range. Although no intensity dependence could be seen, the attenuation may partially be attributed to the resonant process /8/.

In Fig.2 we present internal friction data obtained at low frequencies for PdZr and CuZr. In both materials absorption increases at the lowest temperature in the SC state because $\omega\tau_m > 1$. In PdZr the absorption curve flattens off above 50 mK because relaxation by thermal phonons is sufficient to fulfil the condition $\omega\tau_m = 1$. From the following plateau we deduce $C = 1.9\cdot10^{-4}$, a value considerably larger than that obtained at 720 MHz. This discrepancy between low and high frequency values seems to be a general aspect of metallic glasses /9/. Surprisingly, absorption decreases again at higher temperatures, probably due to the fact that the number of unpaired electrons becomes noticeable. If SC is suppressed internal friction increases steadily with temperature. It is always

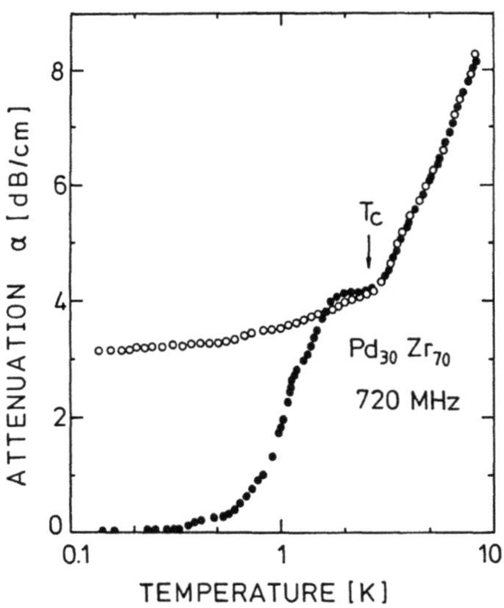

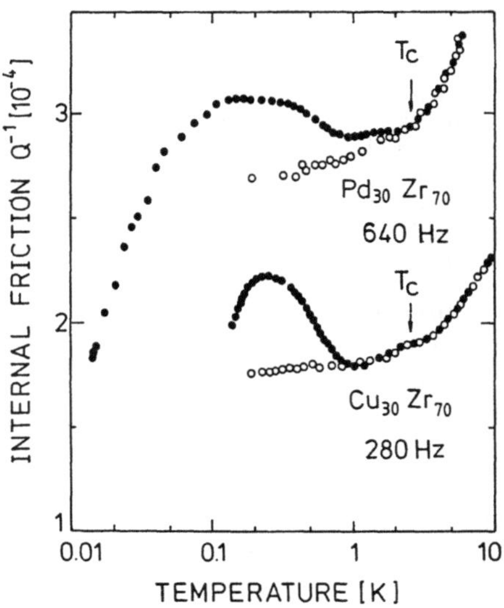

Fig.1 : Ultrasonic absorption in PdZr at 720 MHz
Fig.2 : Internal friction of PdZr and CuZr
Data indicated by open circles were taken after applying a magnetic field of 5.6 T and 6 T, respectively, to suppress SC.

lower than in the superconducting state although $\omega\tau_m << 1$ holds in the whole temperature range, and the nature of relaxation of the TS should be without influence.

For CuZr a well-defined maximum is found at 0.25 K in the SC state. The absence of the plateau cannot be explained by current theories.In the NC state internal friction behaves completely different. At 0.2 K it is 30% smaller than in the SC state, but increases steadily with temperature. Between 0.8 K and 1.6 K the internal friction curve of the NC state crosses two times that of the SC state. Obviously the relaxation mechanism is much more complicated than assumed so far.

According to theory internal friction or absorption provide a direct measure of $\bar{P}\gamma^2$ in the regime $\omega\tau_m << 1$. From our data in the SC state one may conclude that the attenuation decreases when the amount of quasiparticles becomes noticeable. Measurements in the NC state support this interpretation: the observed attenuation is considerably smaller. It seems that the coupling factor $\bar{P}\gamma^2$ is reduced by the presence of quasiparticles. The different low-temperature behaviour of PdZr at 720 MHz is due to the fact that in the SC state the condition $\omega\tau_m < 1$ does not hold to low enough temperatures. No model existing so far explains quantitatively a reduction of $\bar{P}\gamma^2$. Based on the idea of strong coupling between TS and electrons /6/ it has been proposed recently /10/ that the deformation potential γ of the TS to phonons is reduced at low temperatures. It seems, however, that this effect cannot be large enough to explain our results /8/. It seems to be more likely that $\bar{P}$ is changed by the presence of conduction electrons. Measurements of the velocity in these materials /8/ confirm this conclusion.

References:

1. See for example: S. Hunklinger, A.K. Raychaudhuri in Progress in Low-Temperature Physics, Vol.IX, ed. D.F. Brewer, Elsevier Science Publishers (1986),p. 265
2. W.A. Phillips, J. Low-Temp. Phys. 7, 351 (1972); P.W. Anderson, B.I. Halperin, C. Varma, Phil. Mag.25, 1 (1972)
3. V.G. Karpov, M.I. Klinger, F.N. Ignatiev, Sov. Phys. JETP 57, 439 (1983)
4. B. Golding, J.E. Graebner, A.B. Kane, J.L. Black, Phys. Rev. Lett. 41, 1487 (1978)
5. J.L. Black, P. Fulde, Phys. Rev. Lett. 43, 453 (1979)
6. K. Vladar, A. Zawadowsky, Phys. Rev. B28, 1564; B28, 1582; B28, 1596 (1983)
7. J. Jäckle, Z. Phys. 257, 212 (1972)
8. P. Esquinazi, H.M. Ritter, H. Neckel, G. Weiss, S. Hunklinger, Z. Phys.B , to be published
9. A.K. Raychaudhuri, S. Hunklinger, Z. Phys.B 57, 113 (1984)
10. K. Vladar, in Proc. 2. Int. Conf. Phonon Physics, eds. J. Kollar, N. Kroo, N. Menyhard, T. Siklos, World Sci. Publishing, Singapore (1985), p. 96

Evidence for Two Different Kinds of Two Level Systems in Lithium Borate Glasses

J.-Y. Prieur[1], *M. Devaud*[1], *and W.D. Wallace*[2]

[1]Laboratoire d'Ultrasons*, Université Pierre et Marie Curie, Tour 13, 4 place Jussieu, F-75252 Paris Cedex 05, France
[2]Department of Physics, Oakland University, Rochester, MI 48063, USA

1. INTRODUCTION

In Stuttgart [1] we presented echo experiments performed on a family of borate glasses with the general composition : B_2O_3, x Li_2O, y LiCl. We mentioned on that occasion that the echoes we obtained presented some puzzling features which are not in agreement with the expected standard behavior of phonon echoes in glassy systems as observed by other groups [2].

The aim of this communication is twofold : firstly to relate the observation of "standard" phonon echoes on the same borate glasses ; secondly to show how our first "non standard" echoes can be reconciled with the "standard" theory (cf [2]), provided that a power-dependence of the sound velocity and the subsequent phase-*de*matching is taken into account.

2. "STANDARD" PHONON ECHOES IN GLASSES

They take their origin in localized excitations, schematized by Two-Level Systems (TLS) acting like (one-half) pseudospins. The amplitude $\mathcal{E}_{echo}$ of the strain ε associated to the echo is given by :

$$\mathcal{E}_{echo} = e^{-2\Gamma_2\tau}\left(\frac{\omega L}{v}\right)\mathrm{th}\left(\frac{\hbar\omega}{2k_BT}\right)\frac{\gamma\hbar\bar{P}}{\rho v^2}\;\frac{\sin\theta_1\;\sin^2(\theta_2/2)}{\sup\left(\Delta t_1,(\Delta t_2/2)\right)}$$

where τ, Γ_2, L, v, $\bar{P}$, γ, ρ, are respectively the delay time between the two pulses, the transverse relaxation rate of the TLS, the sample length, the sound velocity, the spectral density, the deformation potential, the specific gravity, Δt_i the widths of the excitation pulses (i = 1, 2) ; $\theta_i = \gamma\varepsilon\Delta t_i/\hbar$ is the Rabi rotation angle. The maximum echo amplitude is obtained for $\theta_1 = \theta_2/2 = \pi/2$ (which means $\Delta t_1 = \Delta t_2/2$). Then $\mathcal{E}^{max}_{echo}$ may be written :

$$\mathcal{E}^{max}_{echo} \simeq \mathcal{E}_{in}\; e^{-2\Gamma_2\tau}\left(\frac{\omega L}{v}\right)\mathrm{th}\left(\frac{\hbar\omega}{2k_BT}\right)C$$

where $C = \bar{P}\gamma^2/\rho v^2$ is the dimensionless coupling constant. On the left part (small incident powers) of Fig. 1, P_{echo} is plotted as a function of the incident power P_{in} for two different values of $\Delta t = \Delta t_1 = \Delta t_2/2$; the mean feature shown out by Fig. 1 is the following : the incident power needed to reach the maximum value of $\mathcal{E}_{echo}$ depends on Δt, according to the theoretical relation $\gamma\mathcal{E}_{max}\Delta t/\hbar = \pi/2$.

Measurements of γ by this method on standard echoes in our borate glasses provide values between 0.2 and 0.5 eV, in complete agreement with attenuation-dispersion methods [3]. On the other hand, measuring the exponential decrease

---
*Associated with the Centre National de la Recherche Scientifique

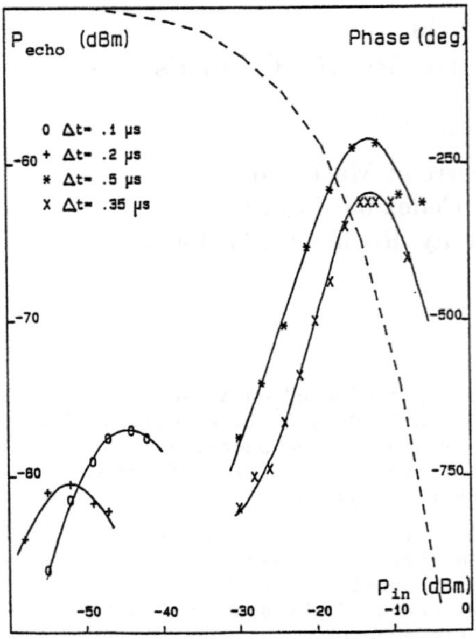

Fig. 1 - P_{echo} versus the input power P_{in} for the standard echoes (left) and non standard echoes (right), each for two different values of Δt. The dashed line shows the variation of the phase of an ultrasonic echo as measured versus P_{in}.

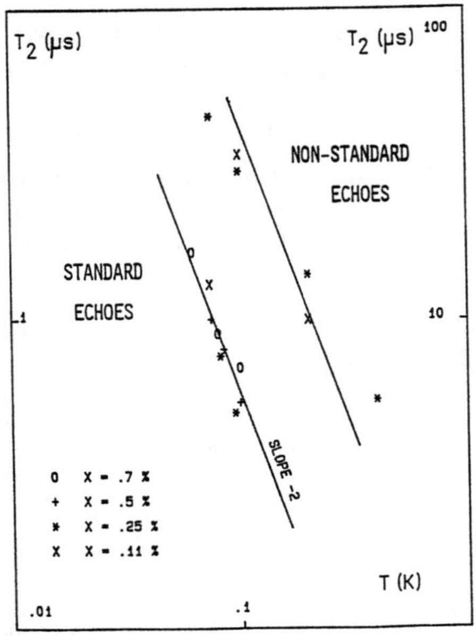

Fig. 2 - Transverse relaxation time T_2 versus temperature T for different concentrations x in Li_2O.
Lower part : standard echoes ;
upper part : non standard echoes.
For both a T^{-2} law is in accordance with experimental data.

$e^{-2\Gamma_2\tau}$ of the echo amplitude as τ increases gives : $T_2 \simeq 1$ μs at T = 75 mK (again in complete agreement with desaturation measurements [3]). Left part of Fig. 2 shows the values found for T_2 as a function of temperature ; accordance with the law : $T_2(T) \propto T^{-2}$ should be noticed.

3. "NON STANDARD" PHONON ECHOES

Using exactly the same experimental device as for standard echoes, but with 40 dB higher input powers and wider pulses (Δt of the order of the μs instead of 100 ns) we could observe forward echoes [1]. Yet, several differences with the "standard" echoes should be pointed out :

- the temperature range of observation of these "non standard" echoes stretches up to 600 mK (only 100 mK for "standard" echoes) ;

- the magnitude of the non standard echoes was always much lower than that of our first reflected ultrasonic echo (- 33 dB at least) ;

- T_2 can be deduced from Fig. 3 (decay of the magnitude of the echo as a function of τ for different input powers). From that figure it is clear that T_2 increases when the input power decreases but tends towards a limit. Those limits are plotted in the upper part of Fig. 2 ;

- on the right part of Fig. I, P_{echo} is plotted versus the input power for two different values of Δt : the position of the maximum of P_{echo} is independent of Δt but depends on temperature.

Fig. 3 - P_{echo} versus the time interval τ between the two pulses for four different input powers.

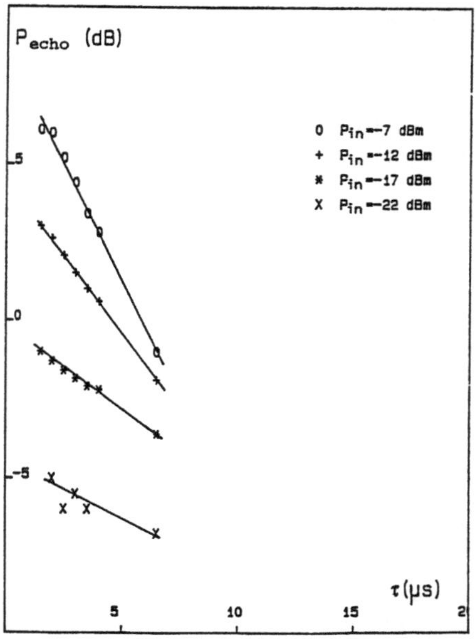

4. HOW NON STANDARD ECHOES CAN BE RECONCILED WITH THE STANDARD THEORY

Let us assume that, in the sample, *two kinds* of TLS do exist : the first ones are strongly coupled to phonons ($\gamma \simeq 0{,}5$ eV) ; they are responsible for attenuation and dispersion of sound [3] and they are at the origin of "standard" echoes of the type involved in [2]. The second ones are weakly coupled to phonons (γ smaller than 10^{-3} eV) : they have no influence on sound attenuation or dispersion or on standard echoes. In contrast, when very high input powers are involved, they are able to undergo a Rabi precession, while the (resonant) strongly coupled TLS are completely saturated over a very large, power-broadened, linewidth.

This saturation is so complete that even the sound velocity is changed (see the dashed curve in Fig. 1). Thus the first and second excitation pulses and the echo propagate with velocities v_1, v_2, v_e. Then the contributions of the different slices of the sample do not reach the output transducer in phase. A naive calculation will show that the overall contribution to the amplitude of this echo is proportional to the "dematching factor" $\sin\left(\frac{\Delta\varphi}{2}\right) / \frac{\Delta\varphi}{2}$ where $\Delta\varphi = \left(\frac{2}{v_1} - \frac{1}{v_2} - \frac{1}{v_e}\right)\omega L$.

5. CONCLUSION

If the above phase-dematching effect is taken into account then the maximum of P_{echo} (Fig. 1) disappears and the curve P_{echo} versus P_{in} keeps increasing along all the range of our experimental incident powers, which proves that the fulfilment of the maximum condition : $\gamma\varepsilon\Delta t/\hbar = \pi/2$ is beyond our experimental possibilities. This effect explains too why T_2 seems to be a function of the incident power, the dematching factor being power-dependent.

REFERENCES

1. M. Devaud, J.-Y. Prieur, W.D. Wallace : in *Phonon Scattering in Condensed Matter*, Springer Ser. Solid-State Sci., Vol. 51 (Springer, Berlin, 1984), p. 449.
2. J.E. Graebner, B. Golding : *Phys. Rev. B* **19**, 964 (1979)
3. M. Devaud, J.-Y. Prieur, W.D. Wallace : *Sol. St. Ionics* **9** & **10**, 593 (1983)

Dielectric Rotary Echoes in Vitreous Silica

M. v. Schickfus and G. Baier

Institut für Angewandte Physik, Universität Heidelberg,
Albert-Überle-Str. 3–5, D-6900 Heidelberg, Fed. Rep. of Germany

Coherent experiments at very low temperatures have proven to be very useful tools in the study of the interaction between the two-level tunneling states in glasses /1/ and external fields. In particular, spontaneous and stimulated echo experiments /2/ have made it possible to directly investigate the resonant coupling of these states to RF electric fields and to ultrasonic phonons. Also the transverse lifetime T_2 and -with limited accuracy- the longitudinal lifetime T_1 of the tunneling states could be determined from these experiments.

Recently the generation of ultrasonic rotary echoes has been demonstrated for vitreous silica /3/. If we use the analogue of a spin 1/2-system in a magnetic field for the two-level states, the rotary echo is brought about by the nutation of the spins which are at resonance with a perturbing external field. Time reversal at time τ is established by inverting the phase of the perturbing field. A rotary echo is then superimposed on this field at $t=2\tau$.

We have investigated dielectric rotary echoes in vitreous silica Suprasil I between 10 mK and 80 mK at a microwave frequency of 710 MHz. The disk-shaped sample was placed in the uniform electric field region of a reentrant resonator. A microwave bridge enabled us to detect the echo signal in the presence of the strong driving field.

If a microwave field $F(t)=F_0 \exp(i\,\omega t)$ is turned on at $t=0$ and phase is reversed at $t=\tau$, the pseudo spin system will respond with a macroscopic polarisation and induce an electrical field /4/:

$$F_e = 2\pi Q \int_0^1 du\; p(E=h\omega_0,u)\, \tanh\left(\frac{h\omega_0}{2kT}\right) p_0 u\, e^{-\alpha t}\, \frac{\omega_1^3}{\beta^3} \sin[\beta(t-2\tau)] \qquad (1)$$

where Q is the Q-factor of the resonator. Polarization passes through zero for $t=2\tau$ and the superposition of pseudo-spins with a distribution of Rabi frequencies $\omega_1=(F_0 p_0 u)/\hbar$ will shape the echo envelope. Here $F_0 p_0 u$ is the off-diagonal interaction matrix element for a tunneling state with dipole moment p_0. $u=(\Delta_0/E)\cos\gamma$ is the ratio between the total energy and the tunnel splitting of the state, multiplied by the direction cosine between the electric field and the dipole. For the density of u we find $p(E,u)=\bar{P}(1/u^2-1)^{1/2}$ after integration over an isotropic orientation of the dipoles. $\bar{P}$ is the constant density of states of the tunneling model. The decay time constant of the echo is $\alpha=(T_1^{-1}+T_2^{-1})/2$, and $\beta^2=\Delta\omega^2+\omega_1^2$ with $\Delta\omega=\omega-\omega_0$.

If weighted with the u-dependence in (1), p(E,u) yields the amplitude spectrum of the Rabi frequencies $g(\omega_1)\propto u(1-u^2)^{1/2}$. This is shown in Fig.1 together with a Fourier transform of our experimental echo envelope. The finite pulse length and the spectrum of $T_2(\approx 20\ \mu s)$ cause the deviations at low Rabi frequencies, whereas the high-frequency excess is probably due to transmitter noise filtered by the cavity. Between 300 kHz and 1.2 MHz our data agree well with the theory and thus confirm the prediction of the tunneling model. With $F_0=108$ V/m we find a dipole moment $p_0=2.25$ Debye, a value which is 40% smaller than that of the spontaneous echo experiment /5/. We believe that the present technique should yield the more reliable results.

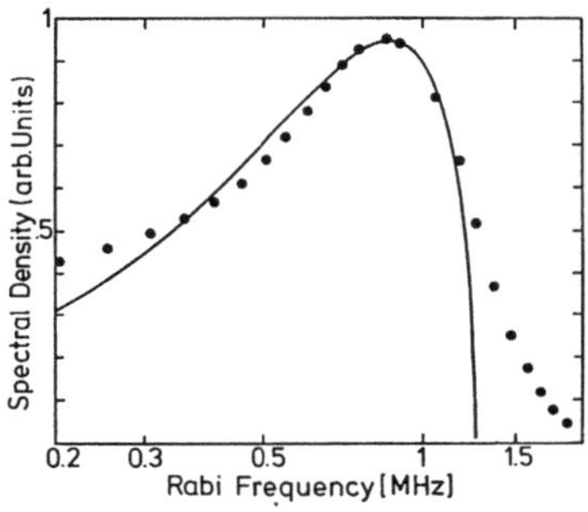

Figure 1

Fourier transform of the rotary echo. The solid line shows the prediction of the tunneling model

Since in glasses $T_2 << T_1$, T can be derived from the τ-dependence of the echo amplitude (1). In practice, however, the driving field heats the sample and the resonator if τ exceeds 20 μs at the lowest temperatures. On the other hand, a modified pulse pattern allows to directly measure the relaxation time T_1 without the need to consider spectral diffusion (Spectral diffusion, caused by thermal fluctuation of the energy of the tunneling states prohibits an immediate derivation of T_1 from stimulated echo experiments). In this experiment the driving field is switched off at $t=\tau$ and back on at $t=\tau+\tau_{12}$. In this case the echo amplitude will be proportional to /4/:

$$F_e \propto \frac{\omega^3}{2\beta^3}\sin\left[\beta(t-\tau_{12}-2\tau)\right] e^{-\alpha(t-\tau_{12})}\left\{e^{-\tau_{12}/T_1}-\cos(\omega_0\tau_{12})\ e^{-\tau_{12}/T_2}\right\} \qquad (2)$$

Since the last term in (2) vanishes for $\tau_{12} >> T_2$, the relaxation time T_1 can directly be derived from the decay of the echo amplitude with τ_{12}. In comparison with a stimulated echo experiment, the influence of spectral diffusion in our experiment is reduced by a factor of the order of $\omega_0/\omega_1 \approx 500$ and can therefore be neglected.

Our experimental result for T=11 mK is shown in Fig.2 together with a fit based on the distribution of T_1 predicted by the tunneling model /6/. At short times we find $T_1=230\ \mu s$, for

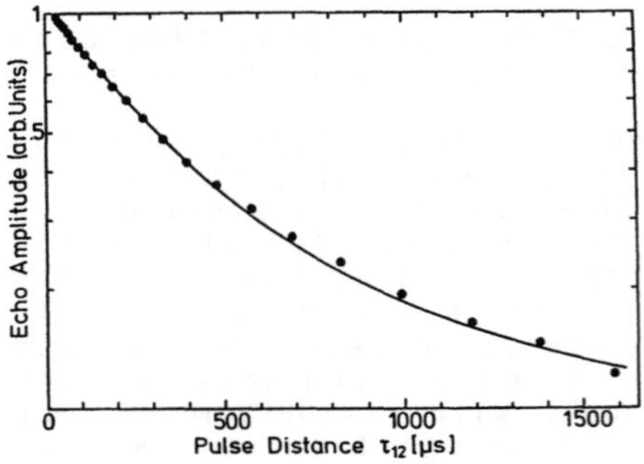

Figure 2
Decay of the two-pulse rotary echo with pulse separation τ_{12}. The solid line represents the theoretical prediction

our longest measuring times the dominant relaxation time is $T_1 \approx 1100$ μs. The temperature dependence of T_1 has been derived from the initial slope of the decay curve, we find the T^{-1}-dependence expected for a one-phonon relaxation process.

Finally we would like to mention that in /7/ a division of the distribution function $p(E,\Delta_0/E)$ into a "slow" and a "fast" branch was suggested. Our $p(E,u)$ is identical with the "slow" branch, indicating that for resonant interaction the density of the strongest coupling states is zero.

1. For a review see: Amorphous Solids: Low-Temperature Properties, ed. by W.A. Phillips, Springer Ser. Topics Curr. Phys., Vol.24 (Springer, Berlin, Heidelberg 1981)
2. B. Golding, J.E. Graebner: In /1/
3. B.Golding, D.L. Fox, W.H. Haemmerle: In Phonon Scattering in Condensed Matter, ed. by W. Eisenmenger, K.Laßmann, and S. Döttinger, Springer Ser. Solid-State Sci., Vol.51 (Springer, Berlin, Heidelberg 1984), p. 446.
4. G. Baier, M. v.Schickfus: To be published
5. B. Golding, M.v. Schickfus, S. Hunklinger, K. Dransfeld: Phys. Rev. Lett. 43, 1817 (1979)
6. J. Jäckle: Z. Phys. 257, 212 (1972)
7. S. Hunklinger: In Phonon Scattering in Condensed Matter, ed. by W. Eisenmenger, K.Laßmann, and S. Döttinger, Springer Ser. Solid-State Sci., Vol.51 (Springer, Berlin, Heidelberg 1984), p. 378.

Hypersonic Phonon Damping in Neutron Irradiated Quartz

R. Vacher, J. Pelous, N. Elkhayati Elidrissi, and M. Boissier

Laboratoire de Science des Matériaux Vitreux, U.A. C.N.R.S. No. 1119, Université des Sciences et Techniques du Languedoc, F-34060 Montpellier Cedex, France

1. Introduction

Among the various materials which are investigated as models of the glassy state, neutron irradiated quartz is one of the most attractive, as it allows continuous structural variations from the perfectly ordered crystal to the amorphous network disorder. A number of studies of this material have been performed, with the purpose of observing "glassy" behaviour and studying its dependence with increasing structural disorder. In particular, the thermal conductivity /1/ and acoustic properties /2-3/ have been widely investigated. An extensive study of the low-temperature properties has also been performed in the past ten years /4-10/, giving clear evidence of signs of anomalies characteristic of the glassy state.

It is well known that the acoustic attenuation in vitreous silica exhibits a strong maximum at a temperature of about 50 K for ultrasonic frequencies. This absorption process is likely to be due to thermally-activated relaxation mechanisms, related to the glassy-network structural disorder. In contrast, our Brillouin scattering results have demonstrated that this process alone is insufficient to explain the observations at hypersonic frequencies /12/. Rather, we suggested that anharmonic phonon-phonon interactions dominate the sound absorption at frequencies higher than 10 GHz. The main purpose of this work is to give support to this assumption by studying the change of the acoustic absorption induced in quartz crystals by neutron irradiation.

2. Experimental details

Two samples with neutron doses of $2.6 \times 10^{19}\,n.cm^{-2}$ and $4.7 \times 10^{19}\,n.cm^{-2}$ - hereafter referred as N_5 and N_6, respectively - were kindly provided by Dr. C. Laermans. Those samples are the same as used in Ref. 13. The doses correspond to neutron energies higher than 0.3 MeV. The sample with the highest neutron dose - referred to as P_3 below - was kindly supplied by Dr. A.M. de Goër. From density measurements, its irradiation dose was evaluated at $7.1\,n.cm^{-2}$ for neutrons of the same energy as above.

Brillouin scattering experiments were performed in the backscattering geometry, using the 488 or 514.4 nm line of an argon-ion laser as the light source. The details of the experimental setup have been described elsewhere /11/. The attenuation of 35 GHz frequency longitudinal phonons propagating in the x-direction was obtained. The refractive index, which is needed for the calculation of the acoustic attenuation, has been measured for all the samples studied.

3. Results and discussion

Figure 1 shows the temperature dependence of the attenuation at 35 GHz of longitudinal phonons propagating in the x-direction for the three samples investigated. The values for N_5 are about 25 percent higher than those in unirradiated quartz. It is evident on this figure that the attenuation strongly increases from N_5 to N_6 and tends to saturate from N_6 to P_3. On the other hand, while the attenuation is beyond the experimental accuracy below 60 K in the unirradiated sample, the values remain sizeable down to the lowest temperature investigated of 4 K for N_6 and P_3, and are very close to those in vitreous silica at the same temperatures. In order to analyse the influence of the different damping mechanisms in the whole frequency-temperature range, it is of interest to compare the attenuation values obtained from ultrasonic and hypersonic experiments. Table 1 shows a comparison of our results to those of Ref. 13 in which ultrasonic measurements on the same samples are reported. The first conclusion to be obtained from this table is that the ω^2-dependence expected for anharmonic phonon-phonon interactions in unirradiated quartz is very well obeyed over two orders of magnitude in frequency.

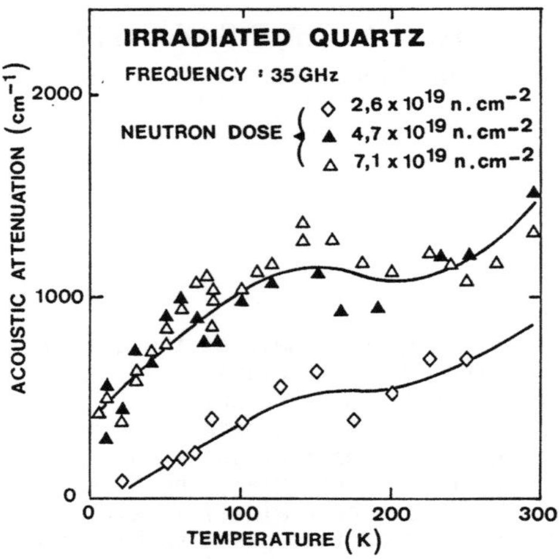

Fig. 1 - Temperature dependence of hypersonic attenuation.

At temperatures around 10 K, ultrasonic experiments /13/ have demonstrated the existence in neutron irradiated quartz of a plateau similar to that observed in glasses. This plateau is usually assumed to originate from relaxation of tunneling systems via one-phonon assisted processes. Its amplitude is proportional to ω. Table 1 shows that this proportionality, which is observed in ultrasonic experiments, holds up to hypersonic frequencies. This demonstrates that the above mechanism dominates the acoustic properties in a very large frequency domain at low temperatures.

Table 1 - Frequency dependence of acoustic attenuation.

		IRRADIATED QUARTZ (dose : 2.6×10^{19} n.cm^{-2})			UNIRRADIATED QUARTZ
Temperature		10 K	100 K	300 K	300 K
Attenuation	U.S. 350 MHz	0.46	0.85	0.51	0.092
$\alpha(cm^{-1})$	H.S. 35 GHz	65	393	896	786
$n = \frac{\log(\alpha_{H.S.}/\alpha_{U.S.})}{\log(\nu_{H.S.}/\nu_{U.S.})}$		1.08	1.33	1.63	1.97

At temperatures in the range from 100 to 300 K, Table 1 shows that the frequency dependence of the attenuation is somewhere in between ω and ω^2. Two mechanisms are expected to dominate phonon damping in this temperature range, namely thermally activated relaxation and anharmonic phonon-phonon interactions. Corresponding to the first process, a peak is observed in the ultrasonic attenuation. The amplitude of this peak must be proportional to ω, while anharmonic damping varies as ω^2. From ultrasonic results, we have estimated the amplitude of the absorption peak to be $\simeq 0.4\ cm^{-1}$ at 100 K and 350 MHz for N_5. The ω-dependence gives a contribution of $\simeq 40\ cm^{-1}$ at the hypersonic frequencies, which is negligible as compared to the experimental values. We are therefore led to the conclusion that anharmonic phonon-phonon interaction is the dominant process at hypersonic frequencies.

Fig. 2 shows the dependence of the above mechanism with the irradiation dose, at room temperature. While the attenuation shows little variation in the first stage of irradiation, a strong increase is observed for doses higher than $3\times 10^{19} n.cm^{-2}$. An irradiation of about $7\times 10^{19} n.cm^{-2}$ gives an attenuation value very close to that observed in vitreous silica. It should be noted that, for the same dose, the relative decrease of the density is only 40 percent of its variation from crystal quartz to vitreous silica. We therefore conclude that the whole anharmonic effect observed in vitreous silica is already obtained with this dose, and that subsequent irradiation will only increase the amplitude of the thermally-activated relaxation process.

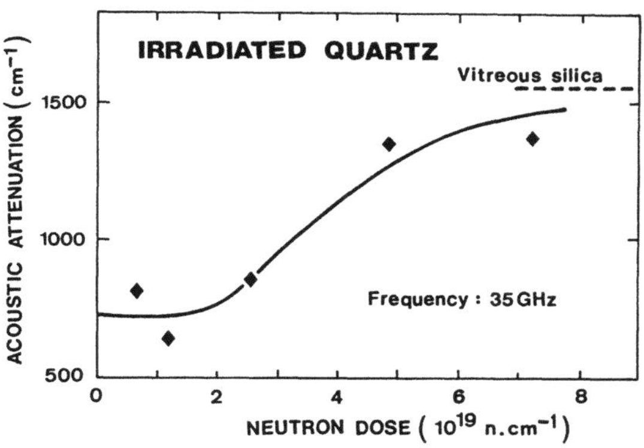

Fig. 2 – Variation of acoustic attenuation with neutron dose.

Acknowledgements - It is a pleasure to acknowledge fruitful discussions with Drs. A.M. de Goër and C. Laermans. We thank SCK/CEN, Mol. Belgium for the irradiation of samples N_5 and N_6, and in particular J. Cornelis. We also thank CEN Grenoble for the loan of the irradiated sample P_3.

1. R. Berman, Proc. Phys. Soc. (London) A208, 90 (1951).
2. H.E. Bömmel and K. Dransfeld, Phys. Rev. 117, 1245 (1960).
3. S. Brawer, Phys. Rev. B 7, 1712 (1973).
4. C. Laermans, Phys. Rev. Lett. 42, 250 (1979).
5. R. Nava, Phys. Rev. B 31, 5497 (1985).
6. B. Golding, J.E. Graebner, W.H. Haemmerle and C. Laermans, Bull. Am. Phys. Soc. 24, 495 (1979).
7. A.C. Anderson, J.A. McMillan and F.J. Walker, Phys. Rev. B24, 1124 (1981).
8. A.M. de Goër, M. Locatelli and C. Laermans, J. Physique 42, C6-78 (1981).
9. M. Saint Paul, J.C. Lasjaunias and M. Locatelli, J. Phys. C15, 2375 (1982).
10. C. Laermans and V. Esteves, Springer Series, Vol.51, p.407 (1984).
11. R. Vacher and J. Pelous, Phys. Rev. B14, 823 (1976).
12. R. Vacher, J. Pelous, F. Plicque and A. Zarembowitch, J. Non-Cryst. Solids 45, 397 (1981).
13. V. Esteves, A. Vanelstraete and C. Laermans in Phonon Physics, Ed. J. Kollar et al (World Sci. Singapore, 1985) p. 45.

Ultrasonic Detection of an Energy Gap Change in the N/S Transition for Trapped H in Nb

K.R. Maschhoff, E. Drescher-Krasicka[1], *and A.V. Granato*

Department of Physics and Materials Research Laboratory, University of Illinois, 104 S. Goodwin, Urbana, IL 61801, USA

[1]Present address: National Bureau of Standards, Metallurgy Div., Rm. A163, Materials Bldg., Gaithersburg, Md. 20899

1. Introduction

For a number of years the properties of low-lying excitations seen in many amorphous materials have been interpreted in terms of a two-level tunneling model[1]. A large number of measurements of specific heat, neutron scattering, and ultrasonic attenuation and velocity have shown that hydrogen trapped at O or N impurities acts as a tunneling system and can be described with a two-level formalism with the impurities providing an internal strain bias. Recently, measurements in both the normal(N) and superconducting(S) states of ultrasonic attenuation in Nb-N-H[2], ultrasonic attenuation and velocity in Nb-O-H[3], and of neutron scattering[4] near 1.5 K and 9 K in Nb-O-H have shown that these systems interact strongly with the conduction electrons of the host material. YU and GRANATO[5] have proposed that the interaction of the conduction electrons with a two-level tunneling system reduces the splitting of the levels in the normal state. The purpose of these ultrasonic velocity measurements was to investigate this predicted effect of the environment on this model tunneling system.

2. Ultrasonic Response of Tunneling Systems

The ultrasonic response of two and four-level systems in crystals has been presented in a form that is particularly useful in describing the results of the measurements presented here[6]. There are two parts to the modulus change caused by tunneling systems in equilibrium: a relaxational term, Δ_R, which is due to the reorientation of the tunneling systems in response to the ultrasonic stress, and a resonance term, Δ_S, which is due to the curvature of the energy levels as a function of strain, ε. The measurements presented here are reasonably well described by a two-level model and, although the defect configuration is not yet known, the data will be discussed in terms of this model. The equilibrium elastic modulus changes are:

$$\Delta_R = -(n\alpha^2/C)(1/kT)[2\alpha\varepsilon/E]^2 \,\mathrm{sech}^2(E/2kT) \tag{1}$$

$$\Delta_S = -(n\alpha^2/C)(1/\Delta_0)[2\Delta_0/E]^3 \tanh(E/2kT) \tag{2}$$

$$\delta C/C = \Delta_R + \Delta_S \tag{3}$$

where n is the defect concentration, α is the strain coupling constant, C is the elastic constant, Δ_0 is the tunneling matrix element, and E, given by $2[\Delta_0^2+(\alpha\varepsilon)^2]^{1/2}$, is the energy splitting. When the temperature is low enough or the relaxation rate is small compared to the ultrasonic frequency, the relaxational term is absent and the elastic modulus change is given by Δ_S. For a real crystal, these relations must be averaged over a suitable strain distribution. For $\alpha\varepsilon \ll \Delta_0$ at T=0 one then obtains $\delta C/C = -n\alpha^2/C\Delta_0$, even for a strain distribution, so that a smaller gap gives a larger elastic constant reduction. One may not increase the size of the effect simply by increasing n because this also increases the internal strain, ε. The relationship between δC and Δ_0 then becomes more complicated and depends on the strain distribution. This has been discussed in detail for a Lorentzian strain distribution of characteristic strain ε_0 by YU and GRANATO[5], who find that optimum conditions for the measurement are obtained when $\alpha\varepsilon_0<\Delta_0$ (low oxygen content), the hydrogen content is comparable to the oxygen content, and the temperature is low($kT<\Delta_0$). This suggests measurements in Nb samples with about 100 ppm at temperatures below 2 K. For this case, they predict an elastic constant change of about 5×10^{-5}.

3. Experimental Proceedure

The Nb specimen used in these measurements was a single crystal provided by H.K. Birnbaum that has been used in previous work[3,7]. Chemical analysis showed that the specimen contained small amounts of the following interstitial impurities: 64 at. ppm oxygen, 49 at. ppm nitrogen, and 70 at. ppm carbon. The sample was doped with approximately 700 at. ppm hydrogen from the gas phase using an equilibrium technique[8]. The measurements were performed in a He^3 cryostat equipped with a magnet.Velocity measurements were done with a standard interferometric technique[9] capable of detecting a velocity change of a few parts in 10^7. The modulus change is given by $\delta C/C=2\delta v/v=2\delta f/f$.

4. Experimental Results

The velocity change for the C'mode was measured for the Nb specimen at temperatures between .5 K and 18 K for both the normal and superconducting states. Data taken at 10.8 MHz are shown in Fig. 1. The important feature in both the normal and superconducting states is the decrease in velocity at low temperatures. This is the resonance part of the tunneling system response. It falls like 1/T above 5 K and saturates at lower temperatures. A dispersion due to the relaxation of the tunneling states is seen near 2.4 K in the superconducting state[7]; no dispersion is seen in the normal state. There is a reduction in the velocity just below Tc (9.26 K for Nb) that is due to the electron-phonon interaction. The velocity decrease at higher temperatures is also a property of the Nb host,and varies as $\beta T^2+\gamma T^4$.

5. Analysis of Results

It is already apparent in Fig. 1 that the two curves, after correcting for the electron-phonon interaction, do not come together at the lowest temperatures, as should be expected if Δ_0 does not change. This is made more clear in Fig. 2, where only the modulus change due to the hydrogen vs 1/T is shown. This curve is obtained by subtracting the host $\beta T^2+\gamma T^4$ and electron-phonon terms from the data. At high temperatures, both the N and S state curves vary as α^2/T, as expected from (1)-(3). This remains true for a strain distribution. Below 2.4 K, in the S state, the relaxation component, Δ_R, is lost and only the resonance component, Δ_S, remains. The deviation from a 1/T dependence in the N state occurs for $kT\approx\Delta_0$, showing directly that $\Delta_0/k\approx1$ K. The relative size of Δ_R/Δ_S is a measure of $\alpha\varepsilon/\Delta_0$ and one sees directly that $\alpha\varepsilon<\Delta_0$. Quantitative values for the three parameters of the model, Δ_0,α, and the Lorentzian strain distribution parameter ε_0 are obtained by fitting the data as shown by the solid lines in Fig. 2. These are: $\Delta_0(S)=0.094$ meV, $\alpha=45$ meV, and $\varepsilon_0=1.1\times10^{-3}$. The elastic constant change, $\delta C(S-N)=5.1\times10^{-5}$, corresponds to a tunneling parameter change of $\{\Delta_0(S)-\Delta_0(N)\}/\Delta_0(S)=36\%$. The middle curved line in Fig. 2 is the calculated N state curve one would expect if Δ_0 does not change, using the parameters Δ_0, α, and ε_0 obtained in fitting the S state data. The value of $\Delta_0(S)$ reported here is in reasonable agreement with the value of 0.085 meV obtained from the neutron scattering measurements[4] for a comparable oxygen

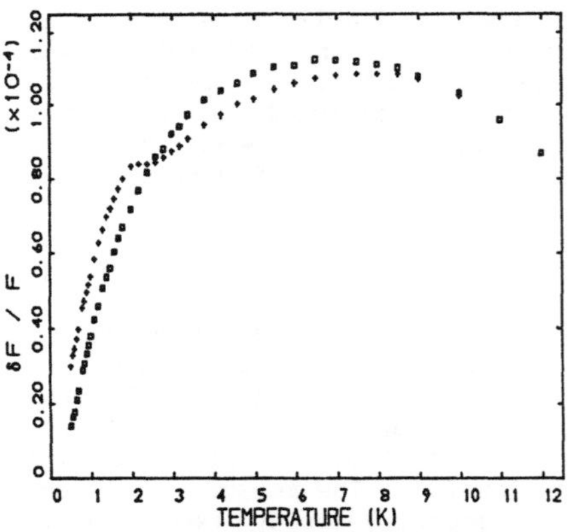

Fig. 1. Relative frequency change vs temperature for the N(□) and S(+) states of Nb-O-H at 10.8 MHz for the C' mode

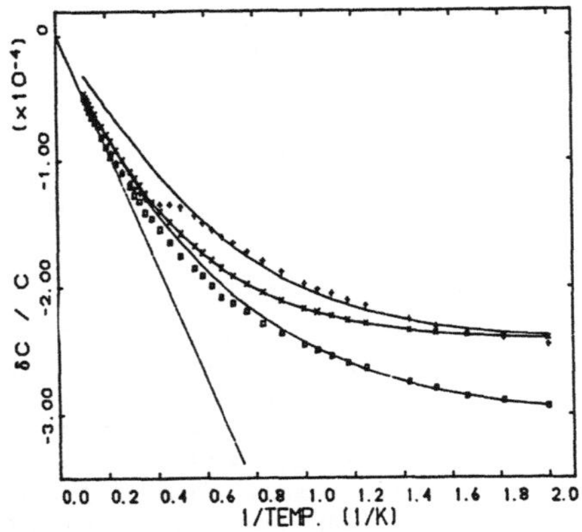

Fig. 2. Relative modulus change vs 1/T after background subtraction for N(□) and S(+) Nb-O-H and fits to the two-level system model(-). The middle curve(-x-) is the computed N state modulus for no change in Δ_0.

concentration. The change in Δ_0 reported here is somewhat larger than the a value of about 20% which can be infered from the neutron scattering measurements at Tc for an OH concentration of 2200 ppm. However, some reduction in the energy gap change is expected with increasing temperature[5].

6. Conclusions

It has been observed that the low-temperature modulus of the Nb-O-H tunneling system is reduced in the normal state. The data can be fit with a two-level model. The tunneling matrix element, Δ_0, is reduced in the normal state by 36%. This is in good agreement with the predicted effect[5], and shows that the effect of the environment on tunneling can be described by a renormalization of the tunneling system parameters.

This work was supported by the National Science Foundation under contract NSF DMR 84-09396.

References

1. P.W. Anderson, B.I. Halperin, and C. M. Varma: Philos. Mag. 25, 1 (1972).
2. J.L. Wang, G. Weiss, H. Wipf, and A. Magerl: in Phonon Scattering in Condensed Matter, eds. W. Eisenmenger, K. Lassmann, and S. Dottinger,Springer-Verlag, Berlin, 1984.
3. E. Drescher-Krasicka and A.V. Granato: J. de Physique Coll. C10,73 (1985).
4. A. Magerl, A.J. Dianoux, H. Wipf, K. Neumaier, and I.S. Anderson: Phys. Rev. Lett. 56, 159 (1986).
5. C.C. Yu and A.V. Granato: Phys. Rev. B32, 4793 (1895).
6. A.V. Granato, K.L. Hultman, and K. F. Huang: J. de Physique Coll. C10, 23 (1985).
7. D.B. Poker, G.G. Setser, A.V. Granato, and H.K. Birnbaum: Phys. Rev. B29, 622 (1984).
8. T.S. Schober and T. Wentzl: in Topics in Applied Physics, Vol. 29, Hydrogen in Metals II, eds. G. Alefeld and J. Volkl, Springer-Verlag, Berlin-Heidelburg-New York, 1978.
9. E.R. Fuller Jr., A.V. Granato, J. Holder, and E.R. Naimon, in Methods of Experimental Physics, Vol. 11, ed. R.V. Coleman, Academic Press, New York and London, 1974.

Fast-Time Heat Pulse Measurements in High- and Low-Diffusivity Materials*

*J. Madsen**, J. Trefny, and R. Yandrofski*

Physics Department, Colorado School of Mines, Golden, CO 80401, USA

1. Introduction

In this paper we discuss two sample geometries which we have used for low-temperature thermal measurements on high- and low-diffusivity materials respectively. In the first geometry, our data reveal features of the response at early times which we ascribe to causality effects. The second geometry gives one the opportunity to investigate both heat capacity and thermal diffusivity on different time-scales in the same sample. Data on vitreous silica, which so far are restricted to temperatures above 0.6 K, do not reveal any large differences from the predictions of standard diffusion theory. However, we discuss an anomaly which may be related to the time-dependent heat capacity of the standard two-level-systems model.

2. High-diffusivity Experiments

The geometry which we have used for studies on crystals is shown in the inset of Fig. 1 [1]. The samples, typically 30 x 10 x 0.1 mm, are clamped at one end and suspended in vacuum with thin-film heater (gold) and thermometer (indium) 10 mm apart as shown. In the boundary scattering regime of our experiment, the response of the thermometer to an instantaneous heater pulse of energy Q can be compared to a one-dimensional solution of the thermal diffusion equation [1]:

$$\delta T = (Q/AC\sqrt{\pi Dt})\ \exp(-x^2/4Dt). \quad (1)$$

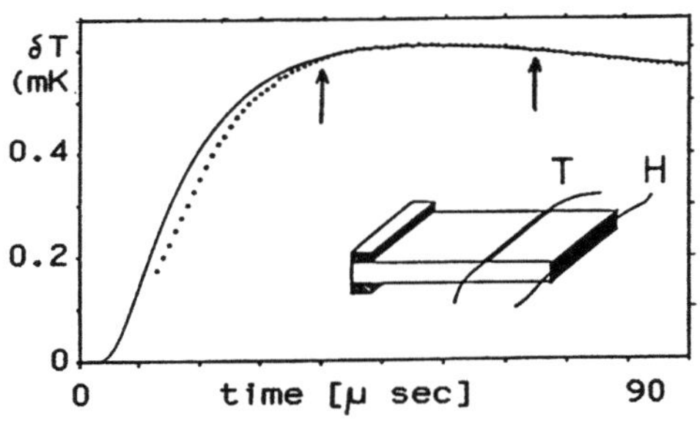

Fig. 1. The geometry for "high-diffusivity" samples is shown in the inset. The temperature pulse, for a roughened sapphire sample at 3.4 K, is fitted to a one-dimensional solution of the diffusion equation in the range indicated by arrows. The early-time discrepancy is discussed in the text

* This material is based upon work supported by the National Science Foundation under Grant No. DMR-8303977.
**IBM graduate fellow.

In (1), L, A and x are the sample length, cross-sectional area, and heater-thermometer separation respectively; the solution is accurate to one percent for times less than $L^2(1-x/L)/5D$. The parameters of the fit are the heat capacity, C, and thermal diffusivity, D, of the sample. We have carried out experiments on single-crystal sapphire with both roughened and smooth surfaces. The values obtained for the heat capacity and for the thermal diffusivity on the roughened samples are in excellent agreement with bulk heat capacity measurements and with steady-state thermal conductivity data on the same specimens. In the case of the smooth samples, specular scattering at the sample surfaces appears to require a more sophisticated analysis, which is presently being developed.

Here we would like to discuss the discrepancy between our data and the predictions of classical diffusion theory at early times. As shown, the data at earlier times always lie low relative to extrapolations of the later-time fits. When measured on faster time scales, the effects are even more pronounced and too large to be explained by thermometer response times, addenda, or other artifacts. However, the observation is consistent with the fact that (1) is acausal, since it predicts that the response to an impulse will begin instantaneously everywhere. The problem has been realized since the beginnings of diffusion theory, and a number of modified equations have been developed to account for it [2]. We are not aware of any previous experiments which have tested these equations. Since the expected minimum propagation time is about two microseconds, our data can provide such a test. We are presently developing a more complete description of fast-time thermal diffusion to explain the observed behavior.

3. Low-Diffusivity Experiments

The geometry which we have described is inappropriate for early-time studies on low-diffusivity materials with slow temporal response such as glasses. Instead, we have adopted the new arrangement shown in the inset of Fig. 2. The sample dimensions are similar to those of Fig. 1 except that the thickness is now typically 1 mm or more. In this case the heat diffusion is initially in two dimensions, and the time scale is determined by the heater-thermometer separation, usually 0.5 mm

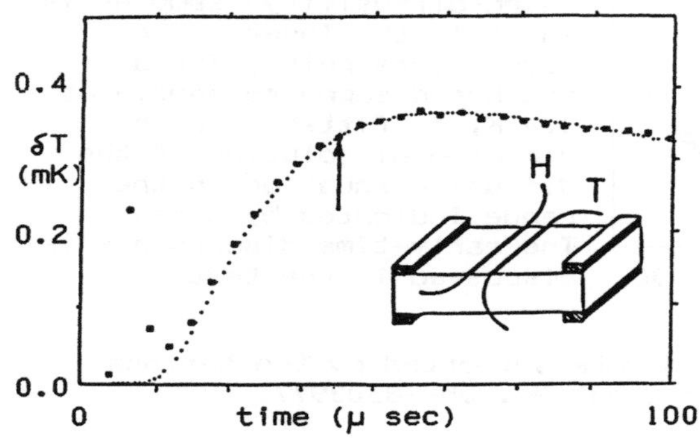

Fig. 2. The geometry for "low-diffusivity" samples with heater and one thermometer shown. The temperature pulse was taken on a sample of Suprasil-W2 at 0.66 K. The small dots indicate the fit to (2)

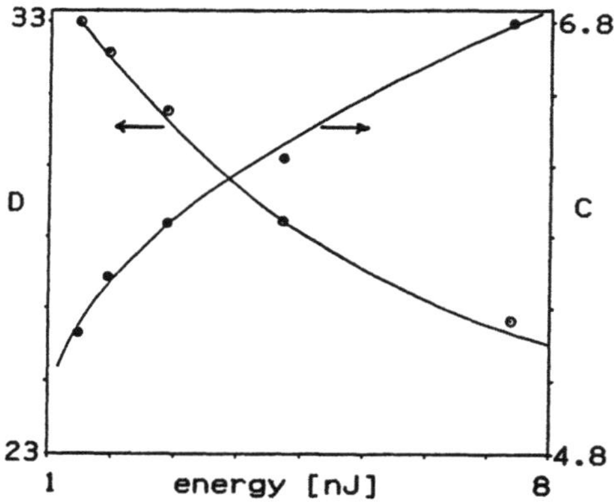

Fig. 3. D [cm^2/sec] and C [J/m^3K] as functions of pulse energy for Suprasil-W2 at 0.66 K

or less. Also shown is a typical response curve for Suprasil-W2 at 0.66 K as well as the fit to an appropriate two-dimensional solution of the diffusion equation [3]:

$$\delta T = (Q/2\pi wCDt) \exp(-x^2/4Dt) \qquad (2)$$

where w is the width of the sample. Fits of similar quality have been obtained at all temperatures so far explored from 0.66 to 3.4 K.

An anomalous behavior which we are investigating is shown in Fig. 3. Contrary to results for crystalline samples, we have discovered that the derived parameters for amorphous systems are sensitive to the energy of the input pulse. Some dependence is to be expected if the specific heat and diffusivity vary with temperature, since the experiment involves an excursion of the sample over a small temperature range. However, the magnitude of the observed dependence is too large in the non-crystalline samples to be explained by such a simple model. We believe that these systematic effects, which have not previously been reported, may be related to the dynamics of equilibration between phonons and two-level systems [4].

References

1. A. Cruz-Uribe and J. Trefny, J. Phys. E: 15, 1054 (1982).
2. A review is given by R. Swenson, Am. J. Phys. 46, 76 (1978).
3. H.S. Carslaw and J.C. Jaeger, Conduction of Heat in Solids (Oxford University Press, Oxford, 1959), p.374.
4. References are summarized by M.T. Laponen, R.C. Dynes, V. Narayanamurti and J.P. Garno, Phys. Rev. B25, 1161 (1982).

Thermal Conductivity of Neutron Irradiated Quartz

A.M. de Goër, N. Devismes, and B. Salce

Département de Recherche Fondamentale, Service des Basses Températures, Laboratoire de Cryophysique, Centre d'Etudes Nucléaires de Grenoble, 85 X F-38041 Grenoble Cedex, France

It is well known that "glassy" thermal and acoustic properties are induced in crystalline quartz by neutron irradiation [1]. In particular, the thermal conductivity K(T) reflects resonant phonon scattering by TLS below 1 K [2][3] and displays a plateau above ≃ 10 K [4]. In view of the unresolved questions about the microscopic nature of the TLS and the physical origin of the plateau, we have carried out a detailed study of K(T) of quartz as a function of neutron dose, in a large temperature range (60 mK - 100 K).

Experiments

Several samples have been cut from the same block of natural Brazilian quartz, and irradiated at CEN-Grenoble with neutron doses from 3.10^{18} to $2.2\ 10^{20}$ n/cm^2 (E > 0.1 MeV). Samples of virgin quartz and Suprasil W have been measured as references. A selection of experimental results is shown in fig.1. The general trend is in qualitative agreement with previous work [3]. The main qualitative results are : (i) below 1 K, phonon scattering increases with dose and saturates above ≃ 6.10^{19} n/cm^2 (ii) in the plateau region, phonon scattering decreases at large doses (> 10^{20} n/cm^2) (iii) for the highest dose (2.2 10^{20} n/cm^2) K(T) is nearly identical to that of neutron-irradiated a-SiO_2 [5], in

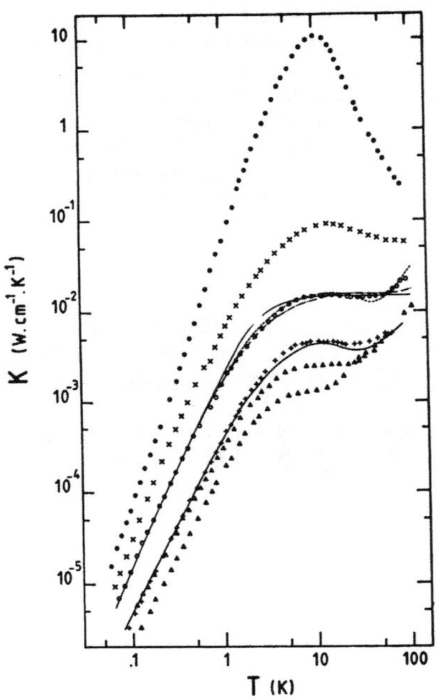

Fig. 1 - Thermal conductivity of neutron-irradiated quartz crystals :

(x) 3.10^{18} n/cm^2 ; (o) 1.8 10^{19} n/cm^2 ;

(Δ) 5.5 10^{19} n/cm^2 ; (+) 2.2 10^{20} n/cm^2.

Results for virgin quartz (•) and Suprasil W (▲) are also shown.

Full curve corresponds to neutron-irradiated a-SiO_2 from [5].

Calculated curves for three models (see text) : -·- model (ii) ; ... (iii) ; --- (iv) the whole temperature range. This is not surprising, as it is known that both crystalline quartz and a-SiO_2 are then transformed in the same highly disordered phase [6].

Analysis

The K(T) curves have been fitted using the simple Debye model, with a total relaxation rate :

$$\tau^{-1} = \tau_V^{-1} + (\pi \bar{P} \gamma^2/\rho V^2)\omega + \tau_D^{-1}$$

τ_V^{-1} includes the ordinary scattering processes needed to fit the curve of the virgin crystal [7]. The second term describes the resonant phonon scattering by TLS, which is dominant below 1 K, and τ_D^{-1} the additional scattering above 1 K. Several phenomenological models have been tested :
(i) Rayleigh scattering, $\tau_D^{-1} = A\omega^4$,
(ii) resonance scattering, $\tau_D^{-1} = D\ \omega^4/(\omega^2 - \omega_0{}^2)^2$
(iii) $\tau_D^{-1} = 0$ but the spectrum of propagating phonons is cut at a frequency $\omega^* = kT^*/\hbar$ well below the Debye frequency. This means that the excitations with $\omega > \omega^*$ are localized and do not contribute to the heat transport, as in the phonon-fracton model [8]. In cases (i) and (iii), it is a priori impossible to describe the increase of K above the plateau, but there is only one adjustable parameter A or T^* instead of two ($T_0 = \hbar\omega_0/k$ and D) in case (ii). With these models, it is also not possible to describe the slight minimum present in most samples in the "plateau" region (fig.1), as an explicit temperature dependence of τ_D^{-1} is needed. Previous suggestions were relaxation scattering [9] or Raman scattering [10] by TLS. We have also tested the relaxation scattering (model (iv)), $\tau_D^{-1} = A\omega^4 + A_1\ T^3$. In that case, it is neces-

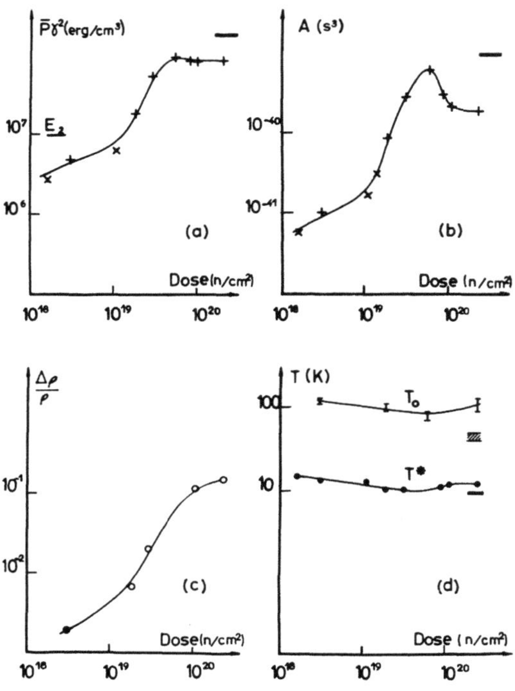

Fig.2 - (a), (b), (d) : Parameters of the fits as a function of dose. The values of $\bar{P}\gamma^2$ are independent of the model used to describe the plateau (the value noted E_2 corresponds to an electron-irradiated sample [7]). — indicates the values for Suprasil W from the same analysis

(c) : relative variation of the mass density as a function of dose. Solid lines are to guide the eye sary to limit the mean free path to some minimum value, so that there are now three adjustable parameters. Fits of comparable quality have been in fact obtained using these models, for all irradiated samples and for Suprasil W, as illustrated in fig.1 for one sample. Models (i) and (iv) give the same results below ≈ 10 K. The evolution of the parameters with neutron dose is shown in fig.2 (a,b,d), and compared to that of the relative variation of mass density (fig.2c). For $\bar{P}\gamma^2$ and A, the three ranges of doses already observed in structural studies [11] are clearly seen. However, the values of A are too large to be physically meaningful. For an intermediate dose (≈ 10^{19} n/cm^2), we have estimated A due to either point defect scattering (using a simple neutron damage model to calculate the number of defects [12]), or scattering by the small defective regions observed by SAXS [13] : they are respectively 10^3 and 10 times smaller than the experimental one. Model (iv) is not satisfactory for another reason : it supposes that TLS are responsible for the "plateau", but experimentally, the plateau is not systematically correlated to the low-temperature resonant scattering, as shown by the present results at high neutron doses, and also by previous work on electron irradiated quartz [7]. On the other hand, T_0 and T^* (fig.2d) depend weakly on dose. The wavelength of the dominant phonons corresponding to T^* varies only from 40 to 60 Å. This result is against the fracton model, as there is no characteristic length related to the state of disorder. Therefore the most likely interpretation of the plateau is a strong resonant interaction of acoustic phonons with localized modes, as suggested many years ago [14] and also quite recently [15].

Acknowledgments

We are grateful to C. Laermans and J.C. Lasjaunias for supplying respectively the quartz block and the Suprasil W sample.

References

1. For a recent review, see C. Laermans : In Structure and Bonding in Non-Crystalline Solids, ed. by G. Walrafen and A. Revesz (Plenum Press, 1985)
2. A.M. de Goër, M. Locatelli, C. Laermans : J. de Physique, Colloque C6 C6-78 (1981)
3. J.W. Gardner, A.C. Anderson : Phys. Rev. B23, 474 (1981)
4. R. Berman, Proc. Roy. Soc. London Ser. A208, 90 (1951)
5. A.K. Raychaudhury, R.O. Pohl : Sol. Stat. Comm. 44, 711 (1982)
6. J.B. Bates, R.W. Hendricks, L.B. Shaffer : J. Chem. Phys. 61, 4163 (1974)
7. C. Laermans, A.M. de Goër, M. Locatelli : Phys. Letters 80A, 331 (1980)
8. S. Alexander, C. Laermans, R. Orbach, H.M. Rosenberg : Phys. Rev. B28, 4615 (1983)
9. M.P. Zaitlin, A.C. Anderson : Phys. Stat. Sol. (b) 71, 323 (1975)
10. A.J. Leadbetter, A.P. Jeapes, C.G. Waterfield, R. Maynard : J. Physique, 38, 95 (1977)
11. R. Comes, M. Lambert, A. Guinier : In Interaction of Radiation with Solids, ed. by A. Bishay (Plenum Press 1967) p. 319
12. D.S. Billington, J.H. Crawford, Jr : Radiation Damage in Solids (Princeton Univ. Press, 1961)
13. D. Grasse, O. Kocar, J. Peisl, S.C. Moss : Rad. Effects 66, 61 (1982)
14. B. Dreyfus, N.C. Fernandes, R. Maynard : Phys. Letters 26A, 347 (1968)
15. U. Buchenau, N. Nücker, A.H. Dianoux : Phys. Rev. Letters 53, 2316 (1984)

Additional Evidence for TLS in Electron-Irradiated Quartz from Ultrasonic Attenuation Measurements

A. Vanelstraete and C. Laermans

Katholieke Universiteit Leuven, Department of Physics, V.S.H.D.
B-3030 Leuven, Belgium

At low temperatures amorphous solids exhibit dynamical properties which are unexpectedly different from these in crystals.[1] They can be attributed to the existence of configurational tunneling states,[2] which are successfully described as two-level systems(TLS).[3] The microscopic origin of these TLS is not clear yet. In an attempt to contribute to this study, crystalline solids with defects, like irradiated crystals, have been investigated. Similar anomalies as in glasses were observed in slightly neutron-irradiated quartz.[4,5,6,7] They were explained by the presence of TLS with a density of states which is smaller than in vitreous silica and which increases with neutron dose. In an earlier stage of our research one of us was involved in low-temperature thermal conductivity studies [8,9] in electron irradiated quartz. These measurements gave evidence for the presence of TLS. However it was not generally accepted that electrons could induce TLS. Here we present conclusive evidence for the existence of TLS in electron-irradiated quartz from ultrasonic studies carried out in a temperature range 1.4 to 300 K. A more detailed discussion can be found in ref.[10].

A single crystal of natural Brazilian quartz, x-cut and of high purity was electron-irradiated up to a dose of 1.0×10^{20} e/cm². (E=2 MeV). The longitudinal ultrasonic attenuation of this sample was measured in the temperature range 1.4 to 300 K at a frequency of 400 and 640 MHz, using the pulse echo technique. The results for 640 MHz are plotted in fig.1 on a log-log-scale. Also data for unirradiated quartz of the same origin are shown. For each sample, a temperature-independent residual attenuation α_o, attributed to geometrical factors, was subtracted from the data. At the lowest temperatures the electron-irradiation causes a remarkable increase of the ultrasonic attenuation compared to the unirradiated crystal. In addition, at the lowest temperatures the attenuation follows a T^3-law. This behaviour is typical for glasses and was also found for neutron-irrad.quartz[7,11], where it is attributed to the interaction of the ultrasonic phonons with TLS. Fig.2 shows the similarity with a neutron irradiated sample (N_2, dose: 1.0×10^{18}n/cm²). In the tunneling model this T^3-behaviour is known as the relaxation absorption where the condition $\omega\tau_m \gg 1$ is fulfilled, with τ_m the smallest relaxation time of the TLS. Fig.3 shows that the T^3-attenuation is frequency-independent, also in agreement with the tunneling model. Another observation that confirms the existence of the TLS in electron-irradiated quartz is the appearance of a shoulder at about 10 K. A similar behaviour was found in vitreous silica and in neutron-irradiated quartz (see fig.2) for comparable frequencies. In the tunneling model, this effect is attributed to the $\omega\tau_m \ll 1$-regime of the relaxation attenuation, for which a T°-attenuation is predicted. Fig. 3 indicates that the height of the plateau varies linearly with frequency, as is also predicted by the tunneling model. However, we remark that only a shoulder is seen and not an extended T°-range. Above 10 K a strong temperature-dependent attenuation [12] due to the anharmonic three phonon-interaction masks the T°- regime of the TLS.

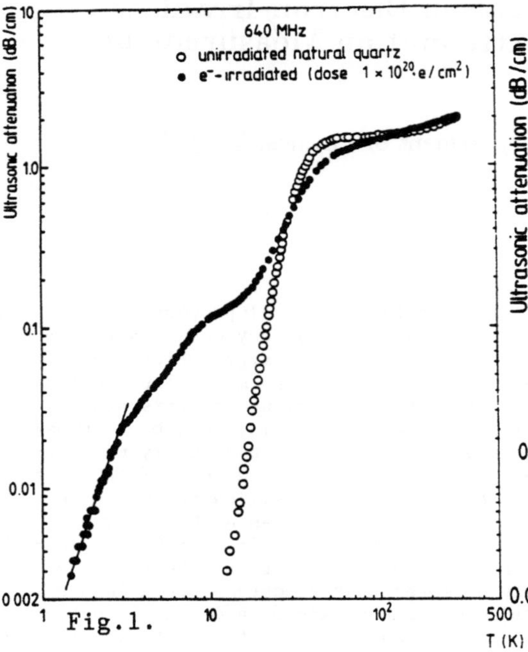

Fig.1.

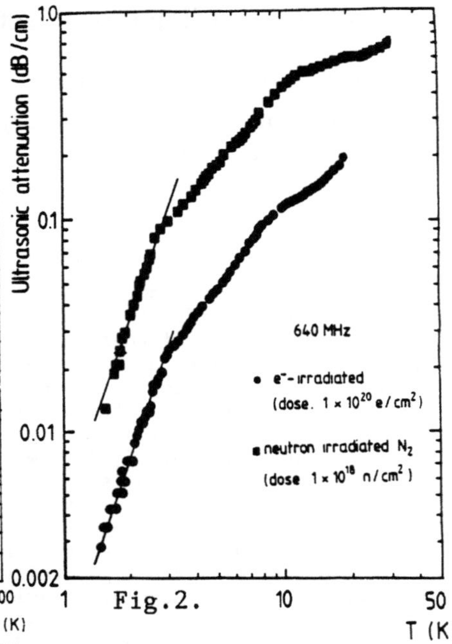

Fig.2.

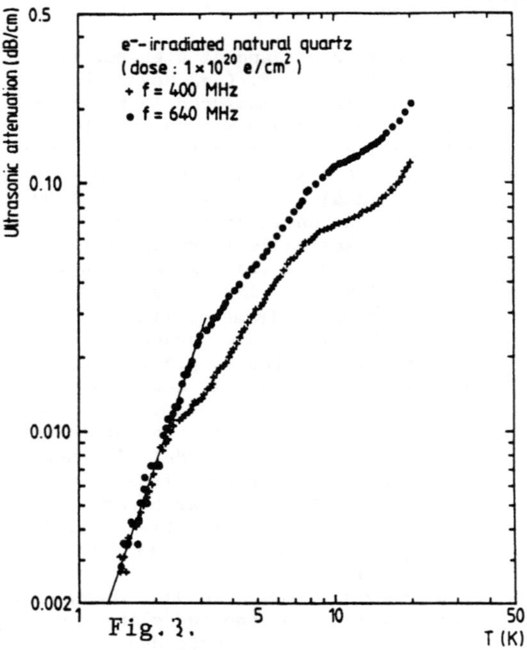

Fig.3.

Fig. 1,2,3 :

Ultrasonic attenuation as a function of temperature.

Between the T^3- behaviour and the shoulder at about 10 K, a transition region is observed which corresponds to the regime $\omega\tau_m \cong 1$. The similarity with the neutron-irradiated quartz is striking (see fig.2).

From the T^3-behaviour and the plateau-regime an estimate can be made for some typical parameters of the TLS, used in the tunneling model. Making use of the expression $\gamma_\ell^2 = 2\gamma_t^2$, we find $\gamma_\ell = 0.50$ eV and $\bar{P} = 3.6 \times 10^{30}$ $cm^{-3}erg^{-1}$. The coupling parameter γ_ℓ is smaller than that found in vitreous silica: $\gamma_\ell = 1.1$ eV if we use a similar method of calculation. Also in neutron-irradiated quartz we had to explain our ultrasonic attenuation data with values for γ_ℓ and γ_t that are smaller than in vitreous silica, although the difference is less.[13] The density of states $\bar{P}$ is found to be about 5% of that in vitreous silica.

As far as the high temperature behaviour is concerned, it is qualitatively similar to that in unirradiated quartz and can mainly be understood in terms of anharmonic three phonon processes. Even the slight reduction of the attenuation around 50 K is a fairly well-known effect and is due to the presence of radiation-induced defects and is discussed elsewhere.[10]
A small increase of attenuation above 100 K however cannot readily be explained in these terms but may be due to the $\omega\tau_{min} \ll 1$-regime of the TLS. Indeed it is striking that this additional attenuation is about the same as that of the 10 K-shoulder. This would indicate that the TLS can be observed up to 300 K.

As can be seen from the above results, MeV-electron-irradiation of quartz does induce TLS similar to those found in glasses and neutron-irradiated quartz. In neutron-irradiated quartz the TLS have been attributed to the presence of amorphous clusters found to be present in the crystalline network. However, because of their small mass, electrons, as opposed to neutrons, cannot cause displacement cascades. Therefore, regions of extensive damage, such as the 20 Å defective clusters found in neutron-irradiated quartz, are not present, as is known from diffuse X-ray studies on one of our samples. [14] Consequently, we believe that the TLS in electron-irradiated quartz are not related to large amorphous clusters.

The authors thank the C.E.N. Grenoble for performing the irradiation and the Belgian IIKW for financial support.

References

1. For a review see: W. A. Phillips(ed.), Amorphous Solids: Low-temperature properties (Springer Verlag,1981).
2. P.W. Anderson, B.I. Halperin, and C.M. Varma, Philos. Mag. 25,1 (1972); W.A. Phillips, J. Low-Temp. Phys. 7, 351 (1972).
3. S. Hunklinger and W. Arnold, in Physical acoustics, edited by W.P.Mason and R.N. Thurston(Academic Press, New York,1976),Vol.12, p. 155.
4. C. Laermans, Phys. Rev. Lett. 42, 250 (1979).
5. B. Golding and J.E. Graebner, in Phonon scattering in condensed matter, ed. H.J. Maris(Plenum Press, 1980), p.11.
6. J.W. Gardner and A.C. Anderson, Phys. Rev. B 23, 474 (1981).
7. For a recent review see: C. Laermans in Structure and Bonding in Non-Crystalline Solids, edited by G. Walrafen and A. Revesz (Plenum Press, 1986, p. 325).
8. C. Laermans, A.M. de Goër and M. Locatelli, Phys. Lett.80A, 331 (1980).

9. A.M. de Goër, M. Locatelli and C. Laermans, J. Phys.(Paris)42,C6-78(1981).
10. C. Laermans, A. Vanelstraete, Phys. Rev. B (1986), to be published.
11. C. Laermans, V. Esteves and A. Vanelstraete, Rad. Eff. 97, 175, (1986).
12. P.G. Klemens in Physical Acoustics IIIB, edited by W.P. Mason and R.N. Thurston (Academic Press, New York,1965),p. 201-223.
13. V. Esteves, Ph. D. Thesis, to be published.
14. D. Grasse, M. Müller, H. Peisl and C. Laermans, J. Phys.(Paris),43, C9-119 (1982).

Measurements of the Low Temperature Sound Velocity at 9.3 GHz in Neutron-Irradiated Quartz Crystals

N. Vanreyten and L. Michiels

Katholieke Universiteit Leuven,
Laboratorium voor Vaste Stof-en Hoge Druk-Fysica,
Celestijnenlaan 200 D, B-3030 Leuven, Belgium

1. Introduction

The thermal [1] and acoustic properties [2] of fast neutron-irradiated crystalline quartz at low temperatures are amorphous-like. Especially of interest in this paper is the variation of sound velocity $\Delta v/v_0$ with temperature for which we have shown that at 9.3 GHz $\Delta v/v_0$ exhibits a logarithmic temperature dependence below 5 K for neutron doses (E > 0.1 MeV) up to 1.6×10^{19} n/cm^2.[3] This temperature dependence can be explained by the existence of localized low-energy excitations, associated with tunneling entities.[4]

The interaction of a hypersonic wave with a two-level tunneling system (TLS) consists of a resonant and relaxation process. The resonant contribution to the relative velocity change for longitudinal acoustical waves is predicted to be [5]

$$\Delta v/v_0 = C_L \; \ell n(T/T_0) \tag{1}$$

with $C_L = \bar{P}\gamma_L^2/\rho v_L^2$. $\bar{P}$ is the spectral density of tunneling states, γ_L is the longitudinal deformation potential describing the coupling between the phonons and the TLS, v_L is the longitudinal sound velocity, ρ is the mass density and T_0 is a reference temperature. From the slope of $\Delta v/v_0$ versus ℓnT one can determine C_L.

In the present contribution we present results from measurements of the variation of sound velocity up to a neutron dose of 8×10^{19} n/cm^2.

2. Results and discussion

The X-cut natural α-quartz crystals were exposed to fast neutrons at SCK Mol, Belgium. The variation of the sound velocity was measured at a frequency of 9.3 GHz and between temperatures of 0.4 K and 6 K using a pulse superposition method.[6] At the highest dose (8×10^{19} n/cm^2) no reflection-echoes were observed because of the deformation of the sample by intense neutron-irradiation. To measure the variation of the sound velocity with temperature in this crystal we used a new technique, described elsewhere [7], based on the occurrence of backward-wave phonon echoes in neutron-irradiated quartz.

Figure 1 shows $\bar{P}\gamma_L^2$, as determined from the slope of $\Delta v/v_0$ against ℓnT as a function of neutron dose. For v_L we used 5.7×10^3 m/s and for the mass density ρ we took interpolated values from ref.8.

The growth of $\bar{P}\gamma_L^2$ on neutron-irradiation for quartz shows three important features. The first is that there is an initial increase which is relatively slow followed by a region of strong growth between 3×10^{19} and 4×10^{19} n/cm^2. At doses higher than 4×10^{19} n/cm^2 $\bar{P}\gamma_L^2$ again increases slower.

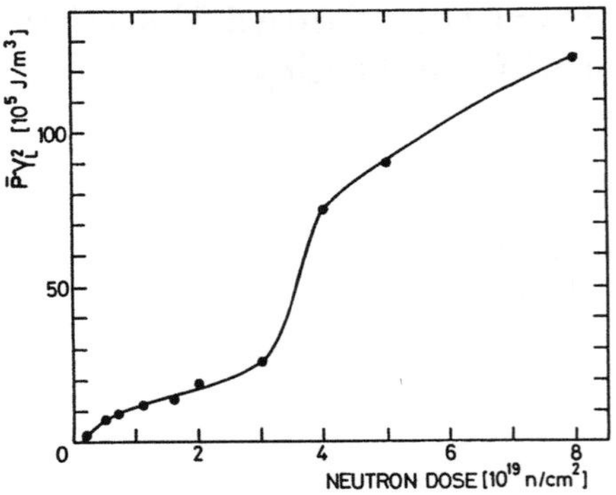

Fig.1.

$\bar{P}\gamma_L^2$ as a function of neutron dose (E > 0.1 MeV).

From the fact that the concentration of amorphous clusters, as reported by GRASSE et al. [9] and the number of TLS do not change in a simultaneous fashion for doses below 1.6 x 10^{19} n/cm² we concluded earlier [3] that there is no simple correlation between the TLS and the amorphous clusters. At higher doses no measurements of the concentration of radiation-induced clusters are available.

It has been proposed by PASCUCCI et al.[10] that the crystalline to non-crystalline transformation in α-quartz under electron-irradiation involves a local rearrangement of individual SiO_4 tetrahedra. The formation of E_1' centers [11] upon irradiation should be responsible for a reduction in the interaction of the tetrahedra so that rearrangement can start. Considerable accumulation of these defects is necessary before substantial freedom for rearrangements of adjacent SiO_4 tetrahedra is permitted. The relative slow increase of $\bar{P}\gamma_L^2$ followed by a strong increase at doses between 3 x 10^{19} n/cm² and 4 x 10^{19}n/cm² can be explained by such an accumulation of defects in neutron-irradiated quartz. In this case the TLS in neutron-irradiated quartz at the higher doses could be connected with motions of these SiO_4 tetrahedra while at lower doses other structural defects can be responsible for the TLS. $\bar{P}\gamma_L^2$ for the latter TLS tends to saturate at 3 x 10^{19}n/cm². BUCHENAU et al. [12] performed a neutron scattering study of the low-frequency vibrations in vitreous silica and to account for their results the authors proposed a model in which there exists coupled rotations of contiguous SiO_4 tetrahedra. In this view some geometric configurations result in two potential minima separated by a small energy barrier. These centers are identified with the tunneling entities which determine the low-temperature properties of amorphous SiO_2. More recently GUTTMAN and RAHMAN [13], using computer simulations of a-SiO_2, proposed that the TLS are attributable to rigid rotations of seven SiO_4 tetrahedra which are mutual connected.

Conclusion

The dependence of $\bar{P}\gamma_L^2$ on neutron dose, as determined from our measurements on the velocity of hypersonic waves,suggests that upon neutron irradiation two kinds of TLS can be created. The strong increase of $\bar{P}\gamma_L^2$ at 3 x 10^{19} n/cm² points to a rearrangement of SiO_4 tetrahedra. This new configuration with two potential minima is the same as in the recent models of a-SiO_2.

Acknowledgement

We are grateful to the Belgian Interuniversitair Instituut for Kernwetenschappen (IIKW) for financial support and SCK, Mol Belgium for irradiating the crystals.

References

1. J.W. Gardner, A.C. Anderson, Phys. Rev. B 23, 474 (1981).
2. M. Rodriguez, R. Oentrich, R. Nava, Physica 107B, 197 (1981).
3. N. Vanreyten, L. Michiels, Solid State Comm. (to be published).
4. P.W. Anderson, B.I. Halperin, C.M. Varma, Phil. Mag. 25, 1 (1972). W.A. Phillips, J. Low. Temp. Phys. 7, 351 (1972).
5. L. Piché, R. Maynard, S. Hünklinger, J. Jäckle, Phys. Rev. Lett.32,1426 (1974).
6. J. Williams, J. Lamb, J. Acoust. Soc.Am. 30, 308 (1958).
7. N. Vanreyten, L. Michiels, (to be published).
8. Mbungu Tsumbu, D. Segers, F. Van Brabander, L. Dorikens-Vanpraet, M. Dorikens, A. Van Den Bosch, Rad. Effects 62, 19 (1982).
9. D. Grasse, O. Kocar, J. Peisl, S.C. Moss, Rad. Effects 66, 61 (1982).
10. M.R. Pascucci, J.L. Hutchison, L.W. Hobbs, Rad. Effects 74, 219 (1983).
11. R.A. Weeks, J. Appl. Phys. 27, 1376 (1956).
12. U. Buchenau, N. Nücker, A.J. Dianoux, Phys. Rev. Lett. 24, 2316 (1984).
13. L. Guttman, S.M. Rahman, Phys. Rev. B. 33, 1506 (1986).

Thermal Properties of Cordierite Glass

E. Bonjour, R. Calemczuk, A.M. de Goër, and B. Salce

Département de Recherche Fondamentale, Service des Basses Températures, Laboratoire de Cryophysique, Centre d'Etudes Nucléaires de Grenoble, 85 X F-38041 Grenoble Cedex, France

The particular features of the thermal properties of noncrystalline solids in the temperature range 2 - 10 K ("plateau" in thermal conductivity K(T) and excess in specific heat c(T)) have tentatively been attributed by several authors to a possible inhomogeneous character of amorphous systems. Following this idea, experimental studies of well characterized inhomogeneous materials have been initiated [1][2]. An interesting system is cordierite glass (52 SiO_2, 34.7 Al_2O_3, 12.5 MgO) in which it has been proved that, by doping with chromium and annealing, it is possible to nucleate clusters and then microcrystallites of $MgAl_2O_4$ spinel, the size of which changing as a function of the heat treatment temperature [3]. K(T) and C(T) measurements have been performed on cordierite (as-received (P), annealed and γ-irradiated) and on different .8 % Cr-doped samples, previously studied by Raman scattering, small angle neutron scattering and electron microscopy, and referred to as A, B and C in [3].

We have measured K(T) from 100 mK to 80 K by the steady state heat flow method, using both a dilution fridge and a classical He^4 cryostat. In fig. 1, we have reported the results for the pure cordierite sample (P) (Data carried out on Cr-doped sample (A) are quite identical). They display the two classical features of disordered systems : a T^2 variation in the low-temperature range and a "plateau" around 10 K. Compared with the standard W-Suprasil glass, the plateau is shifted at higher temperatures and the value of K(T) is increased. However, below 1 K, the two curves merge down to 100 mK. γ-irradiations (performed at nitrogen temperature) do not affect appreciably K(T) for the two samples P and A. The measurements performed on samples B and C show a decrease in K(T), compared with the sample P. The "plateau" tends to disappear and no simple T variation can be evidenced at $T < 1$ K.

Specific heat measurements were performed using a dynamical adiabatic method. The figure 2 presents the logarithmic Cp/T^3 changes as a function of log T.

The cordierite sample (P) displays a maximum, characteristic of the glassy state, around 20 K ; at $T < 4$ K one observes an upward tail which may be associated either with a starting of the low-energy excitations contribution, or with residual magnetic impurities ; the former explanation is more likely.

The Cp/T^3 minimum value is slightly larger than that of the calculated acoustic limit value, deduced from ultrasonic measurements [4]. The Sample A curve is practically superimposed to the latter curve, except in the low-temperature region, $T < 10$ K, where appears the Cr^{3+} ions magnetic contribution.

Following the thermal annealing at 1050 K for 10 minutes, the specific heat of the sample (P) is absolutely unchanged. Contrarily, there is a notable modification of the doped sample curve, after annealing (sample C) :

- firstly the Cp/T^3 maximum is shifted down to 10 K and is located at the same temperature as the pure silica ; the amplitude is strongly increased
- secondly there is a crossing at about 25 K : the sample (C) curve is much lowered above 25 K.

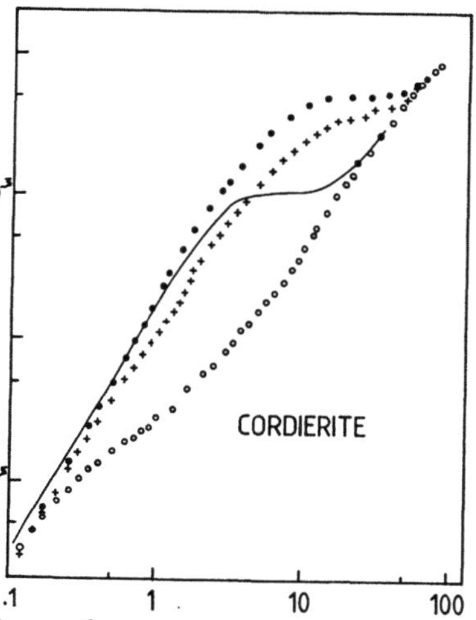

Figure 1

- Thermal conductivity of Cordierite Samples :

  • As-received pure samples (P) and (A)

  + Annealed Sample (B)

  o Annealed Sample (C)

- Thermal conductivity of Silica Glass

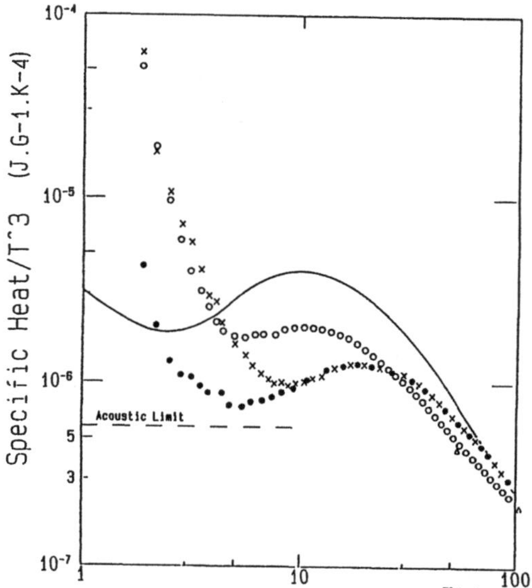

Figure 2

- Specific heat of Cordierite Samples :

  • As-received pure sample (P)

  + As-received Cr-doped Sample (A)

  o Annealed Sample (C)

- Specific Heat of Silica Glass

By applying a magnetic field between 0 and 6 Tesla, there is no modification of the maximum temperature, the only effect is a spreading of the magnetic contribution below 10 K.

It appears that these Cp results well evidence the phase separation which occurs by recrystallization during annealing ; this evolution has been shown by other experiments as mentioned before [3]. These experiments suggest the change from an homogenous glassy state (sample A) to a mixed phase formed with a glassy part near the pure silica and with a crystallized phase, the composition of which being near the spinel $MgAl_2O_4$. Supposing in this sample the maximum $MgAl_2O_4$ crystallization, i.e. 40 %, owing to the cordierite composition, calculated values following the additivity of the two phases Cp contributions, are in good agreement with the experimental curves in the 50-100 K temperature range.

These results do not exclude the possibility of the creation of new modes or excitations by developing the inhomogenity, but this cannot be appreciated, on account of the strong magnetic contribution which presents any observation below 5 K.

It must be emphasized that isolated Cr^{3+} ions do not affect the thermal conductivity of cordierite : This supports the idea that decreases in K(T) are related to the apparition of the crystalline phase, despite a higher

K(T) value in these sample parts. In fact, similar behavior has previously been evidenced in polymers, in which crystallinity was varied from 0 to 50 % [5], and seems directly correlated with the inhomogeneous character of the system. One other important point is that phonon scattering in B sample takes place down to 100 mK. Then, it can be assumed that the same physical source allows for K(T) decrease in the samples B and C. From previous measurements [3], it is known that the numbers of nucleated centers are the same in B and C, the annealed induced change being only on the size of microcrystallites. On the other hand, Raman experiments have evidenced new low-energy excitations ($\hbar\omega \sim 5\ cm^{-1}$) increasing in intensity with annealing. (These excitations cannot be detected in C(T) as their presence is masked by the magnetic Cr^{3+} ion contribution) and which were recently attributed to crystallite surface modes. From these considerations, we can assume that thermal phonons are scattered by these surface modes, the strength of scattering depending on the size of microscrystallites and then on the inhomogeneity of the system. More experimental results are needed to test if part of propagating phonons could be localized.

References

1. R.O. Pohl : in Amorphous Solids low-temperature Properties, ed. by W.A. Phillips, 27 (Springer 1981)
2. A.C. Anderson : in Amorphous Solids low-temperature Properties, ed. by W.A. Phillips, 65 (Springer 1981)
3. A. Boukenter, B. Champagnon, E. Duval and A.F. Wright : Journal de Physique, Colloque C8, C8-443 (1985)
4. J. Pelous : private communication
5. A. Assfalg : J. Phys. Chem. Solids, 36, 1386 (1975)

TLS Spectral Density and Order-Disorder Transition Temperature in $(KBr)_{1-x}(KCN)_x$

J.F. Berret[1] *and M. Meißner*[1,2]

[1]Institut für Festkörperphysik, TU Berlin, D-1000 Berlin 12, Germany
[2]Hahn-Meitner-Institut für Kernforschung, D-1000 Berlin 39, Germany

1. Introduction

Low-temperature measurements ($T \approx 0.1$ to 10K) of the thermal conductivity and the specific heat [1-3] as well as the ultrasonic attenuation and dispersion [4] on $(KBr)_{1-x}(KCN)_x$ single crystals with $0 < x < 1$ have revealed the well-known thermal and acoustic anomalies found in a variety of structurally disordered systems [5]. For x = 0.25, 0.50 and 0.70, the spectral density P of two-level-systems (TLS) has been determined independently from the time-dependence of the specific heat [3, 6] on a time scale $t_{exp} \approx 100\mu$sec to 10sec and from ultrasound data [4] for longitudinal and transverse modes ($\nu \approx 400$MHz). By re-evaluating these ultrasound data we can now show that the P(x)-values obtained by this method are in good agreement with those evaluated from the time-dependence of the specific heat. From this analysis we evidence a variation on the CN^- concentration as approximately $P \propto (1-x)^3$.

Since the glass-like thermal anomalies are not restricted to the orientationally disordered state ($x < 0.56$) but are also present in the range of quadrupolar long-range order ($0.56 < x \leq 0.7$), it has been suggested that the TLS in $(KBr)_{1-x}(KCN)_x$ originate from 180 degree flips of a small fraction of CN^- dipoles [3, 7]. From frequency dependent measurements of the dipolar and quadrupolar susceptibility it is now established that in $(KBr)_{1-x}(KCN)_x$ different freezing processes for the (elastic) quadrupoles and (electric) dipoles give rise to different freezing temperatures, $T_{f,q}(x,\nu)$ and $T_{f,d}(x,\nu)$, respectively [8]. According to the above-mentioned picture of CN^- dipole flips as the origin of TLS (for CN^- concentrations $0.25 \leq x \leq 0.7$) we have connected the TLS spectral density P(x) with the transition temperature T_f where the fast rotations of the CN^- quadrupoles freeze out. For structural glasses attempts have been made in relating the spectral density to the glass transition temperature T_g using different analytical forms, i.e. $P \propto T_g^{-1}$ [9, 10] and $\log(P) \propto T_g^{-1}$ [11]. Based on the dependence $P \propto (1-x)^3$ and an analysis of $T_f(x)$ at $\nu \approx 1$GHZ [8] we find $P \propto (T_{C1}-T_f)^3$ where T_{C1} = 168K is the ferroelastic phase transition temperature of KCN.

2. Results and Discussion

The relative velocity change and attenuation of longitudinal and transverse phonons in $(KBr)_{0.75}(KCN)_{0.25}$ and $(KBr)_{0.5}(KCN)_{0.5}$, respectively, were analyzed in Ref. 4 within the framework of a tunneling model using a distribution function $P(E, r, \mu) = (P/2) \cdot r^{-1}(1-r)^{\mu-0.5}$ where $r = \tau^{-1}/\tau_{min}^{-1}$ (τ_{min} is the fastest relaxation time for given energy E and temperature) and μ is a parameter which decreases the number of fast TLS with respect to slower ones (for $\mu > 0$); $\mu = 0$ turns the above distribution function into the standard form [5]. The use of this modified model gives reasonable good fits to the dispersion and absorption data up to about 5K. However,

Table 1. TLS spectral density P and coupling energy γ for $(KBr)_{1-x}(KCN)_x$ derived from specific heat (C_t), thermal conductivity (K) and ultrasound (us) data; T_f is the order-disorder transition temperature (see text).

x	T_f [K]	P $[10^{44} J^{-1}m^{-3}]$		γ [eV]	
		C_t	us	C_t, K	us
0.25	55	40±4	31±8	0.12	0.16
0.50	82	9±1	14±4	0.18	0.15
0.70	112	2.7±0.3	—	0.48	—

as this spectral density depends on μ any comparison with the P-values derived from the logarithmic time-dependence of the specific heat [3, 6] turns out to be useless. Following the standard tunneling model, we only evaluated data below 1K, where the TLS model predicts a velocity change $\Delta v/v \propto \ln(T/T_0)$ and an attenuation $\alpha \propto T^3$. In Tab. 1 the TLS parameters for x = 0.25 and 0.50 are shown which are consistently determined for both longitudinal and transvere modes. In addition, we also show the same quantities derived from the time-dependent specific heat, $C_t \propto T\cdot\ln(t/\tau_{min})$ and the thermal conductivity, $K \propto T^2$ [3, 6]. Obviously, a fairly good agreement has been obtained for the spectral density P (within 30%) and the averaged coupling energy γ (within 20%).

An analysis of P(x) shows a strong increase as the CN^- concentration decreases, which can be fitted to an empirical power law $P \propto (1-x)^3$. We point out that for $x > 0.7$ this relation can experimentally be tested; however, for $x \leq 0.25$ the low-energy excitations show an increasing tendency to deviate from the standard TLS glass-like behaviour [4, 6]. Hence, we believe that the above relation reflects the increase of the number of TLS in a CN^- concentration range where only tunneling of the CN^- dipoles is possible. In fact, this should be valid for $x > 0.56$ where, below a certain transition temperature T_c, the CN^- quadrupoles arrange into long-range order [12]. For $x < 0.56$ there is no quadrupolar ordering process, although, at high frequencies the quadrupolar freezing can be visualized as a highly cooperative phenomenon [8]. Thus, we conclude that the variation of $P \propto (1-x)^3$ is connected with the transition temperature T_c or T_f, respectively, where the quadrupoles orientate into long-range order (x = 0.70) or into a frozen-in orientationally disordered state (x = 0.50 and 0.25), respectively.

For $(KBr)_{0.3}(KCN)_{0.7}$ we assign T_c = 112K where in a small temperature range two continuous (high order) phase transitions have been observed by X-ray [13] and thermal expansion measurements [14]. For x = 0.50 and 0.25 we use $T_f(\nu)$ of Volkmann et al. [8] who observed the relaxation of dipoles and quadrupoles to behave similar for $\nu > 1$GHz (see Tab. 1). In Fig. 1 we present the variation of the spectral density P with the above defined T_f. The full line indicates a variation $P \propto (T_{c1}-T_f)^3$ which is due to the fact that T_f (for $x < 0.56$) and T_c (for $x > 0.56$) are approximately proportional to the CN^- concentration up to pure KCN, undergoing a ferroelastic (first-order) phase transition at T_{c1} = 168K.

What could be concluded from the above relation between P and T_f or T_c, respectively? First, in a concentration range $0.56 < x < 1$ the dependence on the ferroelastic ordering temperature T_c reflects a change of the potential barrier distribution P(V), hindering dipolar reorientations. This change is connected with the screening of the CN^- interaction by Br^- ions.

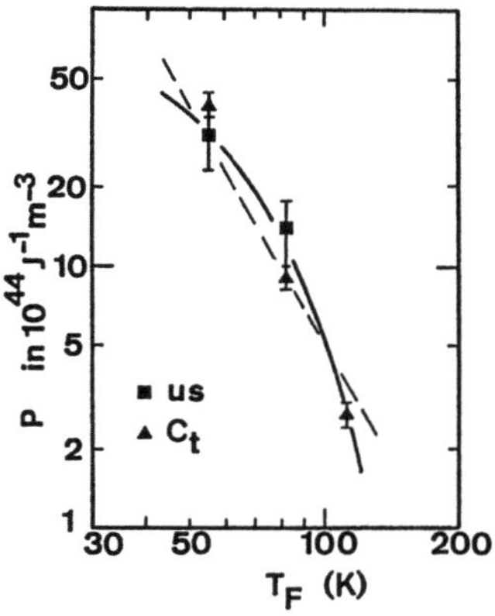

Fig. 1. Variation of the TLS spectral density P with the transition temperature T_f of $(KBr)_{1-x}(KCN)_x$ (as given in Tab. 1). The full line is according to an empirical relation $P \propto (T_{c1}-T_f)^3$ where T_{c1} = 168K is the ferroelastic phase transition temperature of KCN. However, it should be remarked that other types of dependences can be fitted to the data; as an example $P \propto T_f^{-3}$ (broken line) is shown in addition.

Secondly, a comparison between T_f and the glass transition temperature T_g for structural glasses seems to be problematic [8]; we cannot contribute to the recent discussion on a possible relation $P = P(T_g)$. Instead, we suppose that the TLS excitations in CN^- rich $(KBr)_{1-x}(KCN)_x$ are related to the frustration of phase transition ($0.5 < x < 0.9$) whereas for a CN^- range $10^{-3} < x < 0.2$ a situation similar to spin glasses is established.

Acknowledgment

We would like to thank W. Knaak, K. Knorr, A. Loidl, R.O. Pohl and J.J. DeYoreo for stimulating discussions. This work was supported by the Deutsche Forschungsgemeinschaft.

References

1. J.J. DeYoreo, M. Meißner, R.O. Pohl, J.M. Rowe, J.J. Rush and S. Susman: Phys. Rev. Lett. **51**, 1050 (1983)
2. D. Moy, J.N. Dobbs, A.C. Anderson: Phys. Rev. B. **29**, 2160 (1984)
3. M. Meißner, W. Knaak, J.P. Sethna, K.S. Chow, J.J. DeYoreo and R.O. Pohl: Phys. Rev. B **32**, 6091 (1985)
4. J.F. Berret, P. Doussineau, A. Levelut, M. Meißner and W. Schön: Phys. Rev. Lett. **45**, 2013 (1985)
5. For a review see W.A. Phillips (ed.): Amorphous Solids: Low-Temperature Properties (Springer, Berlin, Heidelberg 1981)
6. J.J. DeYoreo, W. Knaak, M. Meißner and R.O. Pohl: to appear in Phys. Rev. B (1986)
7. J.P. Sethna and K.S. Chow: Phase Transitions **5**, 317 (1985)
8. U.G. Volkmann, R. Böhmer, A. Loidl, K. Knorr, U.T. Höchli and S. Haussühl: Phys. Rev. Lett. **56**, 1716 (1986)
9. M.H. Cohen and G.S. Grest: Phys. Rev. Lett. **45**, 1271 (1980)
10. A.K. Raychaudhury and R.O. Pohl: Phys. Rev. B **25**, 1310 (1982)
11. U. Reichert, M. Schmidt and S. Hunklinger: Sol. St. Com. **57**, 315 (1986)
12. K. Knorr and A. Loidl: Phys. Rev. B **31**, 5387 (1985)
13. K. Knorr, A. Loidl and J.K. Kjems: Phys. Rev. Lett. **55**, 2445 (1985)
14. J.F. Berret and M. Meißner: to appear in J. Phys. C (1986)

Part II

Electron-Phonon Interactions and Superconductivity

Phonon-Emission Spectroscopy of a Two-Dimensional Electron Gas

M. Rothenfusser, L. Koester, and W. Dietsche

Physik Department E 10, Technische Universität München, D-8046 Garching, Fed. Rep. of Germany

Electrically heated metal films have long been used as generators of heat pulses. It is well known, that their phonon-frequency spectrum is that of a black-body radiator [1]. If the phonon source is restricted to two dimensions, however, deviations from this behavior can be expected [2]. This situation is realized with the two-dimensional electron gas (2 DEG) in a MOS device [3].

The design of a MOS structure is shown in Fig. 1. The gate electrode (NiCr) is separated from the (100)-Si substrate by 150 nm SiO_2. A positive gate voltage V_G leads to the accumulation of electrons at the interface with a density N_s increasing linearly with V_G. The lowest electron subband is a circle in the k plane [3].

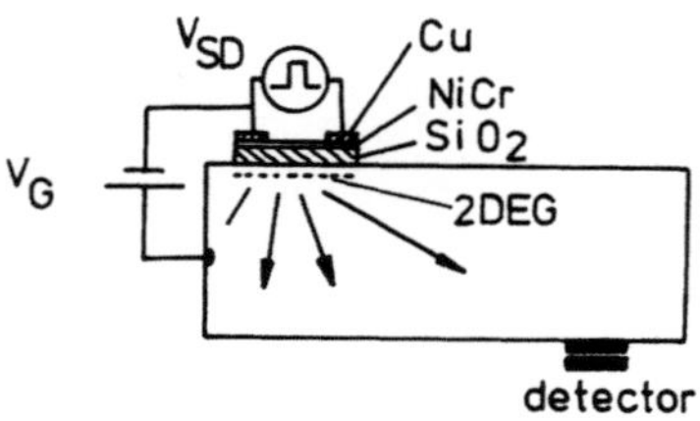

Fig. 1. Sample set-up. Contacts to the 2 DEG (source and drain) were made capacitively with two Cu pads. The source-drain voltage V_{SD} was applied as 100ns pulses. Therefore the impedance of $C_{Ox} \approx 1nF$ could be neglected compared with the resistances of the 2 DEG ($\approx 1k\Omega$) and of the NiCr gate film ($\approx 10k\Omega$). The active area of the 2 DEG was $1mm^2$

The emitted phonons were detected on the opposite surface of the substrate crystal. Two types of tunnel-junction detectors were used. Junctions made of Pb have a detection threshold at $2\Delta/h = 650$ GHz, those of Al at 100 GHz. In either case, there is an upper frequency limit imposed by the isotope scattering in the bulk Si which becomes effective above about 800 GHz. Thus the Pb junction acts as a narrow-band detector (650-800 GHz), while the Al junction behaves as a broad-band detector (100-800 GHz).

The matrix element for phonon emission is determined by the deformation-potential. During an electron transitions from the state $\underline{k}_i$ to $\underline{k}_f$ a phonon is emitted with a momentum component in the 2DEG plane $\underline{q}_p = \underline{k}_i - \underline{k}_f$. The momentum component in the direction normal to the surface follows from energy conservation : $q = \omega/v_s$ with $\hbar\omega = E_1 - E_2$, v_s being the sound velocity. The maximum value of q_p is about $2\ k_F$ (see inset of Fig. 2). Thus, the maximum q is about $2\ k_F/\sin\theta$ were θ is the angle between q and the surface normal. The modification of the phonon propagation by the phonon focusing was taken into account by us. The phonon spectra expected to be emitted by the 2DEG were calculated by summing over all possible transitions [2]. Examples are shown in Fig. 2.

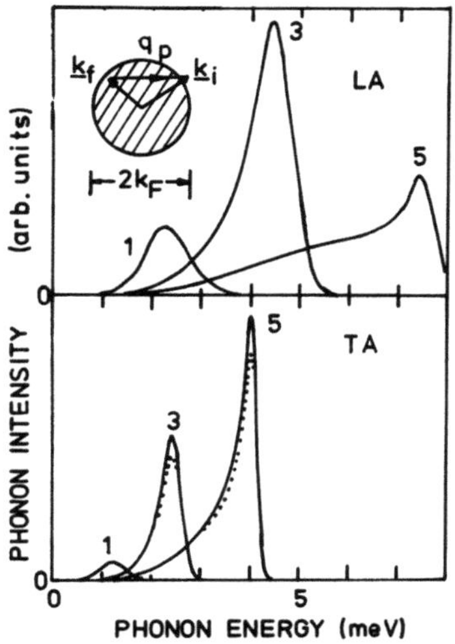

Fig. 2. Theoretical spectra of phonons emitted by a 2DEG. The electron temperature which is required as input parameter was first determined self consistently. It varies between 12 and 18K at an ambient 1K and a source-drain field of 10 V/cm. The phonon spectra are almost monochromatic with a cut-off at a frequency corresponding to 2 $k_F/\sin\theta$. The respective k_F values are given at each trace in 10^6 cm^{-1}. In a few cases, the effect of reversing the electric field direction is also given (dotted lines)

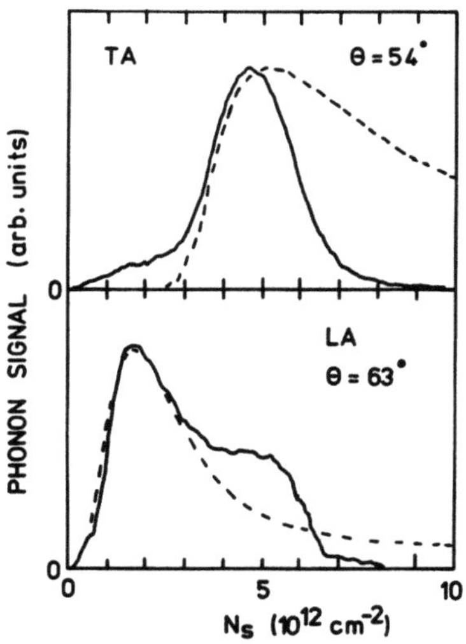

Fig. 3. Phonon intensities measured with the Pb detector for transversely (TA) and longitudinally (LA) polarized phonons. The source-drain field was 10V/cm. The steep increases occurs when the $2k_F$ cut-off is swept over the detector window. Dashed lines: theory

In the experiment the density (or k_F) was varied. With an Al detector, the phonon signal followed roughly the course of the electric power. More structure was observed with the narrow-band Pb junction detector (Fig. 3). The TA phonons observed under $\theta = 54^o$ exhibited a steep increase at a relatively large electron density caused by the coincidence of the 2 k_F-cut-off with the detector window. With LA phonons the cut-off is at higher frequencies. Consequently the signal increase occurred at smaller densities as indeed observed (Fig. 3, bottom). At smaller emission angles the increase shifts to smaller densities with TA phonons too. This is visible from Fig. 4, trace 0 kbar.

The dashed lines in Fig. 3 were calculated from folding the theoretical spectra with the respective detector windows. Good agreement is found near

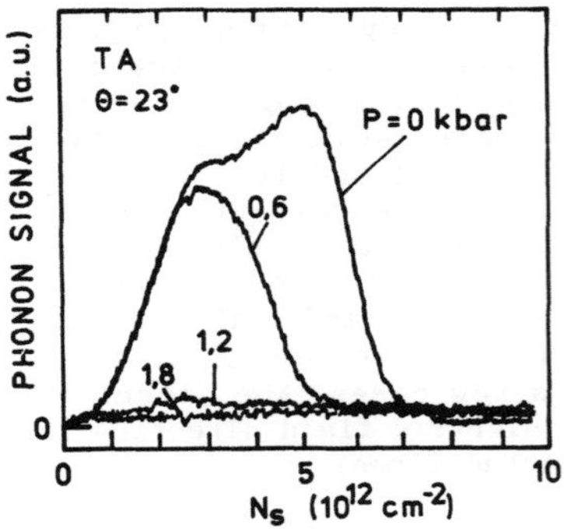

Fig. 4. Phonons detected with a Pb junction. The subbands near the zone boundary are energy-shifted by applying uniaxial stress along the [011] direction. With moderate stress (0.6 kbar), the signal decrease occurs at lower electron densities. At higher stresses, the phonon signal disappears completely. Only very small stress effects were observed with the Al detector

the steep signal increase (the cut-off region). At larger densities, however, there are considerable deviations. With the Pb detector and the LA phonons a maximum around $N_s = 5 \times 10^{12}$ cm^{-2} is observed followed by a steep decrease. A maximum and a steep decrease is also observed with the TA phonons at $\theta = 23^o$ in Fig. 4. The steep decrease is observed throughout.

We suspect that the deviations are caused by the population of higher subbands. The maximum near 5×10^{12} cm^{-2} could be due to transitions between the lowest and the first excited subband at the center of the Brillouin zone. The decrease at higher densities could be interpreted as a cooling of the electrons which were possible if additional phonon emission channels opened up. Such a process could be transitions within the subbands at the zone edge. The additionally emitted phonons would have low frequencies. Indeed, the Al detector signal showed an increase at the same density where the Pb signal decrease occurred [2]. Further evidence comes from observing the phonons with uniaxial stress applied (see Fig. 4).

In conclusion, it has been shown that the most pronounced feature of the phonon-emission spectra of a 2DEG is a 2 k_F cut-off which is not only dependent on the phonon mode but also on the angle. At higher densities the influence of the higher subbands needs to be considered as was demonstrated by the stress experiment. So far, no calculation of the pertinent phonon spectra has been made. They were particularly necessary for the transitions leading to the maximum at $N_s = 5 \times 10^{12}$ cm^{-2}.

1. For reviews see articles in: Nonequilibrium Phonon Dynamics, ed. by W. Bron, Nato Advanced Study Institute Series (Plenum, 1985)
2. M. Rothenfusser, L. Koester, and W. Dietsche, submitted to Phys. Rev.
3. T. Ando, A. B. Fowler, and F. Stern, Rev. Mod. Phys. 54, 437 (1982)

Thermal Conductivity Resonances in Doped Semiconductors and Amorphous Materials

E. Sigmund

Institut für Theoretische Physik, Universität Stuttgart, D-7000 Stuttgart 80, Fed. Rep. of Germany

1. Introduction

The low-temperature thermal conductivity of doped semiconductors (e.g. Si(In), Si(B), and GaAs(Mn)) exhibits a region with resonance-like lowering of the conductivity, which in some cases even leads to a minimum of the curve.[1]

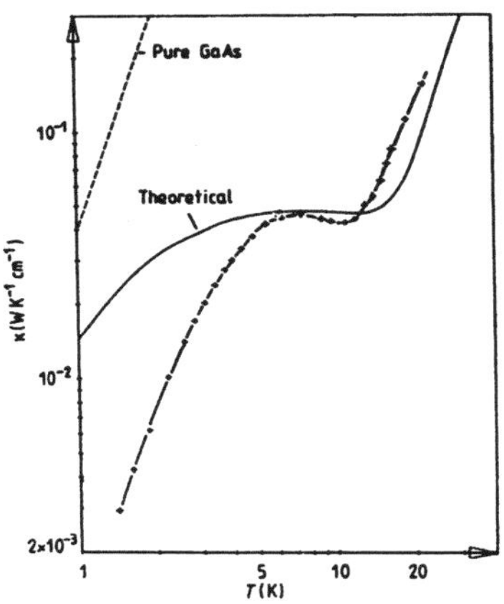

Fig. 1: Thermal conductivity of GaAs(Mn) (3.8×10^{18} cm^{-3}): theoretical values compared with experimental results of de Combarieu and Lassmann (1976).

We show that these resonance structures may be explained by the presence of a dynamical electron-phonon coupling of the defects states creating a new kind of phonon scattering mechanism. This mechanism is strongly temperature-dependent and therefore has a great influence on the thermal conductivity curve. The dynamical scattering mechanism can also explain some features of the thermal conductivity curves of glassy or amorphous materials.

2. The dynamic coupling structure

The interaction scheme of the discussed electron phonon systems consist of the coupling of degenerate or nearly degenerate localized electronic or excitonic levels and low energetic vibrational excitations of the surroundings. In the case when the energetic separation of the high energetic (e.g. electronic) system is of the same order of magnitude as the elementary excitations of the vibrational system, resonance effects in the sense of a direct energy exchange between the subsystems will appear. These systems are of non adiabatic nature since the separation of the motion of the coupled particles is no longer possible.

As an example we consider the case of acceptor states in cubic semiconductors, where the valence-band edge is of Γ_8 type and this symmetry is transferred to the acceptor ground state. Accordingly the Hamiltonian has the form[2]

$$H_{\mathrm{I}} = \sum_{q\lambda} [D^{\varepsilon}(\hat{\rho}_1 r_1^{q\lambda} + \hat{\rho}_2 r_2^{q\lambda}) + D^{\tau}\hat{\rho}_3(\hat{\sigma}_1 s_1^{q\lambda} + \hat{\sigma}_2 s_2^{q\lambda} + \hat{\sigma}_3 s_3^{q\lambda})](b_{q\lambda} + b^{+}_{q\lambda})$$

where $b^{+}_{q\lambda}$, $b_{q\lambda}$ are operators for acoustical phonons of wave vector q in the branch λ, ρ_i, σ_j are two commuting sets of Dirac's 4 x 4 matrices, D^{ε} and D^{τ} are coupling parameters and $s_i^{q\lambda}$ and $r_i^{q\lambda}$ are projectors from lattice phonons to the symmetry coordinates of t_{2g} and e_g vibrations of the defect.

3. The phonon scattering process

The phonon scattering process can be described as follows: Due to the coupling of non-symmetric phonon modes the symmetry of the defect surrounding is lowered, leading to a dynamic splitting of the electronic multiplet. Then, additional phonons can be scattered by the momentary splittings. If these phonons are non-thermal a frequency-selective scattering rate will be observed. The effective resonance energy is related to the statistical distribution of those dynamical splittings. Due to the coupling of the whole frequency spectrum of the crystal the scattering rate is very broad and strongly temperature dependent, which leads to a strong influence on the thermal conductivity curve. The calculations which are performed in a high-order Green function treatment reproduce the resonance-like structure of the thermal conductivity curve which is found in the experiment (Fig. 1).

4. The influence of static splittings

The nonadiabatic and dynamic interaction mechanism can be disturbed by coupling a static elastic or magnetic field to the defect[3]. We confine us to elastic fields which e.g. may be random internal fields and which cause a static splitting of the degenerate electronic levels. As long as the static splitting is much smaller than the phonon-induced oscillation amplitudes of the electronic levels the dynamic behaviour of the system will be unchanged. However, when the static splitting exceeds the dynamic one, the system will lose its dynamic behaviour. This is seen in Fig. 2 where the relaxation rate of phonons scattered at a statically split 2-level system is plotted. For an increasing level splitting the broad dynamic part of the scattering rate decreases drastically.

Fig. 3 shows the scattering rate for a fixed-level splitting in dependence of temperature. As one would expect with increasing temperature the static

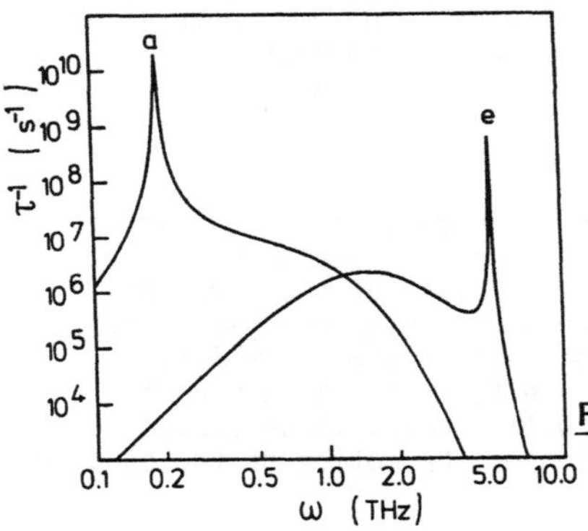

Fig. 2: Mean scattering rate for 2-level system. Two different electronic splittings: a=0,2 THz, e=5.0 THz

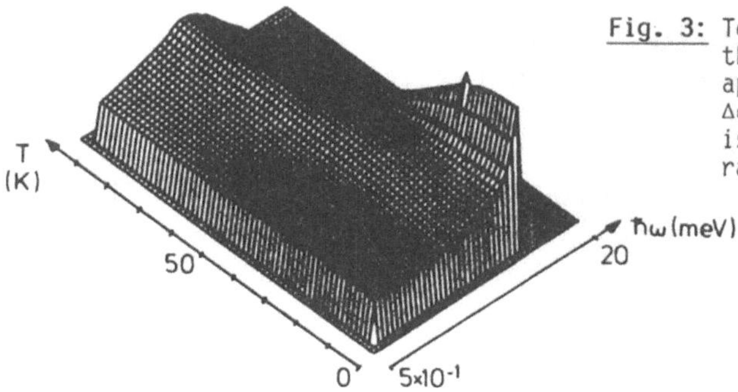

Fig. 3: Temperature dependence of the relaxation rate for an applied elastic field $\Delta\varepsilon=4$ meV. The energy scale is logarithmic, the temperature scale is linear.

part of the scattering rate decreases whereas the dynamic part, which was suppressed by the static one at low temperatures, still survives. This leads to the consequence that the scattering at the dynamic split levels cannot be saturated by increasing the temperature (or the intensity of the phonon pulse) in contrast to the scattering at a statically split 2 level system.

5. Conclusions

The results show that the dynamic behaviour of the electron phonon coupling has an important influence on the phonon scattering process, which can even lead to resonance-like structures (minima) in the thermal conductivity curve. For the appearance of the discussed dynamical behaviour, the origin of the high energetic system need not be necessarily of electronic or excitonic nature, this role can also be played by a localized vibrational degree of freedom. For example, in glassy or amorphous materials when coupling the 2-level systems to the low-energetic vibrations of the surrounding, these dynamic interaction processes will appear which then in a natural way can lead to an explanation of the plateau region in the thermal conductivity curves of these systems.

References

1) de Combarieu, A., Lassmann, K. Phonon Scattering in Solids, Plenum Press, New York, 1976

2) Maier, J, Sigmund, E., J. Phys. C **17**, 4141 (1984)

3) Maier, J., Sigmund, E., Phys. Rev. B (1986) in print

Damping of Phonons by Metal Particles Embedded in an Insulating Matrix

T. Nakayama and K. Yakubo

Department of Applied Physics, Hokkaido University, Sapporo 060, Japan

1. Introduction

The energy spectrum of electrons confined to small region in space becomes discrete due to the quantum-size effect[1]. The problem associated with the observation of such features has been receiving considerable attention using submicron metal particles, particularly from light scattering[2], magnetic susceptibility[3], and specific heat[4]. In this paper we point out that such characteristics can be observed through the damping of phonons by metal particles embedded in an insulating matrix. The mean-level spacing $\bar{\delta}$ depends on the size of particles. For example, $\bar{\delta}$ can be estimated to be about 1K for 100A metal particles using the relation $\bar{\delta} = 4E_f/3N$, where N and E_f are the number of electrons and the Fermi energy of metal particle; i.e., $N \sim 10^4$ and $E_f \sim 10^4$K for a particle with diameter 100A. It should be emphasized here that the level spacings are distributed for an assembly of metal particles owing to different shapes and impurities. The physical quantities must be averaged by the distribution function $P(\delta)$ which should be determined by the random matrix theory[5]. It will be shown in this paper that the remarkable property of energy spectrum of metal particles; i.e., quantum-size effect, can be observable through the frequency and the temperature dependence of the attenuation and the velocity dispersion of phonons. Hereafter a system of units is used in which $k_B = \hbar = 1$.

2. Phonon-metal-particle interaction

Let us consider the case where the wavelength of phonons in a matrix is much larger than the size of metal particles. The volume change of metal particles induced by incident phonons yields the electron-phonon interaction expressed by the deformation coupling as

$$H' = \sum_{n,q,\sigma} A_q a^+_{n+\omega,\sigma} a_{n,\sigma} (b_q + b^+_{-q}) \,, \tag{1}$$

where $a_{n,\sigma}$ and b_q are the annihilation operators of electron and phonon with the wavevector $\vec{q}$, respectively. The symbol n represents the energy level of electron confined in a metal particle. The deformation coupling constant takes the form:

$$A_q = -\frac{2}{3} i\, E_f \left(\frac{q}{2\rho v_s}\right)^{1/2} \,, \tag{2}$$

where ρ and v_s are the mass density and the velocity of phonons in a matrix, respectively.

The interaction Hamiltonian can be separated in a form $H' = -xf$, in which f is the generalized force and x is the generalized displacement, namely, $f = b_q + b^+_{-q}$ and $x = \sum_{n\sigma} A_q a^+_{n+\omega,\sigma} a_{n,\sigma}$. According to the linear response theory, the generalized susceptibility of the imaginary part is expressed as

$$\chi'' = \pi[1-\exp(-\omega\beta)] \sum_{m,n} \rho_n |x_{nm}|^2 \delta(\omega-\omega_{mn}) \quad , \tag{3}$$

where $\beta = 1/T$. The symbol ρ_n is the canonical distribution function, $\omega_{mn} = E_m - E_n$ and x_{nm} is the matrix element of the operator x. The attenuation rate (cm^{-1}) is related to Eq.(3) as

$$\alpha = \chi'' / 2v_s \quad (cm^{-1}) \quad . \tag{4}$$

Using this relation, we have the attenuation rate, for the low temperature regime $\beta\delta > 1$,

$$\alpha = \frac{4\pi E_f^2 n\omega}{9\rho v_f^3} \sum_{m=1}^{\infty} \frac{1-\exp(-m\beta\delta)}{z+4\exp(-\beta\delta)} \delta(\omega - m\delta) \quad . \tag{5}$$

In Eq.(5), z is 1 for even number and 2 for odd number of electrons in metal particle. n is the number density of metal particles. Since the shapes of metal particles are not uniform, the energy-level spacings of electrons for an assembly of metal particles obey the distribution function determined by the random matrix theory. Taking the ensemble average of Eq.(5) using the Wigner distribution function, where the spin-orbit coupling and the magnetic fields are absent, the attenuation rate (cm^{-1}) becomes

$$\alpha = A\omega^2 \sum_{m=1} \frac{1}{m^2} \exp[-\frac{\pi}{4}(\frac{\omega}{m\bar{\delta}})^2] \frac{1-\exp(-\beta\omega)}{z+4\exp(-\beta\omega/m)} \quad , \tag{6}$$

where

$$A = \frac{\pi^2 N^2 n}{8\rho v_s^3} \quad . \tag{7}$$

In the case where a phonon excites an electron from E_f to the first level, we have

$$\alpha = A\omega^2 \exp[-\frac{\pi}{4}(\frac{\omega}{\bar{\delta}})^2] \tanh(\beta\omega/2) \times \begin{cases} 1 - 3\exp(-\beta\omega) & : \text{even} \\ \frac{1}{2}[1 - \exp(-\beta\omega)] & : \text{odd} \end{cases} \tag{8}$$

For $\omega \ll \bar{\delta}$, Eq.(6) shows the frequency dependence proportional to ω^3, and linearly proportional to ω for the case $\omega \gg \bar{\delta}$. This linear dependence is identical with the attenuation rate of phonons for bulk metals. Thus, the ω^3-dependence of the attenuation rate reflects a character of discrete energy level; i.e., quantum-size effect and randomness of energy levels.

From Eq.(8), the velocity change of phonons can be calculated using the relation between the real part of the generalized susceptibility χ' and the velocity v:

$$\chi'(\omega,T) = -\frac{2\omega v_s}{v(\omega,T)} \quad , \tag{9}$$

where v_s is the phonon velocity at $T = 0$. The velocity dispersion is related to α by the following relation,

$$\frac{\Delta v}{v} = \frac{2v\omega}{\pi} P\int_0^{\infty} \frac{\alpha(\omega';T)-\alpha(\omega';T_0)}{\omega'(\omega^2-\omega'^2)}\, d\omega', \tag{10}$$

where P means the principal part of the integral. Since the attenuation rate takes the following form in the case of $\omega\beta >> 1$:

$$\alpha = A\omega^2 \exp[-\frac{\pi}{4}(\frac{\omega}{\bar{\delta}})^2] \tanh(\beta\omega/2) \quad , \tag{11}$$

the velocity change is expressed by

$$\frac{\Delta v}{v} = \frac{2\omega v A}{\pi} \log(\frac{T}{T_0}) \quad . \tag{12}$$

3. Conclusions

It has been predicted[1] that the magnetic susceptibility of metal particles depends on odd or even number of electrons. This odd/even feature is not drastic in the case of the phonon damping as seen from Eq.(5). If we assume D = 100A, $n = 0.9 \times 10^{18} cm^{-3}$ and $\bar{\delta}$ = 1K, the attenuation rate becomes $\alpha = 0.3 cm^{-1}$ for ν = 200MHz and T = 1.5K. Thus the experiments on the attenuation and the velocity dispersion of phonons appear to be quite feasible and would give valuable information on the dynamical properties of metal particles. Especially it should be remarked that the logarithm T-dependence of the velocity dispersion, which comes from the distribution of energy-level spacing, is identical with the T-dependence of the velocity dispersion of phonons in glasses.

References

1. R. Kubo: J. Phys. Soc. Jpn. 14, 975(1962), and L.P. Gor'kov, G.M. Eliashberg, Sov. Phys. JETP 21, 940(1965).
2. R.P. Devaty, A.J. Sievers: Phys.Rev. B32, 1951(1985), W.A. Curtin, N.W. Ashcroft: Phys. Rev. B31, 3287(1985), J. Warnock, D.D. Awschalom: Phys. Rev. B32, 5529(1985), and P.M. Hui, D. Stroud: Phys. Rev. B32, 2163(1986).
3. S. Kobayashi, T. Takahashi, W. Sasaki: J. Phys. Soc. Jpn. 32, 1234(1972), and P. Yee, W.D. Knight: Phys. Rev. B11, 3216(1975).
4. V. Novotny, P.P.M. Meincke, J.H.P. Watson: Phys. Rev. Lett. 28, 901 (1972), V. Novotny, P.P.M. Meincke: Phys. Rev. B8, 4186(1973), and N. Nishiguchi, T. Sakuma: Solid St. Commun. 38, 1073(1981).
5. T.A. Brody, J. Flores, J.B. French, P.A. Mello, A. Pandey, S.S.M. Wong: Rev. Mod. Phys. 53, 385(1981), and R. Denton, B. Muehlshlegel, D.J. Scalapino: Phys. Rev. B7, 3589(1973).

The Role of the Electron-Phonon Interaction in Phonon Anomalies in Transition Metals

K. Schwartzman[1], J.L. Fry[1], and P.C. Pattnaik[2]

[1]Department of Physics, University of Texas at Arlington, Arlington, TX 76019, USA

[2]Texas Instruments Inc., Dallas, TX 75625, USA

1. Introduction

For some time there has been considerable interest in the phonon anomalies in transition metals. A major stimulus for much of the present activity was the work of Varma and co-authors [1,2]. By a judicious separation of the total crystal energy, they isolated the long and short ranged contributions to the phonon dynamical matrix:

$$D_{\alpha\beta}(\vec{q}) = D^{(0)}_{\alpha\beta}(\vec{q}) + D^{(1)}_{\alpha\beta}(\vec{q}) + D^{(2)}_{\alpha\beta}(\vec{q}). \tag{1}$$

The third term is long ranged in space and is primarily responsible for anomalous phonon behavior. The approach taken here is to calculate this quantity from first principles so as to exhibit the source of the anomalies. The first two terms are essentially short ranged and their sum is modeled using four force constants [3]. These constants are determined by requiring the small $\vec{q}$ dependence of $D_{\alpha\beta}$ to agree with observed elastic constants.

It has recently become possible to use the "frozen phonon" method to make fully first principles calculations of phonon spectra [4-6]. This is a more robust approach than the one outlined above since it allows direct computation of force constants. However, for the sake of economy and to focus specifically on phonon anomalies, it is not used here.

2. Theoretical Formulation

The $D^{(2)}_{\alpha\beta}$ contribution to the phonon dynamical matrix may be written diagrammatically as a polarization bubble with dressed electron lines and fully renormalized (Coulomb screened and vertex corrected) electron-phonon vertices at both ends. This is a slight generalization of Ref. 2 and may be proved by explicit expansion of the Coulomb vertex [7]. For computational purposes, $D^{(2)}_{\alpha\beta}$ is written in the random phase approximation. In the tight binding approximation this is

$$D^{(2)}_{\alpha\beta}(\vec{q}) = \sum_{\vec{k}} \sum_{\mu,\mu'} \frac{f(E_{\vec{k}\mu})-f(E_{\vec{k}+\vec{q}\mu'})}{E_{\vec{k}\mu} - E_{\vec{k}+\vec{q}\mu'}} \, g^{\alpha}_{\vec{k}\mu,\vec{k}+\vec{q}\mu'} \, g^{\beta}_{\vec{k}+\vec{q}\mu',\vec{k}\mu} \tag{2}$$

where $E_{\vec{k}\mu}$ is the band energy for wavevector $\vec{k}$ and band index μ. The electron-phonon coupling between electron states $(\vec{k}\mu)$ and $(\vec{k}'\mu')$ for ionic motion in the cartesian direction $\alpha(\alpha{=}x,y,z)$ is

$$g^{\alpha}_{\vec{k}\mu,\vec{k}'\mu'} = \sum_{m,n} A^{+}_{m\mu}(\vec{k}) \, [\gamma^{\alpha}_{mn}(\vec{k}) - \gamma^{\alpha}_{mn}(\vec{k}')] \, A_{n\mu'}(\vec{k}'). \tag{3}$$

The eigenvectors $A_{m\mu}(\vec{k})$ are determined in principle from a self-consistent solution to the electronic Schrodinger equation and the gamma matrices are

$$\gamma^{\alpha}_{mn}(\vec{k}) = \sum_{\vec{R}_{ij}} e^{i\vec{k}\cdot\vec{R}_{ij}} \vec{\nabla}_{\alpha} H_{im,jn} \qquad (4)$$

where H is the crystal Hamiltonian and i and j are lattice site indices. It is known that the tight binding approximation is well suited for treating the d-orbitals of transition metals. The orthogonal representation of the tight binding Hamiltonian is used here (not the non-orthogonal approach of Refs. 1 and 2) as has recently been shown to be appropriate [8].

3. Computational Considerations

Band energies and eigenvectors are obtained from Slater-Koster fits to first principles band structures. The overall fits for the four metals Cr, Mo, W, and Nb are each accurate to 3 mRy. Nine basis vectors (1s, 3p, and 5d) are used in the fitting procedure which give rise to nine energy bands. The angular parts of the gradient in (4) are evaluated analytically and the radial parts are treated using the Harrison scaling scheme [8]. The electron-phonon coupling in (3) is evaluated directly from the eigenvectors and gamma matrices. It is not necessary here to introduce the approximations for the coupling used in Ref. 2. The $\vec{k}$ sum in (2) is evaluated using the analytical tetrahedron method [9] with 55 points in the irreducible (1/48) Brillouin zone (IBZ). These results typically agree to a few percent with calculations using 506 points in the IBZ (for selected $\vec{q}$ points).

4. Results

The calculated phonon dispersion curves for Cr are shown in Fig. 1 together with the experimental spectrum [10]. The dispersion relations obtained by diagonalizing $D^{(0)}_{\alpha\beta} + D^{(1)}_{\alpha\beta}$ and $D^{(2)}_{\alpha\beta}$ individually are shown in Fig. 2. From the figures it can be seen that anomalous phonon behavior arises either from structure in $D^{(2)}_{\alpha\beta}$ or from significant cancellation between the long and short ranged contributions. An example of the former is the sharp dip near the H point along the Δ line, and of the latter the dip in the T_2 mode along the Σ line. Thus, anomalous behavior in general cannot always be associated with Fermi surface extrema or nestings.

Overall agreement between the calculated and observed spectra is good. Absence of the sharp dip along the Δ line in the observed spectrum, however, suggests the presence of the spin density wave [11] may alter the Fermi surface sufficiently to eliminate it.

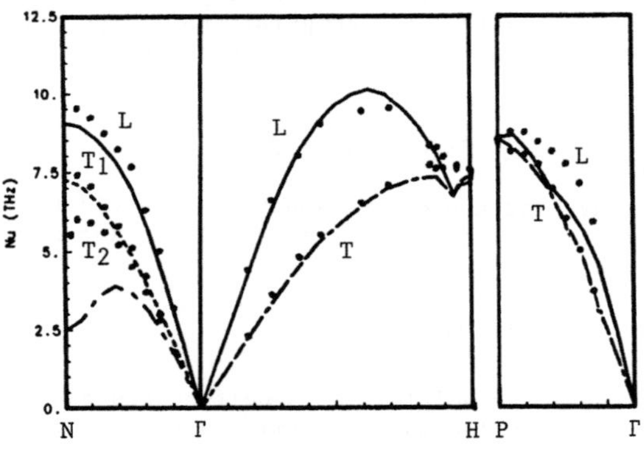

Fig. 1: Calculated and experimental [10] phonon dispersion curves for Cr. Anomalous behavior appears near the N, H, and P points.

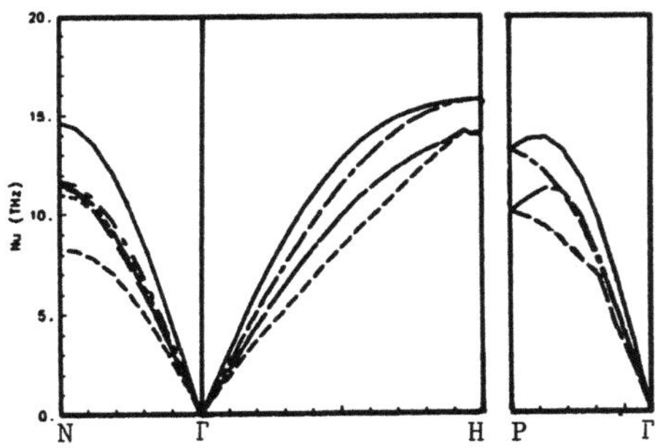

Fig. 2: Individually diagonalized short ranged (upper curves) and long ranged (lower curves) contributions to the phonon dynamical matrix for Cr. The longitudinal modes are shown by solid lines and the transverse modes by dashed lines.

The calculated $D^{(2)}_{\alpha\beta}$ in Mo and W (and so their spectra) show structure much like that found in Cr. This is to be expected since their Fermi surfaces are very similar. Nb has been considered in Ref. 2 and is currently being examined by the authors using the orthogonal tight binding approximation.

Acknowledgements

The authors appreciate enlightening conversations with Dr. Glenn Fletcher and help in running computer programs from Mr. Tin Truong. This work was supported in part by the Robert A. Welch Foundation.

References

1. C. M. Varma, E. I. Blount, P. Vashista, and W. Weber: Phys. Rev. B 19, 6130 (1979)
2. C. M. Varma and W. Weber: Phys. Rev. B 19, 6142 (1979)
3. M. E. Schabes, M.S. Thesis, University of Texas at Arlington, (1983, unpublished); R. C. Rai and M. P. Hemkar: J. Phys. F 8, 45 (1978)
4. K. -M. Ho, C. -L. Fu, B. N. Harmon, W. Weber, and D. R. Hamann: Phys. Rev. Lett. 49, 673 (1982)
5. K. -M. Ho, C. -L. Fu, and B. N. Harmon: Phys. Rev. B 28, 6687 (1983)
6. M. W. Finnis, K. L. Kear, and D. G. Pettifor: Phys. Rev. Lett. 52, 291 (1984)
7. K. Schwartzman: unpublished
8. J. L. Fry, G. Fletcher, P. C. Pattnaik, and D. A. Papaconstantopoulos: Physica 135B, 473 (1985); P. C. Pattnaik, M. E. Schabes, and J. L. Fry: Phys. Rev. B (submitted)
9. J. L. Fry and P. C. Pattnaik: In Integral Methods in Science and Engineering, ed. by Fred R. Payne, et. al. (Springer-Verlag, New York, 1986) p. 27
10. W. M. Shaw and L. D. Muhlestein: Phys. Rev. B 4, 969 (1971)
11. W. C. Koehler, R. M. Moon, A. L. Trego, and A. R. Mackintosh: Phys. Rev. 151, 405 (1966)

An Ultrasonic Study of e$^-$ Induced Defect Complexes in Dilute Al-Ag and Al-Si Alloys

E.C. Johnson and A.V. Granato

Department of Physics and Materials Research Laboratory,
University of Illinois at Urbana-Champaign, Urbana, IL 61801, USA

I - INTRODUCTION

Ultrasonic measurements, though less direct than those of higher frequency techniques, offer several advantages for the study of point defects in dilute metallic alloys. Large effects are often detected for only a few ppm defect concentration where defect-defect interactions may be safely neglected. In addition, for defects produced within a single crystal host, the ultrasonic polarization dependence of the defect response can be exploited in order to answer questions concerning the defect symmety and, as we will see, its kinetics.

In a previous study [1], ultrasonic measurements were made on Al-673 ppm Si following 2.5 MeV e$^-$ radiation with $T < 70$ K (The results were similar to those of a study done earlier [2] on Al-555 ppm Ag). In separate runs, a C' shear mode ($C'=\{C_{11}-C_{12}\}/2$) attenuation peak was detected near 20 K and a C_{44} shear mode peak was detected near 110 K ($f\approx30$ MHz). Both peaks were subjected to the same isochronal annealing program and it appeared that the C' peak annealed several degrees earlier than the C_{44} peak, suggesting the existence of two distinct defect species. The small temperature difference, however, could be a reflection of the fact that it is very difficult to exactly duplicate conditions for two separate irradiation/annealing runs. Both peaks could be the response of a single defect species. Indeed, "peak doublets" of a similar nature have been reported to occur in other alloys [3] and are usually attributed to some type of "frozen free split" defect complex [4].

II - EXPERIMENT

As an alternate approach to this problem, the same irradiation/annealing experiment was performed using <110> propagating longitudinal ultrasound (30.2 MHz). The elastic constant associated with this pure mode may be written [5] $C_L = C_{44} + C'/3 + B$, where B is the bulk modulus. The relaxation strength for a Debye peak at a particular temperature is proportional to $\delta C/C$. Therefore,

$$\text{(Near 20 K)} \qquad \frac{\delta C_L}{C_L} = \frac{1}{3} \times \frac{C'}{C_L} \times \frac{\delta C'}{C'} \approx \frac{1}{13.7}\frac{\delta C'}{C'} \qquad (1)$$

$$\text{(Near 110 K)} \qquad \frac{\delta C_L}{C_L} = \frac{C_{44}}{C_L} \times \frac{\delta C_{44}}{C_{44}} \approx \frac{1}{3.8}\frac{\delta C_{44}}{C_{44}} \qquad (2)$$

so that one would expect to see a reduced version of both shear peaks in this mode simultaneously. No bulk modulus relaxation is expected for a single species of dipolar defects in a cubic lattice [6]. A bulk modulus peak could, however, be the consequence of a reaction between two separate defect species.

Because of the large reduction (≈1/14) in the strength of the low temperature peak in this mode, Al-Si was a better choice for this study than Al-Ag because the decrement of the C' peak was larger.

III - RESULTS

Two peaks were detected (Fig. 1) and as expected, the relaxation strengths were reduced (for example, the ratio of the low-temperature relaxation strength in this mode to that in the shear mode was ≈13.5). No bulk modulus peak was observed. It is difficult to choose a zero reference for the high temperature peak. If one chooses zero reference "A" as indicated in Fig. 1, the C' peak appears to anneal more rapidly than the C_{44} peak. If, however, one chooses zero reference "B" (assuming the defect concentration to be zero after the 125 K anneal), the two peaks appear to anneal together (Fig 2).

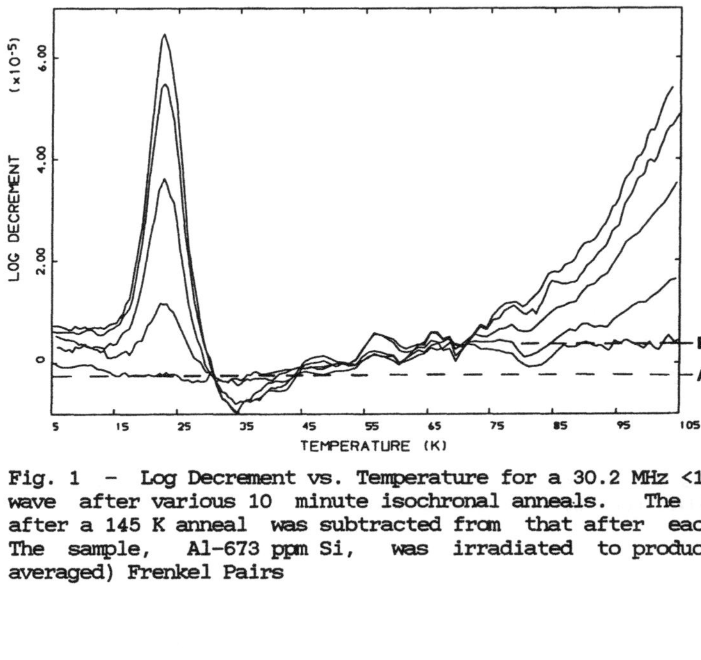

Fig. 1 - Log Decrement vs. Temperature for a 30.2 MHz <110> - longitudinal wave after various 10 minute isochronal anneals. The decrement present after a 145 K anneal was subtracted from that after each anneal plotted. The sample, Al-673 ppm Si, was irradiated to produce 5 ppm (volume averaged) Frenkel Pairs

IV - DISCUSSION

The measurements show that there is indeed a slight increase in the C_{44} peak annealing, as found earlier with separate shear wave measurements, if one supposes the background attenuation has not changed, but that in fact there is no difference if the background has increased slightly to "reference B". The difficulty in choosing a zero reference and apparent lagging of the C_{44} peak annealing might be explained by the existence of another peak at a significantly higher temperature which grows in upon the annealing of this peak, and anneals away before we surmount it in our measurement. The high temperature C_{44} peak of Fig. 1 would then be riding on the tail of this

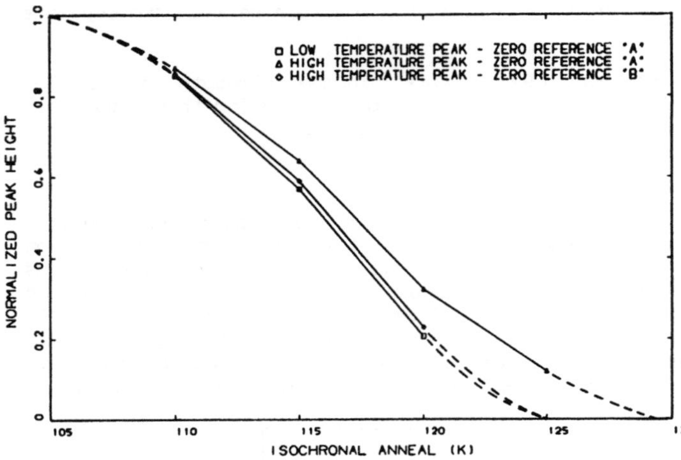

Fig. 2 - Fraction of peaks (Fig. 1) remaining vs. isochronal anneal temperature. For the C_{44} peak, this fraction was measured at 105 K and plotted for the two zero references indicated in Fig. 1

peak. Such a conjecture is supported by an EXAFS [7] study of AL-Ag. Two defect species were detected, only one of which annealed at a temperature consistent with the annealing of the two peaks observed in Al-Ag (corresponding to those of Al-Si shown in Fig. 1). The second defect structure was reported to grow in upon the annealing of the first and to anneal away again at a slightly higher temperature. Since it is most likely that the reference level has changed in this way, the two peaks may be supposed to arise from the same defect.

ACKNOWLEDGEMENTS

This work was supported by the U. S. Department of Energy, Division of Materials Sciences, Contract DE-AC02-76ER091198.

REFERENCES

1. E. C. Johnson and A. V. Granato, Journal de Physique, Colloque C10, 63 (1985).
2. A. V. Granato, J. Holder, K. L. Hultman, D. L. Johnson, G. G. Setzer, P. Wallace, and H. Wong, Point Defects and Defect Interactions in Metals, ed. J. Takamura et. al., (Japan: University of Tokyo Press), 360 (1982).
3. K. H. Robrock, Point Defects and Defect Interactions in Metals, ed. J. Takamura et. al., (Japan: University of Tokyo Press), 353 (1982).
4. A. S. Nowick and B. S. Berry, Anelastic Relaxation in Crystalline Solids, Academic Press, N.Y. and London, 216 (1972).
5. B. A. Auld, Acoustic Fields and Waves in Solids, John Wiley and Sons, N.Y., 83 (1973).
6. A. S. Nowick and W. R. Heller, Advances in Physics, 12, 251 (1963).
7. W. Weber and H. Peisl, Point Defects and Defect Interactions in Metals, ed. J. Takamura *et al.*, (Japan: University of Tokyo Press), 368 (1982).

Influence of the Velocity Operator on Magnetoresistances in Nondegenerate Semiconductors

C.C. Wu

Institute of Electronics, National Chiao Tung University,
Hsinchu, Taiwan, Republic of China

The investigation of the effect of a dc magnetic field on the longitudinal and transverse magnetoresistances yields useful information about the role various scattering interactions are expected to play in solids. Aliev et al [1] investigated the field dependence of the longitudinal and transverse magnetoresistances in the quantum limit in order to determine the influence of the band nonparabolicity and scattering inelasticity on this dependence. Hansen [2] proposed a correct form of the velocity operator derived from the Hamiltonian operator to show that the Hall effect is not influenced by nonparabolicity in the limit of vanishing scattering. However, when we are interested in both low and high frequency regions, the effect of scattering can not be neglected in real crystals, because there are sufficient imperfections to provide plenty of scatterings even at low temperatures. In this paper, we shall investigate the effect of the velocity operator on the magnitudes of longitudinal and transverse magnetoresistances in nondegenerate semiconductors such as n-type InSb by taking into account the effect of an electron relaxation time.

The eigenfunctions and eigenvalues for electrons in a dc magnetic field $\vec{B}$ directed along the z axis for the nonparabolic band model are given by

$$\Psi_{\vec{k}n} = \exp(ik_y y + ik_z z)\phi_n(x - \frac{\hbar c}{eB}k_y) \tag{1}$$

and

$$E_{\vec{k}n} = -\tfrac{1}{2}E_g\{1 - [1 + (4/E_g)\Big((n + \tfrac{1}{2})\hbar\omega_c + \hbar^2 k_z^2/2m^*\Big)]^{\frac{1}{2}}\} , \tag{2}$$

respectively, where m^* is the effective mass of electrons in the conduction band, E_g is the energy gap between the conduction and valence bands, and $\omega_c = |e|B/m^*c$ is the cyclotron frequency of electrons.

The interaction of electrons with acoustic phonons can be taken into account via the vector potential $\vec{A}_1 = \vec{A}_{1o}\exp(i\vec{q}\cdot\vec{r} - i\omega t)$, which arises from the self-consistent field accompanying acoustic phonon waves. The current density $\vec{J}$ can be obtained from

$$\vec{J} = \mathrm{Tr}(\hat{\rho}\cdot\vec{J}_{op}) = \sum_{\vec{k}\vec{k}',nn'} \langle\vec{k}'n'|\hat{\rho}|\vec{k}n\rangle\langle\vec{k}n|\vec{J}_{op}|\vec{k}'n'\rangle . \tag{3}$$

The current density operator $\vec{J}_{op}$ induced by the self-consistent field at a point $\vec{r}_o$ can be expressed by

$$\vec{J}_{op} = -(e/2)[\vec{v} + \vec{v}', \delta(\vec{r} - \vec{r}_o)]_+ \tag{4}$$

with

$$\vec{v} = [\hat{e}_x p_x + \hat{e}_y(p_y - \frac{eBx}{c}) + \hat{e}_z p_z][1 + H_o^{(r)}/E_g + H_o^{(1)}/E_g]^{-1} \tag{5}$$

and

$$\vec{v}' = -(e\vec{A}_1/m^*c)[1 + H_o^{(r)}/E_g + H_o^{(1)}/E_g]^{-1} , \quad (6)$$

where $H_o^{(r)}$ and $H_o^{(1)}$ are the right and left Hamiltonian operators, such as $H_o^{(r)}\psi_{\vec{k}n} = E_{\vec{k}n}\psi_{\vec{k}n}$, and $\psi^*_{\vec{k}'n'}H_o^{(1)} = E_{\vec{k}'n'}\psi^*_{\vec{k}'n'}$, respectively. The density matrix $\overleftrightarrow{\rho}$ should satisfy the quantum Liouville equation [3]

$$\frac{\partial\overleftrightarrow{\rho}}{\partial t} + \frac{i}{\hbar}[H_o + H_1, \overleftrightarrow{\rho}] = -\frac{\overleftrightarrow{\rho} - \overleftrightarrow{\rho}_s}{\tau} , \quad (7)$$

where τ is the electron relaxation time, $\overleftrightarrow{\rho}_s$ is the appropriate equilibrium density matrix, and

$$H_1 = -(e/2c)(\vec{v}\cdot\vec{A}_1 + \vec{A}_1\cdot\vec{v}). \quad (8)$$

From Eqs. (3) - (7), the current density $\vec{J}$ can be expressed by

$$\vec{J} = \overleftrightarrow{\sigma}\cdot\vec{E}, \quad (9)$$

where $\overleftrightarrow{\sigma}$ is the linear conductivity tensors. From this calculation one can obtain the components of linear conductivity tensors. When the acoustic waves are propagating along the dc magnetic field, it can be found that the non-zero linear conductivity tensors are $\sigma_{xx} = \sigma_{yy}$, $\sigma_{xy} = -\sigma_{yx}$ and σ_{zz}.

The longitudinal and transverse resistivities can be expressed from these linear conductivity tensors as

$$\rho_{\parallel} = \frac{1}{\sigma_{zz}} \quad \text{and} \quad (10)$$

$$\rho_{\perp} = \sigma_{xx}/(\sigma_{xx}^2 + \sigma_{xy}^2) . \quad (11)$$

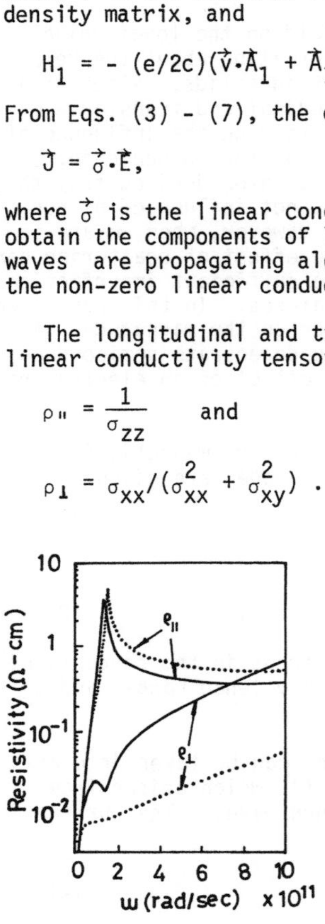

Fig. 1. Longitudinal resistivity $\rho_{\parallel}$ and transverse resistivity $\rho_{\perp}$ as a function of phonon frequency ω in n-type InSb at T = 4.2 K. Solid curves : B = 3 kG; dotted curves : B = 50 kG.

The relevant values of physical parameters are $n_o = 1.75 \times 10^{14}$ cm^{-3} , $m^* = 0.013m_o$, $E_g = 0.2$ eV, $v_s = 4 \times 10^5$ cm/sec, and $\tau = 10^{-12}$ sec. In Fig. 1, it can be seen that the longitudinal resistivity increases rapidly with the phonon frequency up to a cusp maximum point and then decreases slowly. However, for the transverse resistivity no cusp maximum points can be observed. At lower-frequency region, there exist maximum and minimum points in the low magnetic fields. When the field increases, these oscillations of

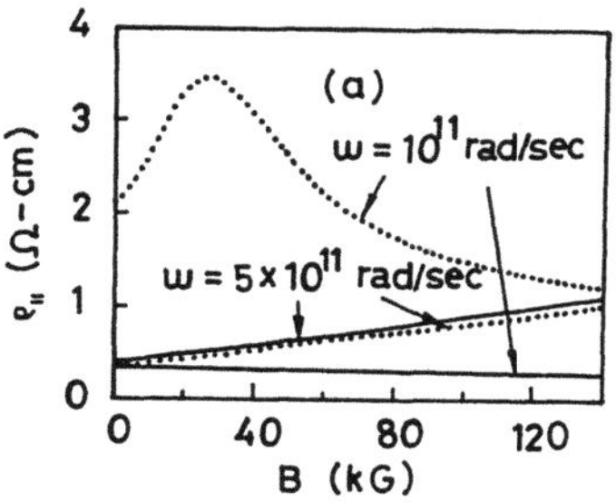

Fig. 2(a). Longitudinal resistivity as a function of dc magnetic field B in n-type InSb. Solid curves : T = 4.2 K; dotted curves : T = 19.7 K.

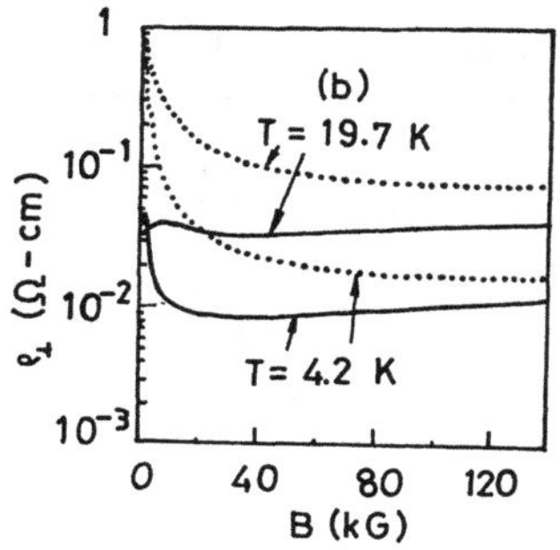

Fig. 2(b). Transverse resistivity as a function of dc magnetic field B in n-type InSb. Solid curves : $\omega = 10^{11}$ rad/sec; dotted curves : $\omega = 5 \times 10^{11}$ rad/sec.

the transverse resistivity with the phonon frequency will diminish. The longitudinal resistivity and transverse resistivity as a function of dc magnetic field for several different frequencies and temperatures are shown in Fig. 2. It can be seen that in the longitudinal resistivity there appears a maximum point around B = 28 kG for T = 19.7 K. This is not the same as our previous work in which the longitudinal resistivity decreases monotonically with the dc magnetic field [4]. It seems that the effects of electron scattering and the velocity operator demonstrate the effect as the temperature increases. For the transverse resistivity, there exist maximum and minimum points at low magnetic fields and low frequencies. These extremum points will be diminished with increasing temperature. This is the same qualitative result as our previous work [4].

References

1. M. I. Aliev, B. M. Askero, R. G. Agaeva, A. Z. Daibov, and I. A. Ismailov, Fiz. Tekh. Poluprovodn. 9, 570 (1975)[Sov. Phys. - Semicond. 9, 8 (1975)].
2. O. P. Hansen, J. Phys. C : Solid State Phys. 14, 5501 (1981).
3. F. R. Sutherland and H. N. Spector, Phys. Rev. B 17, 2728 (1978); 2733 (1978).
4. C. C. Wu, J. Tsai, and C. C. Wu, J. Low Temp. Phys. 19, 269 (1975).

Electron-Phonon Interaction Near the Metal-Insulator Transition of Granular Aluminum

P. Berberich and H. Kinder

Physik-Department E 10, Technische Universität München,
D-8046 Garching, Fed. Rep. of Germany

We have studied the electron phonon interaction of granular aluminum by measuring the phonon emission spectra of Joule heated thin films for various heater powers. The specific resistance of the films covered the range of T_c enhancement as well as the metal-insulator (MI) transition. The spectra of low ohmic specimens ($\rho < 500\ \mu\Omega$cm) are as expected and the electron phonon relaxation time is determined using the model of PERRIN and BUDD [1]. However, near the MI transition the spectra are characterized by a cutoff energy of about 0.8 meV below which phonon emission is strongly reduced. This cutoff is attributed to a discretization of the electronic levels as the metallic grains become more and more isolated near the MI transition.

The granular Al films (thickness d = 30 nm, area A = 1 mm^2) were prepared by thermal evaporation in the presence of oxygen. The phonon emission spectra of the Joule heated films were measured with a stress tuned Si:B spectrometer as shown in the inset of Fig. 1A, the samples being in liquid He at 1 K. Heat pulses of 100 ns duration were detected by a superconducting Al-I-Al tunnel junction. Longitudinal (L) as well as transverse (T) phonons could be observed in [111] direction of Si. When applying uniaxial stress the acceptor ground state splits into a doublet the splitting E being proportional to the stress. As the phonons are resonantly scattered at this two level system the decrease of the transmitted signal $\Delta S(E) = S(0) - S(E)$ gives information about the emitted phonon spectrum. Fig. 1A shows $\Delta S(E)$ of a rather clean Al film ($\rho(4K) = 1.9\ \mu\Omega$cm, $T_c = 1.32$ K) for various heater powers. The spectra are truncated below 0.3 meV by the $2\Delta_D$-threshold. Acoustic thickness resonances are resolved for both the T- and the L-mode. The resulting sound velocities are in fair agreement with those expected for polycrystalline Al. The resonances show that phonon reabsorption is small. In this case the spectra can give information about the electron-phonon relaxation time. Using the model of PERRIN and BUDD [1] we find for the emitted spectral power density per area:

$$p(\Omega) = 9\ N\ d\ (\Omega/\Omega_D)^3\ [n(\Omega,T_e)-n(\Omega,T_o)]/[\tau_{ep}+\tau_{es}] \qquad (1)$$

where Ω is the phonon energy, Ω_D the Debye energy, and N the ion density. T_e and T_o are the electron and the substrate temperature, respectively, and $n(\Omega,T)$ is the Bose-Einstein distribution function. The escape time τ_{es} of phonons from the film is about 50 ps [2]. The reabsorption time τ_{ep} of phonons by electrons can be written in the form [3]

$$1/\tau_{ep}(\Omega) = 4\pi\ N(0)\ \alpha^2\ \Omega/\hbar\ N \qquad (2)$$

where N(0) is the single spin density of states at E_F [3] and α^2 is an effective electron phonon interaction constant. T_e and α^2 were obtained from a fitting procedure. After folding $p(\Omega)$ with a spectrometer weight function [2] we have fitted it to the data to obtain T_e (dashed lines in Fig. 1A). In

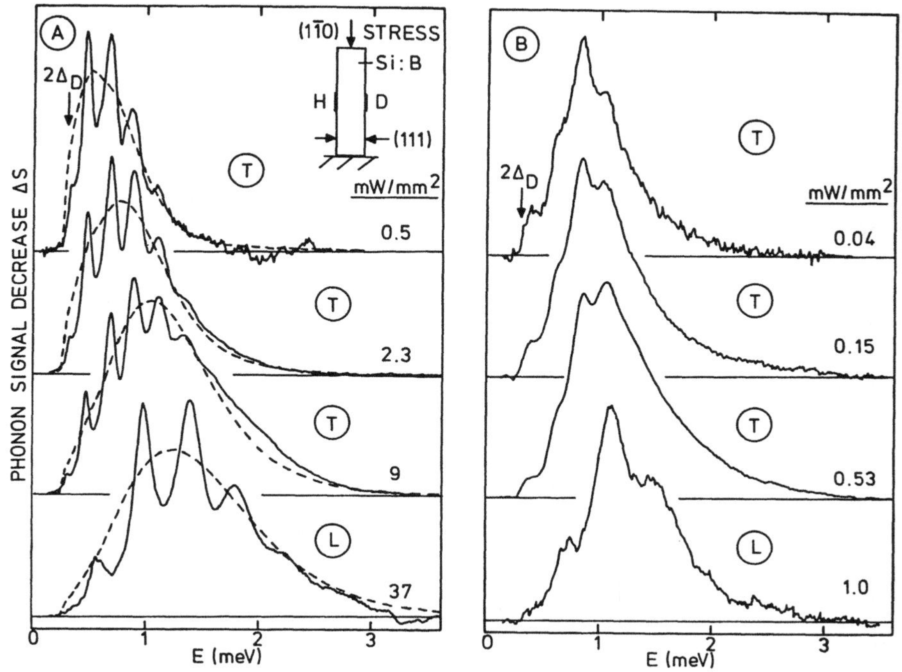

Fig. 1 Phonon emission spectra of a clean (A) and of a granular (B) Al film

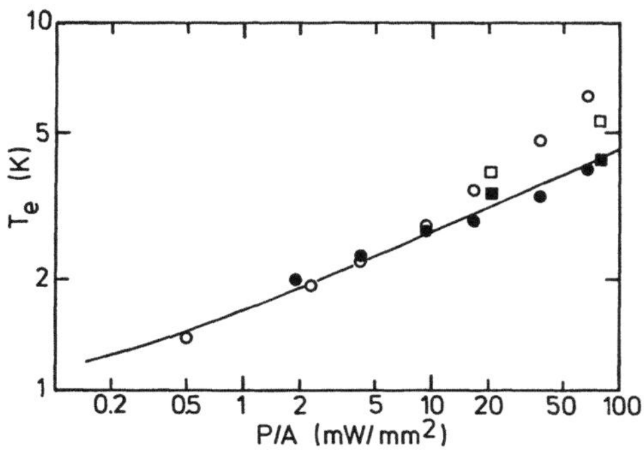

Fig. 2 Electron temperature T_e versus input power per area P/A. Circles/squares: ρ = 1.9/190 μΩcm. Full/open points: L/T-phonons

Fig. 2 we have plotted T_e versus the input power per area (full/open circles: L/T-phonons). The solid line is a fit of the integral of (1) to the data with α^2 as the fitting parameter. (Above 20 mW/mm^2 the data points for the L- and T-phonons differ. As the T-phonons possibly overload the detector these data were not taken into account.) The resulting α^2 = (7±2) meV is in good agreement with previous work on thin films [4,5] but it is a factor of three larger than expected from theory [6].

The results of a similar analysis for a film with a large T_c-enhancement (T_c = 2.3 K, ρ(4 K) = 190 μΩcm) are depicted in Fig. 2 by squares. (The regime of small T_e could not be investigated because of the onset of superconductivity.) The resulting α^2 is (7±3) meV between 1 and 2 meV being the same as for the clean sample. In contrast to this one expects a reduction of the electron-phonon interaction in dirty metals [7] by a factor $q_{ph}l_e = 0.1$ where q_{ph} is the phonon wave vector and l_e is the electron mean free path. This discrepancy is caused by the inhomogeneous structure of the films which are composed of Al grains embedded in an amorphous Al_2O_3 matrix. While electron transport is determined by tunneling between the grains the hot electrons can relax inside the grains. The relevant electron mean free path for relaxation processes is therefore of the order of the grain diameter of about 5 nm leading to $q_{ph}l_e \cong 1$ and no reduction of α^2 is to be expected. Ultrasonic attenuation measurements were interpreted in the same way [8].

Fig. 1B shows the phonon emission spectra of a granular Al film near the MI-transition ($\rho(4K) = 0.03$ Ωcm, $T_c < 1$ K). The sound velocities deduced are within 5 % the same as those of the clean film. We have analysed the spectra using the Perrin-Budd model. An Ω^2-dependence has been assumed for $1/\tau_{ep}$ (Eq.2) to account for the dirty limit of the electron-phonon interaction. The resulting α^2 decrease from 3 meV at 20 mW/mm^2 to 0.4 meV at 0.04 mW/mm^2 indicating a stronger Ω-dependence of $1/\tau_{ep}$. The reason for this behavior becomes clear from Fig. 1B. The maximum of the phonon spectrum does not shift below 0.8 meV even for very small input powers. This is interpreted as a cutoff below which phonon emission is strongly reduced. As the position of the cutoff does not change with the specific resistance between 10^{-3} and 0.1 Ωcm it cannot be caused by localization [9]. Acoustic resonances of the individual metallic grains should lead to rather different cutoff energies for the L- and the T-mode, respectively, in contrast to experiment. As the MI transition is approached, the electronic levels of the individual metallic grains will discretize and consequently the phonon emission spectrum will become discrete, too. However, only a low-energy cutoff will show up in the spectrum for grains being randomly distributed in size and form. The spacing ΔE of the electron levels at E_F is about 1/2N(0)V where V is the volume of a grain. For $\Delta E = 0.8$ meV a diameter of 4 nm results being in good agreement with the average grain diameter of 3 nm determined by electron microscopy [10].

References

1. N. Perrin and H. Budd: J. Physique (Paris) 33, C4-34 (1972)
2. P. Berberich and M. Schwarte: To be published in Z. Phys. B
3. S.B. Kaplan, C.C. Chi, D.N. Langenberg, J.J. Chang, S. Jafarey and D.J. Scalapino: Phys. Rev. B 14, 4854 (1976)
4. A.R. Long: J. Phys. F 3, 2023 (1973)
5. C.C. Chi and J. Clarke: Phys. Rev. B 19, 4495 (1979)
6. W.E. Lawrence and A.B. Meador: Phys. Rev. B 18, 1154 (1978)
7. B. Keck and A. Schmid: J. Low Temp. Phys. 24, 611 (1976)
8. M. Tachiki, H. Salvo jr., D.A. Robinson, and M. Levy: Solid State Comm. 17, 653 (1975)
9. R.C. Dynes and J.P. Garno: Phys. Rev. Lett. 46, 137 (1981)
10. G. Deutscher, H. Fenichel, M. Gershenson, E. Gruenbaum, and Z. Ovadyahu: J. Low Temp. Phys. 10, 231 (1973)

Ultrasonic Attenuation in Superconducting Aluminum Containing Dislocations

Y. Kogure, M. Sekiya, H. Ohtsuka, and Y. Hiki

Faculty of Science, Tokyo Institute of Technology
Oh-okayama, Meguro-ku, Tokyo 152, Japan

1. Introduction

Superconductivity is one of the most important consequences of the electron-phonon interaction. It has generally been believed that the superconductivity was rather a structure-insensitive property, because it was an ordered state of free electrons in k-space. But the circumstance seems to be somewhat altering. Recent studies on a variety of superconducting materials, for example those with the A15 structure and with the disordered state, have suggested the importance of the effect of atomic configuration on the superconductivity. The present study is intended to investigate the effect of crystal dislocations on the superconductivity of very pure aluminum crystals. An ultrasonic method can conveniently be used for the purpose. We have successfully determined the anisotropy of superconducting energy gap in aluminum [1], and the subsequent studies on the dislocated state of the material are described in this paper.

2. Experimental Method

Specimens used were prepared from a zone-refined single crystal of aluminum (99.9999 %). These were spark cut to cylinders 10 [mm] in diameter and 15 [mm] in length. Two specimens with axial crystallographic orientations of [110] and [111] have been used for the ultrasonic measurements. To introduce dislocations, the specimens were successively deformed by compression at a deformation rate of 0.05 [mm/min]. The ultrasonic attenuation in the specimen was measured by the conventional pulse reflection method. The frequency dependence of the attenuation was firstly measured at room temperature to obtain the information on the density and the state of the introduced dislocations. Then the specimen was set in a ^{3}He pumping cryostat to measure the attenuation at the normal ($T_c < T < 2$ K) and at the superconducting ($0.4 \text{ K} \leqq T < T_c$) states. Details of the procedures of the measurement have already been shown [1 - 2].

3. Results and Discussions

(i) Deformation and Dislocation Density

The relations between the applied compressional stress σ and the deformation of the specimen ε are shown in Fig. 1 (a). It is seen that the stresses to deform the [111] specimen are larger than those for the [110] specimen, and the σ-vs-ε shows some irregularity in the case of the [111] specimen. The difference of the stress level can be understood by considering the difference of the Schmid factor m for the deformation modes (m = 0.41 and 0.27 for [110] and [111]), and the mentioned irregularity in the [111] specimen may originate from the instability of the dislocation slip mechanism due to the rotation of the slip planes. The σ-vs-ε can be fitted to a formula

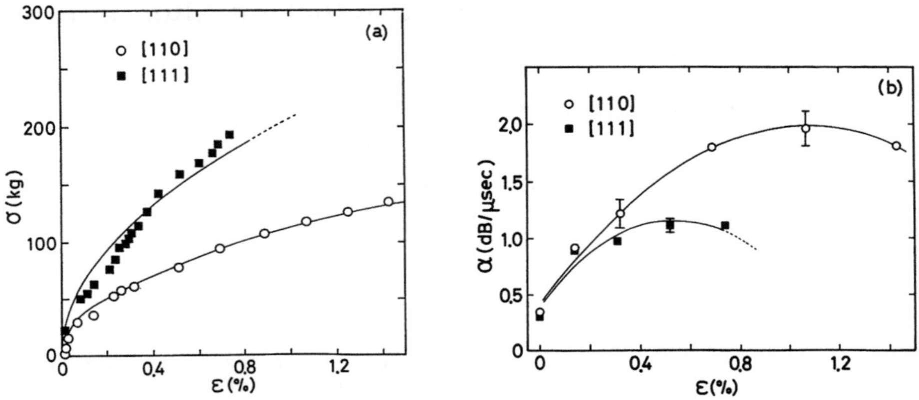

Fig. 1. (a) Applied compressional stress σ versus deformation ε
(b) Room temperature attenuation α versus deformation ε

$\sigma = \beta\sqrt{\varepsilon}$ after considering the Bailey-Hirsch relation [3], and the determined values of β are 1.1 and 2.0×10^8 [dyn/cm^2] for the [110] and the [111] specimens, respectively. The attenuation values at room temperature for the 10 MHz sound are shown in Fig. 1 (b). The increase of the attenuation with the deformation should be attributed to the increasing dislocation damping. The frequency dependence of the attenuation was measured at a low level of the deformation (ε = 0.14 %), and the results were reasonably analyzed on the basis of the theory of overdamped resonance of dislocations [4]. After combining the results of the room temperature attenuation and the σ - ε relation, the density of introduced dislocations was shown to be expressed as $\Lambda = a\varepsilon - b\varepsilon^2$. The determined constant values for the [110] and the [111] specimens are a = 1.1 and 1.7×10^9 [cm^{-2}]; and b = 5.0 and 15.5×10^8 [cm^{-2}], respectively.

(ii) Attenuation in Normal State

As the temperature is lowered, the attenuation decreases due to the decrease of phonon contribution to the dislocation damping. After reaching a minimum, the attenuation increases again due to the interaction of the sound wave with the conduction electrons. Finally, the attenuation becomes temperature independent ($\sim$ 4.2 K - T_c), where the electron mean free path is determined by the impurity- and the defect-scattering. The attenuation in this temperature range is denoted by α_n. The dependence of α_n on the deformation ε is shown in Fig. 2 (a). A theoretical formula for the normal state attenuation has been derived by PIPPARD [5], and the attenuation is a function of the electron mean free path l_e. In the deformed specimens, the electrons are scattered by the impurities and the introduced dislocations. Then the relation $1/l_e = 1/l_i + d\Lambda$ holds, where l_i is the mean free path determined by the impurity-scattering and d is the scattering cross-width of a dislocation. By using this expression and the dislocation density $\Lambda(\varepsilon)$ given in the last section, the experimental data were fitted to the PIPPARD's formula. The determined values of d for the [110] and the [111] specimens are d = 5.1 and 2.7×10^{-8} [cm], respectively. These are consistent with the experimental values determined by the electrical resistivity measurements [6]. However, the origin of the anisotropy observed in the d values is not clear.

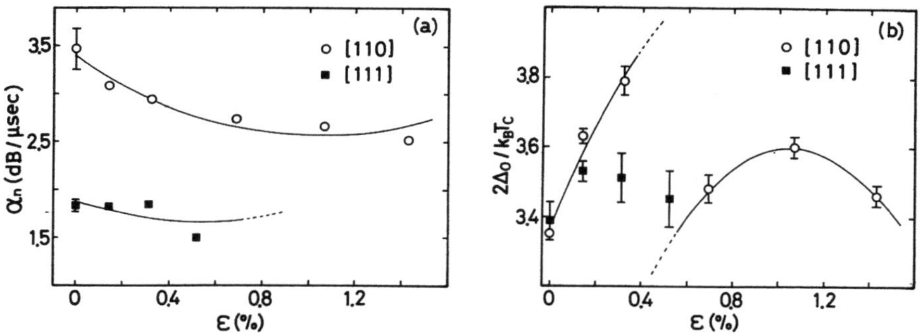

Fig. 2. (a) Normal state attenuation α_n versus deformation ε
(b) Change of superconducting energy gap by deformation

(iii) Change of Energy Gap by Dislocations

Below the superconducting transition temperature T_c, the ultrasonic attenuation rapidly decreases with decreasing temperature because of the condensation of normal electrons into a superconducting ground state. The theoretical form of the temperature dependence of the attenuation in the superconducting state is given by the BCS-theory. The zero-temperature energy gap Δ_o is an important parameter in the theory, and Δ_o can be determined by fitting the attenuation data to the theory. The determined values of $2\Delta_o/k_BT_c$ in several deformed states are shown in Fig. 2 (b). The gap for the [110] specimen initially increases with the deformation, suddenly drops by an appreciable amount, and then increases again. The two solid lines in the figure both represent a relation $2\Delta_o/k_BT_c = A + B\Lambda$. The fitted values of A and B are different in the two lines. The similar behavior is seen in the data for the [111] specimen, but not so clearly as the [110] specimen.

It is noted that all of the present data have shown strong dependences on the orientation of the crystal. There may be two kinds of origin of the anisotropy. First, the directions of the compressional deformation are different, and the configuration and the state of the introduced dislocations might be different. Second, the propagation directions of sound are different, and the sound interacts with the electrons with different k-vectors [1]. It is important that, owing to the high purity of the present specimens, the crystal anisotropy effects can distinctly be observed.

References

1. Y. Kogure, N. Takeuchi, Y. Hiki, K. Mizuno and T. Kino: J. Phys. Soc. Jpn. 54, 3506 (1985)
2. Y. Hiki, H. Ohtsuka, Y. Kogure and T. Kino: Proc. 8th Intern. Conf. on Internal Friction and Ultrasonic Attenuation in Solids, in press
3. J. E. Bailey and P. B. Hirsch: Phil. Mag. 5, 485 (1960)
4. A. V. Granato and K. Lücke: J. Appl. Phys. 27, 583; 789 (1956)
5. A. B. Pippard: Phil. Mag. 46, 1104 (1955)
6. R. A. Brown: Canad. J. Phys. 60, 766 (1982)

Phonons and Thermal Nucleation of the Normal State in Superconducting Indium Granules

N. Perrin

Groupe de Physique des Solides de l'Ecole Normale Supérieure,
24 rue Lhomond, F-75231 Paris Cedex 05, France

As a result of its positive energy surface, which prevents the flux penetration, a type I superconductor submitted to a magnetic field H can remain superconducting up to a value $H_{sh} > H_c(T_b)$ at a temperature $T_b < T_c$ ($H_c(T_b)$ is the critical field at T_b, and T_c is the critical temperature). Small superconducting granules are particularly suitable for the observation of this metastable superheated state because of the small demagnetizing effects, and the division of the superconducting material which prevents the propagation, to the whole sample, of the normal state eventually induced by defects. A thermal nucleation of the normal state can be observed in In granules, embedded in an insulating medium such as paraffin or collodion [1]. The energy required for this nucleation can be provided by the electron of the β^- desintegration of the In nucleus in the collision with a solar neutrino : the thermal nucleation of the normal state is therefore expected to be used for the solar neutrino detection.

The energy deposited in the grain by the impinging particle is shared between the quasiparticle (q.p.) and the phonon systems through the electron-phonon interaction and diffuses through the whole granule ; at the surface energy is lost through the phonon-embedding medium interaction. Each physical process is characterized by a mean relaxation time : τ_e and τ_p respectively for q.p. and phonons in the electron-phonon interaction, τ_s for the phonon relaxation on the embedding medium ; the pure q.p. diffusion is characterized by $t_e(\vec{r})$.

For a 2µ radius (a) granule, the increase of the q.p. thermal energy on a distance about the coherence length ξ from the granule surface, must remain larger than the surface barrier energy for a time $t_h \sim 5$ ns which is the characteristic time for the field penetration in the granule. This time is of the order of magnitude of the smallest expected phonon relaxation time τ_s corresponding to the maximum transmittivity. The actual relaxation time τ_s can be estimated ~ 30 ns, and therefore $\tau_s >> t_h$. Moreover, the mean relevant q.p. and phonon characteristic times τ_e and τ_p, and t_e for the 2 µ radius granule considered here are < 1 ns. We assume that the unknown characteristic time t_n for the propagation of the normal state is larger than t_e. We can therefore study the thermal response of the In granule without any consideration of the electrodynamical aspect.

The time required for the impinging particle to cross over the whole granule being about a few 10^{-14} s, we consider an instantaneous deposition with $P_0 = 30$ keV [2]. Assuming that P_0 is deposited at the center of the granule, we have a spherically symmetric problem. The coupled equations for the temperatures $T_e(r,t)$ and $T_p(r,t)$ respectively characterizing the electron system and the phonon system can therefore be written :

$$\begin{aligned} C_e\, \partial T_e/\partial t - k_e\, \nabla^2 T_e &= -C_e\,(T_e-T_p)/\tau_e + P_0\, \delta(\vec{r})\, \delta(t) \\ C_e\, \partial T_p/\partial t - k_p\, \nabla^2 T_p &= -C_p\,(T_p-T_e)/\tau_p - C_p\,(T_p-T_b)/\tau_s \end{aligned} \qquad (1)$$

with $C_e/\tau_e = C_p/\tau_p$ according to the detailed balance.

The spatial eigenfunctions for this Neumann boundary condition problem $[(\partial T_e/\partial r)_{r=a} = 0]$ are the zero order spherical Bessel function $j_0(k_l r)$, where the k_l are determined from the above boundary condition on T_e ; a Fourier decomposition is used for the time dependence. The most general solution of (1) is given by :

$$T_n - T_b = \sum_l \int_{-\infty}^{+\infty} d\omega \, [A_{l\omega} \exp(i\omega t) \, j_0(k_l r)] \, \tilde{T}_n (l,\omega) \qquad n = e, p \tag{2}$$

where the $\tilde{T}_n$ can be expressed in terms of the eigenfrequencies $\omega_\pm$ of the q.p.–phonon system, the diffusion constants $D_n = k_n/C_n$ and the q.p. specific heat C_e.

The small local variations of $T_e(r,t)$ are considered as fluctuations about the thermodynamical equilibrium state, from which the reduced Gibbs energy difference f between the superconducting and normal state is calculated :

$$f = \varphi^4 - 2\varphi^2 + (3/2)\, h^2 \times \, [1 - 3 \coth x_0/x_0 + 3\, x_0^{-2}] \tag{3}$$

with $f = G_{sh} - G_{n0})/(V\psi_0 H_c^2/2)$, $\varphi = \psi/\psi_0$, $h = H/H_c$, $x_0 = \varphi a/\lambda$ (ψ_0 : bulk order parameter at T_b with $H = 0$, λ : penetration length, V : granule volume). For times $t < 1$ ns, the q.p. and phonon behaviors are nearly independent of $1/\tau_s = 0$ or not. The q.p. temperature obtained for pure q.p. heat diffusion (dashed lines) is shown in Fig.1 for comparison with $T_e(r,t)$ (full lines), solution of (1) with $D_e = 7\times10^{-3}\ m^2 s^{-1}$, $\tau_e = 0.2$ ns, $\tau_p = 0.197$ ns (determined from [3]), $C_e = 629.6\ Jm^{-3}K^{-1}$. The phonon heat diffusion has been neglected. Apart from the expected lower values of T_e due to the fact that part of the input energy is localized in the phonon system, the q.p. time behavior is modified by the phonons as is shown by the larger time required to reach the uniform steady state temperature ($1/\tau_s = 0$) and the maximum in T_e occurring at the granule surface : indeed, for $r = a$, $t_e \sim 10^{-10}$ s is comparable to $\tau = (\tau_e^{-1} + \tau_p^{-1})^{-1} = 10^{-10}$ s and T_e exhibits a behavior corresponding to diffusion damped by the electron–phonon interaction, whereas for $r < a$ and $t << \tau$, T_e exhibits the same kind of behavior for both q.p. heat diffusion : the pure one, and the one including electron–phonon interaction. The decrease of T_e everywhere in the granule at times $t > 0.1$ ns also results from the interaction of the q.p. with the phonons.

A comparison between $T_e(r,t)$ and $T_p(r,t)$ is shown in Fig.2. It is seen that for times $t > 0.1$ ns, the phonon system behaves as a heat reservoir for $r < a/2$ ($\sim (D_e \tau)^{1/2}$) whereas for $r >> a/2$, T_p is always smaller than T_e. This can be understood since the phonons are heated through the q.p. only and their heat diffusion (very small in comparison with the q.p. one) is neglected here. Therefore, the phonons delay the occurrence of a uniform q.p. temperature all over the granule.

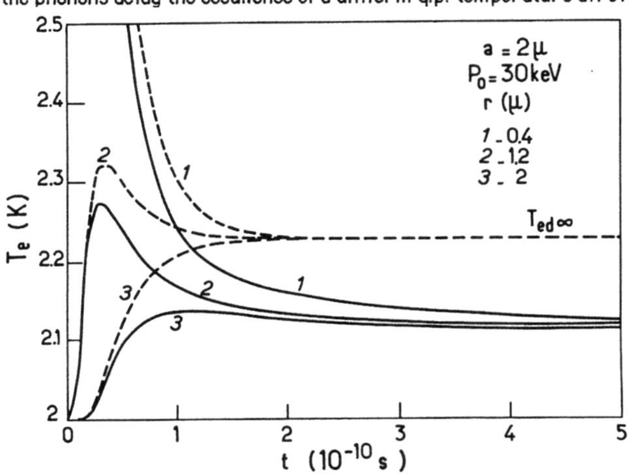

Fig.1 : $T_e(r,t)$ (full lines) compared to $T_e(r,t)$ obtained for pure q.p. heat diffusion (dashed lines)

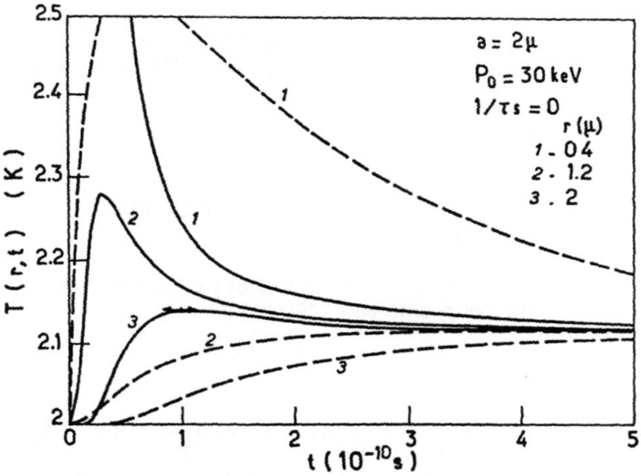

Fig.2 : Comparison between $T_p(r,t)$ (dashed lines) and $T_e(r,t)$ (full lines)

In the case just considered, T_e = 2.115 K at t > 1 ns and the corresponding energy increase (72 J m^{-3}) is sufficient to produce a flipping of the metastable granule for an applied field H > 300 Gauss, even with slow phonon energy loss towards the embedding medium. The superheating critical field, deduced from (3) is 700 Gauss.

This work has been done in the scope of the Neutrino-Indium team program with the financial support of the Centre National de la Recherche Scientifique (C.N.R.S.).

REFERENCES

1. G. Waysand, Comptes rendus de la Rencontre de Cargèse "Astrophysique et Interactions Fondamentales". July 1983, edited by J. Audouze and J.L. Basdevant, Institut d'Astrophysique (Paris), p.166

2. A. de Bellefon and P. Espigat (private communication).

3. S.B. Kaplan, C.C. Chi, D.N. Langenberg, J.J. Chang, S. Jafarey and D.J. Scalapino, Phys. Rev. B 14, 4854 (1976).

Part III

Phonons in Semiconductors

Phonon-Induced Electrical Conductance in Semiconductors

K. Laßmann and W. Burger

Universität Stuttgart, Physikalisches Institut, Teil 1,
Pfaffenwaldring 57, D-7000 Stuttgart 80, Fed. Rep. of Germany

The bath of thermal phonons is a well-known source of impedance to electronic conduction in solids. Turning the tables, a change in electronic conductance may be used for phonon detection, as will be discussed in the following.

Since the phase velocity ω/k in most cases is much smaller for phonons than for electrons in the interaction event the transfer of energy $\hbar\omega$ is negligible for the electron whereas the change of momentum $\hbar k$ may be substantial. Thus, a phonon current will help to direct electron motion. This so-called phonon drag effect shows up in the large thermoelectric power of semiconductors at low temperatures. Activation of electrons by energy transfer in phonon scattering is important in doped semiconductors at low temperatures in situations where the electrons are immobilised in shallow potential wells and is evident in the exponential temperature dependence of conductance due to carrier freeze-out and hopping conductivity. Both aspects of electron-phonon interaction in semiconductors, drag and activation, have been applied for phonon detection by phonon-induced conductance.

The former possibility has been utilized as so-called transmitted phonon drag in Ge and Si by K. HÜBNER and others (/1/ and references therein) in a series of experiments at liquid nitrogen and liquid helium temperatures. A typical set-up is sketched in Fig. 1: A slab of the crystal to be investigated is n-doped on two opposite faces. An electric field E_1 applied to the top n-layer causes an electron current which imparts some of its resultant momentum to the phonon distribution. Those of these perturbed phonons which propagate through the middle layer and penetrate into the bottom n-layer drag the electrons along with them setting up an electric field E_2 opposite in phase to the input field E_1. The variation of the ratio E_2/E_1 with experimental parameters such as thickness or doping of the middle layer, surface roughness, and bath temperature is a measure of the relative transmitted momentum. Though interesting and plausible dependencies on these parameters have been observed, the detailed interpretation suffered from insufficient information on the relevant phonon spectrum involved.

A first example for phonon activation is the technique of thermally stimulated currents /2/ where nonequilibrium distributed electrons are

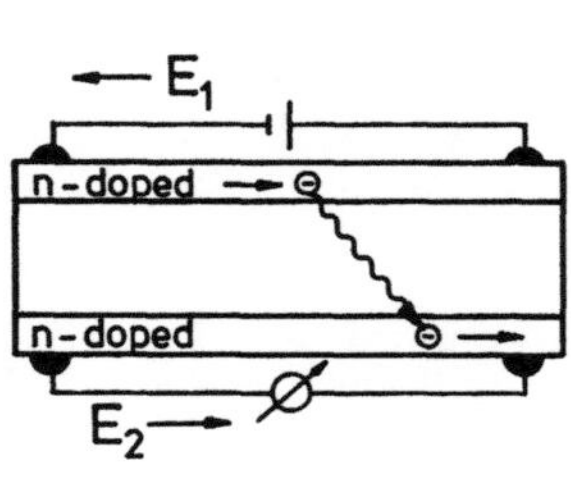

Fig. 1 Scheme of phonon generation and detection in the case of transmitted phonon drag in Si and Ge. Thickness of doped layers several µm, thickness of middle layer ranging from 50 µm to several mm. /1/

liberated from deep traps by thermal phonons during a slow temperature sweep allowing spectral information with resolution of the order of kT. Such an equilibrium phonon technique does not distinguish between one- or multi-phonon processes and the trap depth may be larger than the Debye frequency. (DLTS /3/ may be regarded as a variant of this method.)

Another possibility of phonon-induced conductance is associated with the complementary process where nonequilibrium phonons excite electrons from their equilibrium positions. Three types of phonon-activated current may be envisaged in homogeneous semiconductors: hopping conduction in compensated or highly doped material, carrier liberation from shallow traps, and current due to band-band transitions in narrow bandgap semiconductors.

Phonon spectroscopy might give valuable information in all three cases: the distribution of the hopping states in energy and distance, the Coulomb gap and the Hubbard gap /4/ may be directly accessible, the elementary processes in carrier recombination to shallow traps /5/ may be investigated by the reverse process of activating bound electrons to conducting states and finally the electron-phonon interaction in narrow bandgap semiconductors /6/ may be studied by this type of experiment.

As regards activated conductance due to nonequilibrium phonons, the first experiments have been by ASCARELLI /7/ and ZYLBERSZTEJN /8/ where phonons generated by hot electrons in Ge were detected by a conductivity change in a damaged surface layer biassed below avalanche threshold. However, details of the detection process were not investigated.

A first investigation on the possibility of ionisation from shallow traps has been made by CRANDALL /9/ for shallow donors in GaAs ($E_B \sim 6$ meV). The set-up used was quite analogous to Fig. 1 with 500 μm thick semiinsulating material covered by 40 μm thick, moderately n-doped, top and bottom layers. One of the layers was supposed to generate high-frequency phonons from the recombination of electrons liberated by the avalanche process and the other layer was thought to detect these phonons by the reverse process of ionizing the neutral donors. However, no direct measurement of the phonon energies involved was possible. The same applies to the somewhat modified phonothermal detection scheme (i.e. excitation by a nonequilibrium phonon from ground state to an excited state and from there by phonons of the thermal bath to a current-carrying state) as proposed by ULBRICH /10/ in an experiment with optically generated phonons in GaAs.

WIGMORE /11/ used GaAs doped with acceptors Zn or Cd (E_B about 30 meV) for bolometric detection of heat pulses and microwave ultrasonic pulses. From the temperature dependence of the electrical conductivity /12/ it was concluded that hopping activation was the process responsible for the phonon detection.

Since superconducting tunnel junctions are phonon generators with well established spectral properties /13/ we have used them to analyze the high-frequency phonon detection by an epitaxial layer of n-GaAs (Sn 5×10^{15} cm^{-3}, compensation about 5 %) on a 3 mm thick semiinsulating substrate. From measurement of the temperature dependence of the conductance we find activated behaviour between 6 K and 11 K with E = 5.8 meV corresponding to the donor binding energy and a variation as exp $(-T_o/T)^{1/4}$ below 2.5 K characteristic for variable range hopping with T_o = 0.072 K, so, both types of phonon activated current might be expected in a phonon spectroscopy experiment. Fig. 2 shows the results obtained with a Sn junction generator. Curve A is the

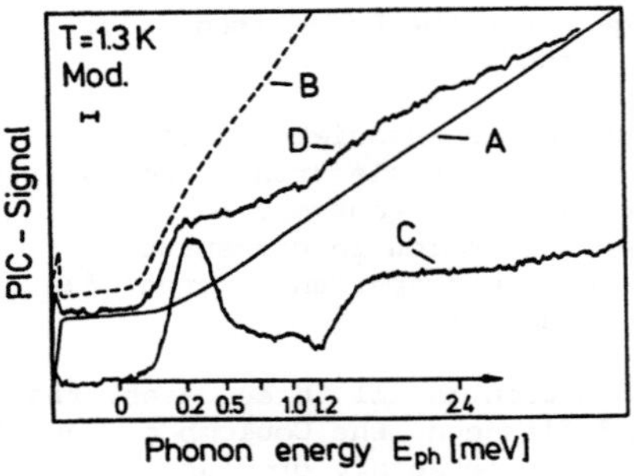

Fig. 2 PIC-signal of an epitaxial layer of n-GaAs (thickness 60 µm, Sn concentr. $5x10^{15}$ cm^{-3}, compensation 0.05) with Sn junction as phonon generator.
A: ↑ junction bias U,
→ junction current I,
I = 155 mA at U = 4Δ /e = 2.36 meV
B: Calculated for constant energy-detection sensitivity
C: PIC-signal, helium contact
D: PIC-signal, oil coverage

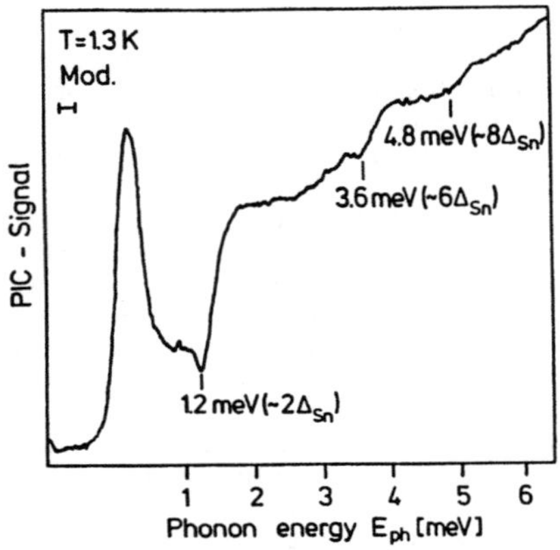

Fig. 3 PIC-signal corresponding to condition of C in Fig. 2,but measured to higher junction bias.
At multiples of the gap down-conversion of high-energy phonons modulo 2Δ by pair breaking increases the number of low-frequency phonons and thus PIC because of the low-frequency sensitivity of hopping. Only few phonons above the gap are emitted from the Sn film.

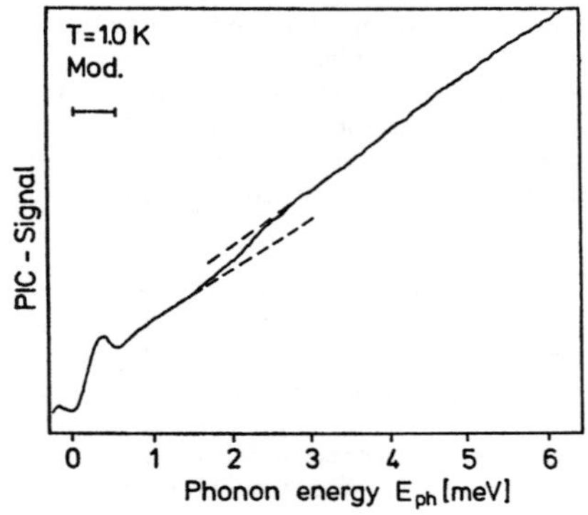

Fig. 4 PIC-signal corresponding to Fig. 3 but with Al junction as phonon generator instead of Sn. The change in slope at 0.2 meV and the decay at 0.35 meV are the same as for Sn. Despite the high-frequency phonons emitted by the Al junction, no threshold is found. The changes in slope at 1.5 meV and 2.8 meV and some small wiggles are reproducible for different Al junctions.

dc-characteristic of the junction, the energy scale being that of the maximum primary relaxation phonons for generator bias beyond the gap voltage. Curve B would be expected qualitatively for energy detection. Curve C shows the phonon -induced current (PIC) signal: After the expected rise beyond the gap voltage there is a change in slope near 0.2 meV followed by a decay starting at 0.35 meV. This latter decay is due to the treshold for increased phonon transmission into liquid helium in contact with the generator junction /14/. This contact can be spoiled by a coverage of the junction with oil which thermalizes the phonons transmitted to it. The effect is shown in curve D where the Kapitza threshold has disappeared, whereas the reduced detector sensitivity starting near 0.2 meV clearly remains. This reduction for high-frequency phonons is manifest also in reduced signal of A in the region of gap-bias where 1.2 meV phonons are generated. After partial down-conversion of these phonons by the oil there is a relative increase of PIC at this bias. Fig. 3 shows the PIC with a Sn generator measured to nominally higher energies. Since reabsorption in Sn prevents sizeable emission of phonons above $2\Delta_{Sn} = 1.18$ meV, the observed structure is astonishing at first glance. Most of it can, however, be understood by this very reabsorption (which leads to a down-conversion of the high-frequency relaxation phonons modulo $2\Delta_{Sn}$) taking into account the Kapitza threshold and the detector preference for low-frequency phonons as is discussed in detail in /15/.

Thin Al generators do not have this reabsorption limit /13/. Fig. 4 shows the corresponding PIC result. Apart from the low-frequency reduction just discussed for Fig. 2 we find distinct slope changes near 1.5 meV and 2.8 meV together with small structures at several frequencies reproducible for different generator junctions. No clear-cut correlation of these structures to any energy differences starting from the ground state of the isolated donor could be made. The overall frequency dependence appears to be rather smooth. It must be kept in mind, however, that part of the nominally high-frequency signal may be made up by low-frequency decay products from phonon scattering in the substrate, in the epitaxial layer, in the junction, and at the interfaces. The details of this have to be sorted out in further experiments. What can be said so far is that the signal is mainly due to hopping activation, whereas ionization of the donors plays no significant role. The reason for this may be either the absorption of the corresponding high-frequency phonons in the substrate or the intrinsic difficulty to excite electrons from the ground state by phonons, because the extended bound state cannot take up the large phonon k-vector connected with the energy difference.

Ionization by phonons has been demonstrated by us /16//17/ for a different class of shallow traps, namely D^- and A^+ centers in Si and Ge first found by FIR-measurements /18/. In complete analogy to atomic hydrogen, neutral effective-mass donors and acceptors can trap a second electron or hole respectively with binding energies of the order of meV (about 1/20 of the neutral impurity binding energy, whereas the extent of the wavefunction is larger by only about a factor of 2 which makes the relation to the involved phonon wavelength more favourable for one-phonon transitions). As an example the variation of the PIC threshold with boron concentration is shown in Fig.5. The high resolution and sensitivity of the method is evident from the curve for the smallest concentration of 5×10^{12} cm^{-3} of boron. The free carriers to be trapped by the neutral impurities were produced by illumination with visible light or, alternatively with room temperature radiation. It is remarkable that already for 5×10^{13} cm^{-3} a broadening of the rise is observed corresponding to a distance of 125 Bohr radii of the neutral boron. Such a broadening and finally a shifting at higher concentrations is attributed to the interaction of two or more neutral impurities in the capture and binding of the extra carrier. The interaction over large distances observed here for the first time may be a consequence of the slowly decaying long-range part of H^--wave function. Similar results have been obtained for Al^+ and Ga^+ (somewhat smaller binding energy than for B^+) and In^+ (Binding energy about 6 meV corresponding to the larger depth of this acceptor).

A phonon energy of 12 meV is quite close to the limiting frequency for some of the TA-branches in Si. So, isotope scattering and anharmonic decay become

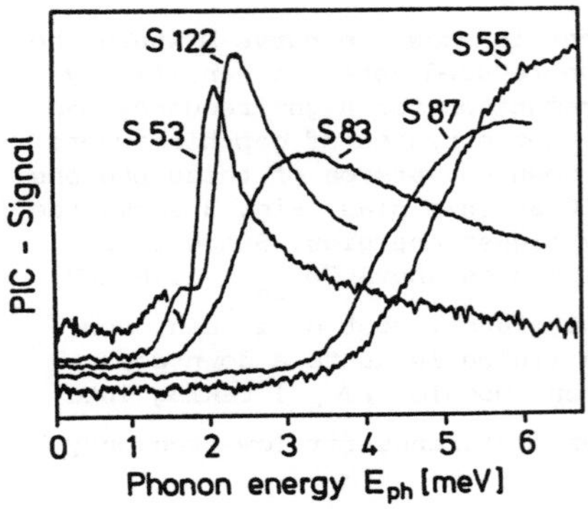

Fig. 5 PIC threshold of B^+ in Si: Dependence on boron concentration.

S53: 5.0×10^{12} cm^{-3}
S122: 5.0×10^{13} cm^{-3}
S83: 9.5×10^{14} cm^{-3}
S87: 5.4×10^{15} cm^{-3}
S55: 1.0×10^{16} cm^{-3}

The precursor resolved for the steeper thresholds of S53 and S122 is a feature of the spectrum emitted by the Al junction.

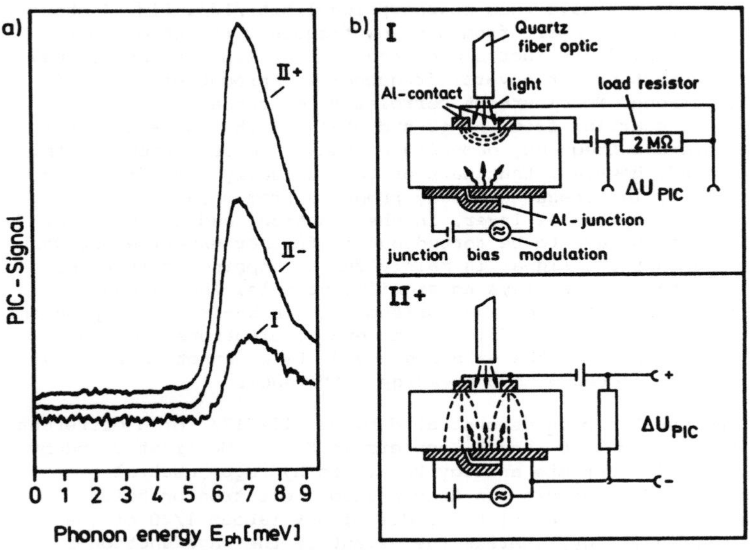

Fig. 6 a) Variation of PIC-signal form and strength for different probing distances from generator junction obtained by different measuring configurations in the case of In^+ in Si.
b) Configuration I: Probing zone opposite to junction (~ 2.5 mm)
Configuration II+: Probing zone near junction since holes produced by illumination are driven there by the bias
Configuration II-: Intermediate

increasingly important. Measuring with different electrode configurations as indicated in Fig. 6b we get a qualitative measure of the phonon mean free path: configuration I probes the region opposite to the phonon generator, whereas II measures the conductivity change along the phonon path. In this case an additional distinction is possible by the polarity of the bias: II+ as indicated in Fig. 6b draws the holes generated by the illumination to the junction so that the A^+ centers produced in this region can probe phonons with short mean free paths. For the reverse bias, the probing region is shifted away from the junction. The effect on the measuring signal is shown in Fig. 6a for the case of In^+: for the largest probing distance we have the smallest signal. Also the decay beyond the threshold is steeper for the longer phonon path which would be expected from the frequency dependence of phonon scattering. So from this type of measurement an estimate of the phonon mean free path and its frequency dependence can be obtained. An example for this is given in /18/.

An important result of these experiments is the reproducible demonstration of the feasibility of phonon spectroscopy with thin Al generators up to at least 12 meV (/17//19/). By this it is ascertained that the negative phono-ionization result in the case of neutral donors in GaAs is not due to the limitations of the Al generator.

Another point to be made is the following: The first excited state(s) above the ground state of the neutral impurity should have a long lifetime if excited at low temperatures because of the large phonon wavevector associated with recombination by single phonon emission. If a long-lived excited state of the neutral impurity were produced by the illumination we might have expected additional thresholds at about 11 meV or below e.g. for Si (P). So far, we could not observe corresponding signals either in silicon or in germanium.

As mentioned above, relaxation phonons with energies larger than the gap are reabsorbed and down-converted within a Sn junction generator because the films cannot easily be made thin enough. Nevertheless, the small amount of beyond-gap phonons emitted is readily detected by a conductivity change at the P^- threshold ($E_{P-} \simeq 2$ meV) as shown by Curve A in Fig. 7. Curve B shows the PIC-signal for an Al junction generator for comparison. Since these high-frequency phonons from a Sn generator could be detected only with difficulty by other methods this result demonstrates the sensitivity for high-frequency phonon analysis. The possibility to shift the threshold by uniaxial pressure might be an additional advantage for phonon spectroscopy.

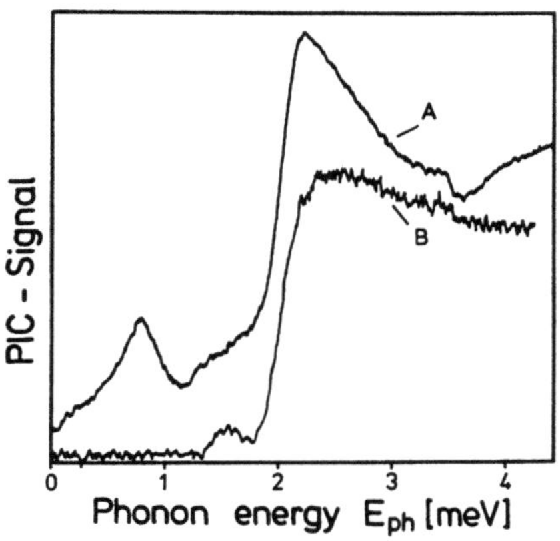

Fig. 7 High-frequency phonon emission from a Sn junction as detected by PIC of Si(P^-) (Curve A) Junction thickness 150 nm,

(P) = 6×10^{14} cm^{-3}
Curve B is obtained with an Al junction generator.
In both cases preceding the threshold at 2 meV a precursor is found at a distance of the corresponding gap which is due to relaxation of excited quasiparticles having tunnelled through the barrier.

For better time and space resolved measurements one needs small and low ohmic structures analogous to the epitaxial CdS layer with finger electrodes evaporated on it (ISHIGURO et al. /20/), used as a detector for heat pulses. The details of the detection mechanism in this case have not been investigated so far. The frequency dependence of the detection sensitivity would be interesting to measure by phonon spectroscopy as described above for the GaAs epitaxial layer.

It is seen from these examples that the combination of tunnel junction phonon generators with the detection by conductivity changes from activated carriers is a powerful experimental tool because of the well-defined and easily controllable spectral properties of the former and the high sensitivity with spectral resolution at high frequencies of the latter. It opens new and direct possibilities to investigate phonon participation in electronic processes in semiconductors at low temperatures in an interesting range of frequencies.

E.Bauser of MPIF in Stuttgart kindly provided us with the GaAs samples. We are grateful to W. Zulehner of Wacker Chemitronic for many well characterized silicon samples.
This work is supported by the Deutsche Forschungsgemeinschaft.

References

1. R. Gereth and K. Hübner: Solid-St. Electron. 10, 935 (1967)
2. R.H. Bube in: Photoelectronic Materials and Devices, ed. by S. Larach, (Van Nostrand, Inc., New York, 1965) p.113
3. D.V. Lang: J. Appl. Phys. 45, 3023 (1974)
4. B.I. Shklovskii and A.I. Efros: Electronic Properties of Doped Semiconductors, Springer Series in Sol. State Science 45 (Springer, Berlin, Heidelberg 1985)
5. F. Beleznay and L. Andor: Solid-St. Electron. 21, 1305 (1978)
6. W. Zawadzki and P. Boguslawski: Phys. Rev. Lett. 31, 1403 (1973)
7. G. Ascarelli : Phys. Rev. Lett. 5, 367 (1960)
8. A. Zylbersztejn : Phys. Rev. Lett. 19, 838 (1967)
9. R.S. Crandall : Solid-St. Commun. 7, 1109 (1969)
10. R. Ulbrich: In Nonequilibrium Phonon Dynamics, ed. by W.E. Bron, NATO ASI Series B124, 101 (1985)
11. J.K. Wigmore: J. Appl. Phys. 41, 1996 (1970)
12. J.K. Wigmore and P. Tlhabologang: Appl. Phys.Lett. 42, 685 (1983)
13. W. Eisenmenger: In Nonequilibrium Superconductivity, Phonons, and Kapitza Boundaries, ed. by E. Gray, NATO ASI Series B65, 73 (1981)
14. O. Koblinger, U. Heim, M. Welte, and W. Eisenmenger: Phys. Rev. Lett. 51, 284 (1983)
15. W. Burger: Thesis, Stuttgart 1986
16. W. Burger and K. Laßmann: Phys. Rev. Lett. 53, 2035 (1984)
17. W. Burger and K. Laßmann: Phys. Rev. B33, 5868 (1986)
18. S. Narita: J. Phys. Soc. Jpn. 49 Suppl. A, 173 (1980)
19. W. Burger and K. Laßmann: These Proceedings
20. T. Ishiguro and S. Morita: Appl. Phys. Lett. 25, 533 (1974)

Phonon Scattering by Cr Ions in GaP and InP

N. Butler[1], *J. Jouglar*[2], *B. Salce*[3], *L.J. Challis*[1], *A. Ramdane*[1], *and P.L. Vuillermoz*[2]

[1]Department of Physics, University of Nottingham, University Park, Nottingham NG7 2RD, UK

[2]Laboratoire de Physique de la Matière, Institut National des Sciences Appliquées de Lyon, F-69621 Villeurbane Cedex, France

[3]Centre d'Etudes Nucléaires de Grenoble, Services des Basses Températures, 85 X, F-38041 Grenoble Cedex, France

$Cr^{2+}(d^4)$ and $Cr^{3+}(d^3)$ both scatter phonons very strongly in GaAs [1]. In Cr^{2+} (present in n-type and semi-insulating - SI - material) the resonant scattering occurs at $\nu_0 \lesssim 5$GHz and is attributable to tunnelling between tetragonally placed Jahn-Teller wells. The scattering by Cr^{3+} (present in p-type and SI material) occurs at higher frequencies and is less well understood. The Cr^{3+} ion is also particularly interesting since the Jahn-Teller effect appears to contain strong non-linear contributions which result in the positioning of the Jahn-Teller wells along the six <110> (orthorhombic) directions. A striking demonstration of this orthorhombic symmetry is that the positions of the energy levels and also the details of the phonon scattering are sensitive to both <100> and <111> stress [2]. This contrasts with tetragonal systems where the <111> stress effects are strongly reduced (quenched), for example [3].

The present work indicates that broadly similar effects occur in Cr doped GaP and InP, see also [4,5]. The reduced thermal resistivity, W/W_0, where W_0 is the resistivity of an undoped sample, is shown in fig 1. The sample cross-sections were all $\sim 2 \times 2\,mm^2$. Evidence of low-frequency scattering in SI InP (sample MIT $\sim 10^6\,\Omega$cm) is seen in the increase in W as the temperature falls below 100mK. This is tentatively assigned to Cr^{2+} in analogy with GaAs:Cr and the strong high-frequency scattering to Cr^{3+}, following Jouglar and Vuillermoz [6]. This latter assignment is supported by measurements on a second InP sample CN1 for which the high-frequency scattering is much weaker. SIMS analysis showed this to contain a similar Cr concentration to sample MIT ($\sim 4 \times 10^{16}\,cm^{-3}$ $\sim$1 ppm) but since its electrical resistivity is much smaller (35 Ωcm) a higher proportion of the Cr should be in the Cr^{2+} state. So this accounts for the much weaker Cr^{3+} scattering but not for the decrease in the low-frequency scattering attributable to Cr^{2+}. We recall however that this

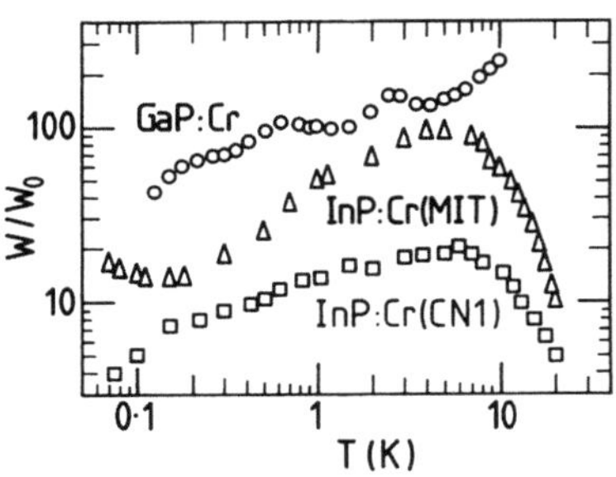

Fig 1. Reduced thermal resistivity, W/W_0, for SI GaP:Cr and InP:Cr. W_0 for InP is that of an undoped sample [6]. W_0 for GaP:Cr is calculated for boundary scattering (K_0 = $0.055T^3\,Wcm^{-1}K^{-1}$ and the apparent rise in W/W_0 above 4K is probably attributable to the neglect of the other instrinsic contributions to W_0.

scattering is very easily quenched by random E-type strains which separate the Jahn-Teller wells by more than the tunnelling frequency [1]. This could indicate then that CN1 is highly strained. It is hoped to repeat these measurements shortly, after annealing and also to make measurements on other n-type InP samples in the hope that they may be less strained and that the Cr^{2+} scattering will appear. The data on SI GaP (sample MCP1) which contains $4 \times 10^{17}\ cm^{-3}$ Cr, show strong high-frequency scattering attributable to Cr^{3+} but do not at present extend to low enough temperatures to reveal any possible Cr^{2+} scattering.

Figure 2 shows data for the stress dependence of the thermal conductivity of SI InP (MIT). The three samples used were cut from the same boule and consequently had very similar zero stress conductivity. Previous measurements were complicated by the fact that the samples had to be cut from different boules [4, 5]. Unfortunately all the samples broke or were destroyed before the highest stresses, ∿2500kg/cm², could be achieved, however, the data clearly show that the conductivity is very sensitive to both <100> and <111> stress and so provide evidence for orthorhombic Jahn-Teller distortions. There are interesting qualitative differences in the <110> stress dependence of GaAs:Cr and InP:Cr which suggest differences in the ratios of the coupling strengths to E and T_2 strains [2,4,5].

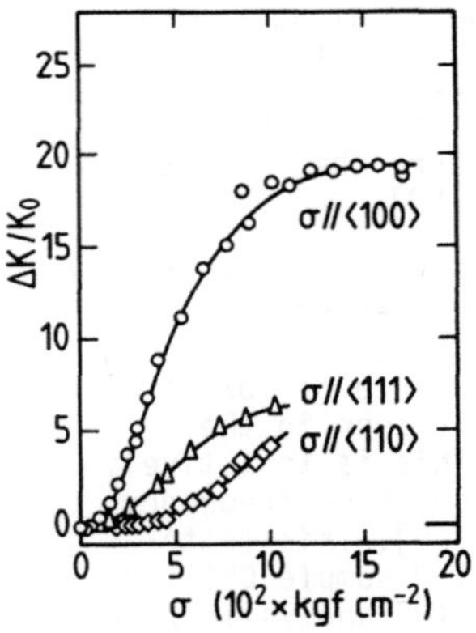

Fig 2. The change in conductivity, $\Delta K/K_0$, with stress, InP:Cr (MIT).

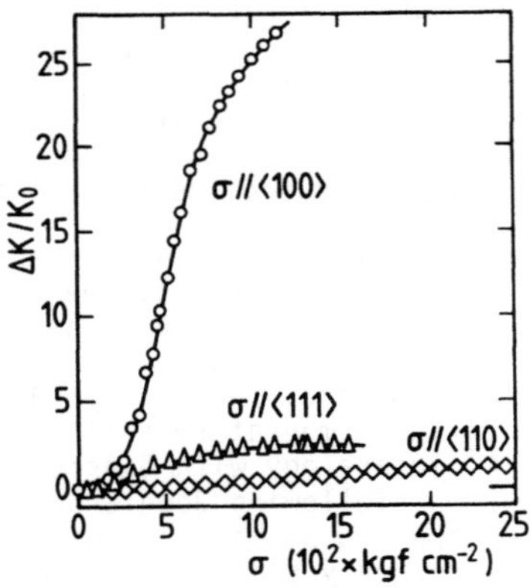

Fig 3. The change in conductivity, $\Delta K/K_0$ with stress, GaP:Cr (MCP1).

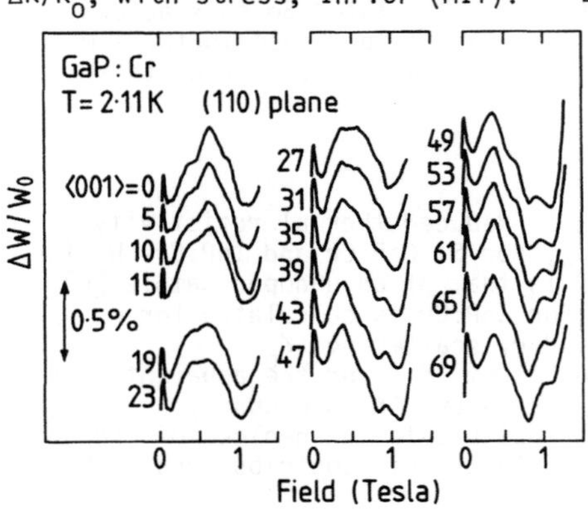

Fig 4. Magnetoresistance of GaP:Cr for fields in the <110> plane at angles ϑ° to the <001>.

Figure 3 shows the data for GaP:Cr (MCP1) [4,5]. In this system the <111> stress dependence is appreciably weaker so the evidence that this system is orthorhombic rather than tetragonal is less clear. Further work on a second sample is in progress.

Finally, in Figure 4, we show magnetoresistivity data for the GaP:Cr sample which shows frequency crossing minima qualitatively similar to that seen for GaAs: Cr [7]. The isotropy seems consistent with a T_d site and the isotropic ground state proposed by Eaves et al [8] for Cr^{3+} in GaP.

References

1. L. J. Challis, M. Locatelli, A. Ramdane and B. Salce: J. Phys. C : Solid State Phys. 15, 1419 (1982)
2. A. Ramdane, B. Salce and L. J. Challis: Phys. Rev. B27, 2554 (1983)
3. L. J. Challis and A. M. de Goër: in The Dynamical Jahn-Teller Effect in Localized Systems, ed. Yu. E. Perlin and M. Wagner (North-Holland, Amsterdam, 1984) p. 533
4. N. Butler, J. Jouglar, B. Salce, L. J. Challis and P. L. Vuillermoz: J. Phys. C : Solid State Phys. 18, L725 (1985)
5. N.Butler, J. Jouglar, B. Salce, L. J. Challis and P. L. Vuillermoz: Proc. 2nd Int. Conf. on Phonon Physics, Budapest, 1985, ed. J. Kollar, N. Kroo, N. Menyhard and T. Siklos (World Scientific, Singapore, 1985) p.720
6. J. Jouglar and P.L. Vuillermoz: J. Physique 45, 791 (1984).
7. L. J. Challis and A. Ramdane: Proc. 3rd Int. Conf. on Phonon Scattering in Condensed Matter, Brown University, 1979, ed. H. J. Maris (Plenum Press, New York, 1980) p.121
8. L. Eaves, D. P. Halliday and C. H. Uhlein: J. Phys. C : Solid State Phys. 18, L449 (1985)

Shallow Impurity Levels in Chalcogen-Doped Silicon as Obtained from Phonon-Induced Electrical Conductivity

W. Burger and K. Laßmann

Universität Stuttgart, Physikalisches Institut, Teil 1,
Pfaffenwaldring 57, D-7000 Stuttgart 80, Fed. Rep. of Germany

At low temperatures shallow neutral donors and acceptors in silicon can bind an extra carrier to form the so-called D^- (donor) and A^+ (acceptor) centers. With the method of **p**honon-**i**nduced electrical **c**onductivity (**PIC**) /1/ we have investigated the binding energy of these centers associated with the shallow impurities P and B /2/, and the deeper acceptors Al, Ga, and In in silicon /3/. In contrast to these hydrogen-like impurities the chalcogens Te, Se, and S in silicon are He-like deep double donors. In the effective mass approximation one would not expect that an extra electron can be trapped by these dopants, since the corresponding He^--state is not stable. However, the central cell correction necessary for the description of such deep impurities might change the situation which is complicated by the tendency of the chalcogens to form complexes. In this paper we show that we have found shallow electron traps associated with chalcogen doping in the case of Te and S.

Sulfur doping was obtained by diffusion. From IR-absorption /4/ it was found that most of the resultant double donors are associated with sulfur pairs. The Te-doped samples consist of Te-doped epitaxial layers (about 0.7 mm thick) grown from the vapour phase /4/ on a Si-substrate, which was about 0.7 mm thick (see inset in Fig. 1). During the growth not only single Te defects are formed but also Te complexes, which can be identified by IR-absorption /4/. These complexes can be dissolved by annealing and quench cooling and formed again by slow cooling /5/.

Figure 1 shows the PIC signals of a Te-doped sample for different contact configurations. In curve A the PIC signal of only the n-doped substrate is obtained. This signal shape is typical for P^- or B^+ states of the residual doping /2/. The absorption structure at 3.6 meV might be due to interstitial oxygen. In curve B the conductivity change is measured through the sample, so the PIC signal of the eptaxial layer as well as of the substrate is obtained. In addition to curve A the signal increases at 4.5 meV and at 7.8 meV followed by structures up to 12 meV. These parts of the PIC signal are due to the Te-doping of the epitaxial layer. The Te signals are observed only when the two Al contacts are negatively biased. In this case electrons are extracted from the illuminated zone to the epitaxial layer and as a result the observed shallow Te states must be due to trapping of an electron and not a hole. Curve C results only from the epitaxial layer, no signal of the substrate is observed but the signal steps at 4.5 mev and 7.8 meV are still there.

The different heights of the PIC steps at 4.5 meV and 7.8 meV in curve B and C result from the different distances of the epitaxial layer to the phonon generator (Al tunnel junction). In B the conductivity change is probed directly under the generator whereas in C the phonons have to cross the substrate before they can ionise the shallow Te traps. Because of the

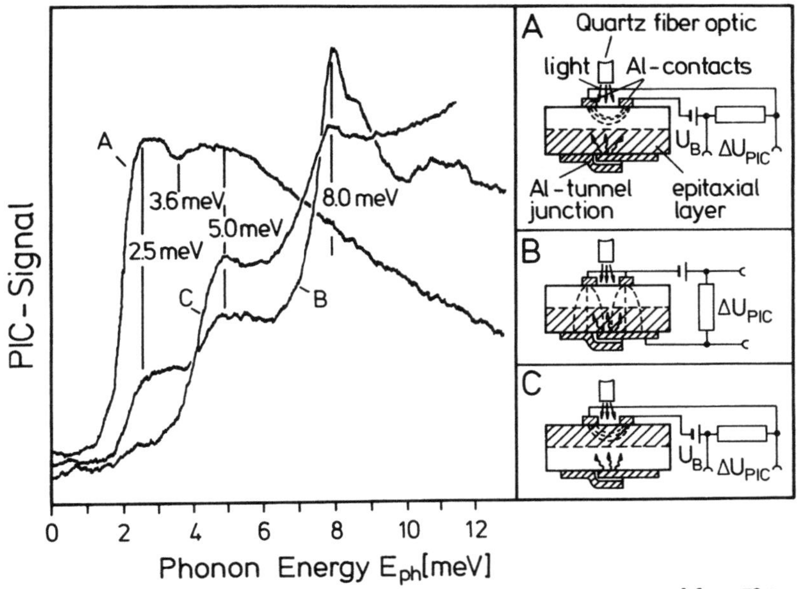

Fig. 1. PIC-signal of a sample with a Te-doped ($5\cdot 10^{16}$ cm^{-3}) epitaxial layer, T = 1.0 K. The inset shows the different contact configurations (the epitaxial layer is marked) of the measured PIC curves. The illumination of the sample with visible light is necessary to produce free electrons in the sample which can then be captured by neutral donors to form the D^- states

strong energy dependence of the anharmonic phonon decay ($\propto E^5$) the phonon mean free path at 4.5 meV is larger by a factor of 15 than that at 7.8 meV leading to the observed dependence of the ratio of the step heights on probing distance and also to the fact that no structures at higher energies are observed in C. Neglecting the effect of elastic scattering one can estimate from the different step heights in curve B and C the constant A for the anharmonic decay ($\tau^{-1} = A\cdot\nu^5$) as $5.4\cdot 10^{-55}$ sec^{-4}, which gives a phonon mean free path of 0.4 mm at 8 meV), in reasonable agreement with calculations for the anharmonic decay in silicon /6/.

As shown in Fig. 2 the PIC-signals depend on the light intensity i.e. on the free carrier concentration produced in the sample. At small light intensities most carriers are trapped within the substrate and only a few reach the epitaxial layer. With increasing light intensity the high energy structures at 8.8 meV and 11.5 meV increase, because more carriers diffuse to the epitaxial layer to be trapped there. At the same time the low energy threshold due to the substrate disappears because the high carrier concentration shortens this part of the resistances in series.

The results reported so far were obtained with two samples containing a large variety of Te-associated complexes. Dissolving the complexes by a baking-quench-cooling cycle the signal completely dissapears apart from the low energy threshold of the substrate. Up to now we cannot identfy the complex(es) causing the signal steps.

In the case of sulfur-pair double donors we find small steps at 4.0 meV and 6.0 meV. Here it was not possible to produce single sulfur by a temperature cycle.

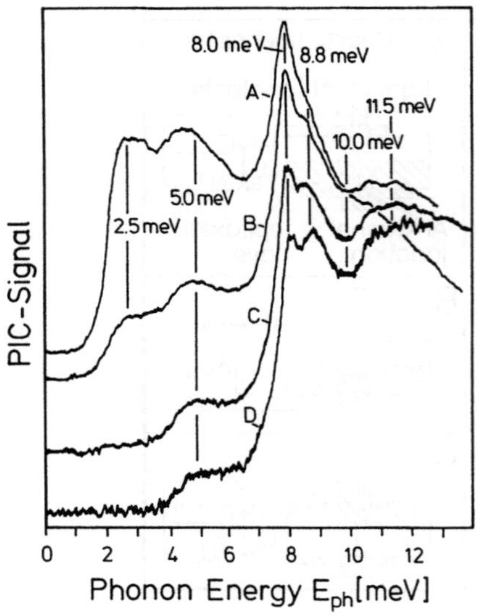

Fig. 2. Dependence of the PIC signal on the intensity of the sample illumination, T = 1.0 K, contact configuration B in Fig. 1, same sample. The light intensity is increasing from A to D

We thank C. Holm and P. Wagner, Heliotronic, Burghausen, FRG for preparing the samples, doing the annealing processes and characterizing the samples by IR-measurements.

References

1. K. Laßmann and W. Burger: These Proceedings
2. W. Burger and K. Laßmann: Phys. Rev. Lett. 53, 2035 (1984)
3. W. Burger and K. Laßmann: Phys. Rev. B 33, 5868 (1986)
4. P. Wagner, C. Holm, E. Sirtl, R. Oeder, and W. Zulehner: Advances in Solid State Science XXIV, p. 191, ed. P.Grosse, Vieweg (1984)
5. P. Wagner: private communication
6. A. Berke, A. P. Mayer, and R. K. Wehner: Solid Stat. Com. 34, 395 (1985) and A. P. Mayer: private communication

Acoustic Losses Due to Vanadium in Gallium Arsenide

P.J. King and I. Atkinson

Department of Physics, University of Nottingham, University Park, Nottingham NG7 2RD, UK

Although it is well established that vanadium doping can yield thermally stable semi-insulating gallium arsenide, the mechanisms by which it does so are unclear [1], and the levels due to the various valence states of vanadium need further elucidation. Acoustic techniques have been very useful in studying the related chromium centres in GaAs and we report here acoustic attenuation measurements made by pulse-echo techniques between 2K and 50K and at frequencies between 200 MHz and 2 GHz on a range of vanadium doped samples. The effect of illumination with either "white" or silicon filtered light ($h\nu < 1.1$ eV) was also investigated; a long wavelength cut-off of 0.78 eV was imposed by the fibre optic light pipe used.

Samples of semi-insulating material manufactured by Wacker Chemitronic and containing 3.4×10^{16} atoms/cm^3 revealed no acoustic losses for longitudinal modes propagating in the [100] or [110] directions under dark conditions except for the expected thermal phonon losses at higher temperatures.

Samples of [110] n-type material were obtained from Professor Ulrici of the Zentralinstitut für Elektronenphysik, Berlin, DDR. SIMS measurements reveal this material to be extremely pure and to be free of detectable transition metal impurities except for 5×10^{16} vanadium atoms/cm^3. Acoustic measurements show a wealth of features for both the longitudinal and fast transverse modes.

Both modes display prominent relaxation peaks in the attenuation as a function of temperature in the region of 10K [Fig. 1]. These peaks can be increased in magnitude by only a few percent by the application of intense white light. The relaxation attenuation appeared to be described by a single Debye expression and on this basis it was separated from the other features, and examined using a least-square fitting technique. This analysis yields the temperature dependence of the single relaxation time (τ), which could be described by a direct one-phonon term plus an Orbach term.

$$\tau^{-1} = AT + B/(\exp(T_0/T) - 1)$$

with $A = 2.3 \times 10^7$, $B = 1.1 \times 10^{11}$ and $T_0 = 34$K. Physically this form would be that expected from a system relaxing between levels in or close to the ground state, partially directly via one-phonon processes and partially via excited state levels 34K higher in energy. Various other forms for the relaxation time including those containing T^7 and T^9 were found to give unacceptable fits to the data. The ratio of the attenuation peaks for the fast transverse and longitudinal modes of 3.6 +/- 0.2 is close to the ratio (3.8) expected for losses involving C_{44} alone and the coupling is therefore of T_2 symmetry.

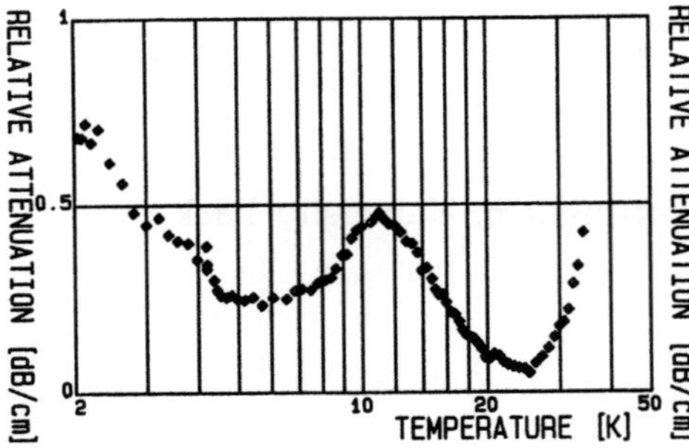

Fig. 1

Acoustic attenuation for the longitudinal mode along [110] in GaAs:V at 1950 MHz. The "10K" relaxation peak and the dark tail discussed in the text are clearly seen.

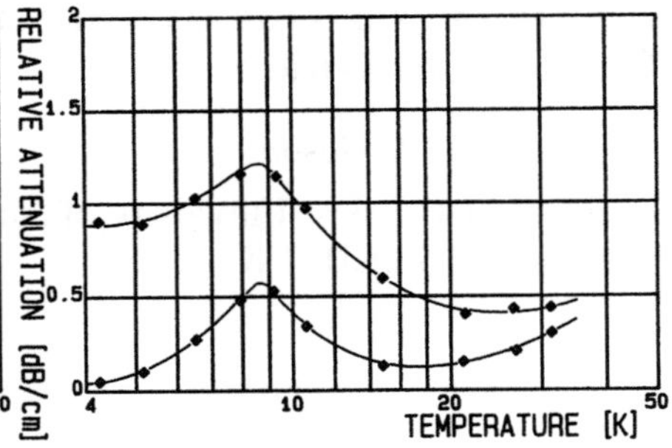

Fig. 2

Acoustic attenuation for the fast transverse mode along [110] in GaAs:V at 665 MHz. The upper trace was obtained under illumination with Si filtered light and the lower trace shows the attenuation under dark conditions.

In addition to the relaxation peak, both modes also exhibit a rising attenuation towards lower temperatures which is present under dark conditions [Fig. 1], and which is somewhat dependent on the thermal history of the sample. The attenuation falls off with temperature more slowly than 1/T in a way which is very frequency dependent, the magnitude being weak below 1 GHz and rising strongly above that frequency. It is possible to reduce the magnitude of this component of the attenuation by the application of intense light in the case of the longitudinal mode. In the case of the transverse modes this effect, if present, would be swamped by the light-induced attenuation about to be described. The dark "tail" is of almost equal magnitude for the two modes and thus involves both E and T_2 coupling.

The third feature found in the n-type samples is a light-induced attenuation which is large at low temperatures and falls as the temperature is increased, becoming unmeasurable by 43K [Fig. 2]. Although the temperature dependence bears a superficial resemblance to the dark "tail" of the previously described feature, the dependence on temperature is different in detail. The dependence on mode is quite distinctive in that the induced attenuation is absent for the longitudinal mode and extremely intense for the fast transverse mode, the echoes of which can be completely quenched by modest intensities of illumination. Simultaneous measurements of the slow transverse mode indicated behaviour similar to that of the fast transverse mode, indicating that this centre is also coupled to E and T_2 modes although clearly with a different dependence to that responsible for the dark "tail". At 4K the attenuation decays with a time constant of order 2-3 minutes while by 40K the time constant is less than 0.1 s.

In attributing the three attenuation components to defect centres we note that the n-type samples are extremely pure and thus the measured phenomena are likely to be vanadium related. The centres which would be expected to couple strongly to phonons are V^{2+}, and the defect complexes (V^{2+}-X) and (V^{+}-X), thought to involve an arsenic vacancy [2]. V^{3+} and its associated complex centre would not be expected to scatter phonons.

We attribute the relaxation peaks to V^{2+} on the grounds that this centre is expected to be the dominant charge state in n-type material [2] and to be absent in semi-insulating material. The V^{2+}/V^{3+} level is reported to lie 0.14 eV below the conduction band. If in our material the Fermi level lay not far above this level residual V^{3+} would be converted to V^{2+} by white but not silicon filtered light, thus slightly enhancing the peaks.

The $(V^{+}\text{-}X)/(V^{2+}\text{-}X)$ level is reported to lie 0.23 eV below the conduction band and thus in n-type material the dominant charge state of the vanadium complexes is expected to be $(V^{+}\text{-}X)$. We attribute the dark "tail" to these centres. Intense illumination is seen to reduce the magnitude of the associated attenuation presumably by inducing a change in valence state of the centres. In the light of this and the reported intense light-induced acoustic paramagnetic resonance [3] and thermally detected electron spin resonance [4] in similar material, attributed to $(V^{2+}\text{-}X)$, the light-induced attenuation seen in the present work may safely be attributed to $(V^{2+}\text{-}X)$ also. Our observation that either white or silicon filtered light induces this attenuation is consistent with this assignment.

Our measurements indicate that three distinct vanadium related centres are responsible for the observed attenuation behaviour and this confirms the existence of defect complexes. The measurements place restrictions upon the coupling symmetry of each centre, which are of great importance in the formation of the theoretical models currently under development.

Acknowledgements

We wish to thank Professor W. Ulrici for providing samples.

References

1. B. Clerjaud: J. Phys. C: Solid State Physics 18, 3615 (1985).
2. W. Ulrici, K. Friedland, L. Eaves, D. P. Halliday: Phys. Status Solidi b131, 719 (1985).
3. V. W. Rampton, M. K. Saker, W. Ulrici: J. Phys. C: Solid State Physics 19, 1037 (1986).
4. A-M Vasson, A. Vasson, C. A. Bates, A. F. Labadz: J. Phys. C: Solid State Physics 17, L837 (1984).

Acoustic Properties of n-Type Ge with Donor Pairs

T. Sota and K. Suzuki*

Department of Electrical Engineering, Waseda University,
Shinjuku, Tokyo 160, Japan

1. Introduction

In a semiconductor lightly doped with shallow donor impurities, donors are isolated and the system is simply a collection of noninteracting donor states which are described by the effective-mass theory. In the case of group V donors in Ge, the donor states consist of the singlet ground state and the triplet excited states which are energetically separated by the valley-orbit splitting 4Δ. The magnitude of 4Δ is sensitive to the species of impurities, i.e., $4\Delta/k_B$=49 K (As), 33 K (P), and 3.7 K (Sb)/1/. Because of the donor states mentioned above, donors strongly interact with the low-frequency phonons.

As the donor concentration N_D is increased, the average separation between donors decreases and donors begin to form various sizes of clusters where interaction between donors plays a role. In this report, considering the simplest donor cluster, i.e., the donor pair, we simulate the electronic states and using the result we study the effect of donor pairs on acoustic properties of Ge:Sb.

2. Procedures of Calculations

Our Hamiltonian for donor pairs in the system Ge:Sb is composed of the kinetic energy K, the Coulomb potential due to donor impurities V, and the electron-electron interaction U, i.e., H=K+V+U. The valley degeneracy is fully taken into account. Donor centers are distributed randomly in a sphere of volume N/N_D where N is the number of donors used in calculating the electronic states, in our case N=2. Then a hydrogenic 1s wave function $\phi_{i,a}$ is attached to each donor center r_i where a denote valley indicies, i.e., a=1 to 4.

We choose an effective one-electron Hamiltonian $\tilde{H}$ as $\tilde{H}$=K+V/2/ to define the one-electron states used in calculations which will be made below. After calculating the matrix elements of $\tilde{H}$, the secular equation $\tilde{H}\Phi_\alpha=\varepsilon_\alpha S\Phi_\alpha$, where S is a matrix whose elements are the overlap integrals, is solved and we obtain the eigenenergies ε_α and the orthonormal one-electron wave functions Φ_α.

The next step is to construct the two-electron wave functions ψ_β using Φ_α and calculate the matrix elements of H in a two-electron configuration space. The total number of configurations for a donor pair with spins is 120. The donor pair's eigenenergies E_n and wave functions Ψ_n are then obtained by diagonalising the Hamiltonian matrix in the configuration space. We have used 100 pairs.

---

* Present address: College of Engineering, Shizuoka University, Hamamatsu, Shizuoka 432, Japan

We have calculated the change of the sound velocity Δv_p due to a donor pair at the long wavelength limit using the following equations,

$$\Delta v_p=\Delta C/(2\rho_m v_0), \qquad \Delta C=\partial^2F/\partial\varepsilon^2,$$
$$F=-(N_D/2k_BT)\ln Z, \qquad Z=\Sigma_n \exp\{-(E_n+C_n\varepsilon)/(k_BT)\}, \tag{1}$$

where ρ_m is the crystal density, v_0 the sound velocity of pure Ge, C_n the deformation potential acting on electrons in the n-th level E_n, ε the strain associated with the acoustic wave. Expressins for the attenuation coefficient are complicated. We do not show them here.

We assume that some donors in the system are isolated and the others form donor pairs. In order to distinguish isolated donors from donor pairs, we introduce the cut-off distance R_C; donors are isolated for $R>R_C$ and form pairs for $R<R_C$ where R is the distance between donors. We treat R_C as a fitting parameter.

3. Results and Discussions

Figures 1(a) and (b) show the comparison between calculated results and experimental ones of the change of the sound velocity Δv for samples with $N_D=3.1\times10^{16}$ cm^{-3} and 5.0×10^{16} cm^{-3}/3/. Broken lines in Fig. 1 show Δv_s calculated using the isolated donor model alone. It can be seen that Δv_s for both samples have steeper temperature dependence than the experimental results. The magnitude of Δv_s is larger than that of the experimental results, in particular at low temperatures. The following has been found from calculations of Δv_p, though they are not shown here. The magnitude of Δv_p becomes smaller than that of Δv_s. The temperature dependence of Δv_p is weaker than that of Δv_s.

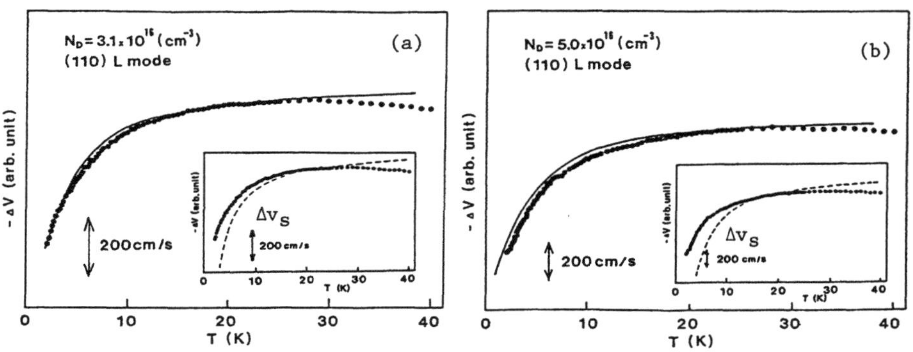

Fig. 1 Comparison between calculations and experiments.

Reasonable agreement between calculated results and the experimental ones has been obtained at $R_C=175$ Å. The values of R_C are smaller and larger than those of the average separation between donors, R_0, for samples with $N_D=3.1\times10^{16}$ cm^{-3} and 5.0×10^{16} cm^{-3}, respectively, where R_0 is given by $R_0=(3/4\pi N_D)^{1/3}$. The proportion of isolated donors to the total number is 50% for the former and 39% for the latter. The contribution from donor pairs to Δv is 10% for the former and 70% for the latter. However, for the latter, the donor pair approximation will be no longer good because of $R_C>R_0$. We should consider larger clusters.

As for the attenuation coefficient α preliminary calculations show the following. α_s calculated using the isolated donor model alone is much smaller than the experimental reslut /4/ at low temperatures. α calculated by choosing R_C=175 Å can qualitatively explain the experimental result.

Acknowledgment

We are greatful to Mr. M. Yawata for his assistance in numerical calculations. This work was supported in part by the Grant-in-Aid for Special Project; Research on Ultrasonic Spectroscopy and Its Application to Material Science from the Ministry of Education, Science, and Culture of Japan.

References

1. J. H. Reuszer and P. Fisher: Phys. Rev. 135, A1125 (1964)
2. T. Takemori and H. Kamimura: J. Phys. C: Solid State Phys. 16, 5167 (1983)
3. M. Kohno, K. Asano, and K. Suzuki: unpublished
4. H. Sakurai and K. Suzuki: unpublished

Heat Pulse Attenuation and Magnetothermal Resistance in Li-doped Si

A.J. Kent[1], *V.W. Rampton*[1], *T. Miyasato*[2], *L.J. Challis*[1], *M.I. Newton*[1], *P.A. Russell*[1], *N.P. Hewett*[1], *and G.A. Hardy*[1]

[1]Department of Physics, University of Nottingham, University Park, Nottingham, NG7 2RD, UK

[2]Institute of Scientific and Industrial Research, Osaka University, Osaka 567, Japan

1. INTRODUCTION

Lithium impurities behave as shallow donors in Si, they are believed to occupy an interstitial site. The six-fold valley degeneracy and the tetrahedral symmetry give rise to the usual ground state levels $^2A_1$, ^2E and $^2T_2$. These levels are inverted with respect to those of a substitutional group V ion. The lower levels are then the approximately degenerate $^2E+^2T_2$ and the excited level is $^2A_1$ at Δ = 440GHz [1]. Theoretically a phonon-induced $T_2 \rightarrow A_1$ transition is symmetry forbidden; the deformation Hamiltonian transforms as A_1 +E [2], the resonant scattering should therefore be from $E \rightarrow A_1$ and only scatter phonons producing E type distortions. We have tested this prediction using heat pulses. The orbital degeneracy, $E+T_2$ should also lead to scattering from within the ground state which is not the case in the group V systems. This scattering is likely to be very sensitive to strain.and it now seems clear that some of the early experiments were significantly affected by random strains probably caused by the relatively high concentrations of C and O present in Si available at that time. This can be seen in the strong scattering observed in the thermal conductivity below 1K which has been attributed to ground state splittings of 50GHz in about 20% of the Li ions [3]. A further interesting aspect is the likelihood of Jahn-Teller effects, which could significantly modify the phonon scattering [4].

A magnetic field can affect the phonon scattering in two ways: For $\nu_{dom} \ll \Delta$, (T<<5K), the dominant effects should be due to Zeeman splitting in the ground state as $g\beta B \sim h\nu_{dom}$. At higher phonon frequencies, changes in the resonant scattering at Δ due to wavefunction shrinkage could become important although, because of the smaller Bohr radius in Si, these should be smaller than those seen in Ge [5] at the same field.

Our present experimental programme is being carried out on Wacker Chemitronic material containing $<5\times10^{15}cm^{-3}$ of O and C. The splittings of the Γ_8 ground state of p-Si of this quality are only 2GHz [6]; APR measurements on Li doped samples reported earlier [7,8] seem broadly consistent with theoretical predictions, assuming the presence of small random internal strains.

2. MAGNETOTHERMAL CONDUCTIVITY RESULTS AND DISCUSSION

For the magnetothermal conductivity measurements, a sample of the above material was Li doped by diffusion to a concentration $9\times10^{15}cm^{-3}$. In some earlier work on this system, thermal contact to the sample was made using indium faced clamps and to test the possibility that this might be introducing strain, we have repeated this work using silver foil glued to the Si with G.E. varnish, an arrangement which is believed to be relatively strain free [9].

The results are shown in fig. 1 for fields in the (110) plane. The maximum change is 2% at 1.2K reducing to 1% at 4K, the changes were independent of field direction. These effects, particularly at the higher temperatures, are much smaller than those seen earlier on the same sample [7] although the zero field thermal conductivity was the same.

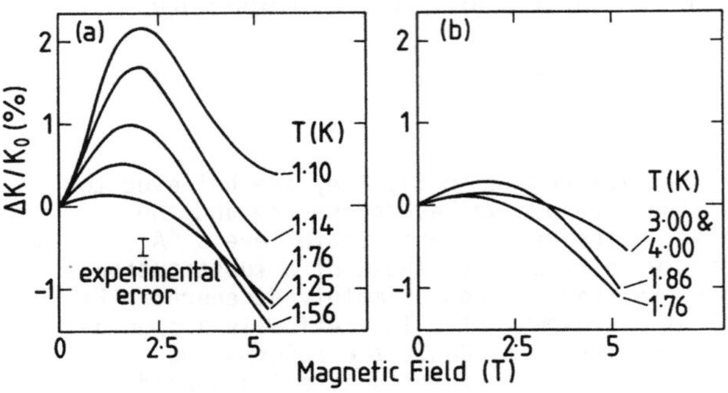

Fig. 1 The increase in conductivity $\Delta K/K_o$ % with magnetic field at various temperatures. The changes at 3.00 and 4.00K are the same to within experimental error.

The increase in conductivity to a maximum observed at low temperatures seems inconsistent with the decrease to a minimum at about 3T ($g\beta B \sim h\nu_{dom}$) expected from Zeeman splitting, although it seems very probable that it results from changes in the $E+T_2$ scattering. An increase could occur from wavefunction shrinkage since this moves Δ to higher frequencies (away from ν_{dom}) but this seems inconsistent with the increasing size of the field effect as T falls. So at present we have no explanation of these data although we have not yet considered if the increase might be explained by field quenching of the Jahn-Teller scattering. The field dependence changes somewhat above 2K and this may be caused by the increasing importance of wavefunction shrinkage effects. The work is being extended to higher temperatures.

3. HEAT PULSE MEASUREMENTS RESULTS AND DISCUSSION

For the heat pulse experiments two samples were cut from the above material, each in the shape of a cube of 6mm side. One was doped with Li by diffusion to a concentration 10^{15} cm^{-3} while the other was left undoped. So that we could make measurements in magnetic fields we used cadmium sulphide bolometers [10], these were fabricated on a (110) face of each sample. On the opposite (110) face Au thin film heaters were deposited.

Heat pulse signals detected after transmission through the Li doped sample are shown in fig.2(a) and compared with signals transmitted by the undoped sample fig.2(b). The heater temperatures T_h shown were calculated using the acoustic mismatch model [11]. We obtain for a gold heater, area A (m^2) and input power P (W) in contact with Si at temperature T_o; $P/A = 24.0\ (T_h^4 - T_o^4)\ Wm^{-2}$. The striking result is the severe attenuation of the slow transverse mode in the Li doped sample, there is also strong attenuation of the longitudinal mode. This is consistent with E mode coupling: a longitudinal acoustic wave propagating along <110>

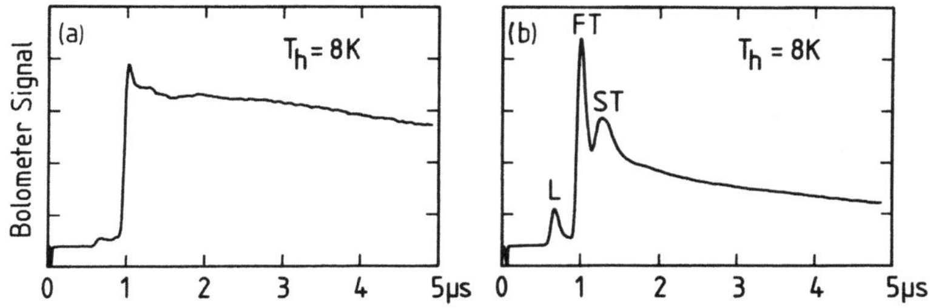

Fig. 2 Heat pulse signal after transmission through, (a) Li-doped sample and (b) undoped sample.

produces A_1, T_2 and E type distortions while the fast transverse wave produces just T_2 type distortions and the slow transverse just E type. The experiment was repeated for different propagation directions, the observations were consistent with this.

An applied magnetic field had no effect on the fast transverse mode in the range of heater temperatures 5-30K. It was difficult to study the other modes in detail as they were severely attenuated, however, preliminary results indicate that the attenuation of the longitudinal mode was increased in an applied field, further work is planned.

We wish to thank Mr W B Roys, Miss S V J Kenmuir and Mr R J Turner for their help in these experiments.

REFERENCES

1. L. Aggarwal, P. Fisher, V. Mourzine and A.K. Ramdas, Phys. Rev. 138, A882, (1965)
2. C. Herring and E. Vogt, Phys. Rev. 101, 944, (1956)
3. A. Adolf, D. Fortier, J. H. Albany and K. Suzuki, Phys. Rev. B21, 5651, (1980)
4. E. Puhl, E. Sigmund and J. Maier, Phys. Rev. B32, 8234, (1985)
5. T. Miyasato, M. Tokumura and F. Akao, J. de Physique, C6, 658, (1981)
6. H. Zeile and K. Lassmann, Phys. stat. sol. (b), 111, 555, (1982)
7. L.J. Challis, A.P. Heraud, V.W. Rampton, M.K. Saker and M.N. Wybourne, Proc. 4th. Int. Conf. Phonon Scattering in Condensed Matter. Stuttgart, 1983, eds. W.Eisenmenger, K.Lassmann and S.Dottinger (Springer, Berlin) p.358
8. V.W. Rampton and M.K. Saker Proc. 2nd. Int. Conf. on Phonon Physics Budapest, 1985, eds. J.Kollar, N.Kroo, N.Menyhard and T.Siklos (World Scientific, Singapore) p.684
9. W.Odoni, P.Fuchs and H.R.Ott, Phys. Rev. B28, 1314, (1983)
10. T.Ishiguro and S.Morita, Appl. Phys. Lett. 25, 533, (1974)
11. W.A. Little, Can. J. Phys. 37, 334, (1959)

Acoustic Paramagnetic Resonance of Ni in GaAs

V.W. Rampton and M.K. Saker

Department of Physics, University of Nottingham, University Park, Nottingham, NG7 2RD, UK

1. INTRODUCTION

Nickel diffuses rapidly into GaAs and acts as a deep acceptor. It has often been used as a contact material on GaAs devices [1,2]. Nickel can exist in the Ni^{3+} [3], Ni^{2+} [4], or Ni^{+} [5] valence states in GaAs depending on the relative concentrations of nickel and other donors and acceptors in the host. It has been suggested that nickel in GaAs has a strong tendency to form pairs with shallow donors such as sulphur, selenium, tellurium, silicon and tin [6], while other workers [7] have found that some results attributed to Ni could be due to copper in GaAs. We have investigated the low-lying energy levels of Ni in GaAs, using ultrasonic waves to excite paramagnetic resonances (a.p.r).

2. EXPERIMENTS AND RESULTS

The two samples of semi-insulating GaAs used were cut from the same boule. Both samples were cleaned and then nickel was evaporated onto the surfaces of one of them. Both samples were then sealed into ampoules and heated at 900°C for 5 hours followed by quenching and washing to remove excess nickel. The samples were then oriented, cut and polished to make specimens for ultrasonic propagation along the [110] direction. Longitudinal ultrasound at 9.6 GHz and at 10.3 GHz was used in a pulse-echo method at a temperature of 1.7K and in magnetic fields up to 2T. The results obtained at 10.3 GHz for the specimen containing nickel are shown in fig. 1 and an isofrequency plot giving the positions of the resonances as a function of the angle between the magnetic field and the crystal axes is shown in fig. 2. The specimen with no nickel gave quite different results. In this control specimen the resonance spectrum appeared to be identical with that found in chromium doped GaAs [8]. SIMS analysis, by Loughborough Consultants, was made of the sample containing nickel. The results are shown in Table 1.

3. DISCUSSION

The sample not treated with nickel showed the a.p.r. due to Cr^{2+} ions implying that the Fermi level in this material was in the upper half of the gap so that the chromium was mainly in the Cr^{2+} state. Nickel is known to act as a deep acceptor with the Ni^{3+}/Ni^{2+} level at about 0.2eV above the valence band. Table 1 shows that sufficient nickel was diffused into the GaAs to more than compensate the chromium initially present. Thus we believe the chromium is entirely transformed to the Cr^{3+} state. This is consistent with the disappearance of the Cr^{2+} a.p.r. No a.p.r. signals have ever been attributed to Cr^{3+} so we believe the resonances we observe in the

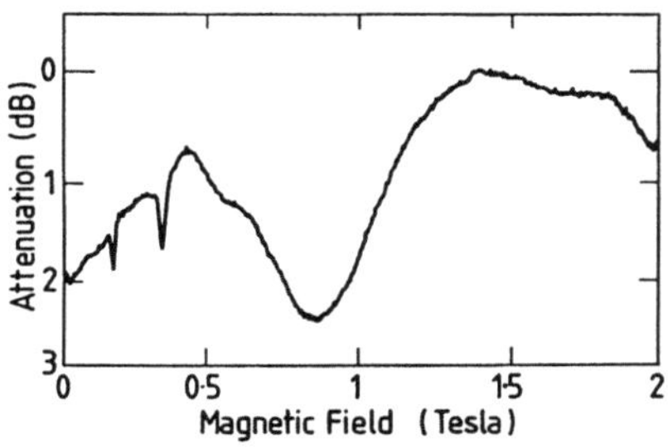

Figure 1 A.p.r. of GaAs:Ni at 10.3GHz and 1.7K

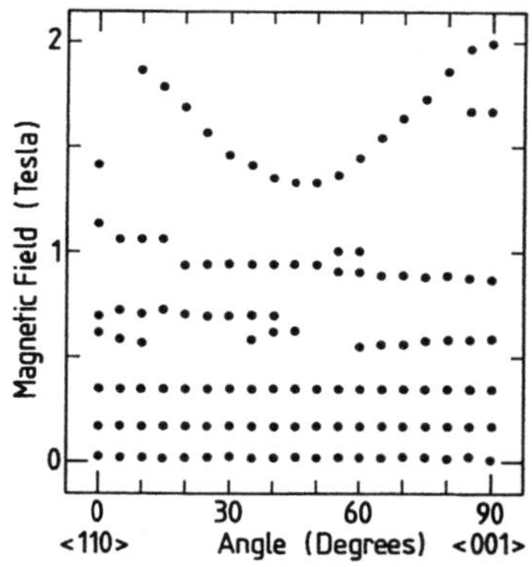

Figure 2 Isofrequency plot of a.p.r. of GaAs:Ni at 10.3GHz

Table 1 SIMS analysis of Ni doped GaAs

Element	Concentration atoms per cm^{-3}
Nickel	1×10^{17}
Manganese	1×10^{15}
Chromium	3×10^{14}

nickel diffused sample are due to nickel. The two isotropic sharp lines at 0.17T and 0.32T at 10.3GHz are attributed to Ni^{3+} ions. Ni^{3+} in a tetrahedral site has an orbital singlet groundstate $^4A_2$ and would be expected to be weakly coupled to the lattice. Hence it is insensitive to random internal strains and will give weak but sharp a.p.r. lines. The $^4A_2$ groundstate splits into four equally spaced levels in a magnetic field. Resonances due to $\Delta M_s = 2$ transitions are found at 0.17T and due to $\Delta M_s = 1$ transitions are found at 0.32T giving a value g = 2.13 ±0.02. This is in good agreement with e.p.r. data [9, 10].

We believe the remaining broad anisotropic resonance absorptions shown in fig. 1 are due to Ni^{2+} ions. Ni^{2+} in a tetrahedral site has a lowest orbital triplet, $^3T_1$, in contrast to the orbital singlet of Ni^{3+} [4]. We expect a much stronger coupling to the phonons. The anistropy of the resonances suggests a sensitivity to other crystal defects and may be due to nickel-donor pairs [6]. We have attempted to explain the results using a strain-stabilised Jahn-Teller model of Ni^{2+}. We used a Hamiltonian for the lowest $^3T_1$ states with a small tetragonal distortion as given by BATES [11], (his equations 7.14 and 7.15 and section 7.3).

$$H = \lambda\gamma k_1^{T_1} \underline{L}.\underline{S} + \beta \underline{H}.(\gamma k_1^{T_1} \underline{L} + 2\underline{S})$$

$$+ \lambda \left[C\{\underline{S}(\underline{S}+1) + E_\vartheta^L E_\vartheta^S + E_\varepsilon^L E_\varepsilon^S\} + b\,(\underline{L}.\underline{S})^2\right] + \tfrac{1}{2}D(3L_z^2 - 2)$$

where the symbols are defined by BATES [11]. We find we can fit the results with parameters as given in table 2.

Table 2 Parameters for Ni^{2+} (cm^{-1})

Frequency	9.6 GHz	10.3 GHz
$\lambda\gamma k_1^{T_1}$	0.458	0.4733
λC^1	-3.647	-2.1157
λb	0.727	0.6786
γ	0.018	0.012
D	10	10

As can be seen, the best fit at each frequency requires slightly different parameters. This is because the isofrequency plots are a different shape at the two frequencies. We also find a further anisotropy in that the resonance line shape and magnitude is not the same when the magnetic field is at the same angle on either side of the [110] direction in the (001) plane. This seems to indicate that there is some overall assymmetry in the sample either due to the original crystal growth direction or due to the direction of Ni diffusion.

ACKNOWLEDGEMENTS

We are pleased to acknowledge the help and advice of Professor C A Bates and Mr C Schiller.

REFERENCES

1. Y. Fujiwara, A. Kojima, T. Nishino and Y. Hamakawa: Jap. J. Appl. Phys. 22, L476 (1983).
2. D.L. Partin, A.G. Milnes and L.F. Vassamillet: J. Elec. Mat. 7, 279 (1978).
3. U. Kaufmann and J. Schneider: Solid State Comm. 25, 1113 (1978).
4. N.I. Suchkova, D.G. Andrianov, E.M. Omel'yanovskii, E.P. Rashevskaya, A.S. Savel'ev, V.I. Fistul' and M.A. Filippov: Sov. Phys. Semicond. 11, 1022 (1977).
5. H. Ennen, U. Kaufmann and J. Schneider: Solid State Comm. 34, 603 (1980).
6. H. Ennen, U. Kaufmann and J. Schneider: Appl. Phys. Lett. 38, 355 (1981).
7. V. Kumar and L-A Ledebo: J. Appl. Phys. 52, 4866 (1981).
8. C.A. Bates, D. Brugel, P.Bury, P.J. King, V.W.Rampton and P.C. Wiscombe: J. Phys. C: Solid State Phys. 17, 6349 (1984).
9. M. de Wit and T.L.Estle: Bull. Am. Phys. Soc. 7, 449 (1962).
10. D.G.Andrianov, N.I.Suchova, A.S. Savel'ev, E.P. Rashevskaya and M.A. Filippov: Sov. Phys. Semicond. 11, 426 (1977).
11. C.A. Bates: Physics Repts. 35, 187 (1978).

Ultrasonic Detection of Interstitial Aluminum in Silicon

W.L. Johnson and A.V. Granato

Department of Physics, University of Illinois at Urbana-Champaign, 1110 W. Green St., Urbana, IL 61801, USA

1. Introduction

In aluminum-doped float-zone silicon the primary defects produced by low-temperature electron irradiation are isolated vacancies and interstitial aluminum. Interstitial silicon, produced along with the vacancy in the initial damage event, is assumed to migrate even at 4.2 K under the influence of ionizing electron irradiation and to displace substitutional aluminum, knocking it into an interstitial position [1].

Interstitial aluminum has been identified in electron paramagnetic resonance studies by Watkins [1] and Brower [2]. The state they observe is doubly positively charged (Al_i^{++}) and is the equilibrium state in p-type silicon. It has tetrahedral symmetry and its energy level is in the valence band.

There has been some speculation concerning the higher charge states of Al_i. In DLTS studies, TROXELL, et al., [3] observed enhanced migration of Al_i under injection conditions at room temperature. The model which they presented to explain this result suggested that Al_i^0 may be distorted from T_d symmetry.

Our ultrasonic measurements reveal a previously unreported defect state which we identify as a non-equilibrium distorted configuration of Al_i.

2. Experimental Results

The samples were float-zone silicon with aluminum concentration of .1-.3 ppm. Two <110> faces were polished flat and parallel within .5 μm, and 16-136 MHz shear waves were propagated using a pulse-echo technique. The transducer was bonded to one face with 3-methylpentane (Aldrich Chem. Co.), which freezes at 120 K and is useful for measurements below 70 K. The samples were irradiated with 2.8 MeV electrons.

Monitoring the attenuation at 5-15 K while the sample is being electron irradiated, a defect relaxation peak appears which grows linearly with dose. It is seen with <1$\bar{1}$0>-polarized (C') waves and not with <001>-polarized (C_{44}) waves, indicating a local tetragonal distortion of the lattice around the defect. When the electron beam is turned off, the peak disappears, but it can be immediately regenerated to the same height by turning the beam back on. The peak can also be regenerated, though not to as great a height, by white light or gamma radiation (Bremsstrahlung from the stopped electron beam). We have not performed detailed studies as a function of wavelength, but have found that .18-.39 eV light is ineffective at generating the peak.

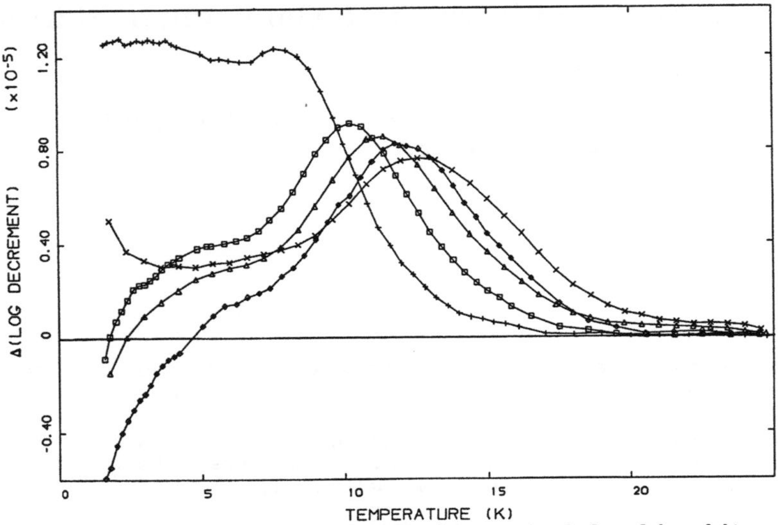

Fig. 1. The change in logarithmic decrement induced by white light. The background measurement (no light) has been subtracted. + 16.0 MHz. □ 45.7 MHz. Δ 75.7 MHz. ◊ 105.7 MHz. × 135.7 MHz

Shown in Fig. 1 are measurements under white light with the background (no light) subtracted. The behavior below ~6 K is not yet well understood and will not be discussed here. The peak displays a classical 1/T dependence of the relaxation strength. The temperature dependence of the relaxation time is found to be Arrhenius in form, with a very low activation energy and frequency factor:

$$\tau^{-1} = (7.7\text{x}10^{10})\exp(-.0051\text{eV}/kT)$$

Thermally activated annealing of the defect occurs well above room temperature, with a time constant of 23 hours at 145°C.

3. Discussion

This relaxation peak is not observed in boron-doped silicon. It grows in proportion to the dose, indicating that it is a primary product of the radiation, and its annealing behavior in the absence of light matches that of the aluminum interstitial [3]. This evidence allows us to conclude that the defect we see is, in fact, the aluminum interstitial.

The charge state is not known. It is not the undistorted equilibrium state, Al_i^{++}. Nor is it likely that it is Al_i^{+}, since that state is ~.17 eV above the valence band [3], and we are unable to generate the peak with .18-.39 eV light.

It may be noted that Al_i appears to behave in a manner similar to the vacancy, which has been studied extensively with EPR and DLTS, and which we have also observed ultrasonically [4] in its positively charged state, V^+. Both defects have equilibrium states of tetrahedral symmetry, and both have tetragonally distorted higher charge states which can be populated by light of appropriate wavelength. The relaxation strengths of the distorted states we see are comparable, and the relaxation rates are such that both produce peaks below 30 K for 100 MHz waves. These similarities suggest that that the symmetry of Al_i charge states may follow one-for-one those of the vacancy as described by Watkins [5]. If this is true, then with the first two localized electrons (Al_i^{++} and Al_i^{+}, corresponding to V^{+++} and V^{++}) the defect maintains cubic symmetry, and with the next two (Al_i^{0} and Al_i^{-}) it becomes tetragonally Jahn-Teller distorted. Higher charge states have an additional trigonal component of distortion. The neutral and singly negatively charged states, therefore, appear to be the best candidates for the relaxation peak reported here.

Acknowledgement

This work has been supported by the U.S. Dept. of Energy, Division of Materials Sciences, Contract DE-AC02-76-ER091198.

References

1. G.D. Watkins: In Radiation Damage in Semiconductors, 7th Intnl. Conf. on the Phys. of Semiconductors, Vol.3 (Dunod,Paris 1965) p.97
2. K.L. Brower: Phys. Rev. B, 1, 1908 (1970)
3. J.R. Troxell, A.P. Chatterjee, G.D. Watkins and L.C. Kimerling: Phys. Rev. B, 19, 5336 (1979)
4. W.L. Johnson and A.V. Granato: Jour. de Physique, 46, Supp. No. 12, C10-537 (1985)
5. G.D. Watkins: In Lattice Defects in Semiconductors, 1974, Inst. Phys. Conf. Ser. No. 23 (Inst. Phys., London, Bristol 1975) p.1

Phonon Spectroscopy of Iron Doped InP

N. Butler[1], *L.J. Challis*[1], *and B. Cockayne*[2]

[1]Department of Physics, University of Nottingham, University Park, Nottingham, NG7 2RD, UK

[2]Royal Signals and Radar Establishment, St. Andrews Road, Great Malvern, Worcs, WR14 3PS, UK

Fe is the most common magnetic impurity in InP and we have examined its contribution to the phonon scattering as part of a programme of investigation of scattering by transition metal ions in the III-V's. In InP, Fe substitutes for In and enters as Fe^{3+} which is expected to be a very weak scattering centre because of its orbital singlet ground state, $^6A_1$. However InP normally contains $\sim 5 \times 10^{16}$ cm^{-3} of shallow donors as impurities (1ppm = 2×10^{16} cm^{-3}) and their electrons are trapped by any Fe^{3+} ions present to give Fe^{2+}. So the concentration of Fe^{2+} ions in InP is equal to that of [Fe] while [Fe] < [D], the shallow donor concentration, but saturates at [D] once [Fe] > [D]. The efficiency of Fe^{3+} as a trap is indicated by the room temperature resistivity of InP:Fe which can be $\sim 10^7\,\Omega$cm (semi-insulating = SI) and the system has been widely studied in recent years [1]. Fe^{2+} has an orbital doublet ^5E ground state and so should scatter phonons more strongly than Fe^{3+}. Spin orbit interaction splits the ^5E state to give the approximately equally spaced energy levels shown in fig. 1 with Δ known experimentally to be $\sim$ 450GHz [2] and the selection rule $\Gamma_i \times \Gamma_f = E$ indicates that resonant phonon scattering should occur predominantly at a frequency $\nu_0 = 2\Delta = 900$GHz.

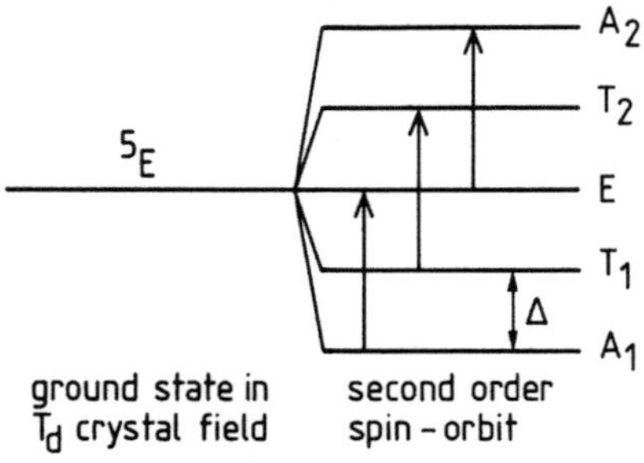

Fig 1.
Ground state energy levels of Fe^{2+} (^5E) in tetrahedral co-ordination showing the allowed phonon transitions for E-mode distortions.

We present here measurements of the thermal conductivity of both n-type and SI samples of InP:Fe whose characteristics are given in the table. The Fe and other impurity concentrations shown were determined by SIMS. The thermal resistivity W_0 of the 'undoped' sample, L973, is used as a reference and the reduced resistivities, W/W_0, of the others are shown in fig. 2. W/W_0 increases broadly with [Fe] but there is no indication of saturation in the SI region at $[Fe^{2+}]$ = [D] which would be expected if the scattering were solely due to Fe^{2+}. Moreover the peak temperature varies from 6K to 12K suggesting that an additional resonant process may be contributing. We believe this is due to Cr present as a trace impurity (table 1) and curve "Cr" shows W/W_0 reduced by a factor of 35 for a SI sample, MIT [3], containing $\sim 5 \times 10^{16}$ cm^{-3} Cr. Curve "Fe" shows values of W/W_0 calculated by adding a resonant scattering term $\tau^{-1} = A\omega^4/(\omega^2 - \omega_0{}^2)^2$ to represent scattering from Fe^{2+} ions to the rate required to describe W_0

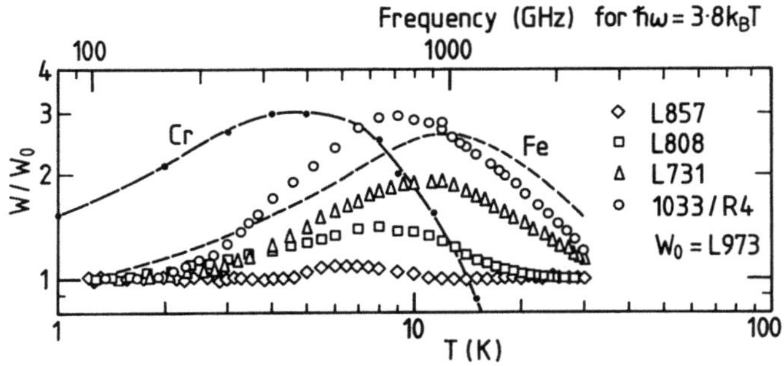

Fig 2 Reduced resistivity, W/W_0, for the InP:Fe samples. Curve "Cr" shows data for InP:Cr reduced by a factor of 35 [3]. Curve "Fe" shows the effect of additional resonant scattering $\tau^{-1} = A\omega^4/(\omega^2 - \omega_0^2)^2$ with ν_0 = 900 GHz

Table 1 Sample characteristics. 1033/R4 was grown at Cambridge Instruments Ltd and the others at RSRE. Their long axes were <110>. Impurity concentrations were determined by SIMS (Loughborough Consultants Ltd) on neighbouring slices

Sample Ref No.	Dimensions mm.	Carrier Type	Resistivity Ωcm (300K)	[Fe] cm^{-3}	[Cr] cm^{-3}	[Mn] cm^{-3}
L973	2.94x2.79 x 20	n	0.3	$3x10^{16}$	$3x10^{14}$	$3x10^{14}$
L857	2.65x2.67 x 14	n	3	$7x10^{16}$	$1x10^{14}$	$4x10^{14}$
L808	2.69x 2.74 x 21	SI	$7x10^4$	$3x10^{16}$	$3x10^{14}$	$3x10^{14}$
L731	2.96x3.07 x 17	SI	10^7	$1x10^{17}$	$2x10^{14}$	$2x10^{14}$
1033/R4	3.18x4.53 x 19	SI	10^6	$2x10^{17}$	$8x10^{14}$	$2x10^{14}$

[4]. The peak for 1033/R4 lies midway between these two curves suggesting approximately equal contributions from Cr and Fe^{2+} while that for L731 lies rather close to that for Fe^{2+} in line with its much smaller Cr concentration. (The Fe concentration is also less but the Fe^{2+} concentrations should = [D] in both samples). The data for L808 and L857 suggest there is less Fe^{2+} present than in L731 and that the scattering in L808 is largely by Cr ions. Cr can in fact exist in InP as Cr^{2+} or Cr^{3+} [1]. Cr^{2+} is likely to have little effect on W above 100mK since any resonant scattering probably occurs at ≤ 5GHz [3] and we attribute the Cr scattering to Cr^{3+}. It is interesting that since the Cr^{2+}/Cr^{3+} level lies above that of Fe^{2+}/Fe^{3+} [1], there should be rather little Cr scattering in n-type samples as the Cr should be largely in its "non-scattering" Cr^{2+} state and this could explain why W(L808) > W (L973). The addition of further Fe converts the Cr into Cr^{3+} and so switches on the Cr scattering. It is also possible that there is a contribution from traces of Mn^{3+} (table 1).

The data for L731 are then very largely consistent with Fe^{2+} scattering and a fit gives $A = 1x10^7\ s^{-1}$ for this Fe^{2+} concentration $\sim 5 \times 10^{16}\ cm^{-3}$ $\sim$2ppm. Fe^{2+} is clearly a very much weaker scattering centre than Cr^{3+} whose concentration in these samples $\lesssim 10^{-2}$ ppm and this relatively weak coupling of ^5E in T_d has been seen in work on other Fe doped systems ($MgAl_2O_4$, CdTe ZnS [5] and provides a striking contrast with the very strong coupling to ^5E in octahedral systems : Cr^{2+} and Mn^{3+} in MgO, $KMgF_3$ and Al_2O_3 [6]. This is presumably directly attributable to a much more effective positioning of the oscillating ligands with respect to the orbital wave functions. From the value A and the expression $A = CN\omega_o{}^2M^4/\pi\rho^2\hbar^2v^7$ where $N = [Fe^{2+}]$, $C \sim 3$ and v is an average sound velocity we obtain an average phonon matrix element $M \sim 1600cm^{-1}$. We hope shortly to use this to determine the spin-phonon coupling constant V_Eb for this system which would in fact be the first measurement for ^5E in tetrahedral systems although there are several in octahedral co-ordination. We are also examining the analyses of the other T_d systems to see if we can use them to obtain further values.

One final point worth emphasising is that the experiments indicate that n-type InP will be more transparent to ballistic phonons than will SI InP. This is not because n-type InP cannot also contain traces of strongly coupled impurities (V^{2+} may be an example [7]) but presumably because their concentration is usually less than that of Cr.

REFERENCES

1. B. Clerjaud: J. Phys. C: Solid State Phys. 18, 3615 (1985); S.G. Bishop: in Deep Centers in Semiconductors, ed. by S. Pantelides (Gordon and Breach, New York, 1985) p 7
2. W. H. Koschel, U. Kaufmann and S.G. Bishop : Solid State Comun. 21, 1069 (1977); V.W. Rampton and P.C. Wiscombe : Acta Phys. Slov. 32, 35, (1982); R.J. Wagner, J.J. Krebs and G.H. Stauss: Bull. Amer. Phys. Soc. 27, 277 (1982).
3. N. Butler, J. Jouglar, B. Salce, L. J. Challis, A. Ramdane and P. L. Vuillermoz: these proceedings
4. N. Butler : Ph.D. Thesis, University of Nottingham (1986).
5. G.A. Slack : Phys. Rev. 134, A1268 (1964) and Phys. Rev. B6, 3791 (1972).
6. L.J. Challis and A.M. de Goër: in The Dynamical Jahn-Teller Effect in Localized Systems, ed. by Yu E. Perlin and M. Wagner, (North-Holland, Amsterdam, 1984) p533
7. V^{2+} appears to scatter phonons strongly in GaAs : N. Butler, L.J. Challis, M. Sahraoui-Tahar and W. Ulrici, to be published.

Ultrasonic Velocity and Electrical Conductivity in p-InSb at Low Temperatures and High Magnetic Fields

G. Quirion, M. Poirier, and J.D.N. Cheeke

Centre de Recherche en Physique du Solide, Département de Physique, Université de Sherbrooke, Sherbrooke, Québec J1K 2R1, Canada

1. Introduction

The piezoelectric interaction in semiconductors connects the ultrasonic velocity and attenuation variations to the electrical conductivity [1]. The ultrasonic techniques offer then the possibility to measure the ac conductivity in a frequency range which is generally hardly accessible for conventional techniques. The ultrasonic investigation of p-InSb at low temperatures is susceptible to give very interesting and important information on the conduction regimes at low temperatures (4-20 K) [2]. We are particularly interested on how the activation and hopping regimes are affected by high magnetic fields and relatively high frequencies. We have thus measured in the low-temperature range the ultrasonic velocity in p-InSb samples having different majority carrier concentration. The results have given the opportunity to evaluate with a good precision the frequency and magnetic field effects on the low-temperature conduction regimes.

2. Samples and Experiment

Three p-InSb samples with different excess impurity concentrations were studied with excess concentration acceptor impurities N_a-N_d, $1\text{x}10^{14}$ cm^{-3} (Ge), $2\text{x}10^{15}$ cm^{-3} (Ge) and $7.5\text{x}10^{15}$ cm^{-3} (Cd). All the samples mentioned above were oriented along the [111] direction in order to analyse the propagation of the longitudinal piezoelectrically active mode. A pulse echo technique has been used to perform the ultrasonic velocity measurements between 4 and 20 K. An automatic acoustic interferometer of the type described by GORODETSKY et al. [3] was used at two frequencies, 90 and 150 MHz. The data for the relative velocity were obtained in this way with an accuracy of 2 ppm.

3. Results and Discussion

In fig. 1 the velocity variations are presented as a function of temperature for three p-InSb samples having different excess majority impurity concentrations. For the weakly doped sample ($1\text{x}10^{14}$ cm^{-3}, fig. 1a), a step-like variation is noticed, as predicted by the theory of HUTSON and WHITE [1]. The velocity variations are observed only for T > 8 K and this means, that the piezoelectric interaction takes place only in the activation regime. When a magnetic field is applied the velocity step is displaced toward higher temperatures because of a decreasing conductivity. The step amplitude is not affected by the field. For the intermediate concentration sample ($2\text{x}10^{15}$ cm^{-3}, fig. 1b), the velocity step begins at a lower temperature (≃ 7 K). It is still displaced toward higher temperatures by a magnetic field but here its amplitude is now an increasing function of the field. At low temperatures, 4 < T < 7 K, the hopping regime is dominating the conductivity which is high enough to partially screen the

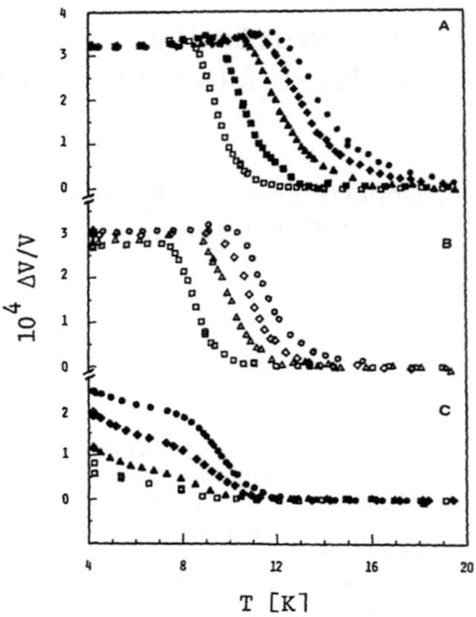

Fig. 1. Relative velocity variation of p-InSb at 90 MHz as a function of temperature for different magnetic field values. a) 1×10^{14} cm^{-3}, b) 2×10^{15} cm^{-3}, c) 7×10^{15} cm^{-3} : B=0 (□), 1 (■), 2 (△), 3 (▲), 4 (◇), 5 (◆), 6 (○), 7 (●) Teslas

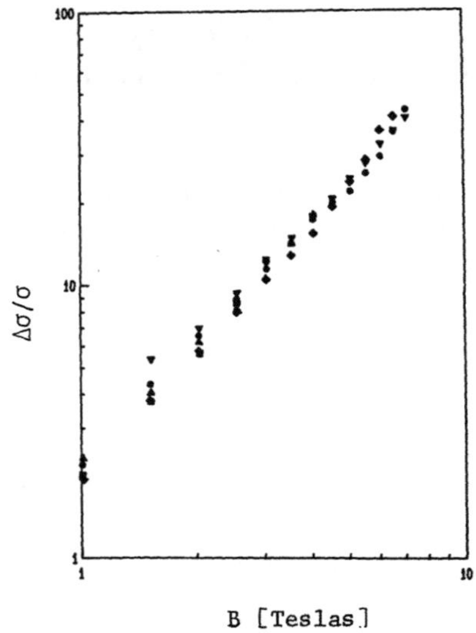

Fig. 2. Relative magnetoconductivity variation of p-InSb at 90 MHz as a function of magnetic field: $N_a - N_d = 1\times10^{14}$ cm^{-3}, T=10.8 (■), 11.3 (▲), 11.8 (◆), 12.5 (●), 13.3 (▼) K

piezoelectric field. When a magnetic field is applied, the conductivity decreases and so does the screening. For sufficiently high fields the velocity variation will saturate at the same value observed for the weakly doped sample. For the sample with the highest concentration (7×10^{15} cm^{-3}, fig. 1c), the variation is small but increases rapidly with increasing field. On the different curves two regions may be distinguished from a change of slope around $T \simeq 8$ K which indicates the passage from one regime of conduction to the other. For $T < 8$ K the hopping conductivity seems to be high enough to efficiently screen the piezoelectric field. The application of a magnetic field decreases the conductivity in both regimes and then restores progressively the piezoelectric interaction.

In fig. 2 we present the relative magnetoconductivity ($\Delta\sigma/\sigma$), deduced from the velocity data, in the activation regime ($10 < T < 14$ K) as a function of magnetic field for different temperatures on a log-log scale. The obtained slope is the same for the two samples studied here and is equal to 1.5±0.2. In the activation regime the relative conductivity variation is given by $\Delta\sigma/\sigma \propto B^2$ if only one type of carrier (heavy hole) is considered. The discrepancy observed here cannot only be explained by a possible stronger contribution of the light holes. In the hopping regime $4 < T < 7$ K, the observed activation energy seems to be nearly independent of the magnetic field intensity, which is consistent with what is generally observed in lightly doped semiconductors [4]. This can be also seen in fig. 3 where the ratio σ_o / σ, σ_o being the zero field conductivity, is presented as a function of the magnetic field for different temperatures. The effects measured here are however too small to be adequately compared to the percolation model of SHKLOVSKII [4].

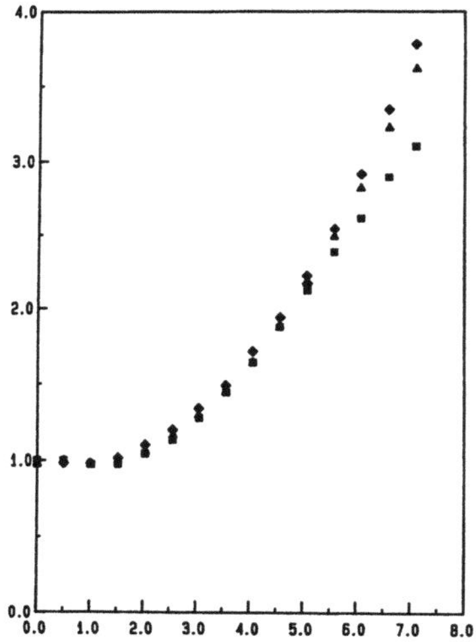

Fig. 3. Relative magnetoconductivity at 90 MHz of p-InSb ($7x10^{15}$ cm^{-3}) as a function of the magnetic field for different temperatures: T = 4.2 (■), 5.6 (▲) and 6.6 (◆) K

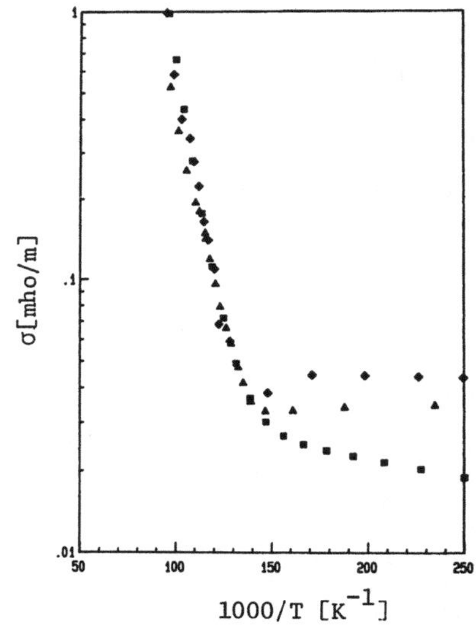

Fig. 4. Electrical conductivity of p-InSb ($2x10^{15}$ cm^{-3}) as a function of 1000/T : dc (■), 90 MHz (▲) and 150 MHz (◆)

In fig. 4, we compare the conductivity curves at 90 and 150 MHz deduced from the velocity data for the $2x10^{15}$ cm^{-3} sample to the dc results. In the activation regime, no frequency effects are observed because $\omega\tau << 1$ as it was expected. In the hopping regime, the frequency effects are important. These effects have already been the object of an analysis [5]. The obtained slope was $s = 0.75\pm0.02$, which is similar to what is generally found for other semiconductors (0.8) in the literature even if the covered frequency range is higher than usual.

References

1. Hutson A.R. and White D.L., J. Appl. Phys. 33, 40 (1962).
2. Madore G. and Cheeke J.D.N., J. Can. Phys. 62, 460 (1984).
3. Gorodetsky G. and Lachterman, Rev. Sci. Instrum. 52, 1386 (1981).
4. Shklovskii B.I., Sov. Phys. Semicond. 6, 1053 (1973).
5. Poirier M., Quirion G. ,Séguin P.-E. and Cheeke J.D.N., Sol. Stat. Comm. 57, 401 (1986).

Magnetic Field Dependence of Sound Velocity in Sb-doped Ge

K. Suzuki, K. Asano, and H. Ogiwara

Department of Electrical Engineering, Waseda University, Shinjuku, Tokyo 160, Japan

1. Introduction

Recently the magnetic field dependence of the ultrasonic attenuation of Sb-doped Ge has been measured over a wide range of impurity concentrations at low temperatures to investigate the concentration dependence of electronic states/1/. The Sb donor has low-lying energy levels and consequently, interacts strongly with low-frequency phonons. Therefore, the system Ge:Sb is suitable to study the concentration dependence of impurity states using the ultrasonic method. It has been found that the magnetic field dependence of the attenuation coefficient, which is determined by that of electronic levels and relaxation rates, depends strongly on the donor concentration in the region less than about 10^{17} cm^{-3}.

It is well known that the velocity change due to the electron-phonon interaction is nearly independent of the relaxation time as far as we are concerned with very low-frequency phonons. The purpose of this paper is to report experimental results of the velocity change in Sb-doped Ge under the magnetic field at low temperatures.

2. Experimental

We have measured the magnetic field dependence of the velocity change in Sb-doped Ge over the temperature range 2.3 to 4.2 K. Measurements have been made using the pulse-echo-overlap method with 60 MHz longitudinal waves propagating along the $(1\bar{1}0)$ direction. The magnetic field up to 5 T was applied along the (111) direction. The impurity concentrations of samples used here are N = 0.31, 0.50, 0.83, 1.2, 2.4, and 3.3 × 10^{17} cm^{-3}.

Figures 1(a)-(d) show the velocity change $\Delta v(B) = v(B) - v(0)$ by the magnetic field /B/ for N = 0.31, 0.83, 1.2, and 2.4 × 10^{17} cm^{-3}. The aspects of $\Delta v(B)$ change with N, but any drastic change does not occur at $N_C = 1.5 \times 10^{17}$ cm^{-3}, the critical concentration for the metal-nonmetal transition.

3. Discussion

In n-type Ge, only the elastic constant C_{44} is sensitive to donor doping/2/. $v - v_0$ (the sound velocity of pure Ge) is proportional to the real part of ΔC_{44}, the change in C_{44} due to the electron-phonon interaction. The latter is determined by the energy levels of electrons in the system under consideration. Therefore, $\Delta v(B)$ reflects the change in electronic states by the magnetic field.

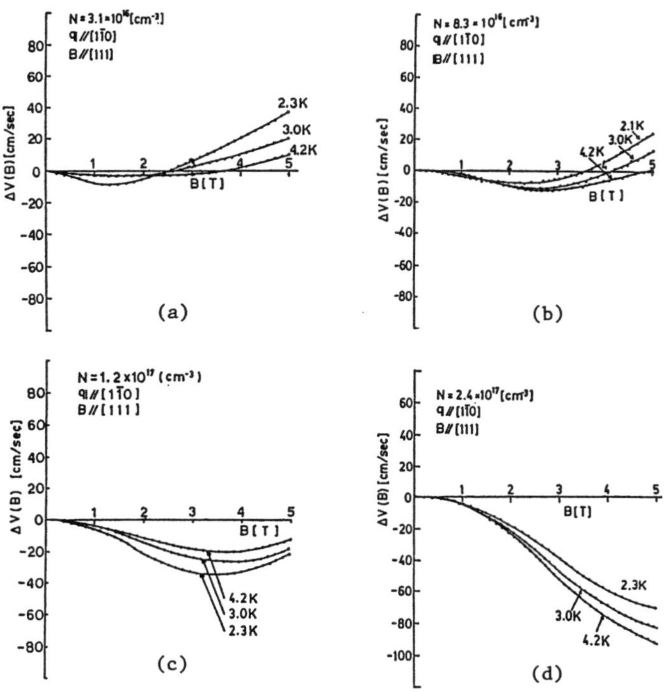

Fig. 1 Δv vs B

For 0.31×10^{17} cm^{-3} where most donors seem isolated/3/, we can semiquantitatively explain Fig. 1(a) based on the level structure of the isolated donor under the magnetic field calculated by Lee et al./4/, though accuracy of their calculation is not so high as to reproduce details. Furthermore we can qualitatively explain the change in the attenuation coefficient by the magnetic field $\Delta\alpha(B) = \alpha(B) - \alpha(0)$.

As N increases, the proportion of isolated donors to the total number decreases. As for a pair, its contribution to $v - v_0$ is small compared with that of an isolated donor at B = 0/3/. The level structure of the pair under B is not known at present. The $\Delta v(B)$ will be small. For $0.50 \times 10^{17} < N < 10^{17}$ cm^{-3}, the presence of electrons in the D^- band must be taken into account even at low temperatures. We conjecture that their contribution to $\Delta v(B)$ is similar to that of electrons in the conduction band.

In the concentration region slightly higher than N_C it is found from a comparison $\Delta\alpha(B)$ with $\Delta v(B)$ that the relaxation time τ increases considerably with B. This means that the dominant relaxation mechanism is not due to intervalley scattering since the intervalley scattering time would be independent of B or shorter. For $N \gtrsim 2 \times 10^{17}$ cm^{-3} $\Delta\alpha(B)$ is proportional to $\Delta v(B)$ where τ is nealy independent of B.

Acknowledgment

This work was supported in part by the Grant-in-Aid for Special Project; Research on Ultrasonic Spectroscopy and Its Application to Material Science from the Ministry of Education, Science, and Culture of Japan.

References

1. H. Sakurai, K. Suzuki, and T. Miyasato: J. Phys. Soc. Jpn. 53, 1356 (1984)
2. R. W. Keyes: IBM J. Res. Develop. 5, 266 (1961)
3. T. Sota and K. Suzuki: this volume
4. N. Lee, D. M. Larsen, and B. Lax: J. Phys. Chem. Solids 34, 1817 (1973)

Ultrasonic Second-Harmonic Generation in n-Type Piezoelectric Semiconductors

C.C. Wu

Institute of Electronics, National Chiao Tung University
Hsinchu, Taiwan, People's Republic of China

The interaction of the fundamental-frequency waves in a non-linear solid gives rise to the polarization at the sum frequency. When large-amplitude ultrasonic flux propagates in a piezoelectric semiconductor, the interaction between the ultrasonic waves and conduction electrons leads to frequency mixing of waves comprising the flux. In high-mobility semiconductors such as n-type InSb, the application of a strong magnetic field can crucially alter the behavior of the electron-phonon interaction. Hansen [1] proposed a correct form of the velocity operator from the Hamiltonian operator to show that the Hall effect can not be influenced by the nonparabolicity of energy bands in the limiting of vanishing scattering. In this paper, we investigate the ultrasonic-harmonic generation in nondegenerate piezoelectric semiconductors such as n-type InSb by taking into account the effect of an electron relaxation time due to the scatterings in solids at the low temperature range when the ultrasonic waves propagate longitudinally. We use the Heisenberg equation of motion to correct the nonlinear effect of energy bands in semiconductors. The major interaction between the electrons and ultrasonic waves is via the piezoelectric coupling in which a self-consistent field is produced accompanying ultrasonic waves.

In a nonparabolic model the eigenfunctions and eigenvalues for electrons in a dc magnetic field $\vec{B}$ directed along the z axis can be expressed as

$$\Psi_{\vec{k}n}=\exp(ik_y y + ik_z z)\phi_n[x - (\hbar c/eB)k_y] \tag{1}$$

and

$$E_{\vec{k}n} = - \tfrac{1}{2}E_g\{1 - [1 + (4/E_g)\left((n+\tfrac{1}{2})\hbar\omega_c + \hbar^2k_z^2/2m^*\right)]^{\frac{1}{2}}\} , \tag{2}$$

respectively, where m* is the effective mass of electrons at the minimum of the conduction band, E_g is the energy gap between the conduction and valenced bands, k_y and k_z are the y and z components of the electron wave vector $\vec{k}$, $\phi_n(x)$ is the harmonic-oscillator wave function, and $\omega_c = |e|B/m^*c$ is the cyclotron frequency of electrons.

The interaction of electrons with ultrasonic waves can be taken into account via the vector potential $\vec{A}_1 = \vec{A}_{1o}\exp(i\vec{q}\cdot\vec{r} - i\omega t)$, which arises from the self-consistent field accompanying waves. Up to second order in $\vec{A}_1$ the Hamiltonian for an electron in the presence of the magnetic field and self-consistent field can be written as

$$H = H_o + H_1 + H_2 \tag{3}$$

where H_o is the unperturbed Hamiltonian of electrons, the perturbed Hamiltonian of first and second orders are given by

$$H_1 = - (e/2c)(\vec{v}\cdot\vec{A}_1 + \vec{A}_1\cdot\vec{v}) \tag{4}$$

and

$$H_2 = \frac{e^2}{2c^2}\left(1 + \frac{H_o^{(r)} + H_o^{(l)}}{E_g}\right)^{-1}\left[\frac{\partial^2 F}{\partial p_i \partial p_j} A_{1i} A_{1j} - \frac{1}{2E_g}\left(1 + \frac{H_o^{(r)} + H_o^{(l)}}{E_g}\right)^{-2}\right.$$

$$\left. \times \left(A_{1i}\frac{\partial F}{\partial p_i} + \frac{\partial F}{\partial p_i}A_{1i}\right)\left(A_{1j}\frac{\partial F}{\partial p_j} + \frac{\partial F}{\partial p_j}A_{1j}\right)\right] \tag{5}$$

respectively, where $\vec{v} = (1/i\hbar)[\vec{r},H_o]$, and $F = (1/2m^*)[p_x^2 + (p_y - eBx/c)^2 + p_z^2]$. $H_o^{(r)}$ and $H_o^{(l)}$ are the right and left Hamiltonian operators such that $H_o^{(r)}\psi_{\vec{k}n} = E_{\vec{k}n}\psi_{\vec{k}n}$ and $\psi^*_{\vec{k}n}H_o^{(l)} = E_{\vec{k}n}\psi^*_{\vec{k}n}$, respectively.

The density matrix $\vec{\rho}$ can be expanded up to second order in amplitude of ultrasonic waves

$$\vec{\rho} = \vec{\rho}_0 + \vec{\rho}_1 + \vec{\rho}_2 \, , \tag{6}$$

where $\vec{\rho}_0$ is independent of time, $\vec{\rho}_1$ varies as $\exp(-i\omega t)$, and $\vec{\rho}_2$ varies as $\exp(-2i\omega t)$. The quantum Liouville equation can be expressed by [2]

$$\frac{\partial \vec{\rho}}{\partial t} + \frac{i}{\hbar}[H,\vec{\rho}] = -\frac{\vec{\rho} - \vec{\rho}_0}{\tau} \, , \tag{7}$$

where τ is the electron relaxation time due to the effect of scatterings in solids. The current density $\vec{J}$ can be expressed as

$$\vec{J} = Tr(\vec{\rho}\cdot\vec{J}_{op}) = \sum_{\vec{k}\vec{k}',nn'} \langle\vec{k}'n'|\vec{\rho}|\vec{k}n\rangle\langle\vec{k}n|\vec{J}_{op}|\vec{k}'n'\rangle \, , \tag{8}$$

where the current density operator induced by the self-consistent field at a point $\vec{r}_o$ is

$$\vec{J}_{op} = -(e/2)[(\vec{v} + \vec{v}'),\delta(\vec{r} - \vec{r}_o)]_+ , \tag{9}$$

where $[,]_+$ denotes the anticommutator. The operator $\vec{v}' = (1/i\hbar)[\vec{r},H_1 + H_2]$ is the velocity operator due to the electron-phonon interaction.

Using the same method as our previous work [3], one can obtain the longitudinal amplitude of displacement for $\vec{q}$ // [111] as

$$\xi''_{2o} = \frac{-4\pi i q\omega\beta_{14}\tau_{zzz}(q,\omega)\xi''^2_{1o}}{\sqrt{3}[2\sigma_{zz}(q,\omega) - \sigma_{zz}(2q,2\omega)][4\pi i\sigma_{zz}(q,\omega) + \omega\varepsilon]} \, . \tag{10}$$

Therefore, the acoustic intensity in the second harmonic can be expressed by

$$\frac{P_2}{P_1^2} = \frac{2}{3dv_s^3}(8\pi\beta_{14}/\varepsilon)^2\left|\frac{\tau_{zzz}(q,\omega)}{[2\sigma_{zz}(q,\omega) - \sigma_{zz}(2q,2\omega)][1 - 4\pi\sigma_{zz}(q,\omega)/i\omega\varepsilon]}\right|^2 \tag{11}$$

where d is the density of materials, v_s is the sound velocity, β_{14} is the piezoelectric constant, ε is the dielectric constant, σ_{zz} is the longitudinal linear conductivity tensor, and τ_{zzz} is the longitudinal nonlinear conductivity tensor.

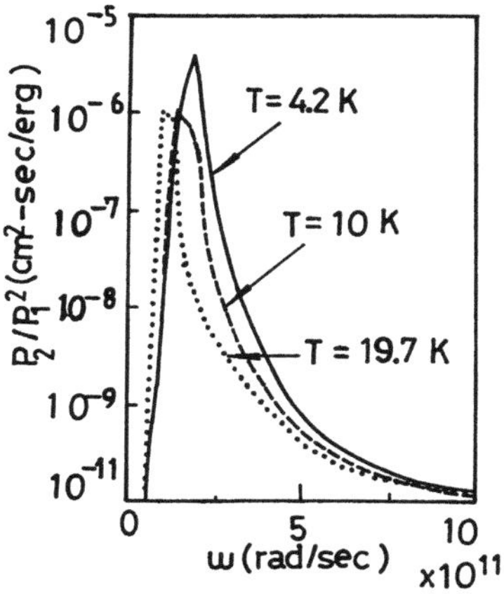

Fig. 1. Ratio of the acoustic intensity in the second harmonic to the square of that in the fundamental as a function of sound frequency.

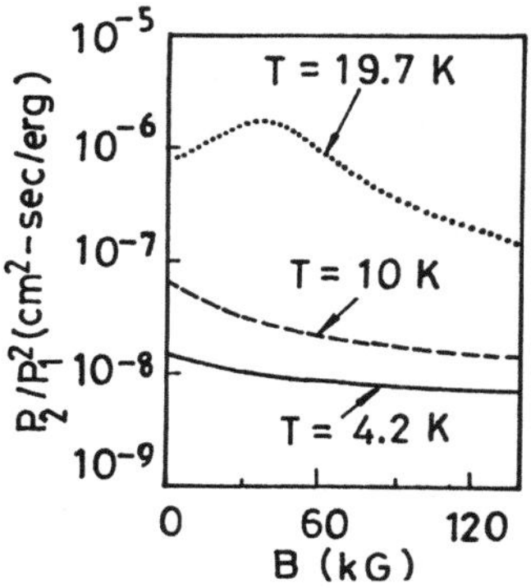

Fig. 2. Ratio of the acoustic intensity in the second harmonic to the square of that in the fundamental as a function of magnetic field.

As a numerical example for n-type InSb, the relevant values of physical parameters are [2,3] n_0=1.75 x 10^{14} cm^{-3}, m* = 0.013m_0 (m_0 is the mass of free electron), ε = 18, β_{14} = 1.8 x 10^4 esu/cm^2, E_g = 0.2 eV, d = 5.8 gm/cm^3, τ = 10^{-12} sec, and v_s = 4 x 10^5 cm/sec. The ratio of the acoustic intensity in the second harmonic to the square of the intensity in the fundamental as a function of sound frequency ω at B = 10 kG is shown in Fig. 1. It shows that the acoustic intensity in the second harmonic increases rapidly with the sound frequency in the microwave region and will peak in the neighborhood of the sound frequency ω = 10^{11} $\sim$ 2 x 10^{11} rad/sec. This maximum point will shift to the higher sound frequency with decreasing temperature. When the sound frequency comes into the high-frequency region, the decrease of the acoustic intensity in the second harmonic becomes slowly with the sound frequency. This is quite different from our previous work [3]. However, the appearance of the maximum point in our present results is in good agreement with those of the phenomenological approach [3,4]. We plot the ratio of the acoustic intensity in the second harmonic to the square of the intensity of the fundamental as a function of the magnetic field at ω = 10^{11} rad/sec as shown in Fig. 2. It can be seen that the acoustic intensity in the second harmonic appears quite complicated relation with the magnetic field and temperature.

References

1. O. P. Hansen, J. Phys. C : Solid State Phys. 14, 550 (1981).
2. F. R. Sutherland and H. N. Spector, Phys. Rev. B 17, 2728 (1978); 2733 (1978).
3. C. C. Wu and H. N. Spector, J. Appl. Phys. 43, 2937 (1972).
4. P. Palanichamy and S. P. Singh, J. Appl. Phys. 54, 3958 (1983).

Phonon Stopbands in Amorphous Superlattices

O. Koblinger, J. Mebert, E. Dittrich, S. Döttinger, and W. Eisenmenger*

Physikalisches Institut, Universität Stuttgart, Pfaffenwaldring 57,
D-7000 Stuttgart 80, Fed. Rep. of Germany
*Now at IBM Deutschland GmbH, Schönaicherstr. 220,
D-7030 Böblingen, Fed. Rep. of Germany

1 Introduction

The development of thin film technology in the last years made it possible to prepare periodic multilayer structures, consisting of alternating very thin crystalline or amorphous films of about nm thicknesses. The periodic structure of such multilayers is the origin of some new properties of those "superlattices" [1]. The acoustic properties of superlattices have been widely studied with resonant raman scattering techniques. Because of the coupling mechanism, in this method only longitudinal phonons are involved. Spectroscopy with superconducting tunnel junctions as generators and detectors of high-frequency acoustic phonons was first used to demonstrate a stopband like behaviour of a $GaAs/Al_xGa_{1-x}As$ superlattice, prepared by using MBE techniques [2].

2 Theory

The calculation of the dispersion relation in layered media was first done by Rytov [3], using a continuuum model. The main result is the existence of band gaps at wavevectors $q=n\cdot\pi/d$, with d the thickness of one period of the superlattice. Calculations of the phonon transmission coefficent can be carried out by using a formalism well known from classical optics [4]. The analogy is made by substituting the impedance for the electromagnetic wave by the acoustic impedance $Z_i = \rho_i \cdot v_i$ of the individual layers with density ρ_i and sound velocity v_i. An example of such a calculation can be seen in Fig. 2.

One important result of a detailed analysis is, that the center frequencies of the stopbands are not sensitive to small statistical variations of the individual film thicknesses. Those variations are expected when using thin film evaporation techniques, but they should only affect the sidestructures [2].

3 Technology

The samples were fabricated in an automatized UHV-evaporation system, which was developed to meet the requirements of preparing very thin, i.e. ~ 10 nm, but uniform metal films. The starting pressure for all process steps was below 10^{-8} mbar. The composition of the residual gas was controlled using a quadrupol mass spectrometer. In-situ measurements of the growing oxide layers were carried out with an automatic ellipsometer. An electron gun was used for the deposition of Si and SiO_2 for the superlattice, as well as Al and Sn for preparation of phonon generators and detectors.

Alternating layers of SiO_2 and Si were deposited on one side of a Si crystal with optical finish. During evaporation, a quartz crystal monitor was used to control the film thicknesses A complete superlattice consisted of seven to fifteen double layers. Without breaking the vacuum an Al-Ox-Al tunnel junction was prepared on top of the superlattice and then a Sn-Ox-Sn tunnel junction was made on the opposite crystal surface.

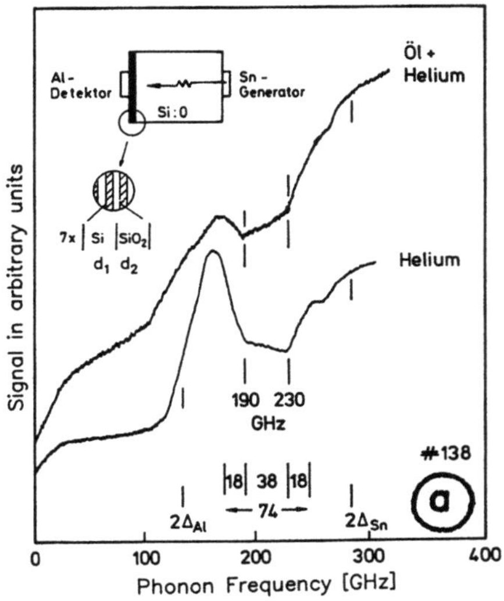

Fig. 1 Phonon spectrum of a Si:0 crystal, covered with a SiO_2/Si superlattice. The useful frequency is 130 to 280 GHz for Fig. 1a, and 280 to 900 GHz for Fig. 1b.

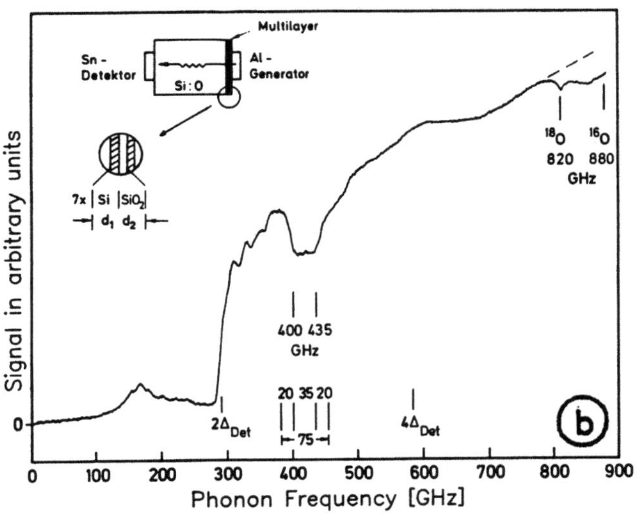

4 Experiments and results

Figure 1 shows a typical spectrum of a phonon transmission experiment with a multilayer structure of seven periods on one crystal surface. In Fig. 1a the phonons were generated in the Sn tunnel junction and detected with the Al junction on top of the multilayer. In Fig. 1b the Al junction was used as a generator, with the phonons first passing the multilayer and then propagating through the crystal to be detected in the Sn junction.

The spectra exhibit strong and broad absorptions with center frequencies around 210 and 420 GHz. A third, but much weaker structure can be seen between 600 and and 700 GHz. At 880 GHz, the well-known resonance of interstital oxygen in Si appears.

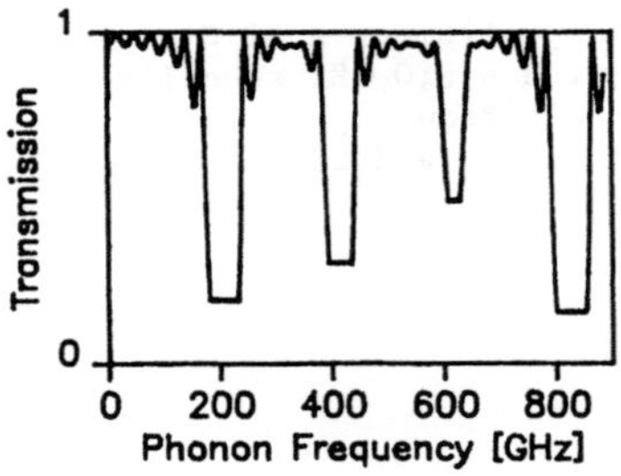

Fig. 2 Calculated phonon transmission for the superlattice of Fig. 1 with the following parameters:

Si: $d_1 = 7.0$ nm
$v_1 = 4700$ m/s
$\rho_1 = 2.2$ g/cm^3

SiO_2: $d_2 = 3.0$ nm
$v_2 = 3330$ m/s
$\rho_2 = 2.3$ g/cm^3

These results may be compared with the calculation for the corresponding superlattice shown in Fig. 2. Here we have taken the value of 3330 m/s for the sound velocity in a-SiO_2, as obtained from phonon spectroscopy [5] for single SiO_2 layers evaporated under the same conditions as ours. Since this value is below that of fused SiO_2 by 10%, we assumed the same reduction for a-Si as compared to the angular average of c-Si. Since it is the ratio of the density of the films which enters into the calculation, we have taken the bulk values for both materials, which means that any influence of the preparation technique is the same for both films. Because of the difficulties in measuring film thicknesses during evaporation, we used the thicknesses of the Si and SiO_2 films d_1 and d_2 as fitting parameters. The best fit with respect to the center frequencies of the stop bands was achievied by setting d_1= 7.0 nm and d_2= 3.0 nm, which is in reasonable agreement with the reading of the quartz crystal monitor. Comparing Fig. 1 and Fig. 2, we can associate the smaller and sharper structures between the main absorptions as side oscillations of the stop band characteristic. In the higher frequency regime, i.e. above 600 GHz, the characteristic is smeared out, which we believe is due to the effect of decreasing phonon wavelength because then our model of uniform, pinhole free layers with well-defined boundaries between the films, no longer holds. On the other hand, in the absorption line of oxygen, even the satellite of the ^{18}O isotope is resolved. This gives us strong evidence that the phonon propagation through the amorphous superlattice is not affected outside the stopbands.

For justification of the phonon stopband behaviour and the reliability of our fabrication process, Fig. 3 shows the spectrum of a second sample with reduced film thicknesses by a factor of 1.7. From the corresponding simulation, we expect an increase of the center frequencies by the same factor of 1.7,

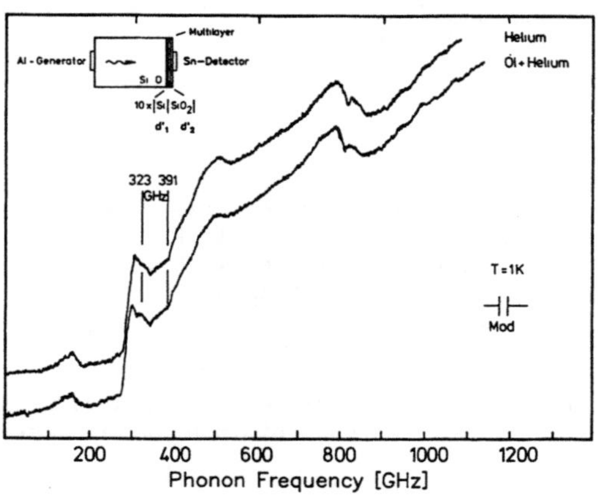

Fig. 3 Phonon spectrum of a superlattice with film thicknesses reduced by a factor of 1.7 compared to those in Fig. 1.

i.e. the first stop band should be shifted from 210 GHz to 360 GHz, which can be seen in Fig. 3. The second stopband appears near 770 GHz, just before the oxygen absorption line, but is broadened by the interference with a phonon resonance in the very thin (10 nm) films of the Al-generator junction.

5 Conclusion

We demonstrated that phonon filters using multilayers as superlattices can be realized very simply by evaporated amorphous SiO_2 and Si films. As few as seven periods resulted in excellent stopband characteristics with steep sides. They are better than MBE grown filters reported previously, which were built up of much larger stacks of crystalline films.

Acknowledgements

We wish to thank Anne Kaup for preparing our samples and Kurt Laßmann for many stimulating discussions.

This work was supported by the Stiftung Volkswagenwerk under AZ I 38/856.

References

1. L. Esaki and R. Tsu, IBM J. Res. Div. 14, 61 (1970)
2. V. Narayanamurti et.al., Phys Rev. Lett. 43, 2012 (1979)
3. S M. Rytov, Zh.Exsp.Teor.Fiz. 29, 605 (1955)[Sov. Phys. JETP 2, 466 (1956)]
4. M. Born and E. Wolf: Principles of Optics (Pergamon, New York, 1964)
5. M. Rothenfusser, W. Dietsche, and H. Kinder in "Phonon Scattering in Condensed Matter" (Springer Series in Solid-State Sciences 51, 1984), 419

Part IV

Thin Films, Surfaces and Thermalization

Phonon Scattering at Crystal Surfaces by *In Situ* Deposited Gas Films

T. Klitsner and R.O. Pohl

Laboratory of Atomic and Solid State Physics, Cornell University, Ithaca, NY 14853, USA

Phonon scattering from surface or interface defects has been a wide ranging problem of interest for many years. Losses at phonon generators and detectors[1], diffuse boundary scattering[2,3] and the anomalous Kapitza resistance[4] are all affected by this type of process, and have led to much speculation about the types of phonon interactions that may be occurring at a surface. Missing from this body of work has been a detailed study on the effects of sytematically introducing well characterized defects onto an otherwise "clean" surface. Here, we report on such a study and briefly discuss its implications for other experiments.

We use thermal conductivity measurements in the boundary scattering regime[2] to study phonon scattering at clean, highly polished Si surfaces. We have shown[5] that in the temperature range of these experiments (.05 - 2.0K) these surfaces are nearly completely specular to thermal phonons and are therefore good substrates for this type of study. Defects are sytematically added to this specular surface in the form of deposited thin films. To avoid any spurious effects due to exposure to air between depositions, thin films of Ne, H_2, or D_2 are deposited and removed in situ without ever warming the cryostat above a few K. A cylindrical, porous Vycor tube surrounds the experiment and acts as our gas reservoir, see Fig. 1. During the cool-down the gas we wish to deposit is adsorbed into the pores of the Vycor. At low temperature, the Vycor is slowly heated until gas desorbs from all sides and condenses onto our sample. Next to the sample is a torsional oscillator

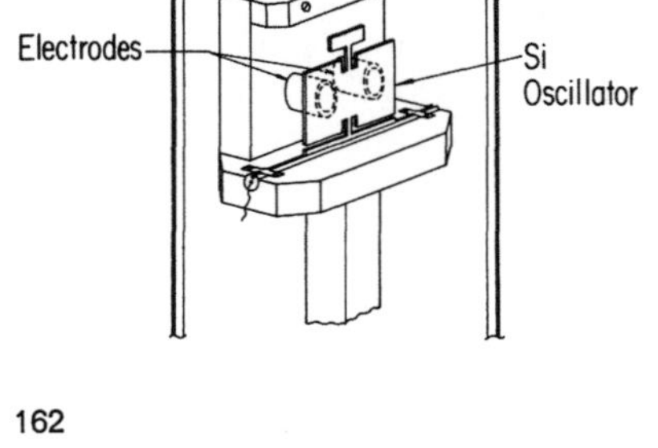

Fig. 1: Experimental set-up. Gas is adsorbed into the pores of the Vycor tube during the cool down. A deposition is made at low temperatures by heating the tube. The film thickness is monitored by the Si torsional oscillator. A two-heater method is used to measure the thermal conductivity.

etched from a polished Si wafer. By monitoring the change in resonant frequency of the oscillator during the deposition, we can measure film thicknesses as low as fractions of a monolayer. We present our data as the reciprocal of the phonon mean free path, $l^{-1}(T)$, which is proportional to the diffuse scattering rate at the surface. We determine l^{-1} from the thermal conductivity, $\Lambda=(1/3)C\cdot v\cdot l$, where C is the specific heat, and v is an average speed of sound.

Figures 2 and 3 show $l^{-1}(T)$ due to various thin films of Ne, H_2, or D_2, bounded by the limits of the clean, polished (specular) surface and a rough (diffuse) surface. Figure 2 has several different coverages of H_2, while Fig. 3 shows a family of coverages of Ne. In both cases, diffuse scattering increases with film thickness up to a maximum scattering rate, whereupon the deposition of thicker films does not increase the scattering rate any further. These "thick" layer limits are different for each condensed gas, see Fig. 2. From these data one can calculate the fraction of specularly reflected phonons[5] at this thick layer limit, and we find that it is equal to the fraction one would calculate on the basis of simple acoustic mismatch (AM) between Si and these films (see arrows in Fig. 2). Clearly, all the diffuse scattering must be occurring in the thin films themselves, while scattering at the interface is specular. In fact, the following observations show that this is the case for all the films we have studied: 1) The H_2 films must be relatively thick (150Å) before any diffuse scattering is observed, and then diffuse scattering increases with film thickness; both observations indicate that the scattering is associated with the bulk film and not with states at the

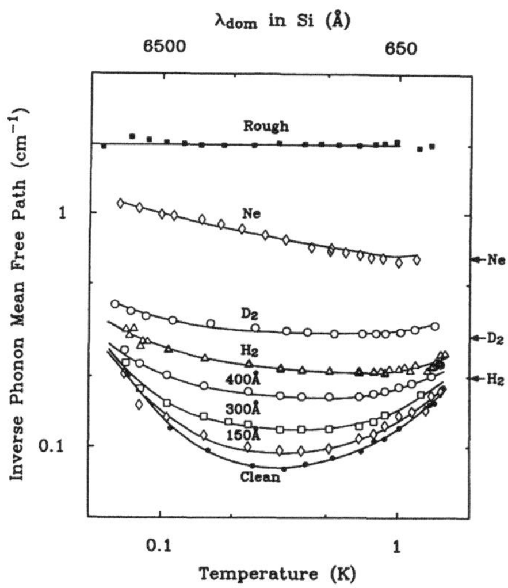

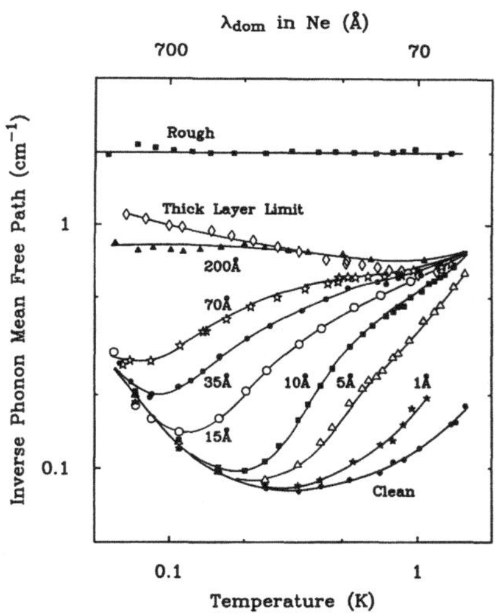

Fig. 2(left): Thick layer scattering limits for thin films of Ne, H_2, and D_2, and coverage dependence of H_2. The arrows on the right show the calculated acoustic mismatch between Si and Ne, D_2, and H_2 for our sample geometry. The rise in the data at the lowest temperatures is not an actual increase in diffuse scattering. It is due to a temperature-dependent boundary resistance at the sample base.

Fig. 3(right): Diffuse scattering rates for different coverages of Ne on Si. The sharp threshold behavior is due to the island structure of the Ne.

interface itself. 2) Both the H_2 and Ne films make a smooth transition from specular scattering to this AM limit as film thickness is increased. Also, each of the thin Ne films has a transition from the specular limit to the AM limit as temperature is increased, see Fig. 3. Such monotonic transitions between the specular and diffuse limits are indicative of the same scattering processes occurring for both thin and thick films.

The scattering profiles for Ne films are characteristically different from those of H_2. The Ne data are characterized by a sharp threshold temperature, below which l^{-1} is equal to the clean, specular limit, and above which diffuse scattering occurs. The H_2 data have a smoother profile, with no sharp threshold temperatures. If we consider the role of film structure, this difference can be explained in terms of our picture that the phonons must enter the thin film in order to scatter diffusely. Films with a discontinuous, island-like structure will have a long wavelength cutoff as temperature is decreased, when the dominant phonon wavelength in the film exceeds the island size. Continuous films, on the other hand, will have no such sharp cutoff, as phonons traveling parallel to the interface will always be able to "fit into" the film. This interpretation is supported by data on films deposited outside the cryostat in an evaporator. Au films also have sharp threshold temperatures[2] similar to the Ne data. Electron micrographs show that these films are indeed discontinuous with island size comparable to the dominant phonon wavelength in Au at the threshold temperature. By contrast, Si films, which appear smooth, do not show sharp threshold temperatures. So, even for small particles (islands) on a surface, there do not appear to be any special (e.g. resonant) surface states. The source of diffuse scattering appears to be the same for both continuous and discontinuous films, i.e. scattering occurs in the film and not by unknown interface states.

The Ne films illustrate an interesting point. A very small amount of Ne (a few Å average thickness) can diffusely scatter high-frequency (>1K) phonons, because the film coalesces into islands that are tens of Å in size (i.e., comparable to the phonon wavelength in the film at this temperature). If the residual "dirt" present on all real surfaces also forms islands, it too will couple strongly to high-frequency phonons in accordance with the scattering process we have presented here. We suggest this type of coupling may be responsible for much of the strong phonon interactions seen at surfaces at high phonon frequencies. This work was supported by the Materials Science Center at Cornell.

1. H.J. Trumpp and W. Eisenmenger, Z. Phys. B 28, 159 (1977).
2. R.O. Pohl and B. Stritzker, Phys. Rev. B 25, 3608 (1982); Tom Klitsner and R.O. Pohl, in Phonon Scattering in Condensed Matter, edited by W. Eisenmenger, K. Lassmann, and S. Dottinger (Springer,Berlin,1984), P.188.
3. D. Marx and W. Eisenmenger, Z. Phys. B 48, 277 (1982).
4. J. Weber, W. Sandmann, W. Dietsche, and H. Kinder, Phys. Rev. Lett. 40, 1469 (1978); E.S. Sabisky and C.H. Anderson, Sol. St. Comm. 17, 1095 (1975).
5. J.E. VanCleve, Tom Klitsner, and R.O. Pohl, in these proceedings.

Investigation of the Non-Equilibrium Distribution Function of Acoustic Phonons and Thermal Conductivity of Very Thin Ge Layers

E. Vass

Institut für Experimentalphysik, Universität Innsbruck,
Technikerstr. 15, A-6020 Innsbruck, Austria

1. Introduction

Heat conduction by phonons in bulk semiconductors [1,2] has been investigated extensively in the past. These investigations imply that the lattice thermal conductivity of a semiconductor is mainly limited by the interaction of the phonons with crystal boundaries, lattice defects, electrons and phonons (N- and U- three phonon processes [3]).

In the last few years it has become possible to realize novel power dissipating semiconductor devices [4] of very small thickness comparable to the de-Broglie wave length λ of the charge carriers which therefore form a quasi two-dimensional (2D) electron gas. In order to optimize the operation of these devices in electronic circuits it is important to know the actual value of the lattice thermal conductivity which can be determined only if the phonon distribution function (DF) is known.

In this work the DF of acoustic phonons in a thin Ge-layer of the thickness $d \approx 30$Å is calculated for the first time and used to determine its thermal conductivity $\varkappa$ in the temperature range $1K \leq T \leq 500K$. The presented analysis is based on the following assumptions:

a) The heat flux in the layer is mainly carried by phonons with the wave vector $\vec{q}$ parallel to the layer.
b) The phonon DF depends on the position vector of the phonon wave packets via the temperature T.
c) The DF of the electron system is given by the Fermi-Dirac function and is not disturbed by the interaction with phonons.

2. Theory and Results

If a temperature gradient $\vec{\nabla}T$ exists along the semiconducting layer the heat current density

$$\vec{w} = \sum_j \int \frac{d^2q}{(2\pi)^2} n_{qj} \hbar\omega_{qj} \vec{v}_{qj} \tag{1}$$

where $\hbar\omega_{qj}$ and $\vec{v}_{qj}$ denote the energy and group velocity of the phonon mode $(\vec{q},j)$ with j being the branch index. The stationary non-equilibrium DF n_{qj} is obtained from the phonon Boltzmann-equation [5]

$$-\vec{v}_{qj}\cdot\vec{\nabla}T \frac{\partial n_{qj}}{\partial T} = \sum_G \left[\frac{\partial n_{qj}}{\partial t}\right]_G + \sum_s \frac{n_{qj}-n^o_{qj}}{\tau_s} \tag{2}$$

where the phonon-phonon collision term is given by

$$\left[\frac{\partial n_{qj}}{\partial t}\right]_{\vec{G}} = \sum_{j'j''} \frac{A^2}{(2\pi)^4} \iint d^2q' d^2q'' \left[(F_{001}-F_{110}) W^{q''}_{qq'} + \frac{1}{2}(F_{011}-F_{100}) W^{q'q''}_{q}\right] \tag{3}$$

with $F_{\alpha\beta\gamma}=(n_{qj}+\alpha)(n_{q'j'}+\beta)(n_{q"j"}+\gamma)$. Using Fermi's golden rule the transition probability rate is calculated to be

$$W_{qq'}^{q"} = \frac{h}{8\rho^3}\left(\frac{C_{pp}}{A}\right)^2 \frac{(qq'q")^3}{\omega_{qj}\omega_{q'h'}\omega_{q"j"}} \delta_{\vec{G},\vec{q}+\vec{q}'+\vec{q}"} \delta(\omega_{q"j"}-\omega_{q'j'}-\omega_{qj}) \tag{4}$$

where C_{pp}, ρ and A denote the phonon-phonon coupling constant, the mass density as well as the area of the layer. $\vec{G}$ is a vector of the reciprocal lattice where $G=0(2\pi/a)$ for N- (U-) processes. The second term on the right side of (2) is the total collision term due to the interaction of the phonon mode (q,j) with boundaries, isotopes as well as 2D free electrons. The relaxation times τ_s limited by these mechanisms have been evaluated in [6] and are not given here in detail. If the deviation of n_{qj} from its equilibrium value is small, (3) can be written as

$$\sum_G \left[\frac{\partial n_{qj}}{\partial t}\right]_{\vec{G}} = \frac{N^o_{qj}-n_{qj}}{\tau_N} + \frac{n^o_{qj}-n_{qj}}{\tau_U} \tag{5}$$

where τ_N, τ_U, n^o_{qj} and N^o_{qj} denote the phonon-phonon relaxation times [6] for N- and U- processes, the Planck-function and the equilibrium distribution function for N-processes. With (5) the Boltzmann-equation (2) reduces to an ordinary differential equation which can be solved exactly provided $\vec{v}_{qj}\cdot\vec{\nabla}T\neq 0$. The result is

$$n_{qj}(T) = \int_0^T dT" \, R_o(T") \exp\left[\int_T^{T"} dT' \, R_1(T')\right] \tag{6}$$

where the initial value $n_{qj}(0)=0$ was used. $R_o=[N^o_{qj}\,\tau_N^{-1}+n^o_{qj}(\tau_U^{-1}+\sum_s\tau_s^{-1})]\tilde{a}^{-1}$, $R_1=[\tau_N^{-1}+\tau_U^{-1}+\sum_s\tau_s^{-1}]\tilde{a}^{-1}$, $\tilde{a}=-\vec{v}_{qj}\cdot\vec{\nabla}T$. If the deviation $n_{qj}-n^o_{qj}$ is large ($\sim n^o_{qj}$) the relaxation time approximation (5) cannot be used. In this case (2) is rewritten with respect to T into an integral equation which was solved iteratively in this work. The result is shown in fig.1 .

The above presented curves imply that the absolute value as well as the slope of the deviation function Δn_{qj} sensitively depend on the limiting scattering mechanism in the layer. For boundary scattering this function lies below

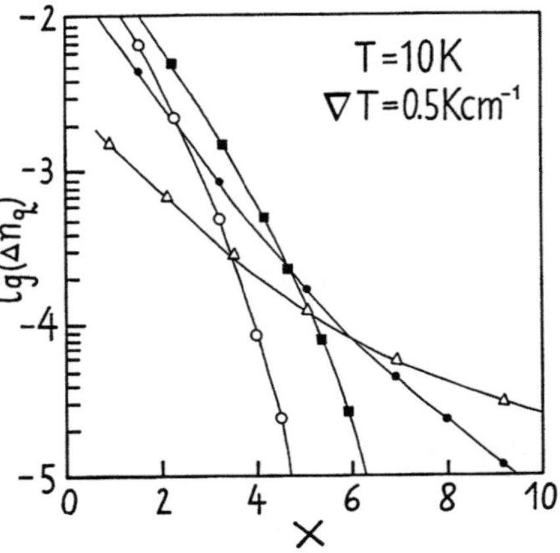

Fig. 1. Deviation $n_{qj}-n^o_{qj}$ vs the ratio $x=\hbar\omega_{qj}/k_BT$ for the interaction of LA-phonons with crystal boundaries (-△-), isotopes (-●-), free electrons (-o-) and phonons (-■-). The presented curves were calculated with $\rho=5.33\text{gcm}^{-3}$, a=5.4Å $m^*_o=0.3m_o$ (effective electron mass), T_D= 370K (Debye-temperature), v_l=5500m/s, v_t=3000m/s (sound velocities). The values of $C_{pp}=10^2$J/m, $\Delta M/M=10^{-4}$ (relative isotope mass difference), Ξ=9.5eV (acoustic deformation potential) and $N_s=1.0\times10^{11}$ cm^{-2} (free carrier density) were chosen arbitrarily. N-processes have been taken into account in all calculations.

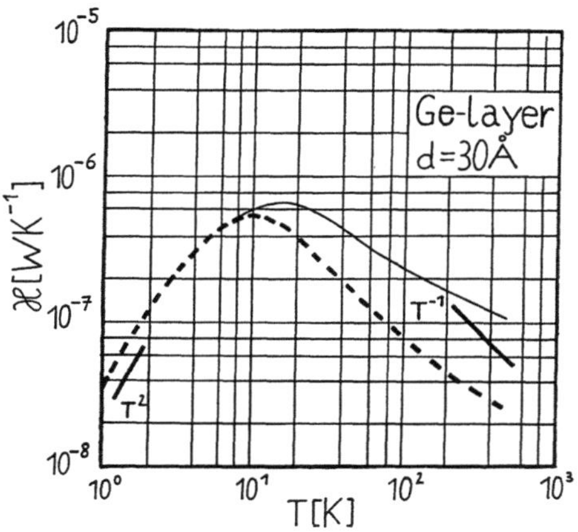

Fig. 2. Surface thermal conductivity of a thin Ge-layer (length L=0.1cm, width W=0.5cm). The dashed curve (---) was calculated with the exact non-equilibrium DF, the full curve (——) with the DF in the relaxation time approximation. If the temperature is very low $\varkappa$ is mainly limited by boundary scattering. When the combined scattering rate due to boundary-, isotope- and electron scattering is comparable to the phonon-phonon scattering rate, $\varkappa$ reaches a maximum.

the Δn_{qj}-curve limited by isotopes if $q \leq 5.1 \times 10^6 \mathrm{cm}^{-1}$. This means that large wave-vector phonons are scattered less effectively by boundaries than by isotopes. With increasing q-values the Δn_{qj}-curve limited by electrons converges stronger to zero than the corresponding curve limited by isotopes because the phonon-electron matrixelement is $\propto q^2$ while the phonon-isotope matrixelement turns out to be $\propto q$ for randomly distributed isotopes [6,7]. Fig.2 shows the theoretical thermal surface conductivity $\varkappa$ (defined by $\vec{w}=\varkappa\vec{\nabla}T$) of a thin Ge-layer vs the temperature T. In the calculation the same physical constants as specified above were used.

The thermal conductivity curve calculated with the collision term (5) considerably deviates from the $\varkappa$-curve calculated with the exact non-equilibrium DF provided T>100K. Therefore we can conclude that the relaxation time approximation (5) cannot be used if the heat current density is mainly limited by phonon-phonon U-processes. In addition it follows that in the boundary scattering range the surface thermal conductivity increases proportional to T^2 due to the surface specific heat. Experimental data which confirm our theoretical results are not available at present in literature.

Acknowledgement

The author thanks Prof. Dr. E. Gornik for critical reading of the manuscript.

References

1. J.A. Carruthers, T.H. Geballe, H.M. Rosenberg and J.M. Ziman, Proc. Roy. Soc. A238, 502 (1957)
2. M.G. Holland in Willardson and Beer Semiconductors and Semimetals Vol.2 p.3-31 (1966)
3. P.G. Klemens, Solid State Physics 7, 1-98 (1958)
4. S. Luryi and A. Kastalsky, Physica 134B+C, 453-465 (1985)
5. J.M. Ziman, Electrons and Phonons, Oxford Press 1979, p.293
6. E. Vass, to be published
7. P. Carruthers, Rev. Mod. Phys. 33, 92 (1961)

Mechanism of Nanosecond Heat Pulse Generation in Metallic Thin Films

M.N. Wybourne[1] *and J.K. Wigmore*[2]

[1]GEC Research Ltd., Long Range Research Laboratory, Wembley, Middlesex, UK

[2]Department of Physics, Lancaster University, Lancaster, UK

In order to understand more clearly the detailed mechanism by which phonons are generated in thin metallic heater films, we have carried out heat pulse experiments with heater excitation pulses only 1 ns long [1]. Previous workers have used pulse lengths typically of the order of 10-100 ns. In this long pulse regime, a steady-state power balance situation is reached where the power absorbed by the electrons in the heater is equal to that radiated into the substrate as phonons. By contrast, the results reported in the present paper correspond to the transient 'warm-up' period existing prior to establishment of the steady state. As will be seen, they differ significantly from observations made in the long pulse limit.

The heater films studied were of nichrome ($Ni_{50}Cr_{50}$), of two different thicknesses 27 and 58 nm, and of $Au_{60}Pd_{40}$, 10 nm thick, having resistivities 210 and 77 μΩcm respectively. They were all excited electrically at power densities between 10 and 2100 Wmm^{-2} from an Avtech pulse generator whose pulse length could be varied between 0.5 and 2 ns. The sample was a 5 mm long single crystal of high purity sapphire (Union Carbide) oriented along the $(22\bar{4}3)$ axis with heater and bolometer on opposite faces. The heaters and bolometers were all made to dimensions of typically 100 μm^2 so that geometrical broadening of the detected pulses was reduced to a fraction of a nanosecond. The surface on which the heaters and bolometers were deposited, after an initial high-grade optical polish, had been annealed at 1200°C under a stream of hydrogen. The effect of this process was to remove many of the outer atomic layers of sapphire including much of the sub-surface damage that scatters phonons strongly [2]. The bolometers used were superconducting PbBi films approximately 30 nm thick. The heater and bolometer were connected directly to the end of coaxial lines and after preamplification by a B&H 3 GHz bandwidth amplifier the bolometer signals were recorded and averaged using an EG&G boxcar integrator having a 350 ps resolution. Assuming that the bolometer integrated the incident signal, we were able to infer approximate pulse lengths from the rise times of the bolometer signals and also to make qualitative observations of changes in pulse shape.

Two separate observations indicated that a steady-state power balance situation did not exist in the heater films. Firstly, the detected heat pulses were considerably broader than the excitation pulse - at power balance the two would be equal. Secondly, for a given excitation energy, the detected pulse length was independent of the excitation pulse length. The broadening effect is illustrated in Figure 1 which shows the transverse mode emitted by the AuPd film excited for 2 ns has a 25 ns rise time and so the heat pulse for this mode is ~25 ns long. We note that the rise time of the corresponding longitudinal modes was approximately half that of the transverse modes. Figure 1 also shows clearly that the length of the emitted heat pulse was strongly power dependent. A decrease in power of

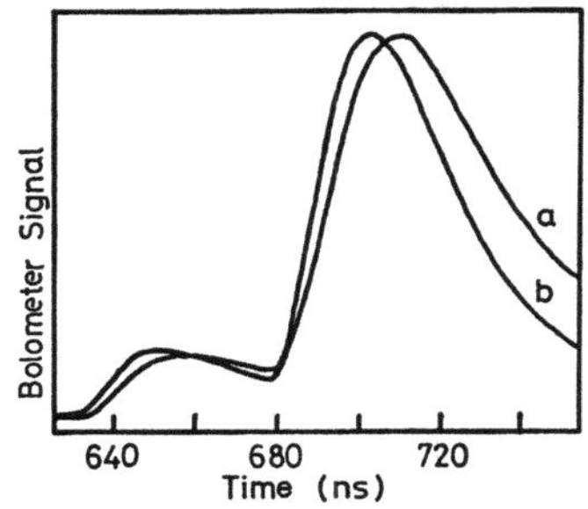

Fig 1 : Transverse mode signals for a AuPd film excited for 2 ns at power densities of (a) 2.07×10^3 Wmm^{-2} and (b) 1.3×10^2 Wmm^{-2}. The signals are normalised to the magnitude of the slow transverse mode.

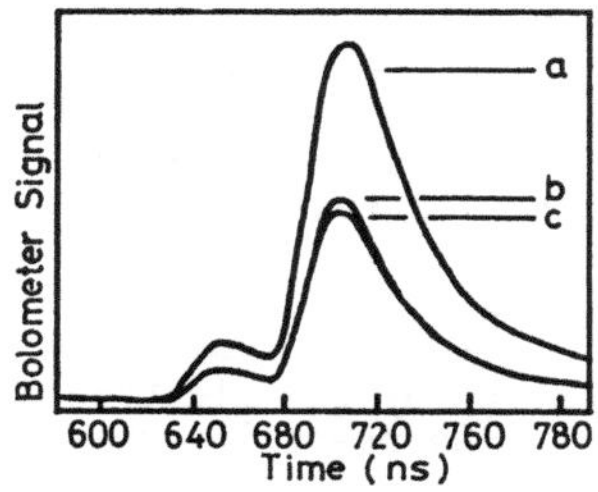

Fig 2 : Transverse mode signals for a AuPd film excited by electrical pulses (a) 2 ns 2.07×10^3 Wmm^{-2} (b) 1 ns, 2.07×10^3 Wmm^{-2} and (c) 2 ns, 1.05×10^3 Wmm^{-2}.

12 dB corresponded to a reduction in length of the slow transverse pulse by 30%. Similar effects were also seen in NiCr but the rise times for a given excitation power density were significantly different from those in AuPd suggesting that dispersion effects within the sapphire were not responsible for the pulse broadening. This observation also indicates that residual surface scattering is unlikely to be responsible for the pulse broadening. An indication of the surface quality is obtained by noting that the rise time of the bolometer signal after three transits across the specimen, that is, involving two reflections, was the same as the direct signal.

Additional evidence that a steady state had not been reached was the fact that the bolometer signals correlated with the energy of the heater excitation pulse and not with the power level, Figure 2. It can be seen that the magnitudes of the transverse modes resulting from 2.07×10^3 Wmm^{-2} 1 ns excitation pulse are almost identical to those produced by a 1.05×10^3 Wmm^{-2} 2 ns excitation pulse and a factor of two smaller than those due to a 2 ns pulse at the higher power density. This effect is not due to integration by the bolometer, indeed all of the effects are observed for many different bolometer bias conditions even when the superconducting transition is broadened by self heating. We draw the inference that even after 2 ns of an excitation pulse, the metal films are still heating up.

The cooling rate of thin films has previously been considered to be determined by the ballistic time-of-flight of phonons across the film combined with an emissivity parameter calculated using the acoustic mismatch model [3]. With an emissivity ~ 0.5 we estimate a characteristic cooling time for our films $\sim 5 \times 10^{-11}$ s and a maximum temperature ~ 30K. The cooling time and consequently the maximum temperature can be considerably increased by diffusive scattering in the substrate [3,4]. To explain the order of magnitude increase in pulse length we observe, the scattering rate in the substrate would need to be $\sim 6.5 \times 10^{10}$ s^{-1} which is much greater than the scattering rate of 29 cm^{-1} phonons determined in less pure sapphire [5]. The diffuse scattering of phonons within the substrate will be highly frequency dependent and so we would expect to see a decrease in the tail of the bolometer signals as the excitation power density was reduced; this was only weakly observed. The decay of the bolometer signal at the highest excitation power levels could be fitted to an exponential decay of time constant ~ 45 ns whilst at the lowest excitation levels the time constant was ~ 35 ns. This value is consistent with the bolometer decay time measured by direct electromagnetic excitation. Our conclusion is that diffuse scattering of phonons within the sapphire is not responsible for the observed effects and we believe the pulse sharpening is due to the details of the electron and phonon dynamics within the heater film itself.

The fact that the length of the emitted heat pulses did not depend on the length of the excitation pulses suggested that absorption of the excitation energy by the electron system was taking place faster than any other stage of the overall process and that the time evolution of the heat pulses had to be determined by the details of the de-excitation, or'cooling' of the films. It is not clear which stage in the chain between the emission of phonons by the electrons followed by their transport to the substrate interface is responsible for the relatively slow release of phonons by the heater. For such thin films, one might expect the Perrin and Budd model to be a good approximation [6]. According to this, the excitation energy goes first to heat up the electron system, which subsequently emits phonons in a characteristic time τ_{ep}. On the assumption that the phonons travelled ballistically to the interface, Perrin and Budd calculated for our power levels heater equilibration times of much less than 1 ns, a result which is clearly at odds with our observations. The weakest point of this model, we believe, is the assumption of ballistic propagation. The fact that we observed longitudinal and transverse mode pulses of different lengths suggested that phonons travelled diffusively within the heater. Finally, we note that the electron-phonon coupling used by Perrin and Budd was for the limit of short phonon wavelength compared with the mean free path of the electrons. For our disordered films the opposite limit, in which the electron-phonon interaction is weaker, may be more appropriate. Detailed calculations are in hand.

References

1 M.N. Wybourne, C.G. Eddison and J.K. Wigmore, Sol. Stat. Comm. 56, 755 (1985)

2 C.G. Eddison and M.N. Wybourne, J. Phys. C: Sol. State Phys. 18, 5225 (1985)

3 D.V. Kazakovtsev and Y.B. Levinson, J. Low Temp. Phys. 45, 49 (1981)

4 W.L. Schaich, J. Phys. C: Sol. Stat Phys. 11, 4341 (1978)

5 M. Engelhardt, U. Happek and K.F. Renk, Phys. Rev. Lett. 50, 116 (1983)

6 N. Perrin and H. Budd, J. Physique 33, C4-33 (1972)

Phonon Spectroscopy of Adsorbed H_2O Molecules

L. Koester, S. Wurdack, W. Dietsche, and H. Kinder

Physik Department E 10, Technische Universität München,
D-8046 Garching, Fed. Rep. of Germany

It is now well established from cleaving [1] and laser annealing [2,3] experiments that the anomalous Kapitza transmission is caused by surface defects. Furthermore, apparently all experimentally observed phenomena can be described with a phenomonological defect model [4]. This model requires the presence of two-level states at the surface with splittings in the range of typical phonon energies, i.e. about 1 meV.

Recently, we showed that submonolayers of Au cause the reentry of the anomalous Kapitza transmission [3]. The data could be well described by the defect model assuming an energy-independent density of two-level states. The goal of this work was to extend that work to a completely different system, namely adsorbed H_2O molecules. These molecules are particularly interesting because they are present at all "real" surfaces.

We used laser annealing to prepare perfect Si surfaces. The experimental set-up is shown in Fig. 1. The 3 mm thick Si (100) crystal was part of a vacuum chamber. The inner side of the Si could be laser annealed through the window by ruby-laser pulses scanned over the surface. The phonon generator ($Pb_{0.85}Bi_{0.15}$-tunnel junction) and the detector (Al-junction) were placed onto the outer side. They were positioned along the [010] direction and were either 1 mm or 6 mm apart from each other. In the first case, the observed phonon pulses were incident on the test surface near the normal direction. With a large distance between the junctions more obliquely reflected phonons were observed. The latter had the advantage that the specularly and the diffusively scattered phonons could be separated more easily [5].

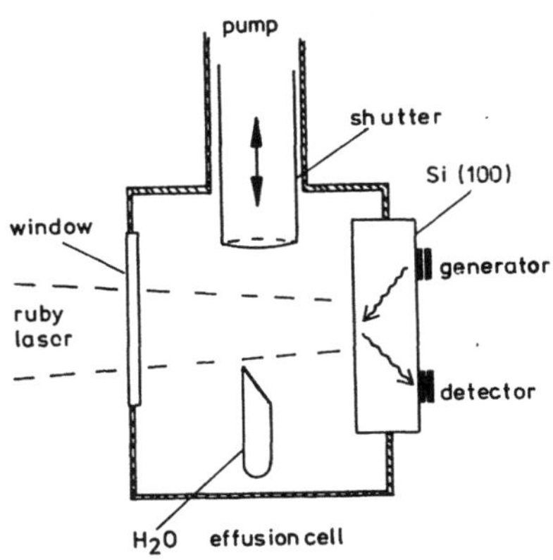

Fig. 1. Experimental set-up. At the bottom of the vacuum chamber there was a stainless steel tube, 10 mm in length and 1 mm in diameter, serving as effusion cell for H_2O molecules. At an ambient temperature of 1 K, it was heated to temperatures around 200 K while the shutter tube was closed. With a quartz balance (not shown) it was checked if a constant evaporation rate was reached. After opening the shutter, the molecules could be deposited onto the test surface with coverages as low as a 1/100 of a monolayer. The evaporation rate was calibrated by evaporating thick H_2O layers and measuring their thicknesses interferometrically

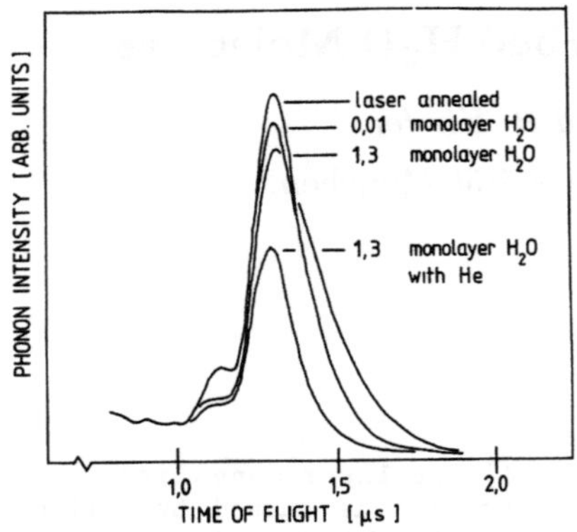

Fig. 2. Phonon echoes observed under near-normal incidence

In Fig. 2, reflected phonon pulses are shown for near-normal incidence. After the annealing, there was no signal change after the chamber was filled with He. Thus there was no anomalous Kapitza transmission at this surface. The deposition of only 0.01 molecular layers of H_2O, however, led to a signal decrease. With 1.3 monolayers a diffusive tail showed up also. Furthermore, an anomalous Kapitza transmission into the He was induced. Thus, adsorbed H_2O molecules show similar effects as Au adatoms [3]. A quantitative analysis yielded a deformation potential of about 1 eV if a constant density of states was assumed. With Au the potential was 0.4 eV under the same assumption.

Results for the oblique incidence are shown in Fig. 3. The echo pulse measured with the as-received sample was relatively small and had a long diffusive tail. After annealing an 10 x 10 mm^2 area the specularly reflected part of the pulse almost doubled while the diffusive tail diminished. At the same time, the anomalous transmission into the He was reduced to less than 10 % (not shown in Fig. 3). After laser annealing, H_2O molecules were again deposited. In Fig. 3, the effect of 1 monolayer is shown.

The dependence of the phonon-echo intensities on frequency is shown in Fig. 4. The predominant feature of the "as-received" trace is the broad minimum around 280 GHz. Such a minimum was always observed under the oblique angle but not for near-normal incidence. We suspect that this minimum is of similar origin as the onset behavior reported bei Koblinger et al. [6]. The frequency scales are different, but this may be caused by the different substrates, i.e. Si vs. Sn.

After laser annealing, the minimum disappeared. This proves that it must have been a surface effect. The deposition of H_2O molecules did not cause any sharp structures. This came as a surprise because H_2O molecules should be the prevailing impurity on a "real" surfce and, thus, they were the prime suspect for causing the minimum at the as-received surface. The featureless trace in Fig. 4 is instead compatible with a constant density of states as it was found with Au [3]. On the other hand, the minimum at the as received trace demonstrates that sharp and more interesting structures exist. We will continue our efforts and expect that phonon spectroscopy of surface defects will soon be a reality.

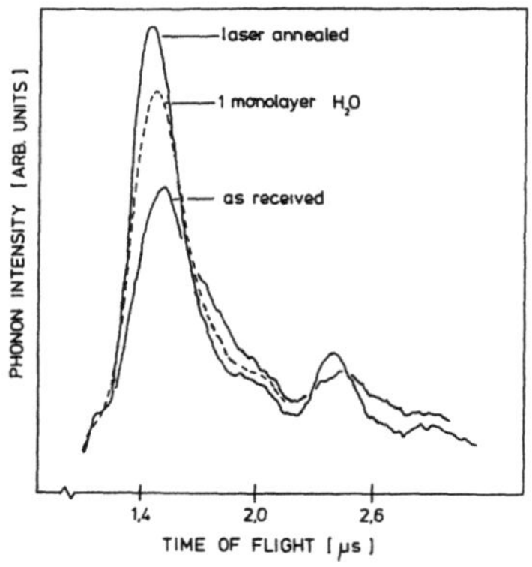

Fig. 3. Phonon echoes observed under an oblique angle

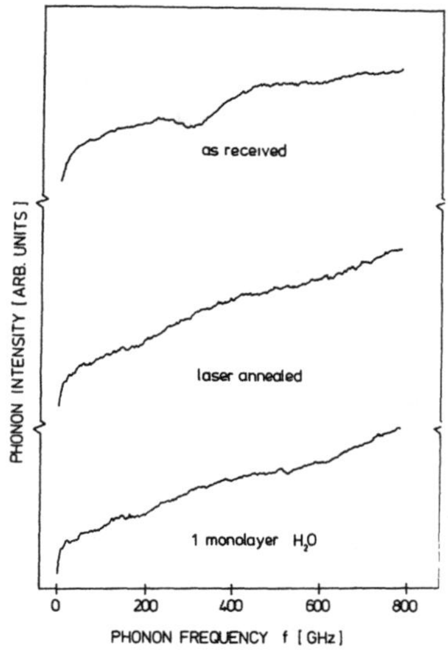

Fig. 4. Frequency dependence of the pulses of Fig.3

1. J. Weber, W.Sandmann, W. Dietsche, and H. Kinder: Phys. Rev. Lett. 40, 1469 (1978)
2. H.C. Basso, W. Dietsche, H. Kinder, and P. Leiderer: in Phonon Scattering in Condensed-Matter, ed. by W. Eisenmenger et al. (Springer, 1984) pg. 212
3. H.C. Basso, W. Dietsche, and H. Kinder: submitted to J. Low Temp. Phys.
4. H. Kinder, Physica 107B 549 (1981); H. Kinder and K.Weiss: contribution G 3 this volume
5. S. Burger, K. Lassmann, and W.Eisenmenger: J. Low Temp. Phys. 61, 401 (1985)
6. O. Koblinger, U. Heim, M. Welte, and W. Eisenmenger: Phys.Rev. Lett.51, 284 (1983)

Thermalization of Ballistic Phonon Pulses in Dielectric Crystals Below 1K Using Time-Resolved Thermometry

W. Knaak[1,2], *T. Hauß*[1], *M. Kummrow*[1], *and M. Meißner*[1,2]

[1]Institut für Festkörperphysik, TU Berlin, D-1000 Berlin 12, Germany
[2]Hahn-Meitner-Institut für Kernforschung, D-1000 Berlin 39, Germany

Time-resolved measurements of the specific heat can characterize low-energy excitations with respect to their coupling to the phonons, as has been shown e.g. for vitreous SiO_2 [1] and KBr:KCN [2]. As these experiments have achieved a μsec-time resolution, the questions arise, at what times do the phonons reach a new equilibrium after the application of a heat pulse and what are the processes involved. These questions have now been further studied on two dielectric crystals (Si, NaF) which we used earlier [3] to test the absolute determination of the specific heat by this method. It turned out that in pure crystals below 1K, thermalization by surface scattering is the dominant process, so that this work is also related to questions concerning the anomalous Kapitza-resistance, like the problem of specular and diffusive phonon reflection at a free surface [4,5].

The method we used for short-time-calorimetry is a quasiadiabatic heat pulse experiment, i.e. the sample is connected to a dilution cryostat via a weak link, and a heat pulse is applied by a thin Au-film-heater. The response of the sample is measured with a thin carbonfilm bolometer, made by rubbing pencil material ($m \approx 20\mu g$) onto a roughened spot of the crystal surface. The heat capacity of this carbon material is about $c \approx 10^{-4} \cdot T[K]^{1/2}$ J/gK. The time-resolution of the bolometer was measured with short hf- and/ or light pulses and was found to be $\tau \approx 0.1\mu sK^2 \cdot T^{-2}$ when referred to the actual bolometer temperature. Earlier experiments [3] have demonstrated, that this set-up gives correct specific heat values for t>1ms and down to 50mK. However, at shorter times several transient effects occur in the temperature-vs.-time profiles (Fig.1): for T>0.1K, where the thermometer is

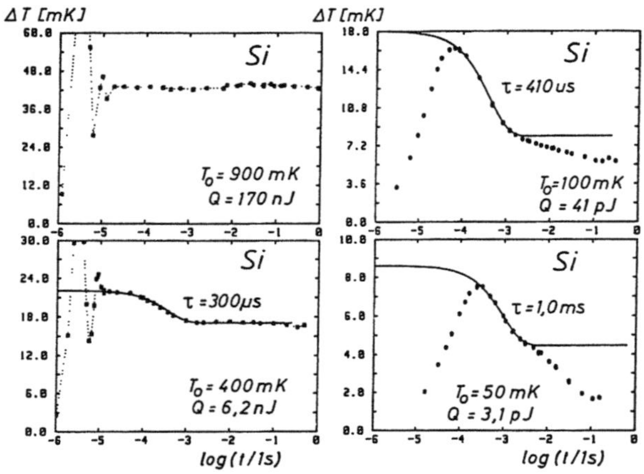

Fig. 1. Temperature-vs.-time profiles for an adiabatic heat pulse experiment on pure Si (16g, dim. $10 \times 28 \times 25 mm^3$, zone-refined and VPE-purified, $B \approx 5 \cdot 10^{12} cm^{-3}$; $O,C \approx 10^{15} cm^{-3}$). The solid lines are exponential fits showing the time constants of the thermalization of the injected phonons. The decrease of the signal for T<0.1K and t>2ms is due to the added heat capacities.

fast enough, the ballistic phonon signal is clearly resolved after some microsec. But, except for 900mK, the "temperature rise" at about 20μs is too large to be explained by the Debye elastic specific heat of that crystal. The expected value is reached only after a subsequent, exponential decay of the signal to its long-time value. The data points shown have been corrected for the heat flow out of the sample due to the non-adiabatic mounting.

This overshoot, which is only visible below 0.5K in this experiment and which rapidly grows below 0.2K, reflects the frequency-down conversion of the injected phonons, whose frequencies correspond to a much higher temperature than the base temperature. The fact, that this process can be observed with the bolometer, is due to its frequency sensitivity. In situations where the incoming phonon distribution is nonthermal, the bolometer-signal should depend on the energy flux *and* the phonon absorption coefficient for the frequencies involved. However, the absorption coefficient drops substantially for frequencies below the fundamental mode across the bolometer thickness. The value of this threshold will be between 6 and 12 GHz with an effective thermometer-thickness of 0.1μm and a Debye-velocity between 1.2 and 2.4 km/s of pure graphite [6]. If the dominant frequencies of the injected phonons cross this threshold during thermalization the bolometer-signal will decrease as is shown in Fig.1.

Next, we focus the question whether volume or surface inelastic scattering is the dominant process below 1K. Here, the only inelastic volume process present in perfect crystals is the anharmonic decay which is proportional to ω^a, a=4...5. Therefore, lifetimes of 10GHz-phonons against anharmonic decay are larger than sec. for Si [7] and for NaF as estimated from second-sound data [8]. Even resonant scatterers introduced into the sample volume like OH^- in NaF do not contribute to the thermalization of phonons. Fig. 2 shows the $\Delta T(t)$-profiles for two NaF crystals, mainly different in the OH^--content by a factor of 130. Doping with 0.8ppm OH^- leads in thermal conductivity to a mean free path of ≈0.3 mm at 0.2K [8], but the mean free path against inelastic scattering at these centers is much larger than $50\mu s \cdot v_D$=18cm, because the small increase of the thermalization can be explained by the somewhat smaller dimensions of the NaF:OH-sample. Therefore, these resonant scattering centers are very ineffective in thermalization, the probability for an inelastic event being less than 10^{-4}. Experiments are prepared to put further limits to the volume inelastic scattering in Si by changing the volume-to-surface ratio but not the surface quality.

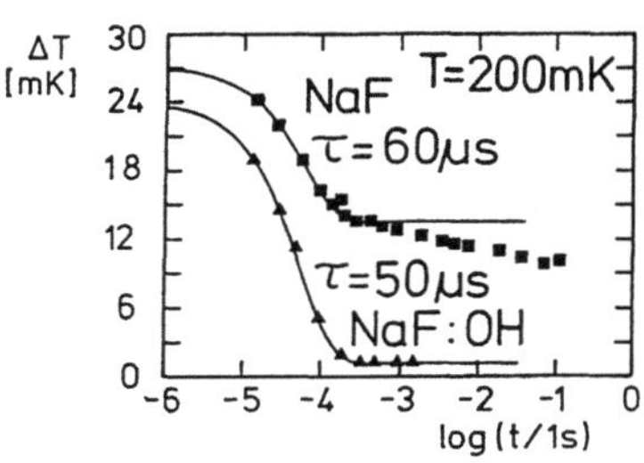

Fig. 2. Thermalization of heater phonons in two different NaF crystals; **pure:** $[OH^-] < 5$ppb, dimensions = 12x12x12mm³, surface scattering rate $\tau_S^{-1}=2.2\cdot10^5 s^{-1}$;
NaF:OH⁻: $[OH^-]$=0.8ppm, dimensions = 6.8x12x11mm³, $\tau_S^{-1}=2.8\cdot10^5 s^{-1}$.

Thermal conductivity measurements have shown that, except at very perfect surfaces, scattering is mainly diffusive [4], but, as can be seen in Fig.1, the probability for inelastic scattering at the surface of this sample is only 1%, concluded from a comparison of the thermalization rate 1/(300μs) to the total surface scattering rate of 1/(3μs). This sample was diamond polished and if thermalization is a surface process, the rates should depend on surface treatment. The largest effect in increasing the thermalization

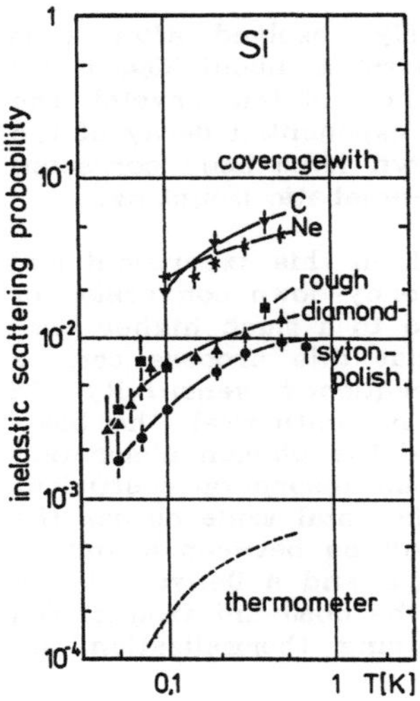

Fig. 3. Inelastic scattering probability at different Si-crystal surfaces as calculated from the ratio of thermalization rate to total surface scattering rate. Dimensions of syton polished and with Ne-covered crystal $5\times5\times50mm^3$, $\tau_s^{-1}=8\cdot10^5 s^{-1}$; other $10\times28\times25mm^3$, $\tau_s^{-1}=3.3\cdot10^5 s^{-1}$. The thermalization effect at the thermometer is estimated from the C-covered sample by scaling down the areas.

rates was achieved by covering the surfaces with films of neon ($d<0.1\mu m$) or carbon. The neon film was simply frozen onto the cold Si-crystal (syton polished) while the carbon film was made by also rubbing pencil material onto the whole, rough surface of the crystal. This addendum clearly contributed to the specific heat, which however coupled only at relatively long times ($t>10ms$) to the sample capacity, i.e. the actual thermometric capacity is much lower than the value given above.

The measured thermalization rates of the different surfaces are shown in Fig. 3, where the rates are normalized to the total surface scattering rates τ_s^{-1}. The results show, that there is no large difference between the diamond polished (or rough) surface, which is supposed to contain a lot of subcutaneus dislocations, and the syton polished surface on the other hand. The thermalization at the small area of the thermometer is estimated to be one order of magnitude lower, so that the syton polished surface is still not ideal because of inelastic scattering being introduced presumably by adsorbates from the air. Remarkably the absolute values of the inelastic scattering probability of the covered surfaces are larger than those of the totally diffusively scattering rough surface, but are still below 10%, so that in the Kapitza problem, where a phonon hits the surface once, inelastic processes should not represent a major part.

Summarizing, we have shown, that the phonon absorption of thin film thermometers has a frequency dependence for low frequencies which resembles that of tunneling junctions and which makes it possible to observe directly the frequency-down conversion, i.e. thermalization of 3-30GHz-phonons. This thermalization occurs at the surfaces after about 100 reflections even for rough surfaces. Resonant scatterers like OH^- in NaF do not contribute to thermalization so that even in doped crystals or amorphous systems thermalization might also extend to msec-time scales below 1K.

References

1. M. Meißner, K. Spitzmann: Phys. Rev. Lett. **46**, 265 (1981)
2. M. Meißner, W. Knaak, J.P. Sethna, K.S. Chow, J.J. DeYoreo, R.O. Pohl: Phys. Rev. **B32**, 6091 (1985)
3. W. Knaak, M. Meißner in Proceedings of the 17th Int. Conf. on Low Temp. Phys., U. Eckern et al., ed. (North Holland, Amsterdam 1984)
4. T. Klitsner, R.O. Pohl: these proceedings
5. D. Marx, W. Eisenmenger: Z. Phys. **B48**, 277 (1982)
6. B.J.C. van der Hoeven, jr. P.H. Keesom: Phys. Rev. **130**, 1318 (1963)
7. estimated from data of W. Burger, K. Lassman: these proceedings
8. T.F. McNelly: Ph.D. thesis, Cornell University (1974)

How Specular are Polished Crystal Surfaces?

J.E. VanCleve, T. Klitsner, and R.O. Pohl

Laboratory of Atomic and Solid State Physics, Cornell University, Ithaca, NY 14853, USA

1. Introduction

In pure crystals at low (<1K) temperatures, bulk phonon scattering vanishes, and all scattering takes place at the surface. The fraction of phonons which are diffusely scattered, f, may be substantially reduced by polishing and cleaning. Here, we report measurements of heat conduction in samples with highly polished, specular surfaces (see also Klitsner and Pohl in these Proceedings), and will describe how to obtain f from the measured temperature drop ΔT_e between two thermometers clamped to the crystal a distance d apart. A detailed treatment will be presented elsewhere[1]. We define the mean free path (mfp) $l \equiv 3\dot{Q}/CvA(dT/dx)$, where $\dot{Q}$ is the heat input, A is the sample cross-section, v the Debye speed of sound, C the Debye specific heat, x the distance from the top of the sample, and dT/dx is the temperature gradient. Figure 1 shows the data (with $dT/dx=\Delta T_e/d$). We see that l^{-1} depends on the sample length L, and in addition on the size of the clamps used to attach the thermometers to the sample. We will outline how to interpret this remarkable observation, and show that the three curves of Fig. 1 are consistent with nearly the same f once one accounts for the actual temperature profile along the surface of the sample and for the effect of the clamps on the measurement. Previously, CASIMIR[2] considered an infinitely long cylinder with f=1 and found $l=2R=l_c$, where R is the sample radius. Henceforth all lengths will be normalized so that $l_c=1$. ZIMAN (BSZ)[3] considered cylindrical samples of arbitrary f and finite L, assuming dT/dx to be constant. A numerical solution of the same problem by PERLMUTTER and SIEGEL(PS)[4] shows this to be a particularly poor approximation when the surface is highly specular. Because of the nature of radiative heat transfer, T(x) must jump discontinuously at the ends of the sample. Ziman's solution actually gives a $\dot{Q}$ which varies with x, and thus is not the steady-state solution. The effect of clamps has never been considered.

2. One-Dimensional Model

In order to illuminate the effect of the clamps and the sample ends on the temperature profile, we consider a cylinder with perfectly specular walls, with its base held at T and a power input of $\dot{Q}$ at the other end. A series of "grey" sheets perpendicular to the axis is used to replace the scattering by the walls with a 1D transfer model.

The ith sheet intercepts a fraction α_i of all radiation incident on it, reaching the temperature $(T + \Delta T_i)$, and thermally radiates $2\alpha_i \cdot 4\sigma T^3 \cdot \Delta T_i \cdot A$, where σ is the phonon equivalent of the Stefan-Boltzmann factor, $\sigma = (1/16)vC/T^3$. The top of the sample is "black" ($\alpha_1=1$). Two other α_i are equal to α_c, the fraction of phonons inter-

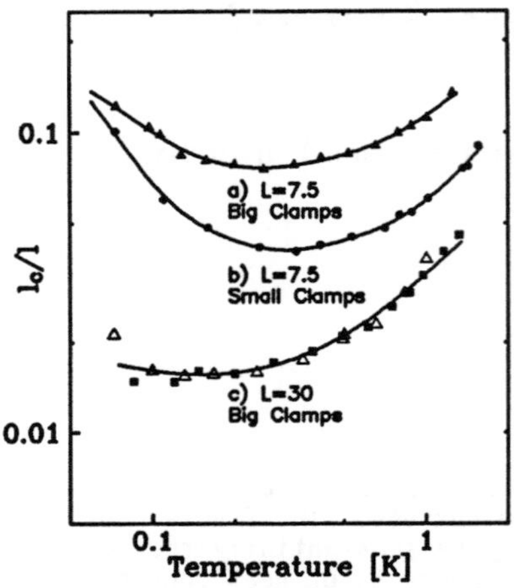

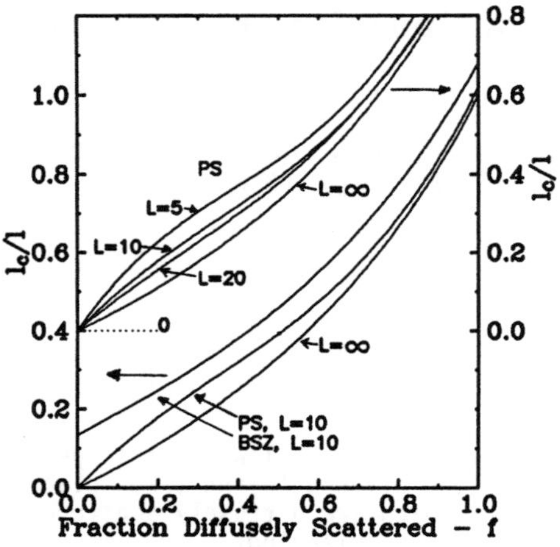

Fig. 1(left) Data on clean, specular surfaces for Si crystals of different length or clamp size (see text). The upturn of l_c/l at low temperatures is an experimental artifact, not an actual increase in diffuse scattering (see Klitsner and Pohl, these Proceedings).

Fig. 2(right) Theoretical curves for various sample lengths. Upper curves are using PS[4]; lower curves are comparison of BSZ[3] to PS for the same sample size.

cepted by a clamp, and the rest equal α_s, representing the scattering at the surface. Analysis of this system[1,5] shows that as the number of sheets N is increased so that $\alpha_s \to 0$ as $N \to \infty$ with $\alpha_s \cdot N$ constrained to be constant, the following temperature profile is obtained: The temperature drop per unit length is $(N \cdot \alpha_s/L) \cdot \dot{Q}/(8 \cdot \sigma \cdot T^3)$ except at the ends of the sample and the clamps, where there are discontinuous jumps in temperature of $(1/2) \cdot \dot{Q}/(4 \cdot \sigma \cdot T^3)$ and $(\alpha_c/(2-\alpha_c)) \cdot (\dot{Q}/(4 \cdot \sigma \cdot T^3))$ respectively. Since each clamp measures the average of the discontinuous temperatures on either side of it, the additional temperature drop between the clamps due to the area of the clamps themselves is $\Delta T_c = (\alpha_c/(2-\alpha_c)) \cdot (\dot{Q}/(4 \cdot \sigma \cdot T^3))$, and the total temperature drop across the clamps is $\Delta T_e = \Delta T_c + \Delta T_s$, where ΔT_s is the temperature drop due to diffuse scattering at the surface itself, which one would measure if the clamps were infinitesimally small. We define $l_e^{-1} = l_c^{-1} + l_s^{-1}$, with $l_c^{-1} = (4/3d) \cdot \alpha_c/(2-\alpha_c)$, l_e the experimentally measured l, and l_s the mfp in the case of infinitesimal clamps. In our experiments, we assume ΔT_c is as above and find f from $\Delta T_s/d$ as explained below. ΔT_c may be measured for a given clamp size by comparing two samples of the same specularity and length, one of which has very small clamps (giving ΔT_s). α_c may also be calculated from geometrical factors. The experimental α_c may be somewhat smaller than one calculates, since the entire nominal clamped area may not be an ideal thermalizing surface.

3. Interpretation of ΔT_s

Temperature profiles for various lengths and specularities are found in PS[4]. As in our 1D model, there's a temperature jump of roughly $(1/2) \cdot \dot{Q}/(4 \cdot \sigma \cdot T^3)$ at each end, unlike the BSZ model, and a nearly linear region around the center of the sample. The linear part of the PS temperature profile for a given f and L is $(dT/dx)_0$. By plotting $(dT/dx)_0 \cdot (4/3) \cdot (4 \cdot \sigma \cdot T^3 \cdot A)/\dot{Q} = l^{-1}$ vs. f for various L, we can interpolate our measured $dT/dx = \Delta T_s/d$ to find f, given L, see Fig. 2. The lower set of curves shows the disagreement between BSZ and PS for L=10; in the limit f=0, BSZ would predict $l^{-1}=4/3L$ while PS predict $l^{-1}=0$. Hence BSZ predict a minimum temperature gradient, an artifact of their assumption that the temperature gradient is constant, while PS agrees with the elementary consideration that $dT/dx \rightarrow 0$ as $f \rightarrow 0$. For a given $\Delta T_s/d$, BSZ yields a substantially lower f than the PS, primarily because it does not account for the temperature jumps at the ends of the sample. The upper set of curves shows the PS results for L=5,10, and 20. Given fixed f, l^{-1} decreases as L increases, reaching a limit of $f/(2-f)$; conversely, given l^{-1}, the value of f calculated increases substantially with L.

To interpret Fig. 1, we take points at T=.3K, $l_e^{-1}=.075$, for sample a). The clamps have $\alpha_c=.25$; with d=6.0, we obtain $l_c^{-1}=.032$, and $l_s^{-1}=.043$. Curve b) has $l_e^{-1}=.045$, $\alpha_c=.05$, giving $l_c^{-1}=.006$, and $l_s^{-1}=.039$, close to what was found for the large clamps. Hence curves a) and b) can be brought into agreement merely by considering the clamp size in the way described. For $l_s^{-1}=.04$ and L=7.5, interpolation of the PS results gives f=.036. For sample c), $l_e^{-1}=.016$ and the same big clamps as before. Here d=22, yielding $l_c^{-1}=.009$ and $l_s^{-1}=.007$. Since L=30, we obtain from PS $.009<f<.013$. Similar values of f are obtained for other cleaned and polished samples of this kind.

Acknowledgements

One of us (J.V.C.) has been supported by an NSF graduate fellowship. We also thank Dr. Robert Buxbaum for interesting discussions. This work was supported by the NSF, Grant DMR-8417557, and the Materials Science Center at Cornell.

References

1. J. E. VanCleve, to be published.
2. H.B.G. Casimir: Physica 5, 495 (1938).
3. R. Berman, F. E. Simon, and J. M. Ziman: Proc. Roy. Soc. A220, 171 (1953); R. Berman, E. L. Foster, and J. M. Ziman: Proc. Roy. Soc. A231, 130 (1955).
4. M. Perlmutter and R. Siegel: J. Heat Transfer 85C, 55 (1963).
5. A thorough discussion of radiative heat transfer in general is given by M. N. Ozisik: Radiative Transfer (Wiley-Interscience, New York (1973).

Phonon Thermalization at Al_2O_3 Surfaces

L.J. Challis, S.V.J. Kenmuir, A.P. Heraud, and P.A. Russell

Department of Physics, University of Nottingham, University Park, Nottingham, NG7 2RD, UK

Studies of polished Al_2O_3 surfaces by techniques such as LEED show that vacuum annealing at temperatures above 1400°C and H_2 annealing at 1200°C, a chemical etching process, are both effective in producing well ordered surfaces [1]. Further evidence for this is that phonon reflection at H_2 annealed surfaces has recently been shown to be very largely specular [2]. In the present work we have investigated whether similar treatments reduce the probability of inelastic scattering at Al_2O_3 surfaces and studied the decay and growth of holes arising from it.

The technique used is to inject into a crystal a heat current containing a narrow 'hole' at a frequency ν_i. As the current propagates down the crystal the hole fills up because of inelastic scattering. The size of the hole can be seen using the frequency crossing technique. The work is an extension of earlier experiments [3]. The sample used is a cylindrical a-axis Al_2O_3 bicrystal 4mm in diameter, 40mm long, grown by Hrand Djevahirdjian S.A. (fig 1). One half (V) is vanadium doped and contains ~70ppm of V^{3+} and 0.3ppm Fe^{2+}; the other (Fe) is iron doped and contains ~2ppm of Fe^{2+} and < 1ppm of V^{3+}. A heat current is passed along the sample and the temperature gradients measured at different distances from the interface using pairs of thermometers T_n (near) and T_f (far) as shown. Heat can be injected at either H_{Fe} or H_V. When the heat is injected at H_{Fe}, the phonon current entering V contains 3 holes due to the 3 Fe^{2+} transitions. These are tuned by a magnetic field parallel to the c-axis and crossing signals C, D and H [3] are produced when each Fe^{2+} hole crosses, in turn, the V^{3+} hole due to the $|0\rangle \rightarrow |-1\rangle$ transition.

Two treatments were applied to the sample. It was annealed in argon for 30 hours at 1775°C then Syton polished and annealed in H_2 for 5 hours at 1200°C. The surfaces were not touched after this second treatment and contacts were made using a special jig [4]. The table shows examples of the signal sizes after each treatment. The signals are seen to be much less 12mm from the interface (H_{Fe}-T_f) than 6mm (H_{Fe} - T_n) and at 12mm the signal was unchanged, to within experimental error, when the heat

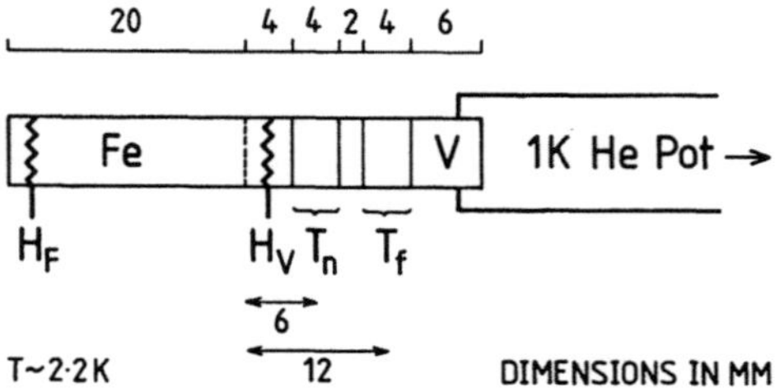

Figure 1
The bicrystal arrangement. The contact positions were varied for different experiments

injection was switched from H_{Fe} to H_V. This shows that the Fe^{2+} hole had completely decayed to a remanent value caused by traces of Fe^{2+} impurities in V. The thermal conductivity was increased ∿ 40 ± 20% by argon annealing (the absolute error is large because of the small contact separation) but was changed very little by polish plus H_2 anneal. So these treatments had no very significant effect on the speculär reflection nor on the probability of inelastic scattering. This may be associated with the fact that the surfaces were cylindrical and not flat as in earlier work [2, 5]. Attempts to obtain bicrystals with square sections have been unsuccessful, however. Both samples supplied were too highly doped to be useful.

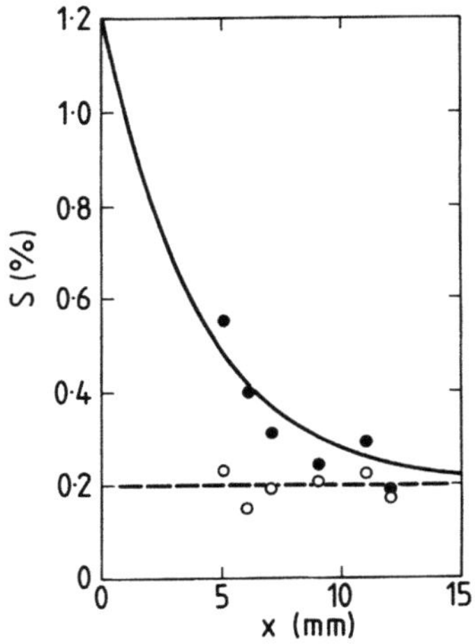

Fig 2. The decay in size ($S=\Delta W/W_o$) of line D (H_{Fe}) with distance from the interface (solid circles). The open circles show the sizes of the remanent signals obtained using H_V. The solid line is an exponential with ℓ_I = 4mm. (S=0.196+Aexp(-x/4)).

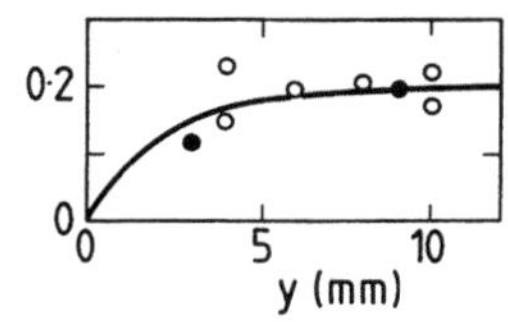

Fig 3. The growth in size of S of line D (H_V) with distance from the heater. The line is an exponential S=0.196(1-exp(-y/2)).

Fig. 2 shows the signal decay for line D (before the polish) in more detail. Measurements made at 6 distances, x, from the interface are broadly consistent with an exponential decay with ℓ_I=4mm to a remanent value given by the line passing through the points obtained using heater H_V. Since the crystal diameter is 4 mm these data are consistent with surface thermalization and similar results were obtained for the other lines. Fig. 3 provides information on the growth of holes in the injected current as it passes along the V half of the bicrystal. The points show the size of line D measured at distances y from the heater H_V and the line is an exponential S(y)=0.196(1-exp(-y/2)) suggesting a growth length ~2 mm. The closed circles show signals measured 3 and 9 mm from the heater (S(3),S(9)) after H_2 anneal. This increased the Fe^{2+} concentration and to correct for this the signals have been reduced so that S(9)=0.196. Values of S(3) appreciably less than S(9) were also seen for lines C, H, ans F(V^{3+}/V^{3+}) (The data for lines C and H are shown in the last column of the table. S(3) and S(9) are given in rows 3 and 4 respectively). These data suggest that the growth length is less than the inelastic mean free path which is in line with the initial theoretical calculations of Sheard et al [6].

TABLE

The sizes of frequency crossing signals C, D and H, ($S = \Delta W/W_o\%$) at 2K. The signals were measured at 6mm (T_n) and 12mm (T_f) from the interface using heater H_{Fe} or H_v. The increases in size after the argon and H_2 anneal are attributed to increases in the Fe^{2+} concentrations.

	Fine ground + air(1200°C)[3]			Fine ground + argon (1775°C)			Syton polish $+H_2$ (1200°C)		
Heater - Thermometer	C	D	H	C	D	H	C	D	H
H_{Fe} - T_n	0.24	0.21	0.30	0.44	0.40	0.46	0.34	0.41	0.61
H_{Fe} - T_f	0.10	0.10	-	0.15	0.18	0.19	0.22	0.28	0.40
H_v - T_f	0.10	0.10	-	0.13	0.17	0.20	0.22	0.30	0.37
H_v - T_n	0.09	0.10	0.13	0.14	0.15	0.20	0.11	0.18	0.22

REFERENCES

1. G.W. Cullen: 1978 in Heteroepitaxial semiconductors for electronic devices, (Springer-Verlag, New York) eds. G.W. Cullen and C.C. Wang, Chapter 2.
2. M.N. Wybourne, C.G. Eddison, M.J. Kelly: J. Phys. C : Solid State Phys. 17, L607 (1984).
3. L.J. Challis, A.A. Ghazi and M.N. Wybourne: Phys. Rev. Letts. 48, 759 (1982).
4. We are very grateful to Mr W.B. Roys and to Mr C.G. Eddison and Dr M.N. Wybourne (GEC Hirst Research Laboratory) for their help with these treatments.
5. The samples also contain a small concentration of bubble defects although these do not appear to produce significant phonon scattering.
6. F.W. Sheard, G.A. Toombs and S.R. Williams: 1985. Proc 2nd Int. Conf. on Phonon Physics (Budapest, 1985) eds. J.Kollar, N.Kroo, K.Menyhard & T.Silkos (World Scientific Pubs. Co., Singapore) p435 and these proceedings.

Investigation of the Sound Velocity of Very Thin Metal Films

P. Berberich, P. Hiergeist, and J. Jahrstorfer

Physik-Department E 10, Technische Universität München, D-8046 Garching, Fed. Rep. of Germany

Recently, several high-frequency phonon experiments have revealed a weak acoustic coupling of evaporated metal films to the substrate. For instance, MARX and EISENMENGER [1] found anomalously large phonon reflection coefficients for silicon-metal interfaces. EVERY et al. [2] observed a so-called halo in phonon imaging experiments which they attributed to a disordered boundary layer at the interface to the substrate. We have studied the sound velocity of ultra thin metal films. As the measured velocities strongly decrease below 10 nm the results give further evidence for the formation of a boundary layer with rather different acoustical properties.

The sound velocities were determined from the acoustic thickness resonances observed in the phonon emission spectra of Joule heated metal films. For phonon spectroscopy we used a stress tuned Si:B spectrometer as described in more detail in Ref. 3. As shown in the inset of Fig. 1 a Pb-I-Pb tunnel junction was on the same crystal close to the heater film H. The monochromatic phonons emitted by this junction were used to calibrate the energy scale of the spectrometer. The Al films were evaporated in a vacuum of 2×10^{-4} Pa onto chemically polished Si substrates (Wacker-Chemitronic). Prior to evaporation the Si specimens were cleaned in HF acid to remove the oxide layer. The film thickness varied between 4 nm and 30 nm. It was monitored during evaporation by a quartz oscillator and was checked afterwards by a surface profiler. With decreasing thickness the sheet resistance $R_\square$ at 4 K increased from 1 to 66 Ω, the resistance ratio R(300 K)/R(4 K) decreased from 3 to 1.2 and T_c varied from 1.34 K to 1.85 K indicating dirty but homogeneous metal films. The Pt films were kindly prepared by J. Vancea, Univ. of Regensburg, using electron beam deposition in a UHV of 10^{-6} Pa [4]. For film thicknesses between 4 nm and 1.2 nm, the temperature dependence of the resistivity at low temperatures revealed weak localization [5] for the thicker films (0.27 kΩ < $R_\square$ (4 K) < 7.2 kΩ) and strong localization for the 1.2 nm thin film ($R_\square$ (4K) = 70 kΩ).

The phonon emission spectra of a 30 nm Al film are shown in Fig. 1A of Ref. 3. The positions of the acoustic thickness resonances $\Omega_n = (n - \alpha)hv/2d$ (n = 1,2,3..) were used to determine the sound velocity v and the phase factor α. The resulting sound velocities both for the L- and the T-mode agree fairly well with the Voigt average phase velocities of Al [6]. However, the phase factor is not 1/2 as expected from the acoustic mismatch model but varies between 1/2 and 1. In addition, the modulation amplitude of the resonances is surprisingly large. Calculations show that the intensity reflection coefficient at the Al/Si interface is 0.2 instead of about 0.01 predicted by the acoustic mismatch model. Fig. 1 shows some interferograms for Al films with d = 4, 7.5 and 10 nm, respectively. The position of the resonances are marked by dashed vertical lines. Irrespective of the film thickness the resonances are resolved only up to about 2 meV for the T-phonons and up to 3 meV for the L-phonons. This is attributable to the

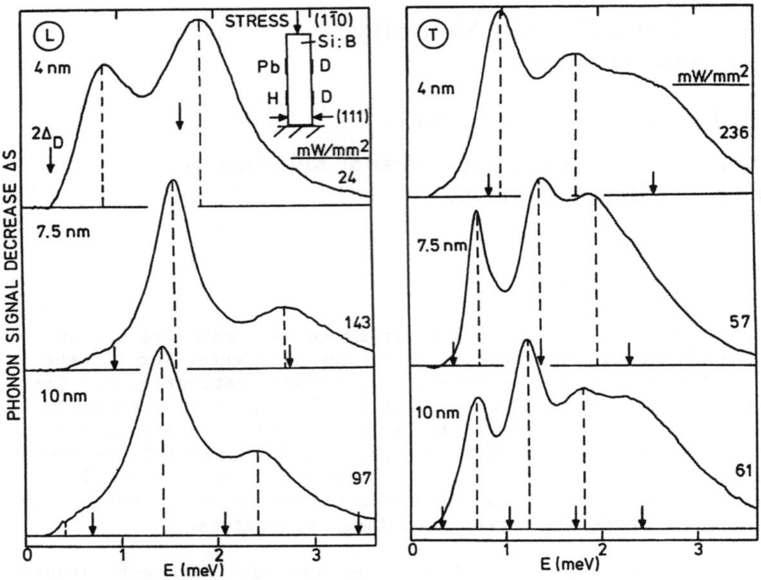

Fig. 1 Phonon emission spectra of Al films for various thicknesses.

roughness of the film surfaces. It should come into play when $2hq_{ph}>1$ where q_{ph} is the phonon wavevector and h the roughness amplitude. A calculation of the interferograms analogous to optical interference filters [7] results in h = 0.5 nm. The arrows mark the positions of the resonances for bulk polycrystalline Al with boundary conditions according to the acoustic mismatch model. Obviously, the spacing between the peaks is smaller than expected, indicating a reduced sound velocity.

In Fig. 2, we have plotted the measured sound velocities normalized to their bulk values $v(\infty)$ as a function of film thickness. Both, the longitudinal (open circles) and the transverse (full circles) sound velocities of Al continuously decrease down to 50 % with the thickness decreasing from 15 nm to 4 nm. A similar behavior has been observed for the transverse sound velocity of Pt films. Unfortunately, only one resonance showed up in the spectra. (As the sound velocity of Pt is smaller, surface roughness already gets important at lower frequencies than for Al.) Therefore we had to fix

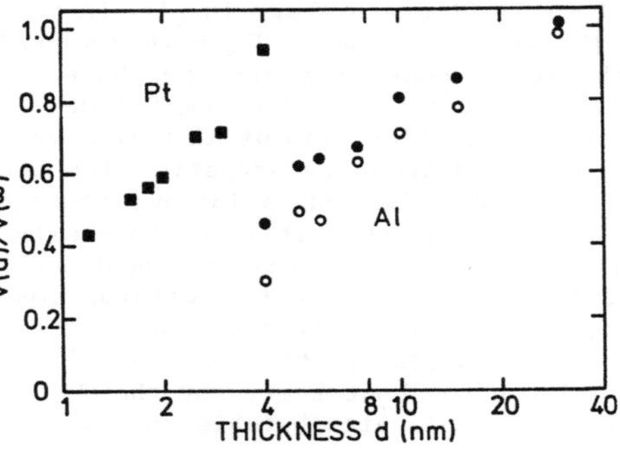

Fig. 2 Normalized sound velocities $v(d)/v(\infty)$ vs. thickness d. Circles/squares: Al/Pt Full/open: T-/L-phonons Al: $v_L(\infty)$ = 6.6 km/s $v_T(\infty)$ = 3.3 km/s Pt: $v_T(\infty)$ = 1.77 km/s [6]

the phase factor α to determine v. As the acoustic impedance of Pt is much larger than that of Si we assumed clamped boundary conditions at both interfaces (α = 0). The resulting sound velocities which are plotted in Fig. 2 as squares decrease between 4 nm and 1.2 nm in a similar way as for Al.

The Pt films which were prepared in UHV are of high purity. In addition, thick granular Al films which are evaporated in the presence of oxygen have the same sound velocity as clean Al films [3]. We have evaporated Pb-I-Pb junctions onto Al films to eliminate a possible influence of adsorbed layers of water or air. However, the same reduction of v was found as for the unprotected films. All these observations show that the phonon softening is not caused by impurities inside the films or by surface contaminations. In the case of inhomogeneous film growth percolation, theory predicts a correlation between the electrical conductivity σ and v as both scale with $(p-p_c)^t$ near the percolation threshold p_c where p is the fraction of sites or bonds occupied. This leads to the relation $v = A\,\sigma^x$ with $x = t_v/t_\sigma = 0.26$ [8,9]. Our data yield x = 0.23 for Pt(T), x = 0.36 for Al(T) and x = 0.52 for Al(L) deviating considerably from the predicted value. There are other observations which indicate that island formation cannot play the dominant role: (i) we found no film discontinuities by scanning electron microscopy within the resolution of 5 nm; the roughness amplitude of 0.5 nm deduced from the interferograms is small. A similar phonon softening has been reported for metallic Ni/Mo superlattices [10] where the sound velocity decreases by about 30 % for modulation wavelengths below 10 nm. This softening has been explained by surface strains which produce a disordered transition layer of about 1 nm in thickness [11]. In this respect, our results demonstrate that such a boundary layer already forms at a single interface.

References

1. D. Marx and W. Eisenmenger: Z. Phys. B 48, 277 (1982)
2. A.G. Every, G.L. Koos, and J.P. Wolfe: Phys. Rev. B 29, 2190 (1984)
3. P. Berberich and H. Kinder: These conference proceedings
4. G. Fischer, H. Hoffmann, and J. Vancea: Phys. Rev. B 22, 6065 (1980)
5. R.S. Markiewicz and L.A. Harris: Phys. Rev. Lett. 46, 1149 (1981)
6. G. Simmons and H. Wang: Single Crystal Elastic Constants and Calculated Aggregate Properties (MIT Press, Cambridge Mass., 1971)
7. J.M. Eastman: In Physics of Thin Films, Vol 10, ed. by G. Haas and M.H. Francombe (Academic Press New York 1978) p. 167
8. B. Derrida, D. Stauffer, H.J. Herrmann, and J. Vannimenus: J.Physique 44, L 701 (1983)
9. O. Entin-Wohlmann, S. Alexander, R. Orbach, and K. Yu: Phys. Rev. B 29, 4588 (1984)
10. M.R. Khan, C.S.L. Chun, G.P. Felcher, M. Grimsditch, A. Kueny, C.M. Falco, and I. K. Schuller: Phys. Rev. B 27, 7186 (1983)
11. I.K. Schuller and A. Rahman: Phys. Rev. Lett. 50, 1377 (1983)

Surface Modes and Surface Reconstruction in Diamond-Type Crystal Lattices

W. Goldammer, W. Ludwig, and W. Zierau

Institute for Theoretical Physics II, University Münster, Domagkstr. 75, D-4400 Münster, Fed. Rep. of Germany

1. Introduction

Surface modes and surface reconstructions of various crystals have been investigated in many papers during the last decades. Especially for the diamond structure crystals such modes have been investigated for a (111)-surface. In the calculations done up till now we have used force-constant models including interactions up to 4.th neighbors /1,2/. Apart from the surface modes, these models lead to a soft mode behavior of the lowest surface mode if the force constants at the surface are changed in an appropriate way. Interpreting this soft mode as a lattice instability which leads to a reconstruction, statements on this are possible. Without any further assumptions than a certain change of the surface constants, we find a 7×7 reconstruction for the Si-(111)-surface, an 8×8 one for Ge and a 3×3 one for α-Sn, whereas diamond exhibits no reconstruction. In the present paper we compare the former results with those of a dipole-model proposed by R.F.Wallis /3/ using Si as an example.

2. Force constant and dipole model

The force constant model starts with the interaction

$$\delta\Phi = \Sigma\tfrac{1}{2}f(\delta r)^2 + \Sigma\tfrac{1}{2}g(\delta r)^2 + \Sigma\tfrac{1}{2}h(\delta r)^2 + \Sigma\tfrac{1}{2}\ell(\delta r)^2 + \Sigma\tfrac{1}{2}\sigma r_o^2(\delta\theta)^2 + \Sigma\varphi r_o(\delta\theta\delta r) \qquad (1)$$

where f,g,h,ℓ describe central forces between 1.st,... neighbours and σ,φ angle bending and mixed forces. The atoms in this model are looked upon as rigid ones. - - The dipole model considers the fact, that the ion cores and the electronic distributions in a phonon mode undergo relative displacements which build up dipole moments. The induced dipoles then interact by the wellknown potential. Thus if we put

$$d_i\binom{\vec m}{o} = \sum_j P_{ij}\binom{\vec m\,\vec m}{o\,o}\, u_j\binom{\vec m}{o} + \sum_j \sum_{k=1}^{4} P_{ij}\binom{\vec m\;\;\vec m+\vec\delta_k}{o\;\;1}\, u_j\binom{\vec m+\vec\delta_k}{1} \qquad (2)$$

for the dipole-moment induced on particle $\binom{\vec m}{o}$ by the displacements of $\vec u(1,\vec m+\vec\delta_k)$ ($\vec\delta_k$ restricted to nearest neighbours here), we obtain for the total dipole-dipole interaction

$$\delta\Phi = \frac{1}{2\varepsilon} \sum_{\substack{mn\\ \alpha\beta ij}}{}' \left\{ \frac{\delta_{ij}}{|\vec R^{\vec m}_{\alpha} - \vec R^{\vec n}_{\beta}|^3} - 3\,\frac{(X^{\vec m}_{\alpha i} - X^{\vec n}_{\beta i})(X^{\vec m}_{\alpha j} - X^{\vec n}_{\beta j})}{|\vec R^{\vec m}_{\alpha} - \vec R^{\vec n}_{\beta}|^5} \right\} d_i\binom{\vec m}{\alpha}\, d_j\binom{\vec n}{\beta} \qquad (3)$$

ε is the dielectric constant. The lattice sums occurring in this connection are evaluated with the Ewald method. The parameters of the interaction are obtained by a least square fit to the bulk phonon modes (Fig.1) and are given in Table 1. For tetrahedral symmetry there are only two independent dipole parameters. The surface is introduced by cutting all the interactions between two half-infinite pieces of the crystal and treating the missing interaction by a perturbation theory using the Green function method as the appropriate one.

TABLE 1	f	g	h	ℓ	σ	$\varphi\sqrt{2}$	P_1	P_2
f.c. model	8,8	1,44	-o,125	o,o4	-o,o2	-1,125	—	—
dipole mod.	11,38	o,71	o,18	-o,o6	o,69	-o,o3	o,24	-o,56
	in $1o^4$ dyn/cm						in units e	

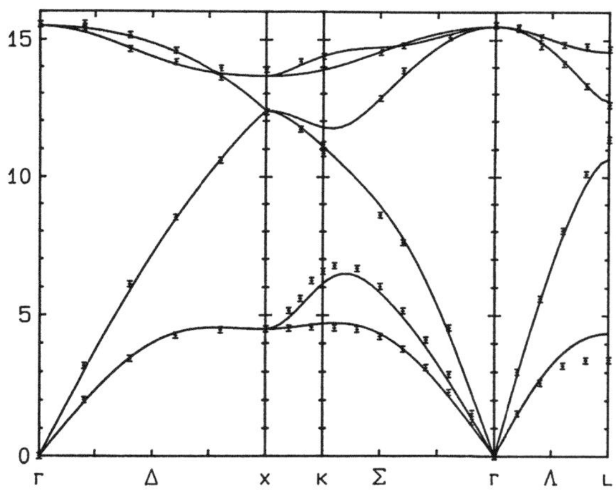

Fig.1. Bulk phonon modes according to the dipole model compared to experimental points. Frequencies in units of THz.

One set of relative force constant changes for each model leading to soft modes is given in Table 2. There are other possible sets of such changes, which also lead to soft modes. The essential point, however is that all these sets give the soft modes at the same point in $\vec{k}$-space.

TABLE 2	
f.c. model	$f'/f = 1$; $f''/f = o,5$; $g'/g=o,62$; $g''/g=g'''/g=o,5$ $h'/h=1$; $\sigma'/\sigma=\sigma''/\sigma=1,5$; $\varphi'/\varphi=o,5$; $\varphi''/\varphi=1,5$
dipole model	$P_1(o,o)=1,o5P_1$; $P_1^*(o,1)=P_1^*(1,o)=1,4oP_1$; $P_1(1,1)=1,3oP_1$; $P_2(o,o)=P_2^*(o,1)=P_2^*(1,o)=1,4oP_2$; $P_2(1,1)=o,4oP_2$

The primes and numbers in brackets denote different positions of the atoms relative to the surface, where the symmetry is less than in the bulk. In the dipole-model the short-range forces are left unchanged; only the dipole parameters marked with an asterisk are independent, the others follow from the acoustic sum rule (translational invariance).

3. Results and conclusions

The results are shown and compared in Figs.2 and 3. In the force constant model the soft mode always occurs at a point which corresponds to an instability leading to a 7×7 reconstruction and this is independent from the special set of force constants. On the other hand, the dipole model exhibits a soft mode at the $\overline{M}$-point, which corresponds to a 2×1 reconstruction.

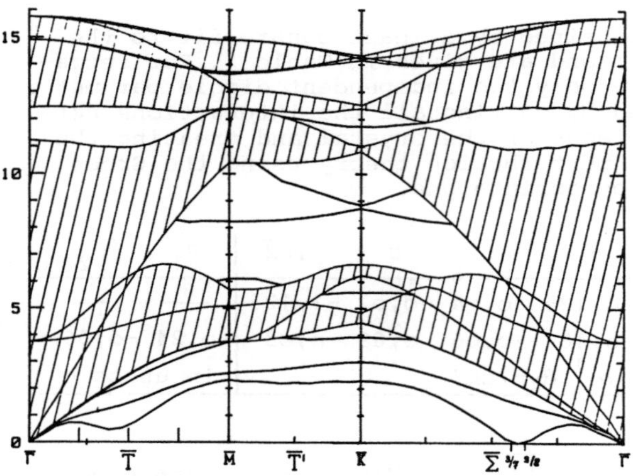

Fig.2. Surface modes according to the force constant model with surface constants from Table 2.

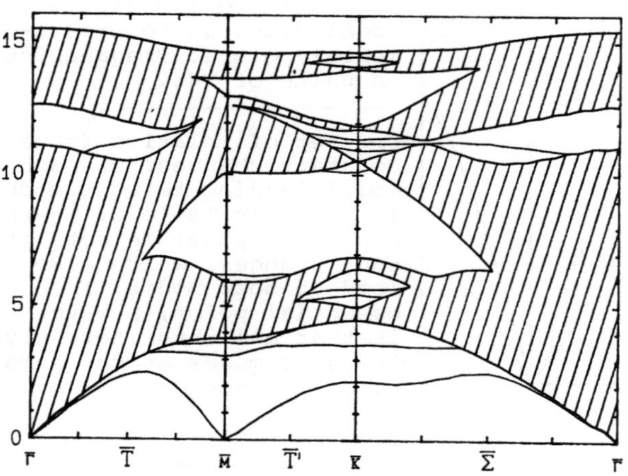

Fig.3. Surface modes according to the dipole-model with surface constants from Table 2.

From the soft mode displacements we can construct reconstruction patterns but we have to see , that these patterns are due to the instabilities and give no information on new equilibrium positions; this would need a calculation taking into account nonlinear parts of the interaction. Furthermore, such a displacement pattern is not unique; because of the degeneracy of the soft mode, different linear combinations are possible. Nevertheless, there is a strong hint to the symmetry of the reconstruction, namely 7×7 or 2×1, resp.
Since the instabilities for different sets of force constants always occur at the same points in $\vec{k}$-space (this is also valid for the other diamond structure crystals) we conclude that the bulk properties determine the dynamical behaviour of the surface to a large extent, and that only certain details are related to the special surface constants.

Furthermore, short-range forces seem to favour the 7×7 reconstruction with a large length of periodicity in real space giving a small site symmetry to a surface particle, whereas long-range forces favour the 2×1 reconstruction with a small length of periodicity in real space and a larger site symmetry of a surface particle. Thus in Si (and possibly the other similar structures) there seems to be a competition between short- and long-range forces, which makes it possible, that both reconstructions are energetically close together.

1. W.Goldammer, et al., Surf. Sci 141 (1984) 139.
2. W.Goldammer, et al., Surf. Sci, in press
3. R.F.Wallis, private communication

Measurements of Relaxation Time of Sodium Ions in Sputtered Thin Film Beta-Alumina

K. Nobugai, Y. Nakagiri, F. Kanamaru, M. Tokumura, and T. Miyasato

The Institute of Scientific and Industrial Research, Osaka University, Mihogaoka 8–1, Ibaraki, Osaka 567, Japan

It has been reported by Almond and West that the temperature dependence curve of the acoustic attenuation in sodium-beta-alumina shows a broad peak compared with Debye theory. In the present experiment, the relaxation attenuation of rf-electric field accompanied by the surface elastic waves on a lithium-niobate transducer which is adjacent to the sputtered thin film sodium-beta-alumina deposited on a substrate with an air gap of a few μm has been measured (i.e.,by means of acousto-electric effect) as a function of temperature at frequencies 200, 400 and 600 MHz, respectively, and it has been clarified that the temperature dependence curve fits quite well with Debye theory in this system, and the activation energy was estimated to be approximately 0.13 eV. It should be emphasized that this method is very useful for the investigation of such thin film material.

1. INTRODUCTION

The superionic material draws attention in this decade because of its peculiar property, a high ionic conductivity, and from the industrial interest, as this material is expected for the solid electrolyte in secondary battery. This material is classified into three groups, a. Silver and/or copper compounds, b. Material with fluorite structure and c. with beta-alumina structure.

Many discussions were held for the mechanism of ionic conductivity in this material, and first of all, the interaction between conductive ion and the lattice system or phonon is extensively discussed. Some of them were given by Hayes/1/. Five years ago, Almond and West /2/ had reported on the experimental results of the temperature dependence of acoustic attenuation in a single crystal of sodium-beta-alumina, and clarified that the data are inconsistent with simple Debye theory, and found to be in good agreement with each other when analyzed with use of "universal" dielectric theory.

On the other hand, two of us (K.N. and F.K. /3/) have succeeded in the preparation of a thin film sodium-beta-alumina onto alpha-Al_2O_3 substrate by means of rf-sputtering method from a sintered target of mixed powders of Na_2CO_3 and Al_2O_3, in Ar and O_2 mixture sputtering gas, and the deposited film was annealed at 1300°C for 24 hours in the air. This film has a polycrystalline structure with c-planes oriented parallel with the film surface, and has a similar high ionic conductivity along the surface plane(c-plane) with that of single crystal one.

In order to investigate the interaction between the sodium ion and the lattice system or phonon, we carried out the measurement of the acoustic attenuation on the lithium niobate transducer (SAW device) which is adjacent to the deposited sodium-beta-alumina film with an air gap, because we considered that if there were coupling between sodium ion and the rf-electric field which was induced by SAW on the lithium niobate transducer, the electric energy might be absorbed by the relaxation of sodium ion in the film, the relaxation time of which is determined by the interaction between the sodium ion and the thermal phonon in the film.

2. EXPERIMENTS

2.1 Sample Preparation

This is partially mentioned at the bottom of "INTRODUCTION". A target with an appropriate Al_2O_3/Na_2O ratio was sputtered with Ar(50%)+O_2(50%) sputter gas at the pressure of 4×10^{-3} Torr onto columnar single crystal substrate. The concentration of sodium ion in the film depends on the sputtering conditions, especially on the target temperature. The thickness of the film was estimated to be 0.5-2.3 μm.

2.2 $LiNbO_3$ Transducer

A 128° rotated Y-cut lithium niobate plate with interdigital electrode (the fundamental frequency is 200 MHz, and higher modes 400 MHz and 600 MHz were also excited) was used. The temperature dependence of the transducer itself was measured without the sample and it could be ignored.

2.3 Sample Setting

A sample film deposited on the substrate was set on the SAW device with an air gap of 2-9 μm, and the SAW device was also set on the copper block, and they were contained in a copper can, inside of which was evacuated at first to 10^{-6} Torr, and then helium gas was introduced till 260 Torr at room temperature to keep the temperature uniform.

2.4 Measurement

The can with the sample of the cryostat was cooled down to liquid nitrogen temperature at first, and then warmed up by heater at a rate of one degree per minute or so during the measurement. The echo signal was gated and amplified by a Box-Car Integrator, and fed to the Y-axis of the X-Y recorder. The temperature was measured by means of Au:Fe-Chromel thermo-couple, and its out-put was fed to X-axis of the recorder.

3.EXPERIMENTAL RESULTS AND DISCUSSIONS

The temperature dependence of the attenuation of the signal height for 200 MHz and 600 MHz are shown in Fig.1 and Fig.2. The filled circles show experimental results, and broken lines are by Debye theory. As seen from these figures, the theoretical curves fit quite well with experiments. We can get the activation energy from the temperature at peak point and the frequency, estimated to be approximtely 0.13 eV, which is very close to the theoretical value by Wang et al./4/ for a single crystal, they got 0.14 eV. When we plot each peak point for 200 MHz, 400 MHz and 600 MHz on a T^{-1} vs. $\log\omega$ plane, they show linear relationship, and furthermore the peak point for 240 MHz by Almond and West also lies on this line as shown in Fig.3 (shown by the open circle). The activation energies obtained by various methods are listed on Table 1. At this point we do not know the reason for the difference between the present results and those of Almond and West, because of the following. 1. The coupling between the wave and the ion is very simple in our case, namely acoustic wave on the device couples with ion through the electric field, and does not couple directly with thermal phonon in the film. But in case of bulk wave in a crystal, the cou-

Table 1. Activation energy for sodium-beta-alumina

	Ionic Conductivity	Relaxation Time	Calculation
Thin Film	0.25 eV	0.13 eV	-
Single Crystal	0.16 eV	-	0.14 eV

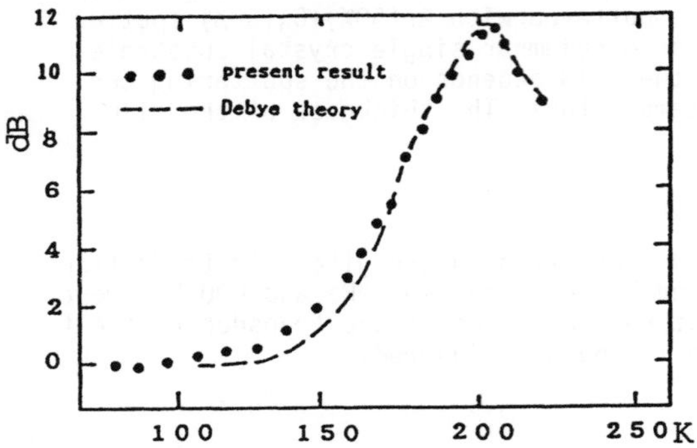

Fig.1. SAW signal attenuation for 200 MHz.

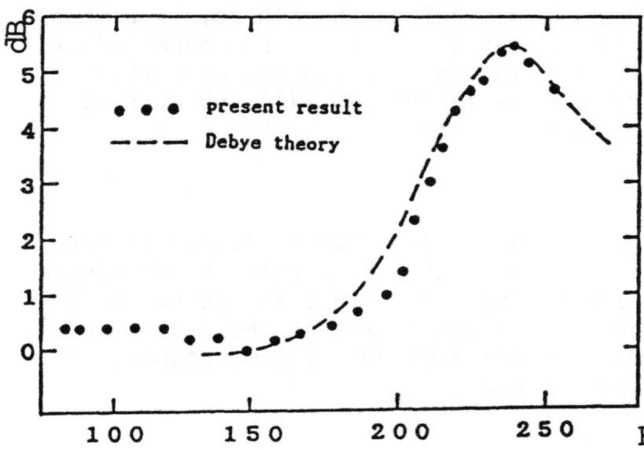

Fig.2. SAW signal attenuation for 600 MHz.

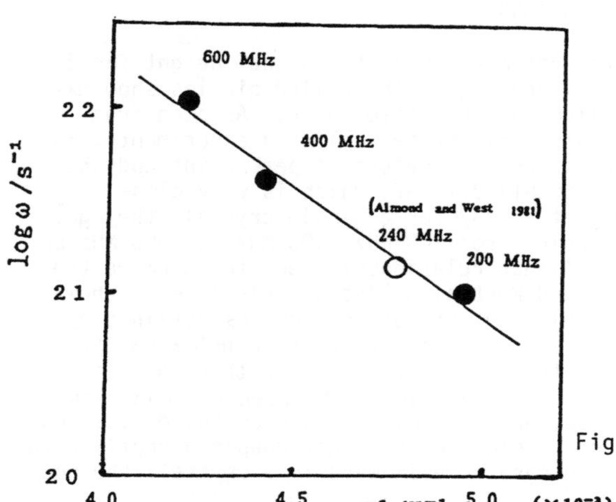

Fig.3. T^{-1} vs frequency relationship of attenuation peaks.

pling mechanism between the ion and acoustic wave is not simple, for example by deformation potential, "piezoelectric" coupling due to local electric fluctuation, due to the Akhieser effect or due to other non-linear effect. 2. The sample preparation is quite different, ours is a sputtered polycrystalline film and the other is a single crystal. Finally, it should be noticed that the concentration of sodium ion is not discussed in both cases. Further experiment is to be continued, for example using a single crystal.

References

/1/ W.Hayes:Journal de Physique, C6-167(1981).
/2/ D.P.Almond and A.R.West:Phys.Rev.Lett.47(1981)431.
/3/ K.Nobugai and F.Kanamaru:Reactivity of Solids.(ed. by P.Barret and L. C.Dufour) Amsterdam 1985. p.811-816.
/4/ J.C.Wang et al.:J.Chem.Phys.63(1975)772.

Part V

Quantum Matter and Kapitza Resistance

Phonons in Quantum Liquids

A.F.G. Wyatt

Department of Physics, Unversity of Exeter, Exeter, Devon EX4 4QL, UK

The excitation picture of liquid ^{4}He has not only stood the test of time but has improved with the years. Introduced by Landau [1] it envisages the atoms in a ground state at T = 0. For T > 0 we have excitations plus the zero entropy ground state; all the energy of the liquid is associated with the excitations. Density waves involving all the atoms have much smaller energy quanta associated with them than excited single atoms moving in the confines of neighbouring atoms [2].

The ground state of liquid ^{4}He is a Bose condensate with all the atoms in the same state. For non interacting bosons each particle would have zero momentum (k) but, for ^{4}He the wavefunction reflects each atom's confining cage of other atoms and there is only ~ 15 % probability of k = 0 [3]. The elementary excitations are phonons and rotons (see fig 1). The rotons have no angular momentum associated with them despite their name and their spectrum joins continuously onto the phonon spectrum. A phonon in liquid ^{4}He is more similar in many respects to a longitudinal phonon in a crystal than to a sound wave in a classical gas, except that the lack of periodicity means there are no umklapp processes. A roton for $q \geq 2.1$ Å^{-1} is pictured as a moving ^{4}He atom together with the back flow of fluid around it [2,4]. This fluid dipole enables roton-roton scattering to be modelled by dipole-dipole interactions.

For T < 2 K we consider liquid ^{4}He as a gas of phonons and rotons. The thermodynamic quantites can be calculated to within a few % with this model [5,6]. It also gives a coherent account of viscosity, attenuation of sound and second sound [7]. For 1 < T < 2 K there are more rotons than phonons and for T < 0.6 K there are essentially no rotons. At the higher temperatures the excitations increasingly mutually scatter as their density increases. Although this leads to a shortening of lifetimes, the excitations retain their essential character and the excitation model is still useful up to T_λ though not beyond it [8].

This short review is mainly concerned with the dynamics of the excitations and its variation with energy, pressure and temperature. We shall consider the dispersion curve $\omega(q)$ for both ^{3}He and ^{4}He. However we shall mainly concentrate on ^{4}He as the collective excitations are longer lived than in ^{3}He and have been studied more widely. We shall discuss the intrinsic lifetime of an excitation when it is essentially isolated and finally the lifetime when it is scattered by a thermal population of excitations.

2 The Dispersion Curve

The dispersion curve for liquid ^{4}He has been measured by neutron scattering (see review by Woods and Cowley [9]) and the results are shown schematically in fig 1. In the phonon, maxon and roton regions, for $q < 2.6\ \text{Å}^{-1}$ the excitations have a well-defined energy and momentum. However, energetic neutrons can create 2 rotons and their combined momentum runs continuously from $q = 0$ so these form a broad band of excitations across the diagram for $\omega \geqslant 2\Delta/\hbar$, where Δ is the energy of the roton minimum. The line in this region shows the position of the maximum of the broad measured line shape. There is evidence for bound roton pairs with energy $\leqslant 2\Delta$ [10,11,12] and for roton-maxon pairs [12]. The single roton curve runs into the 2Δ continuum at $q \sim 2.6\ \text{Å}^{-1}$ and the measured line broadens. The excitation states with energy $> 2\Delta$ are insignificantly populated at $T < 2K$ and will not concern us further.

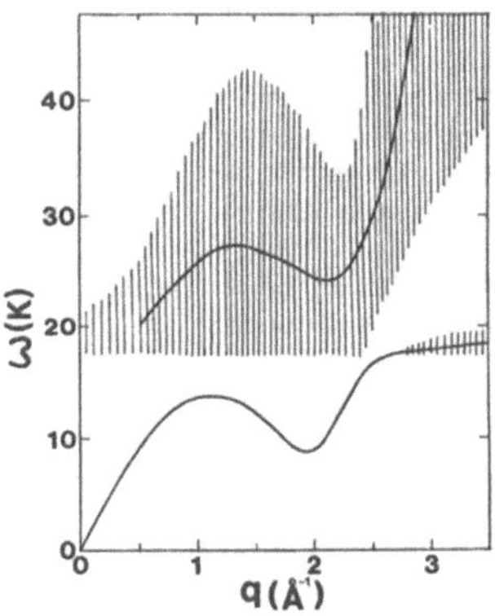

Fig 1. Dispersion curve for ^4He at $T \sim 1$ K [9]. Excitations for $\omega < 17$ K are well defined

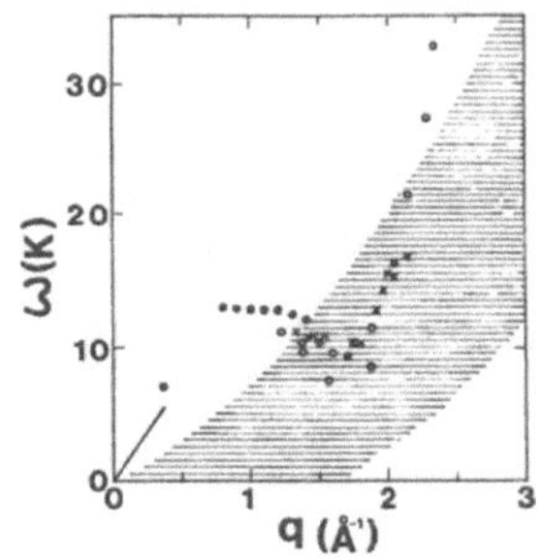

Fig 2. Dispersion curve for ^3He at $T \sim 0.015 - 0.7$ K [14–17]. The line shows the sound velocity

The only other quantum liquid, ^3He, which is a fermi system, has a dispersion curve with similarities to that of ^4He. The different statistics leads to quite different superfluid transition temperatures, however it is argued that the long-range interactions between the atoms is similar in both cases and so the excitation spectrum of density fluctuations should also be similar [4,13]. It is suggested that the interatomic potential can be split into short and long-range parts as is done for electrons in a metal and the density fluctuations are akin to plasmons. The shorter range of the van der Waals forces compared to coulomb forces makes $\omega(q) \to 0$ as $q \to 0$ for the He liquids. These excitations are called zero sound modes to emphasise that they are in the collisionless regime, however we shall continue to refer to them as phonons and rotons in liquid ^4He. This concept is a major extension of the zero sound in a condensed Fermi liquid where the restoring force to a density fluctuation is the change in Fermi energy. Such a mode has been seen at $T \ll T_F$, SKOLD et al [14]. Their existence for $T \gg T_F$ [15] greatly supports this approach. The $\omega(q)$ for these modes as shown in fig 2 [14–17] shows broadly the same shape as the dispersion durve for ^4He.

However the major difference is indicated by the much shorter lifetime of these modes in ^3He. This arises as there are other excitations in ^3He into which the zero sound modes can decay. For $T \ll T_F$, there is a well-defined fermi surface and particle-hole pairs can be created, these form a continuous band of excitations shown in fig 2 for $m^* = 2.88\ m_3$. If two or more particle-hole pairs can be created then this fills the entire ω-q plane. For $T > T_F$ the concept of a hole disappears but pairs of ^3He atoms can be similarly excited. The neutron scattering at T = 15 mK [14] clearly shows the lifetime of the zero sound mode becoming very short as the dispersion curve enters the particle-hole continuum. The relaxation to multipairs and atoms is much less and allows the zero sound to be seen outside the continuum even at 1.2 K [15]. The actual lifetime has not been measured because of the difficulties of doing neutron scattering with such a strong absorber. In liquid ^4He the excitations are only well defined for $T < T_\lambda$ [8] which indicates that their lifetime is intimately connected with the Bose condensation.

Due to the large zero point motion and the weak van der Waals forces, the liquids are extremely compressible. The dispersion curve for liquid ^4He under 25 bar pressure is shown in fig 3 [9]. The sound velocity increases by ~ 50 %, the maxon peak increases, the roton minimum decreases and shifts to higher wavevectors. For liquid ^3He the 'roton' region also decreases with pressure [18]. Besides these large changes there is a subtle alteration to the shape of the phonon dispersion in ^4He. At 0 bar the dispersion curve bends upwards initially before bending back to the maxon peak, but at 25 bar the dispersion curve bends continuously downwards. This has dramatic effects on the phonon lifetime and ultrasonic attenuation. Recently the neutron scattering resolution has dramatically improved and STIRLING's [19] results clearly show upward dispersion in fig 4. We shall return to this point in the next section.

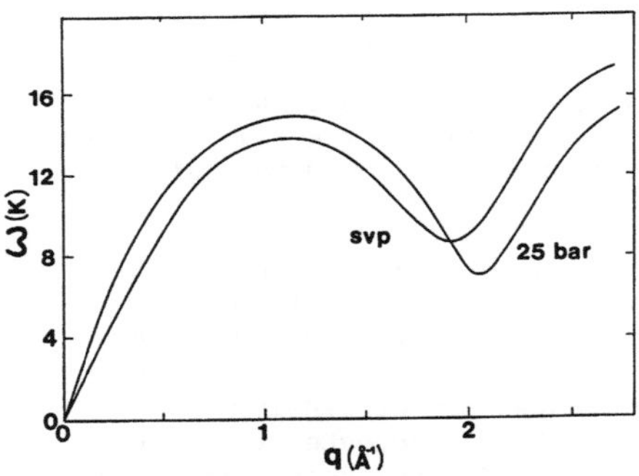

Fig 3. Dispersion curves for ^4He at svp and 25 bar [9]

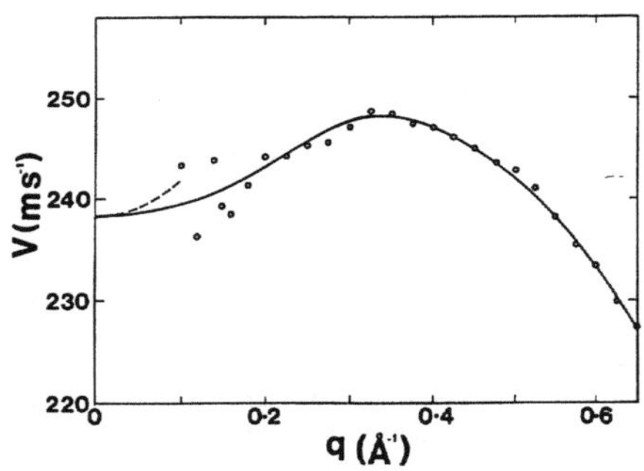

Fig 4. The phase velocity of phonons [19]. The broken line is $\omega/q = C_0(1 + 1.49q^2)$ [30]

The energy of the excitations are weakly temperature dependent. As the number of excitations increases there is a slight downward shift in their energy. Recently, MEZEI [20] has shown that $\Delta(0) - \Delta(T) = 19\ T^{1/2} \exp(-\Delta(T)/T)$ (where Δ and T are in Kelvin) which indicates that the decrease in energy is proportional to the thermal population of rotons. As there is only a slight mass density change in this temperature range, the effect of a higher density of excitations is to weaken the long-range force. Other regions have not been studied in detail but the maxon peak appears to have a small downward shift in the data of SVENSSON et al [8] and the sound velocity decreases by 4 % between 1 and 2 K [21]. Below 1 K the dispersion curve is essentially independent of temperature.

3 Intrinsic Lifetimes of Excitations in ^{4}He

For the excitation picture to be sensible the excitations must have intrinsically long lifetimes. In the ideal limit, an excitation injected into liquid ^{4}He at T = 0 should not decay into two or more other excitations. For most regions of the dispersion curve with $\hbar\omega < 2\Delta$, such decay is impossible as energy and momentum cannot be conserved. There are two regions where this might not be so, the first is the phonon region when there is upward dispersion and the second is the roton region around q ~ 2.2 $Å^{-1}$ if the local gradient of the dispersion curve is greater than the sound velocity. In this case a roton could decay into a lower energy roton and a low-energy phonon. A direct measurement of lifetimes around q = 2.2 $Å^{-1}$ indicates that it does not occur at ~ 0 bar [20]. At 25 bar the measured $\omega(q)$ shows that this process is not possible [9] and it probably does not occur at any pressure.

Upward dispersion in the phonon region has been thought about for some time. Clearly it could not be a large deviation from linear behaviour as otherwise the specific heat would not vary as T^3 ($T \leq 0.6$) and early measurements of the structure factor were not sensitive enough to show any deviation from linearity. It was realised that upward dispersion would have a strong effect on ultrasonic attenuation as it would allow 3 phonon process (3pp) rather than the less probable 4 phonon processes. It was the inability of 4pp to account for the large ultrasonic attenuation and the possibility of explaining it with the 3pp [22] that gave impetus to search for upward dispersion. To account for the change of ultrasonic attenuation with pressure it was suggested that the upward dispersion decreased with pressure [23]

Clear evidence for upward dispersion came from angualar spreading of a phonon beam [24]. Phonons can be injected into liquid ^{4}He at T < 0.1 K where there is a negligible chance of scattering from thermal excitations. If these are collinated into a beam then any angular broadening of the beam beyond its geometric width must be due to decay processes. In the 3pp, the two created phonons make an angle with the original phonon so that momentum is conserved. The measurements are shown in fig 5. The pressure dependence shows that the upward curvature decreases with increasing pressure and above 19 bar there is only downward curvature.

There will be a critical energy, ω_c, such that phonons with $\omega > \omega_c$ do not decay via the 3pp while those below do. This is clearly demonstrated in two different experiments using superconducting tunnel junctions. DYNES and NORAYANAMURTI [25] used one as a tuneable phonon generator in a

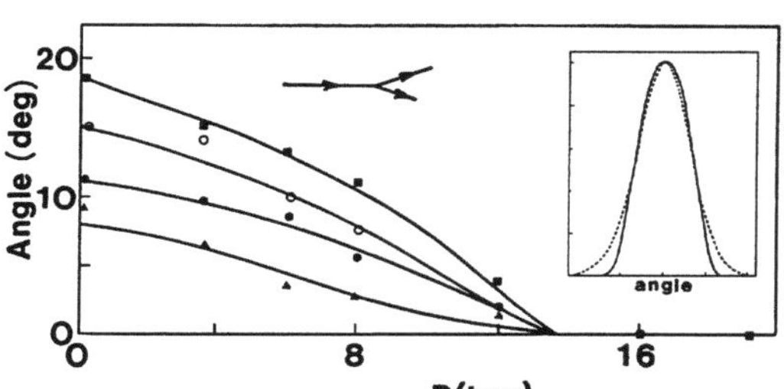

Fig 5. Increase in angular width of a beam of phonons as a function of pressure for heater temperatures 1.2, 1.6, 2.2, 2.9 K [24]

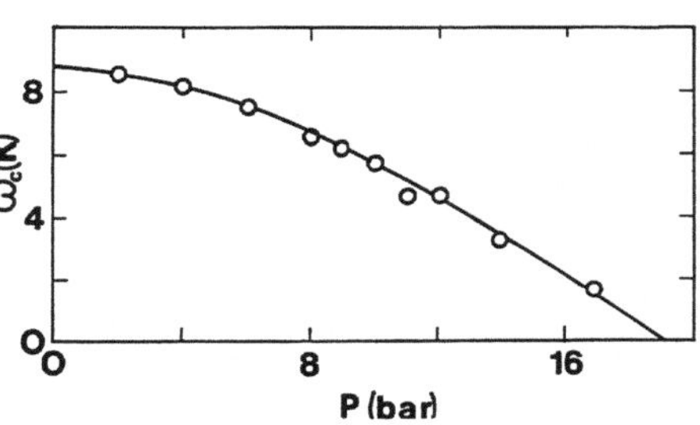

Fig 6. The critical energy for the 3 phonon spontaneous decay process as a function of pressure [25]

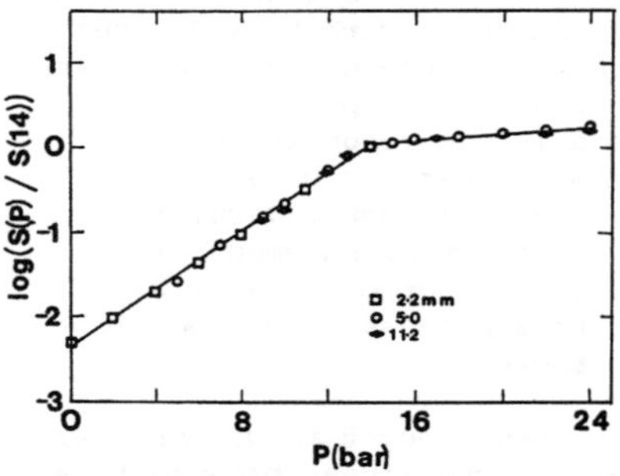

Fig 7. The signal due to phonons with $\omega > \omega_c(P)$ normalised at 14 bar as a function of pressure [26]

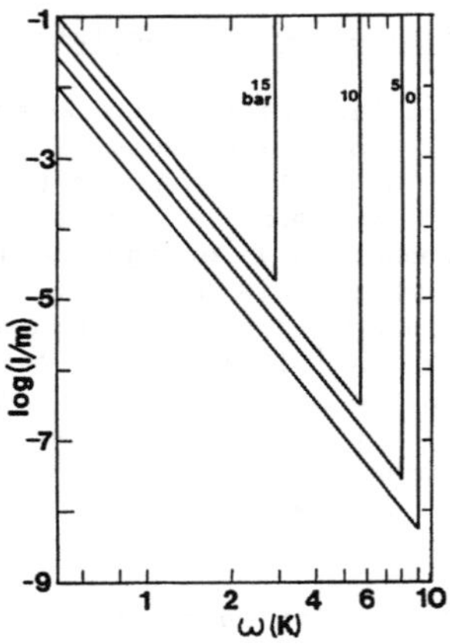

Fig 8. The calculated mean free path, ℓ, for phonons as a function of energy [27,28]

steady-state experiment. The 3pp decay is so strong that the phonons cascade down in energy to less than the cut off energy of the detector. The measured values of ω_c as a function of pressure are shown in fig 6. WYATT, LOCKERBIE and SHERLOCK [26] measured the magnitude of the response of a superconducting tunnel junction to a phonon pulse. The results are shown in fig 7. The signal depends on the pressure of the liquid ^{4}He and it was shown that this is exactly what one would expect from the exponential tail of a Planck distribution of phonons and a critical cut-off that decreased linearly with pressure.

The mean free path (ℓ) from the 3pp is expected to vary at ω^{-5} [27]. In fig 8 is shown the expected dependence of ℓ on ω and pressure [28]. The qualitative features of this figure have been confirmed experimentally. Further evidence for the 3pp comes from the nonlinear interaction of two phonon beams. Two low-energy phonons both with $\hbar\omega < 2\Delta_{Al}$ so that they are not detected by an Al superconducting tunnel junction, can combine to create one high-energy phonon with $\hbar\omega > 2\Delta_{Al}$ [29]. The results of such an experiment are shown in fig 9. In a quite different experiment with two coherent sound waves with ω_1, q_1 and ω_2, q_2, the non linear interaction allowed by the 3pp gives a spatial variation with a repetition length = $2\pi/|2q_1 - q_2|$. This gives a good measure of the initial upward dispersion, $\omega = cq(1 - \gamma q^2)$, $\gamma = 1.49 \pm .04$ Å^2 [30].

Apart from the phonon region, where $\omega < \omega_c(P)$, the excitations are expected to have very long intrinsic lifetimes. To show this, it is necessary to measure mean free paths rather than linewidths. The technique of quantum evaporation [31] has shown that the mean free path for some phonon and roton energies is macroscopic. It can only be done at P = 0, but it is expected that the lifetimes will be similar at higher pressures, as it only depends on the 3pp being not allowed. Only quantum evaporation allows the high-energy excitations to be studied. If an excitation is injected into liquid ^{4}He with $T \leqslant 0.1$ K and it reaches the liquid/vacuum interface, it has a probability of evaporating a single atom if $\hbar\omega > E_B$ where $E_B/k_B = 7.15$ K, the binding energy of a ^{4}He atom. The excess energy is carried away as kinetic energy, $mv^2/2 = \hbar\omega - E_B$. From time-of-flight measurements and the excitation group velocity, found from $\omega(q)$, the energy of the excitation can be found. The results for phonons at normal incidence are shown in fig 10. The energy of the phonons responsible for the fastest times at a given liquid depth are marked on the curves. From this experiment it is clear that phonons with energy 9 K < $\hbar\omega/k_B$ < 11 K can travel several millimetres in the liquid.

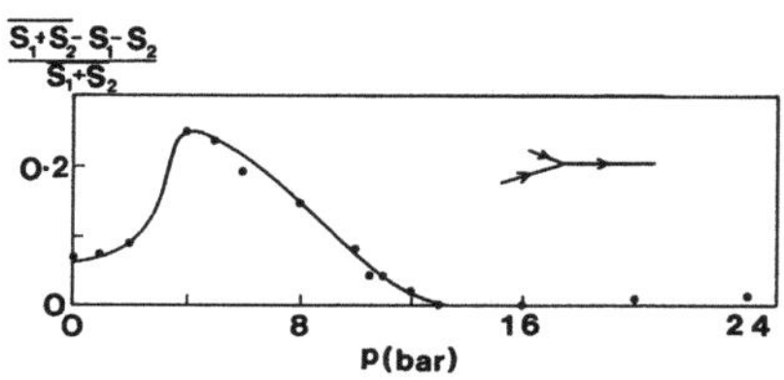

Fig 9. Phonon up processes shown by the excess signal for two interacting phonon beams ($S_1 + S_2$) over the two beams detected separately [29]

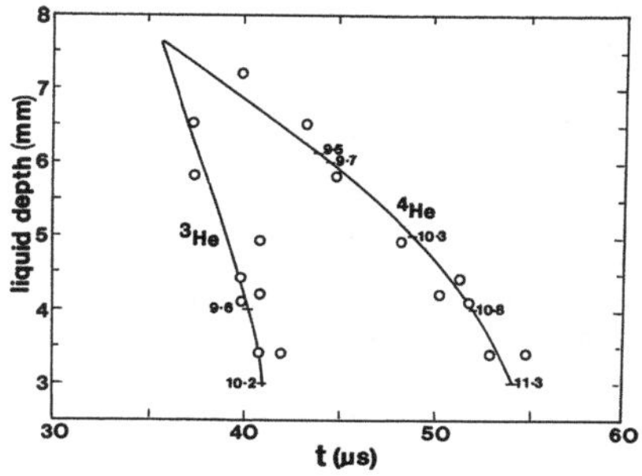

Fig 10. The minimum time for phonon-atom signals shown as a function of liquid depth. The solid lines are calculated [31]

The evaporation from rotons can be distinguished from phonons by using the angular dispersion at the interface [32]. The boundary condition that the parallel component of momentum is conserved means that phonons and rotons at the same angle of incidence create two atom beams at different angles. Rotons give a larger angle than phonons. This is shown in figure 11, where again the energies corresponding to particular times and angles are indicated.

A roton-atom signal as a function of time separates the contributions from different energy rotons. This simplicity arises from the negligible

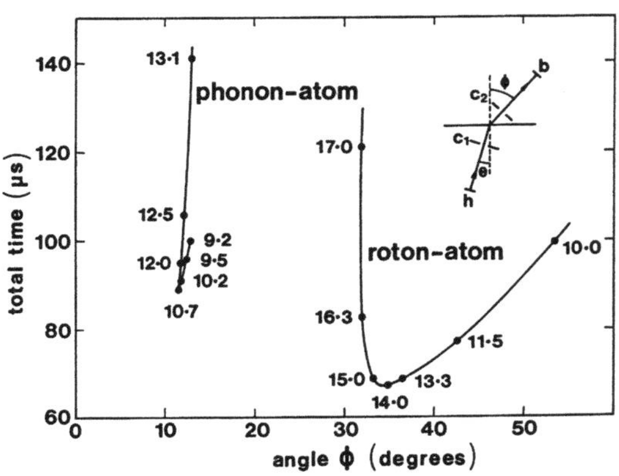

Fig 11. Calculated times and angles for phonons with angle of incidence 15°. The curves are annotated with the energies of the excitations [32]

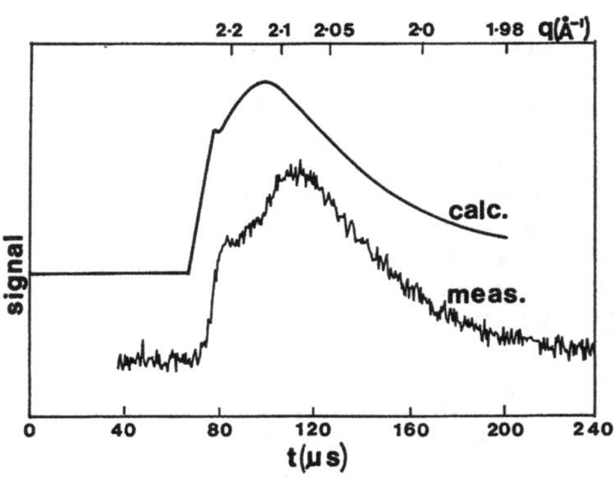

Fig 12. Roton-atom signals as a function of time. For liquid and vacuum path lengths of 6.5 mm [32,33]

occupation of the high-energy roton states with low group velocity. In figure 12 is shown a measured [32] and calculated [33] roton-atom signal. The calculation assumes that a Planck spectrum ($T = 1$ K) of rotons are injected into the liquid ^{4}He and their mean free path is much longer than the propagation distance. The roton wave vectors that give the signal at different times are shown. In particular, the narrow spike on the rising edge of the signal comes from the linear region of the dispersion curve around 2.2 $Å^{-1}$. This agreement shows very clearly that these rotons have a long mean free path.

4 Thermal Lifetimes

At finite temperatures $0.1 \leq T < 2.2$ K at SVP the lifetimes of phonons and rotons are governed by collisions with the thermal equilibrium density of phonons and rotons. For $T \geq 1$ K both the phonon and roton lifetimes are dominated by interactions with rotons. The lifetime of rotons with $q \sim 2$ $Å^{-1}$ is inversely proportional to the thermal roton density, $\tau^{-1} = 94\ \hbar\ T^{1/2} \exp(-\Delta/T)$. The measured line width [20] of the neutron spin echo and the line width of the light scattered from bound roton pairs [34] follow this temperature dependence closely. The strength of the scattering is the same as that from viscosity values [7] which similarly arises from the 4 roton process. The phonon lifetime, in this temperature range follows approximately the same temperature dependence for the higher wavevectors [35]. This suggests that at the higher temperatures ≥ 1.4 K that roton scattering of the phonons is dominant for $q > 0.5$ $Å^{-1}$. However, for the lower temperatures and $q < q_c$ there appears to be a contribution from phonon-phonon scattering via the 3pp.

At lower temperatures $T < 0.6$ K only phonons have a significant population and phonon lifetimes are governed by phonon-phonon interactions. For injected phonons with $q < q_c$ these are scattered by 3pp while for larger q, the scattering will be by 4pp. So an injected phonon with $q > q_c$ will decay slowly until $q \leq q_c$ and then more rapidly. The viscosity for the temperature $0.4 < T < 0.6$ K has been explained by 3pp [27]. The 4pp lifetime for $T \leq 0.3$ K should be measurable using quantum evaporation.

5 Conclusions

$\omega(q)$ is much better known for ^{4}He than for ^{3}He, however, there is enough evidence to see broad similarities in $\omega(q)$. For ^{3}He the lifetimes are broadened by decays to particle-hole pairs and probably multipairs. In the Bose system ^{4}He this cannot happen, and the intrinsic lifetimes are mostly very long and the excitation picture correspondingly good. At pressures $P < 19$ bar the phonon lifetimes are shortened by 3 phonon processes. At higher temperatures interactions with the thermal population reduce the lifetimes. At low temperatures the lifetimes are so long that only lower bounds can be found from time-of-flight and detection by quantum evaporation.

References

1. L.D. Landau: J. Phys. Moscow 5, 71 (1941)
2. R.P.Feynman: Phys. Rev. 94, 262 (1954)
3. V.F. Sears, E.C. Svensson, P. Martel, A.D.B. Woods: Phys. Rev. Lett. 49, 279 (1982)
4. D. Pines: Quantum Fluids, ed. D.F. Brewer, (N. Holland, 1966) p.257
5. P.J. Bendt, R.D. Cowan, J.L. Yarnell: Phys. Rev. 113, 1386 (1959)
6. M. Sudrand, E. Varoquaux: J. de Physique, C6, 21 (1976)

7. I.M. Khalatnikov: An Introduction to the Theory of Superfluidity (W. A.Benjamin Inc. N.Y. Amsterdam 1965)
8. E.C. Svensson, R. Scherm, A.D.B. Woods: J. de Physique C6, 211 (1976)
9. A.D.B. Woods, R.A. Cowley: Repts in Progress in Physics 36, 1145 (1973)
10. T.J. Greytak, R. Woerner, J. Yan, R. Benjamin: Phys. Rev. Lett. 25, 1547 (1970)
11. J. Ruvalds, A. Zawadowski: Phys. Rev. Lett. 25, 333 (1970)
12. W.G. Stirling, Proc. 2nd Internat. Conf. on Phonon Physics, eds. J. Koltar, N. Kroo, N. Menyhard, T. Siklos (World Scientific 1985) p.829
13. C.H. Aldrich, C.J. Pethic, D. Pines: Phys. Rev. Lett. 37, 245 (1976)
14. K. Skold, C.A. Pelizzari, R. Kleb, G.E. Ostrowski: Phys. Rev. Lett. 37, 842 (1976)
15. K. Skold, C.A. Pelizzari: J. Phys. C11, L589 (1978)
16. R. Scherm, W.G. Stirling, A.D.B. Woods, R.A. Cowley, G.J. Coombs: J. Phys. C7, L341 (1974)
17. W.G. Stirling, R. Scherm, P.A. Hilton, R.A. Cowley: J. Phys. C9, 1643 (1976)
18. P.A. Hilton, R.A. Cowley, W.G. Stirling, R. Scherm: J. de Physique C6, 208 (1978)
19. W. Stirling: 75th Conf on ^{4}He, ed. J.G.M. Armitage, (World Scientific Pub. Co., Singapore 1983) p.109
20. F. Mezei: Phys. Rev. Lett. 44, 1601 (1980)
21. J. Wilks: Properties of Liquid and Solid Helium (Clarendon Press 1967)
22. H.J. Maris, W.E. Massey: Phys. Rev. Lett. 25, 220 (1970)
23. J. Jackle, K.W. Kehr: Phys. Rev. A9, 1757 (1974)
24. N.G. Mills, R.A. Sherlock, A.F.G. Wyatt: Phys. Rev. Lett. 32, 978 (1974)
25. R.C. Dynes, V. Narayanamurti: Phys. Rev. Lett. 33, 1195 (1974)
26. A.F.G. Wyatt, N.A. Lockerbie, R.A. Sherlock: Phys. Rev. Lett. 33, 1425 (1974)
27. H.J. Maris: Rev. Mod. Phys. 49, 341 (1977)
28. A.F.G. Wyatt, D.R. Allum: J. Phys. C15, 1917 (1982)
29. Y. Korczynskyj, A.F.G. Wyatt: J. de Physique C6, 230 (1978)
30. D. Rugar: Phonon Scattering in Condensed Matter, eds. W. Eisenmenger, K. Lassmann, S. Dottinger (Springer, Berlin, Heidelberg, N.Y., Tokyo 1984) p.26
31. M.J. Baird, F.R. Hope, A.F.G. Wyatt: Nature 304, 325 (1983)
32. F.R. Hope, M.J. Baird, A.F.G. Wyatt: Phys. Rev. Lett. 52, 1528 (1984)
33. A.F.G. Wyatt, M. Brown: Proc. 2nd Internat. Conf. on Phonon Physics, eds. J. Kollar, H. Kroo, N. Menyhard, T. Siklos, (World Scientific 1985) p.440
34. T.J. Greytak, R. Woerner, J. Yan, R. Benjamin: Phys. Rev. Lett. 25, 1547 (1970)
35. F. Mezei, W.G. Stirling: 75th Jubilee Conf. on ^{4}He, ed. J.G.M. Armitage (World Scientific 1983) p.111

Phonon Scattering at Surfaces and Interfaces

W. Eisenmenger

Physikalisches Institut, Universität Stuttgart, Pfaffenwaldring 57, D-7000 Stuttgart 80, Fed. Rep. of Germany

1. Introduction

Since the discovery of the thermal boundary resistance by Kapitza /1/ in 1941, there have been extensive experimental and theoretical studies on the underlying phonon transmission, reflection and scattering processes at surfaces and interfaces. The field has been recently reviewed by A.C. Anderson /2/ and A.F.G. Wyatt /3/. The reader is referred to these authors, since the present contribution cannot give a complete account of all the relevant work done. For ideal solid-liquid or solid-solid boundary conditions the theoretical treatment according to the acoustic mismatch model by Khalatnikov /4/ and Little /5/ predicts strong phonon reflection at the solid-liquid ^4He interface, but weak phonon reflection for most solid-metal interfaces. In the latter case the acoustic model was especially successful for the calculation of phonon spectra and intensity distributions emitted by metallic heaters into dielectric substrates as performed by Weis and co-workers /6/ taking also account of phonon focussing /7/. Experimental evidence for the applicability of the acoustic model to the solid-liquid ^4He interface for phonon frequencies ranging from 100 to about 700 GHz was provided by the famous UHV cleaving experiments of Kinder and co-workers /8/. More recently Basso et al. /9/ demonstrated that phonons of 100 GHz are almost ideally reflected from Si-liquid ^4He interfaces after laser annealing the Si-crystal under UHV conditions. Even multiple solid-solid interfaces as in the phonon transmission experiment through a GaAs:AlGaAs MBE superlattice by Narayanamurti et al. /10/ exhibit ideal acoustic properties in the frequency regime of 200 GHz. Similar results have been obtained with amorphous Si/SiO_2 superlattices by Koblinger et al. /11/.

In contrast to these examples of "ideal" acoustic behaviour, many "real" solid-liquid or solid-solid interfaces deviate significantly from the acoustic model, as well known by the anomalous high phonon transmission /Kapitza anomaly/ from solids into liquid He cf. /3/ or by phonon losses /12/ and anomalous high phonon backscattering /13/ at solid-solid interfaces. Phonon transmission and phonon reflection experiments with nonideal acoustic interface systems demonstrated in addition to specular diffraction or reflection large contributions of diffusely transmitted and or scattered phonons cf. /3/, /14/, /15/. This is consistent with the fact that violation of the k-parallel conservation leads to additional channels of phonon transport cf. /3/. The diffuse backscattering from uncovered crystal surfaces is strongly reduced by contact with liquid He /16/, /17/, whereas the specular reflected components remain almost unchanged. Therefore, it appears natural to assume that the same nonideal interface properties leading to diffuse backscattering from the uncovered surface are mainly responsible for anomalous high phonon transmission into the ^4He bath or for enhanced backscattering at solid-solid interfaces.

The following report on experiments with frequency-dependent phonon backscattering at differently prepared Si and Al_2O_3 surfaces demonstrates the variety of phenomena in phonon backscattering of real surfaces to be explained by appropriate models. This work has been performed by S. Burger in co-operation with K. Laßmann, S. Döttinger, E. Mok, R. Schneider and the author.

2. Experiment

In the experimental arrangement of Fig. 1 a Si-crystal is sealed against an UHV chamber, carrying also a window for laser annealing experiments /18/ to be described later. The main phonon propagation path for 90° specular reflection corresponds to maximum phonon focussing of the ft-mode, the same holds for the paths of diffuse scattering and mode conversion. The geometry is well suited for differentiating between various phonon paths by reflection and scattering in pulse experiments. For comparison it is possible to perform theoretical Monte Carlo simulations of the expected pulse signal /13/. For phonon generation mostly constantan heaters have been used with frequency variation by changing the heater power. The main frequency range could be selected with Sn-tunneling junctions at 280 GHz and Al-junctions with a lower frequency limit of roughly 80 GHz /19/ as detectors.

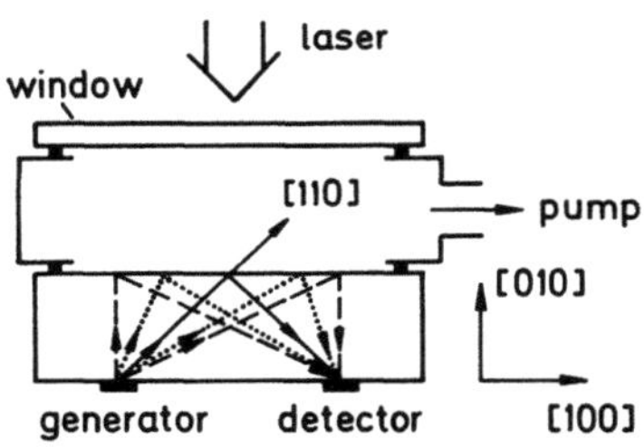

Fig. 1 UHV sample chamber with crystal and optical window for laser annealing. The propagation paths are: Full lines: specular and diffuse longitudinal mode, first peak in Fig.2. Specular and diffuse fast transverse mode, third peak in Fig.2. Dotted lines: mode conversion longitudinal - fast transverse phonons, second peak in Fig.2. Dashed lines: diffuse scattered fast transverse phonons along phonon focussing direction, fourth peak in Fig.2.

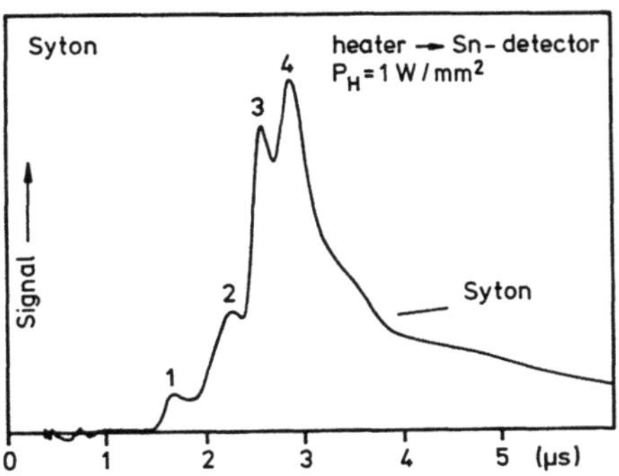

Fig. 2 Phonon backscattering signal from a Syton-polished, uncovered Si-surface at 280 GHz. The scattering is almost completely diffuse, see text. For peak identification cf. Fig.1.

3. Phonon backscattering from mechanical and chemical polished Silicon

The crystals were prepared by polishing with 9 μm diamond and finally with Syton. Fig. 2 shows the Sn-detector pulse response /19/ corresponding to 285 GHz with high heater generator power. This agrees with earlier results of Marx /20/ indicating almost complete diffuse phonon backscattering as evidenced by comparison with the computer simulated pulse response. A more direct check was the preparation of real "rough" surfaces with 0,25 μm or even 15 μm diamond showing agreement with the result of the Syton polished surface.

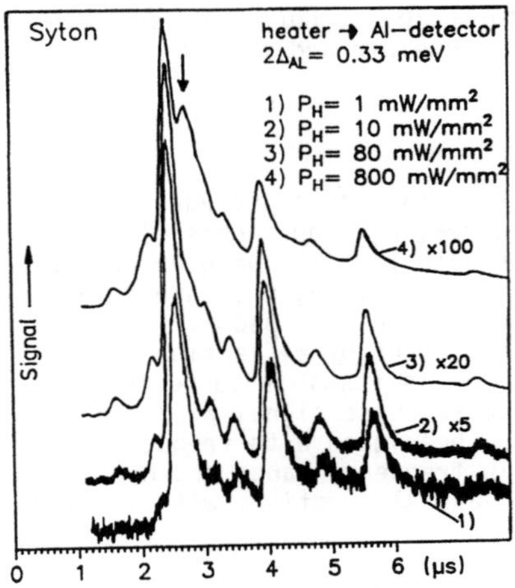

Fig. 3 Backscattering signal changing from specular to diffuse of a Syton-polished Si-surface for different heater power with an Al-junction as detector. The arrow at curve 4 marks the diffuse scattering contribution at high heater power.

A dramatic change of the backscattering signal (Fig. 3) is observed with an Al-detector sensitive also for low frequencies in the 80 GHz range. The observed echoes result from oblique multiple phonon reflection between the crystal face carrying generator and detector and the opposing surface. This gives direct evidence for specular phonon reflection especially at low heater power. With increasing heater power and the resulting spectral shift to higher frequencies again the most prominent diffuse scattering peak no. 4 (cf. Fig.1) marked by the arrow is clearly observed. This is supported by the influence of ^{4}He coverage shown in Fig. 4 and 5. With low heater power the signal reduction in Fig. 4 is almost unobservable, whereas at high heater power the significant signal reduction in Fig. 5 corresponds to a strong diffuse backscattering contribution. Polishing only with diamond 0,25 μm grain size reduces the specular reflection as expected, but at low heater power the influence of the ^{4}He coverage is surprisingly small. Apparently the polishing grain size of 0,25 μm is too small for strong diffuse scattering at this frequency. The

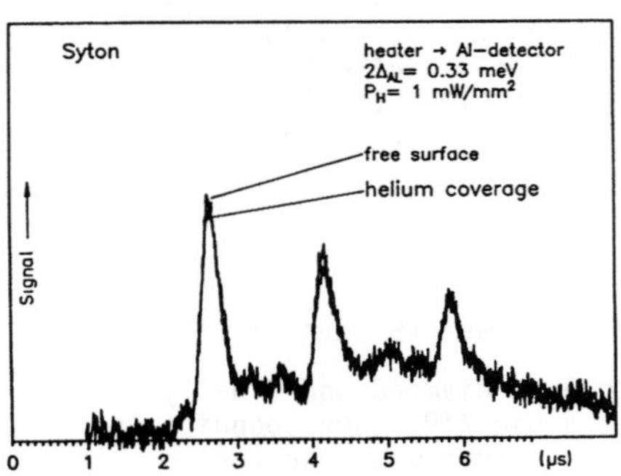

Fig. 4 Backscattering signal for a Syton-polished Si-surface at low heater power without and with liquid ^{4}He-contact.

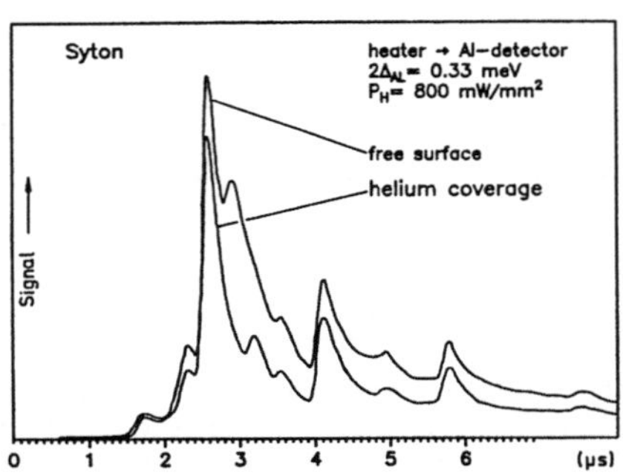

Fig. 5 Backscattering signal for a Syton-polished Si-surface at high heater power without and with liquid ^{4}He-contact.

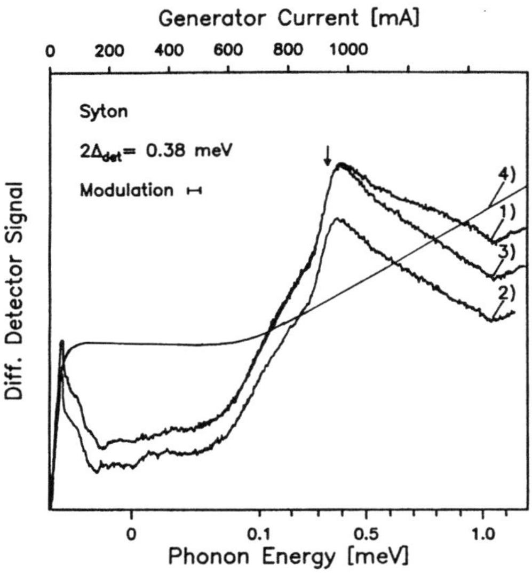

Fig. 6 Phonon energy dependence of the differential backscattering signal from a Syton-polished Si-surface:
curve 1: uncovered surface
curve 2: helium-covered surface
curve 3: curve 2 fitted to the lower part (from zero voltage to 0,35 meV) of curve 1
curve 4: Sn-generator I-V dependence.
The arrow marks the position of the detector gap.

crystal surface showed residual fine scratches under high optical resolution microscopy, whereas the Syton polished surface appeared optically smooth. Increasing the heater power to 1 W/mm^2 the diamond polished surface showed with Al-detection strong diffuse scattering almost identical to Fig. 2 with corresponding signal reduction by liquid ^{4}He contact.

More detailed information on the frequency dependence of the phonon scattering has been obtained with a Sn-generator (and an Al-detector) as a tunable phonon source using the modulation technique. From intensity reasons pulse resolution was not obtained and the result for a Syton polished Si-surface in Fig. 6 represents a signal-time average. Curve 1 and curve 2 correspond to the frequency dependence of phonon backscattering without and with liquid ^{4}He coverage respectively. Curve 3 is the same signal as in curve 2, but normalized to the maximal value of curve 1. Both curves being almost identical below phonon energies of 0,3 meV show a signal reduction of curve 3 relative to curve 1 above 0,35 meV. This indicates an increasing phonon escape from the solid into the ^{4}He liquid with a threshold of 0,35 meV in agreement with the frequency dependence /21/ of the phonon escape from a Sn-tunnel junction into liquid ^{4}He. This substrate independence is an indication for the more general nature of the 0,35 meV escape threshold which consequently may be related to surface impurities or defects in combination with liquid ^{4}He boundary states /22/, /23/.

4. Phonon backscattering from Silicon treated by laser annealing

Since mechanical, chemical and other polishing procedures did not change the almost complete diffuse backscattering of 285 GHz phonons into at least partly specular reflection, the laser annealing method /9/ has been applied also to our sample configuration /18/ (see Fig. 1) with a Sn-detector. In using a Q-switch ruby laser with 1 J pulse power and 50 ns duration, it was sufficient to treat the effective scattering surface area of the Si-crytal with 5 laser pulses until reaching the maximal and final change to specular reflection shown in Fig. 7. Again covering the scattering surface with liquid ^{4}He results

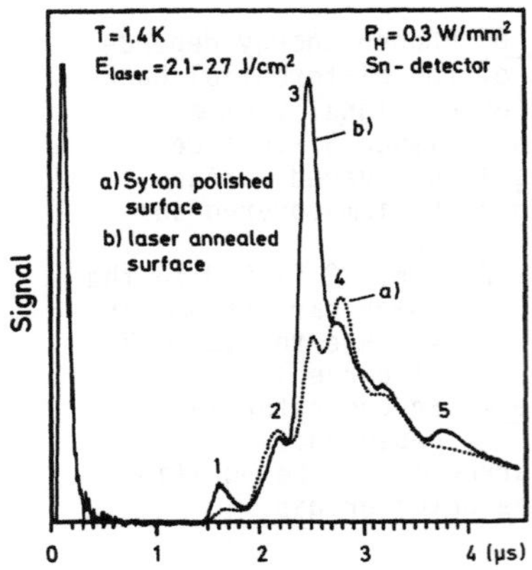

Fig. 7 Phonon backscattering signal of a polished Si-surface before (curve a) and after (curve b) laser annealing. Besides the increase of pulse 3 by specular reflection also a double specular reflection signal 5 is observed.

in only a 13% ft signal reduction. This demonstrates the high degree of specular phonon reflection of the Si-surface treated by laser annealing even at the frequency of 285 GHz. Exposing the reflecting surface to air (at room temperature) changed the reflection again back into complete diffuse scattering with corresponding signal reduction by liquid ^4He contact. Reflection could only be obtained anew with pulse powers of the original value, i.e. surface melting. The laser annealing method /9/ also in our experiments at 285 GHz proved as a very efficient means of ideal acoustic surface preparation.

The residual diffuse scattering observed at 285 GHz decreases as to be expected with frequency /24/. The experimental accuracy, however, has to be improved to resolve the 0,35 meV /21/ phonon escape threshold.

5. Phonon backscattering from Sapphire surfaces

Specular and diffuse reflection at sapphire surfaces has been already observed by Taborek and Goodstein /17/. Experiments with the sample arrangement of Fig.1 and the sapphire surface in the (001) plane /25/ indicated by comparision with Monte Carlo computations a ratio of 3:4 between specular and diffuse scattering in the 80 GHz frequency range. The scattering surface was diamond polished with a final Syton treatment. Surprisingly a change to almost complete specular reflection (Fig. 8) could be obtained by a 10 min. polishing

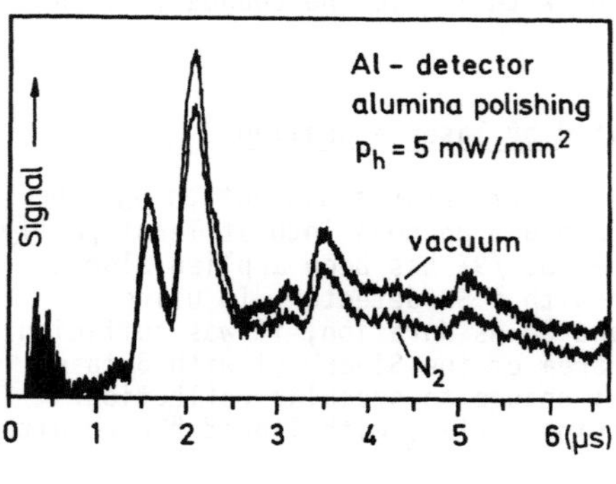

Fig 8 Specular backscattering of a Sapphire surface polished with cubic Al_2O_3 without and with condensed N_2.

with cubic Al_2O_3 of 50 nm grain size. This is demonstrated by multiple reflection signals and by only a small signal reduction with condensed N_2 or liquid ^4He contact. If the surface is treated again a few minutes with Syton, the original contribution of diffuse phonons is observed, correspondingly the specular reflection is found by a succeeding Al_2O_3 polish. This has been repeated several times. The corresponding changes of the surface properties are possibly related to structural or chemical differences on the nanometer scale. But electron micrographs of surface replicas did not show different surface structures within 10 nm resolution.

6. Phonon scattering from condensates on ideal surfaces

With primarily ideal reflecting surfaces it is now possible to study the phonon scattering by controlled adsorption of atoms or molecules /26/ or condensed layers. The result of an experiment /27/ with a 324 nm constantan film evaporated on a Syton polished Si-crystal is shown in Fig. 9. At a frequency of 80 GHz the specular reflection of the uncovered surface is almost completely changed to strongly reduced diffuse backscattering. This is in accord with the acoustic model. Almost perfect acoustic matching results in high transmission to the metal film. Phonon absorption and emission by electron interactions results in thermalization and subsequent diffuse emission back into the substrate. Similar experiments by Klitzner and Pohl /28/ on the temperature dependence of heat transport in perfect single crystal rods covered by different adsorbates and thin filmsexhibit a significant frequency (temperature) dependence related in a characteristic way to the film structure.

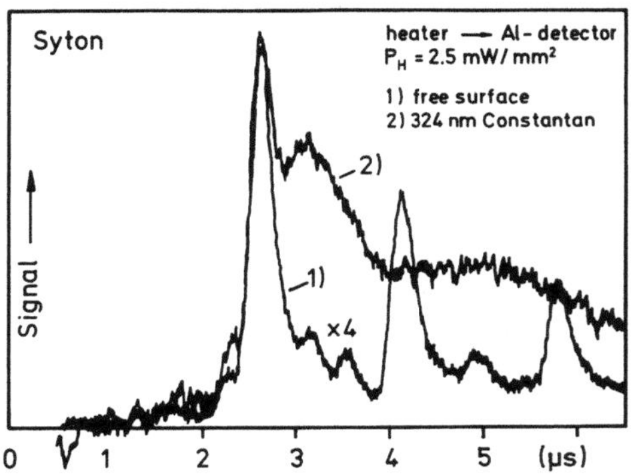

Fig. 9 Phonon backscattering from a Syton-polished Si-surface before (curve 1) and after (curve 2) evaporation of 324 nm thick constantan film. Curve 1 is reduced in amplitude by a factor of four.

7. Conclusion

Several experimental techniques as UHV cleaving /8/, laser annealing /9/, MBE /10/ or other UHV evaporation techniques /11/ or even the simple polishing procedures (100 GHz range) reported here are presently available for the preparation of acoustically ideal surfaces and interfaces. For these systems the classical acoustic models appear to be applicable up to several 100 GHz. Deviations from the acoustic model as the Kapitza anomaly are accompanied by diffuse phonon backscattering and or transmission. Only little is known under what conditions these scattering processes are elastic or inelastic. For uncovered Si-surfaces /12/ elastic diffuse scattering has been observed. Diffuse phonon transmission into liquid ^4He appears to be accompanied by inelastic decay, cf. ref. /3/, as has been found also in phonon loss processes /12/ at the interface between different solids as e.g. tunneling junctions and substrates or other systems.

A large variety of models cf. also ref. /3/ has been proposed in order to explain anomalous or nonacoustical properties. These models are based in principle on diffuse scattering (violation of the $k_=$ conservation) by surface roughness /31/,surface,subsurface or interface defects,adsorbed atoms,molecules and corresponding phonon states in the form of two-level systems, either at the surface /22/ or in the first ajoining ^{4}He layer /23/. Similar phonon states or nonlinear resonant systems are possible in adsorbed atomic or molecular clusters or islands of adsorbates /21/,/19/,/26/,/28/,/29/,/32/, for which already experimental evidence has been given by a characteristic frequency dependence of the diffuse scattering processes. It is to be hoped that different mechanisms for diffuse acoustic phonon scattering at surfaces and interfaces can in future be identified by their frequency dependence and by other significant properties, as for example the polarisation and angle dependence /30/.

I should like to thank all my collaborators for their contributions and numerous stimulating discussions. I am also grateful to Tom Klitzner and R.O. Pohl as well as to A. Khater for supplying preprints of their recent results. This work was supported by the Deutsche Forschungsgemeinschaft.

References

/1/ P.L. Kapitza, J. Phys., USSR, 4, 181 (1941)
/2/ A.C. Anderson in Nonequilibrium, Superconductivity, Phonons and Kapitza Boundaries; K.E. Gray Ed., Nato Adv.Study, Inst. Series B, Physics, Vol.65, Plenum Press, New York, 1981
/3/ A.F.G. Wyatt, same source as /2/
/4/ I.M. Khalatnikov, Zh. eksp. theor. Fiz. (USSR) 22:687 (1951)
/5/ W.A. Little, Can. J. Phys. 37, 334 (1959)
/6/ F. Rösch, O. Weis, Z. f. Phys., B 27, 33 (1977)
/7/ B. Taylor, H.J. Maris, C. Elbaum, Phys. Rev. Lett. 23, 416 (1969)
/8/ J. Weber, W. Sandmann, W. Dietsche, H. Kinder, Phys.Rev.Lett 40, 1469 (1978)
/9/ H.C. Basso, W. Dietsche and H. Kinder, P. Leiderer in Phonon Scattering in Condensed Matter (W. Eisenmenger, K. Laßmann, S. Döttinger Ed.) Springer Series in Solid State Sciences, 51
/10/ V. Narayanamurti, H.L. Störmer, M.A. Chin, A.C. Gossard. W. Wiegmann, Phys. Rev. Lett. 43, 2012 (1979)
/11/ O. Koblinger, J. Mebert, E. Dittrich, S. Döttinger, W. Eisenmenger, this volume and Verhandlungen d. DPG 5, 1481 (1986)
/12/ L.J. Challis, in Phonon Scattering in Condensed Matter, W. Eisenmenger, K. Laßmann, S. Döttinger Ed., Springer Series in Solid State Sciences, Vol 51, 2, (1984)
H.J. Trumpp, W. Eisenmenger, Z. Phys. B 28, 159 (1977)
/13/ D. Marx, W. Eisenmenger, Phys. Lett. 82 A, 291 (1981)
/14/ C.J. Guo, H.J. Maris, Phys. Rev. Lett 29, 855 (1972)
/15/ R.E. Horstmann, J. Wolter, Phys. Lett. 62A, 279 (1977)
/16/ J.T. Folinsbee, J.P. Harrison, J. of Low Temp. Phys. 32, 469 (1978)
D. Marx, J. Buck, K. Laßmann, W. Eisenmenger, J. Phys. C6, suppl. to No. 8, 1015 (1978)
/17/ P. Taborek, D.L. Goodstein, J. Phys. C12, 4737 (1979);
Sol. State Comm. 38, 215 (1981) ; Phys. Rev. B22, 1550 (1980)
/18/ E. Mok, S. Burger, S. Döttinger, K. Laßmann, W. Eisenmenger, Phys. Lett. 114, 473 (1986)

/19/ S. Burger, K. Laßmann, W. Eisenmenger, J. of Low Temp.Phys.61,401 (1985)
/20/ D. Marx, W. Eisenmenger, Phys. Lett. 93A, 152 (1983);
Z. Phys. B48, 277 (1982)
/21/ O. Koblinger, U. Heim, M. Welte, W. Eisenmenger, Phys.Rev.Lett.51, 284 (1983); a slightly different frequency dependence was earlier found by E.S. Sabisky, C.H. Anderson, Sol. State Comm. 17, 1095 (1975)
/22/ H. Kinder, Physica 107B, 549 (1981)
/23/ T. Nakayama, Phys. Rev. B32, 777, (1985); J. Phys. C18, 667 (1985)
/24/ R. Schneider, Diplomarbeit, Universität Stuttgart, 1986
/25/ S. Burger, W. Eisenmenger, K. Laßmann, Proc.LT17, Vol. 9A, North Holland (1984)
/26/ L. Köster, St. Wurdack, W. Dietsche, H. Kinder, this volume and Verhandlungen d. DPG 5, 1477 (1986)
/27/ S. Burger, Dissertation, Universität Stuttgart (1986)
/28/ R.O. Pohl and B. Stritzker, Phys. Rev. B25, 3608 (1982);
T. Klitzner and R.O. Pohl in Phonon Scattering in Condensed Matter ed. by W. Eisenmenger, K. Laßmann and S. Döttinger (Springer, Berlin,1984) p. 188; T. Klitzner and R.O. Pohl this volume
/29/ A. Khater, Phys. Rev. B (submitted)
/30/ H. Kinder, A. de Ninno, D. Goodstein, G. Paterno, F. Scaramuzzi and S. Cunsolo,, Phys. Rev. Lett. 55, 2441 (1985) and this volume
/31/ N.S. Shiren J. Phys. Suppl. C6, 42, 816, (1981),
Phys. Rev. Lett. 47, 1466, (1982)
/32/ L.J. Challis, J. Phys. C7, 481, (1974)

Three-Phonon Processes in Second Sound, Poiseuille Flow, and Thermal Conduction in Solid ^{4}He

S.J. Rogers and G.N. Gamini

Physics Laboratory, University of Kent, Canterbury, Kent CT2 7NR, UK

Poiseuille flow and second sound have provided for solid ^{4}He a body of evidence concerning τ_N^{-1}, the N-process scattering rate. τ_N^{-1} has also been inferred from the thermal resistivity of isotopic point defects. However, the various estimates of τ_N^{-1} differ both as regards temperature dependence and the absolute magnitude [1]. Recent Poiseuille flow experiments of Golub et al [2] point to differences in dislocation concentration in the helium crystals as a source for these disparities. In their analysis, allowing for dislocations increases the overall estimate of τ_N^{-1} and changes its apparent temperature dependence from $(T/\theta)^3$ to $(T/\theta)^5$. We have looked again at the pulse propagation of second sound in solid ^{4}He to see what evidence there might be for the effects of dislocations.

Our observations were in the pressure range from 40-100 atmospheres. To simplify the analysis of the data we chose to use a planar geometry with the dimensions of both heater and detector (3 × 3 mm^2) large compared with the crystal thickness, L, (0.4-1.5 mm). Pulse propagation in this situation should be well characterised by the propagating Gaussian solution of the one-dimensional second sound equation - for a δ function input of heat at x = 0 at time zero, the temperature excursion at x = L at time t is

$$\Delta T(L,t) = \frac{A\exp(-t/2\tau_R)}{(\tau_N t)^{\frac{1}{2}}} \exp\left\{-\frac{(L - c_2 t)^2}{0.4 c_2^2 \tau_N t}\right\}, \qquad (1)$$

where τ_R^{-1} is the scattering rate for resistive processes, c_2 is the second sound velocity, and A is proportional to the input pulse. This form of solution does not include the effects of second viscosity, and resistive scattering only appears as an attenuation factor. For large defect concentrations the resistive scattering has the additional effect of broadening the received second sound pulse. This can be seen in Fig. 1, where second sound transmission in a crystal before and after annealing is compared. The crystal, as initially grown by rapid cooling, has a large concentration of defects, possibly dislocations, which give rise to a broadening of the detected pulse. At 1.175 K the pulse width for the unannealed crystal is nearly 40% more than for the annealed crystal. If Eq. (1) is used to determine τ_N from the two pulse profiles, the estimates obtained differ by almost a factor of 2. The resistive scattering has the effect of *increasing* the apparent value of τ_N.

The strength of the resistive scattering for the unannealed crystal may be inferred in two ways: firstly the factor $\exp(-t/2\tau_R)$ when applied to the small echo pulse at 1.175 K yields τ_R = 2.2 µs; secondly, computer modelling of the effect of τ_R on the pulse profile [4] suggests $\tau_R \cong 1.5$ µs at 1.175 K. For the unannealed crystal of Fig. 1a the resistive scattering appears to be at a minimum at about 1.1 K. We expect τ_R^{-1} to increase at

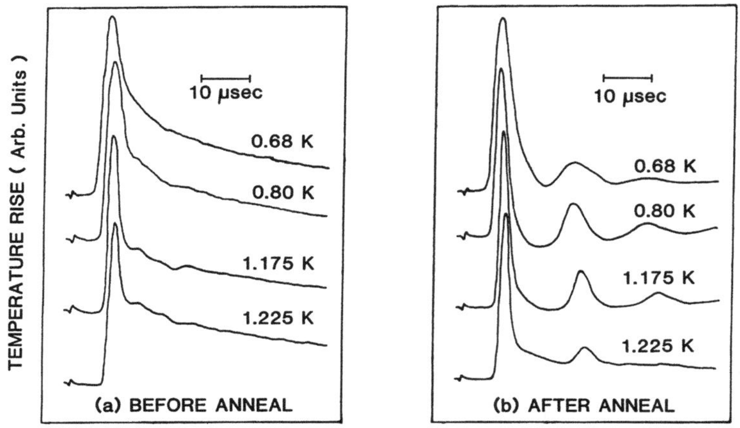

Fig. 1. The effect of defects on second sound propagation in solid ^4He at a pressure of 99.7 atm. L = 1.52 mm.

higher temperatures because of the increase in the resistive 3-phonon processes; the apparent increase at lower temperatures, where the second sound echo disappears, is consistent with scattering by the *flutter* (vibration) of dislocations. Whereas the strain field scattering for dislocations varies linearly with the phonon wave vector q, the flutter scattering varies as q^{-1} and so, in the dominant phonon approximation, varies at T^{-1}. In the analysis of their work, Golub et al identify such scattering by dislocation flutter as important. For a dislocation concentration of 3×10^6 cm^{-2}, they represent the scattering rate, averaged over the phonon distribution, by

$$\tau_R^{-1} = 1.4 \times 10^4\, T + 3.5 \times 10^4/T \ s^{-1} \quad , \qquad (2)$$

where the second term represents the flutter scattering. It is only necessary to increase this τ_R^{-1} by a factor of 10 to account for what we see in Fig. 1a.

It is safe to assume that annealing will not remove all such defects, but at the lowest temperature of our observations, we are able to account for all of the apparent attenuation in our best crystals in terms of second viscosity and the change in τ_N. Fig. 2 shows data for such a crystal; the relative amplitude of the first echo and the primary pulse, which provide a measure of the attenuation, are plotted as a function of T. Making reasonable assumptions about the dominant frequency in the propagating pulses, we have used the computed dispersion relations of Ref. [4] for defect free crystals to estimate the pulse attenuation due to τ_N. The assumption $\tau_N = 9.4 \times 10^{-15}\ (\theta/T)^4$ s leads to the estimates represented by the open circles. Although the agreement with the observations is perhaps fortuitously good, it is clear that the variation in τ_N alone can account for the apparent attenuation below the temperature at which it is a minimum. The choice of τ_N is also consistent with the observed increase in the second sound velocity at the lowest temperatures. At higher temperatures where resistive scattering is significant, τ_N must be derived from the pulse width data. Such values for the crystal of Fig. 2 are shown as solid triangles in Fig. 3 together with similar data for some other crystals. The solid line is the form for τ_N assumed above. The τ_N values in Fig. 3 are derived from the forward half widths of the pulses.

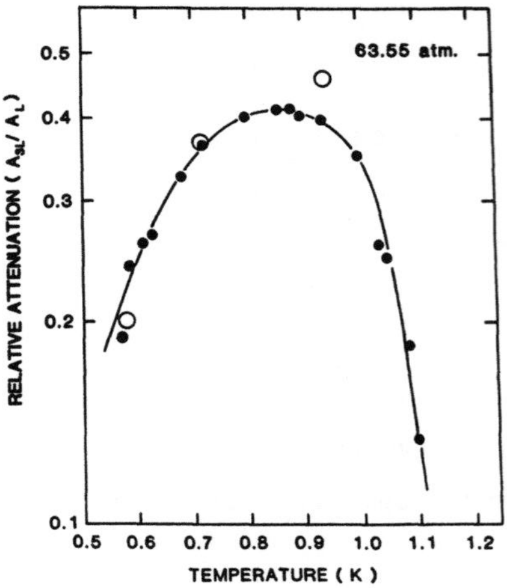

Fig. 2. Relative attenuation data. L = 0.43 mm.

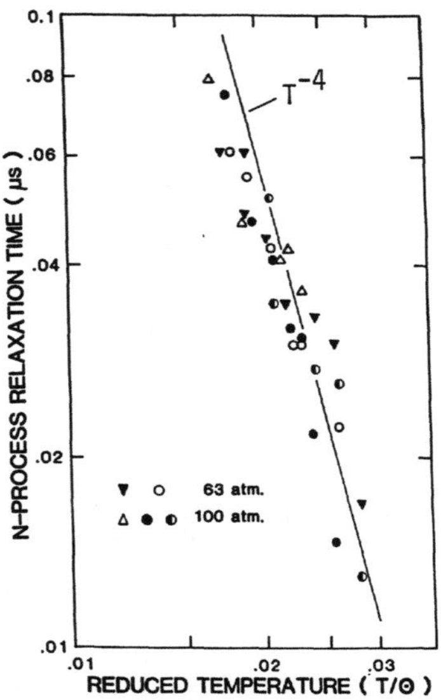

Fig. 3. τ_N values for various crystals. L = 0.43 mm.

Although a T^{-4} variation for τ_N fits the overall pattern of the data, the apparent temperature variation for τ_N for a given crystal can be quite different from this. We believe that resistive scattering is responsible for much of this variation, and that consistent values for τ_N will only be obtained when the role of the resistive scattering is taken fully into account. The preliminary analysis of our results presented here encourages us to have confidence in the findings of Golub et al as regards the strength of τ_N. In most cases the data available for each crystal spans only a limited temperature range and the precise form of the T dependence is still an open question. The T^4 form for τ_N used above has the virtue of being consistent with the values deduced from point defect scattering.

1. See references cited in [2-4] below.

2. Golub, A.A. and Svatko, S.V., Sov. J. Low Temp. Phys., 6, 465 (1980), and Golub, A.A., Zuev, N.V. and Mikhailov, G.A., Sov. J. Low Temp. Phys., 9, 229 (1983).

3. Trefny, J.U., Guo, C.J. and Fox, J.N., J. Low Temp. Phys., 29, 533 (1977).

4. Rogers, S.J., Phys. Rev., B3, 1440 (1971).

One-Dimensional Second Sound in Superfluid Helium

J.P. Eisenstein and V. Narayanamurti

AT & T Bell Laboratories, 600 Mountain Ave.,
Murray Hill, NJ 07974, USA

In the low temperature, phonon dominated, regime the velocity of second sound in superfluid helium approaches $c_0/\sqrt{3}$, where c_0 is the ordinary acoustic sound velocity. Observation of this limiting behavior is difficult owing to the rapid decrease of the wide-angle scattering rate $\tau_\perp^{-1}$ as the temperature is reduced. At a given frequency ω, the product $\omega\tau_\perp$ soon exceeds unity and the mode collapses. Although wide-angle phonon scattering is required for the propagation of second sound, at low pressures and temperatures it is the nearly collinear ($\sim 5°$) 3-phonon process that dominates the mean free path[1]. This is due to the phonon dispersion being anomalous or "upward" at small wavevector. Maris[2] showed that for times intermediate between $\tau_{\|}$, the small-angle time and $\tau_\perp$ a pseudo-temperature can be assigned to groups of phonons propagating in a given direction and equilibrating amongst themselves via the 3-phonon process. He further predicted that a new type of second sound would propagate in this one-dimensional regime with velocity close to c_0. It is the study of this new mode that we report on here.

Heat pulses are generated by a thin-film heater and detected by a novel 2-D electron gas bolometer which has a demonstrated sensitivity to low energy phonons[3]. The glass plate on which the heater is evaporated and the MBE-grown GaAs heterostructure form a parallel plate "resonator" with 1.31mm spacing. Both the pulsed and cw response of this device have been studied. Figure 1 shows typical pulse trains resulting from a single short (100ns) heater pulse at both low and high pressure. At low pressure the time-of-flight gives, within experimental uncertainty, a speed of propagation equal to c_0 as expected for ballistic propagation. Uncharacteristic of ballistic phonons is the long train of strong, well-defined echoes usually associated with a collective mode. At high pressure the speed falls about 3% below c_0 and only one echo is significant. The ratio of first

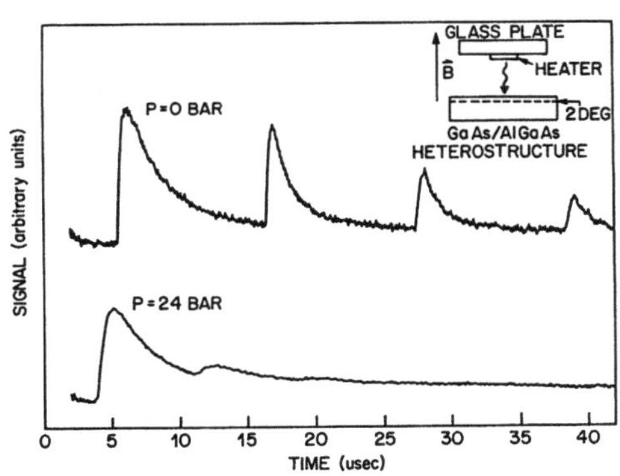

Fig. 1. Typical pulse trains at low and high pressure at 150mK. Inset shows experimental arrangement.

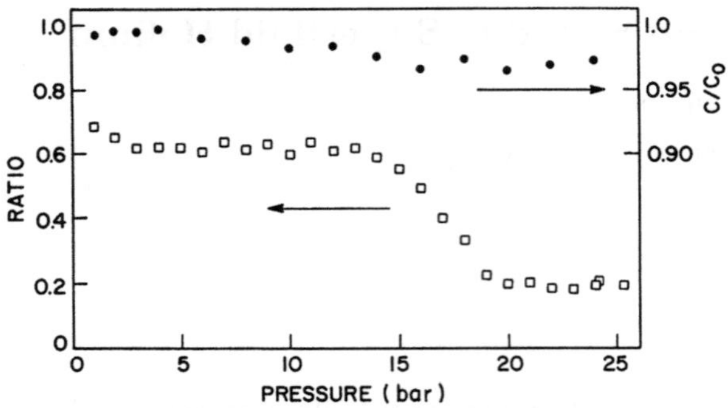

Fig. 2. Open squares show ratio of first echo to main pulse amplitude vs. pressure. Solid dots give propagation speed in units of acoustic sound velocity.

echo to main pulse amplitude versus pressure is shown in Fig. 2 along with the speed of propagation. Up to about 13 bar this ratio is roughly constant but then falls sharply to a lower plateau reached by 19 to 20 bar. We believe these data represent the first observation of one-dimensional second sound propagating at speeds close to c_0. The data in Fig. 2 show that the mode exists only below about 20 bar where the phonon dispersion is anomalous and the essential 3-phonon process is operative. Above 20 bar the dispersion becomes normal (downward) and the 3-phonon process shuts off and the collective mode is replaced by ballistic propagation.

At pressures below 20 bar not all of the phonons participate in the collective mode. While the phonon dispersion at low energy is anomalous in this pressure range, there exists a series of closely-spaced cut-off energies above which all multi-phonon decay processes cease and mean free path becomes very large[1]. It suffices to consider only the highest of these cut-offs and to also consider only the 3-phonon events since they dominate the mean free path at low energy. A given heat pulse ($T_h \sim 1$K) will contain ballistic components above cut-off as well as lower energy phonons which initially down-convert via 3-phonon decays but eventually settle down as a low-energy mass propagating collectively as 1-D second sound and equilibrating through 3-phonon processes. As the pressure is increased the cut-off drops from about 9.85K at P=0 to zero at around 20 bar[4]. At low pressure essentially all phonons participate in the collective mode but as the pressure is raised an increasing fraction are above cut-off and propagating ballistically. The pressure variation of the echo ratio plotted in Fig. 2 depends slightly on the magnetic field applied to operate the detector, through the changing spectral response of the 2-D electron gas.

To verify the existence of a collective mode with macroscopic wavelength we have also studied 1-D second sound standing waves in this parallel plate geometry. These cw observations, in the frequency range below 250kHz, have revealed as many as three resonances in nearly harmonic sequence. The frequencies are given, to within a few percent by $f_n = nc_0/2d$ where n = 1,2,3 etc. and d is the plate spacing, 1.31mm. A typical frequency spectrum is shown in Fig. 3 along with a depiction of the inferred temperature variation across the cell.

Boltzmann equation calculations[2,5] have yielded a complete picture of the transition between ordinary second sound at low frequencies and high temperatures and this new mode in the opposite limits. In the temperature (T<0.25K) and frequency (f>70kHz) range of our measurements the calculations show the

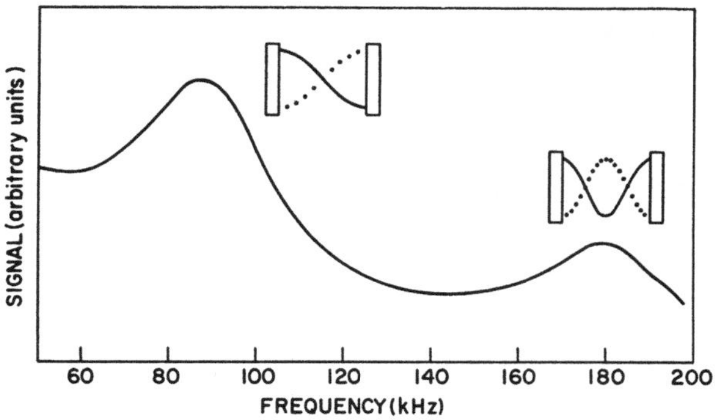

Fig. 3. The two lowest standing wave resonances at P=0 and 120mK. Insets depict temperature variation across cell.

propagation speed differing from c_0 by no more than about 5%. For the cw measurements we have observed systematic variations of the speed with both frequency and temperature that are close to the predictions but the very narrow window studied prohibits a thorough comparison. Below about 0.2K we find the Q of these resonances, typically 5, to be temperature independent indicating the probable dominance of reflection losses over the bulk attenuation Maris[2] calculates.

An additional interesting feature of the theory has been pointed out by Benin[5]. He predicts that the 1-D second sound mode will propagate well into the "collisionless" regime $\omega\tau_{\|} > 1$ and shows, with a simple geometric picture, that this is due to the small angle of the 3-phonon process. At P=0 and T=0.25K he finds the mode survives till $\omega\tau_{\|} \sim 2700$. If we use the pulse width of 100ns to define a frequency, and Benin's formula for $\tau_{\|}$, the pulsed measurements at 150mK give $\omega\tau_{\|} \sim 50$. Given the uncertainties inherent in the pulsed method, this number is only a rough estimate. The more reliable cw resonance measurements span the range $.06 < \omega\tau_{\|} < 2.6$ at P=0. Although very high values of $\omega\tau_{\|}$ have not yet been studied it is already clear that this mode propagates into the collisionless regime, in clear contrast to ordinary second sound[6].

In summary, we have made the first observations of 1-D second sound in superfluid helium. This new collective mode, propagating with speed close to the acoustic sound velocity, has been studied with both pulsed and cw methods. Above 20 bar, where the phonon dispersion is no longer anomalous, the mode collapses. It is the small scattering angle of the 3-phonon process provided by anomalous dispersion that distinguishes this mode from ordinary second sound.

It is a pleasure to thank R. C. Dynes for several essential comments and H. L. Stormer for useful discussions. We are indebted to J. C. M. Hwang for growing the GaAs heterostructure and to M. A. Chin and K. Baldwin for fabricating the heaters and bolometer.

1. H. J. Maris, Rev. Mod. Phys. *49*, 341 (1977).
2. H. J. Maris, Phys. Rev. A, *9*, 1412 (1974).
3. J. P. Eisenstein, V. Narayanamurti, H. L. Stormer, and A. Y. Cho Bull. Am. Phys. Soc. *31*, 606 (1986) and these Proceedings.
4. T. Haavasoja, V. Narayanamurti and M. A. Chin, J. Low Temp. Phys. *57*, 55 (1984).
5. David Benin, Phys. Rev. B, *13*, 1105 (1976).
6. V. Narayanamurti, R. C. Dynes and K. Andres, Phys. Rev. B *11*, 2500 (1975).

Angular Dependence of Kapitza Transmission by Thin Defect Layers

H. Kinder[1] *and K. Weiss*[2]

[1]Physik Department, Technische Universität München, D-8046 Garching, Fed. Rep. of Germany
[2]Hilti AG, Schaan, Liechtenstein

To understand the anomalous Kapitza conductance, the presence of defects, e.g., 2-level-systems with random excitation energies on the surface was assumed. If these defects couple directly to both, the phonons of the solid and the excitations of the helium (phonons, rotons), they open a new channel of heat transfer which can be much more effective than acoustic transmission. [1] Here we present a quantitative theory of the angular dependence and frequency dependence of this mechanism which is not limited to small transmission coefficients. Explicit calculations are made for the case of thin defect layers on smooth surfaces.

The simultaneous coupling of the defects to solid-phonons and helium-excitations is described by

$$H = H_o + M_S \varepsilon_S + M_H \varepsilon_H \tag{1}$$

where H_o is the unperturbed Hamiltonian of the defect, ε_S the strain of the phonon, ε_H the density fluctuation associated with a helium excitation, and M_S, M_H the corresponding deformation potentials. Following JAECKLE et al. [2], this coupling leads to an imaginary part of the elastic constants, δc_α, within the interaction ranges:

$$\delta c_\alpha(\omega) = 2\pi i \; M_\alpha^2 \; N(\omega) \; \tanh\,(\hbar\omega/2k_B T) \tag{2}$$

where α refers to either solid or helium, and $N(\omega)$ is the density of states of the defects per unit volume. Equ. (2) implies a continuum picture where the individual defect positions do not enter. This is justified for transmission coefficients of order unity which require mean distances between resonant defects of the order of the wavelength, indeed. Using the complex elastic constants, we first solve the wave equation for each interaction range.

For a smooth surface, it is appropriate to Fourier transform the wave equation in parallel (x,y) directions, but not in normal (z) direction:

$$-\,\omega^2 \;\rho\; u_i(\vec{q}_\parallel,z) = (i\; q_{\parallel j} + \delta_{3j}\; \partial/\partial z)\; \sigma_{ij}\; (\vec{q}_\parallel,z). \tag{3}$$

Here, ρ is the mass density, u_i the displacement, and σ_{ij} the stress tensor. Further, Hooke's law has the form:

$$\sigma_{ij}(\vec{q}_\parallel,z) = c_{ijk\ell}(z)\; (iq_{\parallel k} + \delta_{3k}\; (\partial/\partial z)\;)\; u_\ell\; (\vec{q}_\parallel,z) \tag{4}$$

where $c_{ijk\ell}$ are the elastic constants including the imaginary parts induced by the defects. Equ. (3) and (4) can be cast in a linear first order differential equation system in $u_i(z)$ and $\sigma_{i3}(z)$:

$$\partial u_i/\partial z = -i\; c^{-1}_{i3j3}\; c_{j3k\ell}\; q_{\parallel k}\; u_\ell + c^{-1}_{i3j3}\; \sigma_{j3} \tag{5}$$

$$\partial\sigma_{i3}/\partial z = \{q_{\parallel j}\,(c_{ijk\ell} - c_{ijm3}\,c^{-1}_{m3n3}\,c_{n3k\ell})\,q_{\parallel k} - \rho\omega^2\delta_{i\ell}\}u_\ell$$
$$- i\,q_{\parallel j}\,c_{ijk3}\,c^{-1}_{k3\ell3}\,\sigma_{\ell 3}. \tag{6}$$

Here, c^{-1}_{i3j3} is the matrix inverse defined by $c^{-1}_{i3j3}\,c_{j3k3} = \delta_{ik}$. We solve the system by iteration because we are interested in the case where the interaction range is shorter than the phonon wave length. For the initial values we use $\sigma_{i3}(z{=}-0){=}0$ at the free surface of the solid, and $u_i(z{=}+0){=}0$ at the hard wall seen by the helium. This amounts to a neglect of acoustic transmission, which is well justified because of the celebrated impedance mismatch. We find a linear relation between stress and displacement of the form:

$$\sigma_{i3}(z_o) = -i\,\omega\,Z_{ij}u_j(z_o) \tag{7}$$

where the Z_{ij} can be viewed as "input impedance" of the interaction ranges in the solid ($z_o = -a_S$) or in the helium ($z_o = a_H$), respectively. Only the Hermitean part of the impedance contributes to absorption or emission of phonons. Its leading terms are for the solid side:

$$Z^S_{ij} = \omega a_S\,s_{\parallel k}\,s_{\parallel \ell}\,\mathrm{Im}\,\{c_{ik\ell j} - c_{ikm3}\,c^{-1}_{m3n3}\,c_{n3\ell j}\}$$
$$+ i\,\omega^2\,a_S^2\,s_{\parallel k}\,\rho\,\mathrm{Im}\,\{c_{jk\ell 3}\,c^{-1}_{\ell 3i3} - c_{ik\ell 3}\,c^{-1}_{\ell 3j3}\}\,/\,2$$
$$- \omega^3\,a_S^3\,\rho^2\,\mathrm{Im}\,\{c^{-1}_{i3j3}\}\,/\,3 \tag{8}$$

with the slowness $s_i = q_i/\omega$. Here, we have assumed an average homogeneous elastic constant of the layer. The terms differ by their dependence on $\vec{s}_\parallel$, i.e. the angle of incidence. For normal incidence, only the third order term in ωa_s survives. For a range of small angles, the second order term can be important. Only for sufficiently large angles, the first order term takes over.

For the helium side, assuming isotropy, the leading term is

$$Z^H_{33} = -\,\mathrm{Im}\,\{c_{1111}\}\,/\,(\omega\,a_H) \tag{9}$$

and $Z^H_{ij} = 0$ otherwise.

The solutions of the bulk wave equation, $(\rho\,\delta_{i\ell} - s_j\,c_{ijk\ell}\,s_k)\,u_\ell = 0$, for given s_1 and s_2 are determined by the six roots, $s_3^{(\lambda)}$, of det $(\rho\delta_{i\ell} - s_j c_{ijk\ell} s_k){=}0$ with respect to s_3. They can always be grouped in 3 incident and 3 reflected modes (possibly evanescent). Their relative amplitudes, $b^{(\lambda)}$, must be chosen so as to satisfy the boundary conditions at $z{=}-a_s$ or $z{=}a_H$, respectively:

$$\sum_{\lambda=1}^{6} (c_{i3jk}\,s_k^{(\lambda)} + Z_{ij})\,e_j^{(\lambda)}\,b^{(\lambda)} = 0 \tag{10}$$

Here, $e_i^{(\lambda)} = u_i^{(\lambda)}\,/\,u^{(\lambda)}$ are the polarization vectors as given by the wave equation.

By inversion of (10) we find for each incident mode (λ_o) the three amplitude reflection coefficients $b^{(\lambda)}/b^{(\lambda_o)}$ for the general case of mode conversion. Then, the energy flux absorption coefficients are

$$A_{\lambda o} = 1 - \sum_{\lambda=1}^{3} |b^{(\lambda)}\,/\,b^{(\lambda_o)}|^2\,\mathrm{Re}\,\{v_3^{(\lambda)}\}\,/\,v_3^{(\lambda_o)} \tag{11}$$

with the z components of the group velocities $v_3^{(\lambda)} = e_i^{(\lambda)} c_{i3jk} s_j^{(\lambda)} e_k^{(\lambda)} / \rho$.

For phonons in bulk liquid He, (11) boils down (using (9)) to

$$A_H = 4\, Z_{33}\, e_3\, \rho_H\, v_H \,/\, (\rho_H\, v_H + Z_{33}\, e_H)^2 \tag{12}$$

where $e_3 = \cos\Theta$ for longitudinal waves. The angular distribution of phonons emitted out into helium is given by the energy flux density, i.e. by $v_3\, A_H$ in our continuum description. For weak interaction (Z_{33} large, low frequency), the flux density is independent of angle (isotropic), but it falls off for angles near 90°. For stronger interaction (Z_{33} smaller, higher frequencies), the distribution falls off at smaller angles already, thereby attaining a shape similar to the cosine law found experimentally by SHERLOCK, MILLS and WYATT for the larger heater powers.

Typical numerical results for the $A_{\lambda o}$ as functions of $\vec{s}_\parallel$ are shown for the Si (001) surface along [100] direction in Fig. 1.

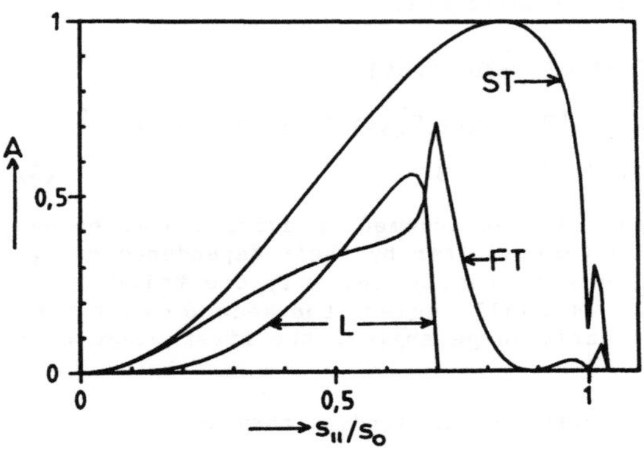

Fig. 1. Absorption coefficient as a function of parallel slowness (units $\sqrt{(\rho/c_{44})}$) for small thickness of the defect layer. For normal incidence,($s_\parallel$=0) the absorption vanishes for all polarizations. This was studied experimentally by KINDER et al.[4]. There is a peak ("halo" [5]) in the fast transverse mode (FT). A Brewster angle appears in the slow transverse mode (ST). There are no effects due to phonon focusing.

References

1. H. Kinder, Physica (Amsterdam) 107, 549 (1981)
2. J. Jäckle, L. Piché, W. Arnold and S. Hunklinger, J. Noncryst. Sol. 20, 365 (1976)
3. R.A. Sherlock, N.G. Mills and A.F.G. Wyatt, J. Phys. C: Sol. State 8, 300 (1975)
4. H. Kinder, A. De Ninno, D. Goodstein, G. Games Paternò, and F. Scaramuzzi, Phys. Rev. Letters 55, 2441 (1985)
5. G. L. Koos, A. G. Every, G. A. Northrop, and J. P. Wolfe, Phys. Rev. Lett. 51, 276 (1983)

Phase-Insensitive Measurements of Ultrasonic Attenuation Near the Phonon-Fracton Crossover in Sintered Powder

*J.H. Page and R.D. McCulloch**

Physics Department, University of Manitoba,
Winnipeg, Manitoba, R3T 2N2, Canada

During the past few years, the nature of the vibrational spectrum in sintered powders has been a subject of both practical and fundamental importance. The initial motivation for our work was the possibility that new vibrational modes in sinter are responsible for the large enhancement of the thermal boundary conductance between liquid helium and sintered metal heat exchangers at millikelvin temperatures [1, 2]. More recently we have also realized that sinters may be model systems for testing new fractal and percolation theories of the vibrational modes of disordered materials [3-6]. Evidence that sintered metal powder is well described as a percolating structure above the percolation threshold has come from measurements of the electrical conductivity, Young's modulus and ultrasonic velocity [2,7]. Such percolating systems are believed to be homogeneous (Euclidean) at long length scales but self-similar (fractal) at short length scales, with the transition between the two regimes occurring at the percolation correlation length ξ. Orbach and co-workers [3-5] have used scaling arguments and effective-medium-approximation calculations to predict a crossover in the vibrational spectrum of percolating networks from extended (phonon) excitations for wave-lengths $\lambda > \xi$ to localized excitations, known as fractons, for $\lambda < \xi$. In this paper, we report ultrasonic attenuation measurements, made using a phase-insensitive acoustoelectric transducer, that give evidence for such a phonon-fracton crossover in sintered metal powder.

One of the difficulties in measuring the ultrasonic attenuation in inhomogeneous materials is that the non-uniformity of the elastic properties can give rise to wavefront distortion, leading to spurious attenuation readings when conventional piezoelectric transducers are used. To avoid this problem, the ultrasonic signals from our specimens were detected using a CdS acoustoelectric transducer [8] which is well suited to phase-insensitive attenuation measurements in our frequency range (1-20 MHz). This transducer makes use of the acoustoelectric effect in piezoelectric semiconductors: because of the piezoelectric coupling, the ultrasonic wave propagating through the transducer gives rise to a local electric field which in turn sets up a current density j given by [8]

$$j = \sigma E_{AE} = \frac{\mu}{v} \alpha \Phi. \qquad (1)$$

Here μ is the electron mobility, v the ultrasonic velocity, α the ultrasonic attenuation, and Φ the ultrasonic energy flux. Equation (1) shows that the acoustoelectric signal, detected in our experiments as the voltage $V_{AE} = \int E_{AE} dx$, depends on the ultrasonic intensity and is therefore insensitive to phase distortion. One advantage of using CdS as transducer material is that its conductivity can be varied over many orders of magnitude by light illumination, enabling the attenuation in (1) due to the electron-phonon interaction [9] to be adjusted for optimum transducer sensitivity.

*Present address: Atlantis Scientific, Ottawa, Ontario, Canada

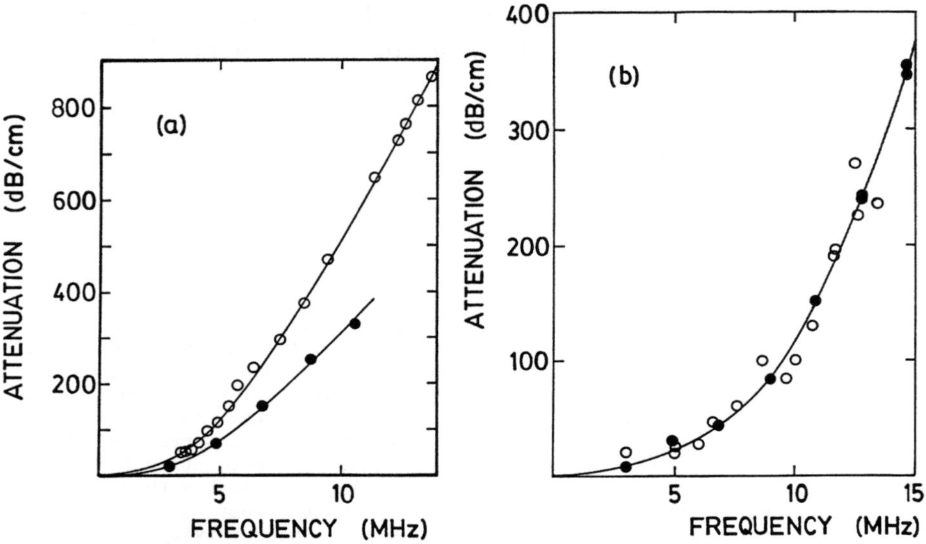

Fig. 1 Attenuation measured by phase-insensitive (•) and piezoelectric (o) transducers in copper sinters. In (a), the particle diameter d = 0.5 μm and the occupied volume fraction f = 0.37. In (b), d = 10 μm and f = 0.56

Figure 1 shows a comparison of the attenuation of longitudinal ultrasonic waves measured by the acoustoelectric and piezoelectric transducers for two representative sinter samples. In Fig. 1(a), the apparent attenuation measured by piezoelectric detection greatly exceeds the actual attenuation measured by the acoustoelectric transducer, indicating considerable wavefront distortion by inhomogeneities in this sample. By contrast, the two techniques gave similar results for a more homogeneous sinter as shown in Fig. 1(b). In general, however, it is clear that piezoelectric detection cannot be relied upon to give accurate attenuation measurements in porous sinters, necessitating the use of the phase-insensitive transducer in our measurements.

The data in Fig. 1 indicate that the attenuation in the sinter samples increases rapidly with frequency and is very much larger than the attenuation in bulk copper. This implies that the attenuation is not caused by the usual sound absorption mechanisms in metals but arises predominantly from elastic scattering of inhomogeneities in the structure. In this case, the inverse of the attenuation can be interpreted as a frequency-dependent localization length for vibrations, $\ell(\omega)$; this follows since the intensity of a sound wave travelling in the x direction falls off as $I_o\ e^{-x/\ell(\omega)} = I_o\ e^{-\alpha(\omega)x}$. The localization length measured in two sinters made from 10 μm diameter copper powder is shown in Fig. 2(a). As the frequency is increased, the localization length drops precipitously, reaching values smaller than the ultrasonic wavelength at the maximum frequencies for which data could be taken in each sample. Such small values of $\ell(\omega)$ indicate the breakdown of wave propagation at high frequencies, implying a crossover to localized vibrational modes. Note that the condition $\ell(\omega) < \lambda$ is similar in spirit to the Ioffe-Regel criterion [10], extensively invoked in the context of electron localization, that modes are localized when the mean free path is less than the wavelength.

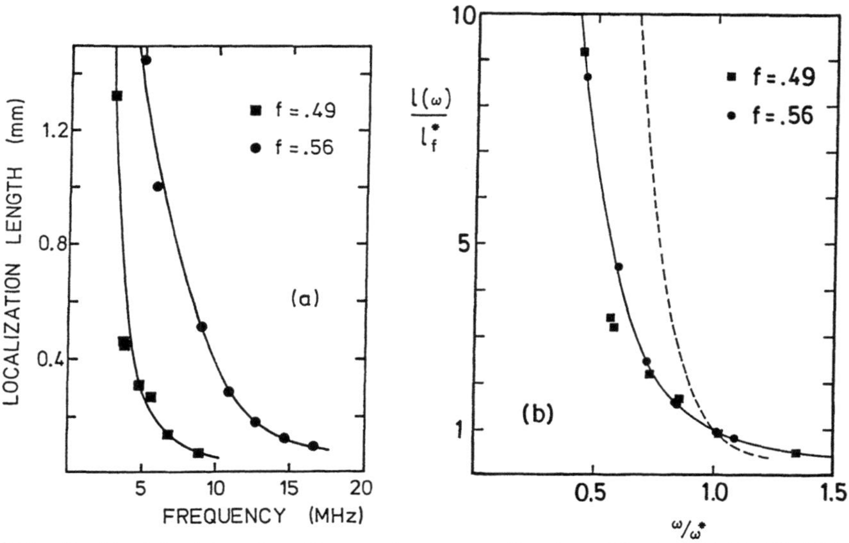

Fig. 2 Localization length vs. frequency for two 10 μm copper sinters

Using this criterion to define a "critical" value of the localization length, $\ell_f^* = \ell(\omega^*)$ at the frequency ω^* such that $\ell(\omega^*) = \lambda = 2\pi v/\omega^*$, it can be shown (Fig. 2(b)) that the localization lengths in sinters with different occupied volume fractions f scale with ℓ_f^* as expected for single-parameter scaling [11]. Furthermore the values of ℓ_f^* determined from the data are proportional to the percolation correlation lengths $\xi(f)$ that mark the boundary between the Euclidean and fractal regions, implying that localized modes in sinter can be correctly referred to as fractons. The dashed curve in Fig. 2(b) represents effective-medium-approximation calculations [4] of the localization length near the phonon-fracton crossover in a simple bond-percolation network, showing qualitative agreement with our results. A more extensive analysis of these data, along with similar results for submicron sinter, will be discussed in another publication [11].

Acknowledgements

Research support from NSERC is gratefully acknowledged. Much of this work was performed while the authors were at Queen's University, Canada.

References

1. A.R. Rutherford, J.P. Harrison and M.J. Stott: J. Low Temp. Phys. 55, 157 (1984)
2. M.C. Maliepaard, J.H. Page, J.P. Harrison and R.J. Stubbs: Phys. Rev. B 32, 6261 (1985)
3. S. Alexander and R. Orbach: J. Phys. Lett. (Paris) 43, 625 (1982)
4. O. Entin-Wohlman, S. Alexander, R. Orbach and K.-W. Yu: Phys. Rev. B 29, 4588 (1984)
5. A. Aharony, S. Alexander, O. Entin-Wohlman and R. Orbach: Phys. Rev. B 28, 4615 (1983)
6. Y. Kantor and I. Webman: Phys. Rev. Lett. 52, 1891 (1984)
7. D. Deptuck, J.P. Harrison and P. Zawadski: Phys. Rev. Lett. 54, 913 (1985)

8. L.J. Busse and J.G. Miller: J. Acoust. Soc. Am. 70, 1370 (1981); J.S. Heyman: J. Acoust. Soc. Am. 64, 243 (1978)
9. A.R. Hutson and D.L. White: J. Appl. Phys. 33 40 (1962)
10. A.F. Ioffe and A.R. Regel: Prog. Semicond. 4, 237 (1960)
11. J.H. Page and R.D. McCulloch: to be published

Percolation and Kapitza Resistance

J.P. Harrison and P. Zawadzki

Queen's University, Kingston, Ontario, Canada

1. Introduction

The millikelvin temperature Kapitza resistance problem remains unresolved at present. Experiment has demonstrated that for sintered metal powder and pure ^3He the resistance varies as T^{-1} (see reference 1) and for dilute ^3He varies as T^{-2} [2-4]. Measurements have also shown both an increase [5] and decrease [6] of the resistance with increase in magnetic field, an increase [5], no change and decrease [4] with pressure, and a variation with T_F^{-1} for dilute ^3He mixtures [3,4]. Below the superfluid transition temperature in liquid ^3He the resistance increases exponentially [7,8], reflecting the exponentially decreasing number of quasiparticle excitations. Many, but not all, of the experimental results can be described by the modified acoustic ("shaking-box") model of Rutherford et al [9], together with estimates for the confined ^3He and sinter thermal resistances that are in series with the boundary or Kapitza resistance. Figure 1 illustrates a comparison of the calculated resistances with the various pure ^3He results. A similar comparison for dilute mixtures shows agreement to within a factor of 2. Note however that the model is not able to describe some of the observed variations with pressure [5], or magnetic field [5,6].

For the shaking-box model it was postulated that the pores of a sinter could be treated as vibrating boxes containing a gas of ^3He quasiparticles [9]. The heat transfer was then a result of the coupling between the ^3He quasiparticle excitations and the S.H.O. modes of the boxes. There are

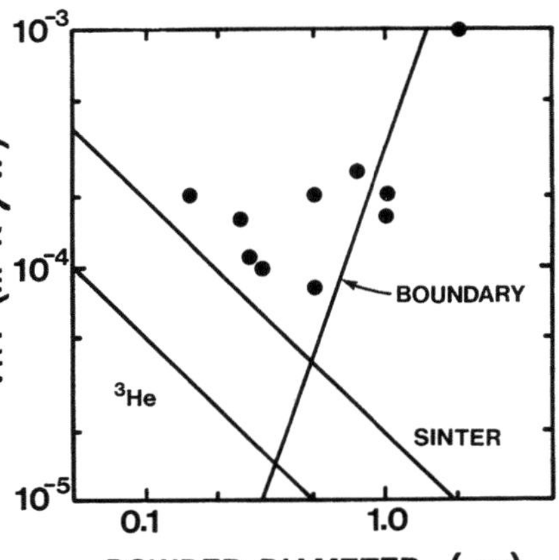

Fig. 1 Estimates of the three distributed thermal resistances that limit heat transfer between bulk ^3He and bulk metal in a sintered metal powder heat exchanger, as a function of the powder diameter. The resistances are of the sinter from electrical resistivity measurements, of the ^3He in the pores from a size effect calculation and of the boundary from the shaking box model. The thickness of sinter was 5 mm. The data points represent the experimental results from many laboratories (see Ref. 9 for original references).

no adjustable parameters but there are two hypotheses underlying the calculation:
(1) There are localised vibrational modes of a sinter or packed powder with a frequency-independant density of states.
(2) For submicron powder sinter these modes extend over the frequency range (20 - 400 MHz) that is thermally excited in the temperature range 1- 20 mK. These hypotheses have been studied by identifying sintered metal powder as a lattice-based percolation system. In turn this identification assists in choosing the sinter parameters for optimum heat exchanger design.

2. Sinter as a Percolation System

The bond percolation system consists of a lattice of points with neighbouring points connected by a bond with a probability p. At a threshold value (p_c), there is a continuous connection across the system [10]. Above p_c, if the bonds are conducting, the conductivity of the system varies as $(p - p_c)^t$ and if the bonds are elastic, the elasticity varies as $(p - p_c)^\tau$ where t and τ are scaling exponents. Sinter can be considered an analogue of this system with the powder particles as the lattice points and the necks between the particles, formed by the heat treatment, as the bonds [11]. Experiment has shown that the conductivity and elasticity of sinter do scale with the occupied volume fraction of the sinter, with $t = 2.15 \pm 0.25$ and $\tau = 3.8 \pm 0.5$ [12]. These exponents are in agreement with the best estimates from percolation theory, provided that the elasticity is determined by the bending of the bonds (necks) at the lattice points (particles).

A study of the dynamics of the percolation system has shown two things that relate directly to the hypotheses of the shaking box model:
(1) There are localised vibrational modes (fractons) with a density of states that depends only weakly upon frequency, as $\nu^{1/3}$ by applying the Alexander-Orbach conjecture [13] and as $\nu^{-0.1}$ by applying scaling arguments to the bond-bending model of elasticity [14].
(2) At a length scale comparable with the correlation length (size of the largest cluster or hole) there is a cross-over from the localised excitations to phonon-like excitations [15].

The existence of the cross-over has been supported by ultrasonics experiments [16, 17] which showed both phonon-like propagation at low frequencies and a threshold wavelength, below which there was no propagation, which scaled with packing fraction in the same way as the theoretical correlation length.

3. Application of Percolation Ideas to Kapitza Resistance

An understanding of the millikelvin Kapitza resistance is of inherent interest, and of practical interest for the design of heat exchangers. Osheroff and Richardson have demonstrated a magnetic coupling for oxidised silver that is quenched by a field of 20 mT. Since a similar magnetic field is required for Pt n.m.r. thermometry, very much larger fields are in use for spin polarised ^3He studies and no magnetic coupling is expected for ^3He - ^4He mixtures, therefore most experiments depend upon the acoustic coupling for heat exchange. This theoretical resistance varies as d^3 where d is the power diameter and can be minimised by using smaller diameter powder. However, the confined ^3He and sinter resistances that are in series with the boundary resistance increase with decrease in d [18]. Furthermore, the percolation study showed that the localised modes had a low-frequency band edge at ~ 10 M Hz for 0.5 μm sinter with occupied volume fraction f = 40%. This sets the low-temperature limit of the shaking box model at $T_e \sim h\nu_e/k_B \sim 0.5$ mK. Scaling arguments show that $T_e \propto f^2 d^{-1}$ (approximately) and that the temperature can be reduced by decreasing f; but that in turn increases the boundary and sinter resistances. However, the optimum diameter, occupied volume fraction and geometry can be found by numerical analysis.

Below the band edge temperature, acoustic heat transfer is from ^{3}He quasiparticle excitations to sinter phonons. The model of Toombs et al [19] should describe this regime, so that a T^{-3} dependence is expected. For most pure ^{3}He results this dependence was not observed either because the particle size was large and T_e was lower than the measured temperature range or because the particle size was small and the sinter resistance dominated. However, for dilute ^{3}He the sinter resistance is not important, and it is possible that the measured T^{-2} dependances are reflecting the cross-over of the boundary resistance from T^{-1} to T^{-3}.

Acknowledgements

This work has been supported by NSERC and the Killam Research Foundation. We wish to thank Dr. John Page for many valuable discussions.

References

1. J.P. Harrison, J. Low Temp. Phys. 37, 467 (1979).
2. D.D. Osheroff and L.R. Corruccini, Phys. Lett. 82A, 38 (1981).
3. D.A. Ritchie, J. Saunders and D.F. Brewer in Proc. LT17, p. 743 (Elsevier, New York, 1984).
4. H.C. Chocholacs, R.M. Mueller, J.R. Owers-Bradley, Ch. Buchal, M. Kubota and F. Pobell, unpublished report at Ultralow Temperature Symposium, Bayreuth, 1984.
5. D.D. Osheroff and R.C. Richardson, Phys. Rev. Lett. 54, 1178 (1985).
6. T. Perry, K. De Conde, D.L. Stein and J. A. Sauls, Phys. Rev. Lett. 48, 1831 (1982).
7. C.A.M. Castelijns, K.F. Coates, A.M. Guenault, S.G. Mussett, and G.R. Pickett, Phys. Rev. Lett. 55, 2021 (1985).
8. J.M. Parpia, Phys. Rev. B32, 7564 (1985).
9. A.R. Rutherford, J.P. Harrison and M.J. Stott, J. Low Temp. Phys. 55, 157 (1984).
10. D. Stauffer, in Disordered Systems and Localisation p. 1, (Springer, Berlin (1981)).
11. Gravity prevents the sinter existing with $p < p_c$ and even with $p > p_c$ there will be isolated clusters.
12. D. Deptuck, J.P. Harrison and P. Zawadzki, Phys. Rev. Lett. 54, 913 (1985).
13. S. Alexander and R. Orbach, Phys. Lett. 98A, 357 (1983).
14. I. Webman and G.S. Grest, Phys. Rev. B31, 1689 (1985).
15. O. Entin-Wohlman, S. Alexander, R. Orbach, and K.-W. Yu, Phys. Rev. B29, 4588 (1984).
16. M.C. Maliepaard, J.H. Page, J.P. Harrison, and R.J. Stubbs, Phys. Rev. B32, 6261 (1985).
17. J.H. Page and R.D. McCullough (to be published).
18. R.J. Robertson, F. Guillon and J.P. Harrison, Canad. J. Phys. 61, 164 (1983).
19. G.A. Toombs, F.W. Sheard, and M.J. Rice, J. Low Temp. Phys. 39, 272 (1980).

Thermal Boundary Resistance from 0.5–300 K

E.T. Swartz and R.O. Pohl

Laboratory of Atomic and Solid State Physics, Cornell University, Ithaca, NY 14853, USA

Many experiments[1] have shown conclusively that phonons scatter diffusively at interfaces when the temperature exceeds a few K. On the other hand, heat pulse transmission experiments[2] have shown that the acoustic mismatch model, which assumes that the phonons are not diffusively scattered, approximately predicts the transmission of phonons through an interface even at temperatures above several tens of K. The thermal boundary resistance should depend on whether or not the phonons are diffusively scattered at the interface. In order to see the effect of diffuse scattering on the thermal boundary resistance, one must be able to accurately measure the thermal boundary resistance over a broad temperature range. There are two major difficulties in measuring the thermal boundary resistance: temperature extrapolation from the thermometers to the interface becomes much more difficult as the interface temperature is raised, and the power required to produce a temperature discontinuity increases rapidly with increasing temperature. We have developed a technique for accurately measuring the thermal boundary resistance from below 1K to near room temperature.

Our technique utilizes photolithography to pattern thin metal films on dielectric substrates. We pattern the metal into two parallel narrow lines, each about 2 microns wide, separated by about 2 microns. Each line acts as thermometer, as its resistance is temperature dependent. The interface that we are studying is the interface between one of the thermometers (thermometer 1) and the substrate. A relatively large D.C. current is passed through thermometer 1, and the power and resistance are calculated. The temperature on the thermometer side of the interface is thus determined from the resistance of thermometer 1. A relatively small current is passed through the other thermometer (thermometer 2), and its temperature is determined. In thermometer 2 there is negligible self-heating because the power used to measure the resistance is negligible. Therefore thermometer 2 measures the temperature of the substrate just under thermometer 2. The temperature under thermometer 1, and therefore at the interface of interest, is determined by extrapolating the temperature over the very short distance between the thermometers, a distance more than 1000 times shorter than the distance extrapolated in conventional measurements. The problem of the large power required to produce a temperature discontinuity at higher temperatures is also less severe with our technique because the area of the interface is typically less than 10^{-4} cm^2, again more than 1000 times smaller than in a conventional experiment. Thus the key to our technique is the ability to shrink the experiment to the micron scale.

The first substrate/metal film pairs we measured were pure metals on sapphire or quartz. The resistivity of pure metals is temperature dependent only at temperatures higher than about 5K. Therefore, if we use

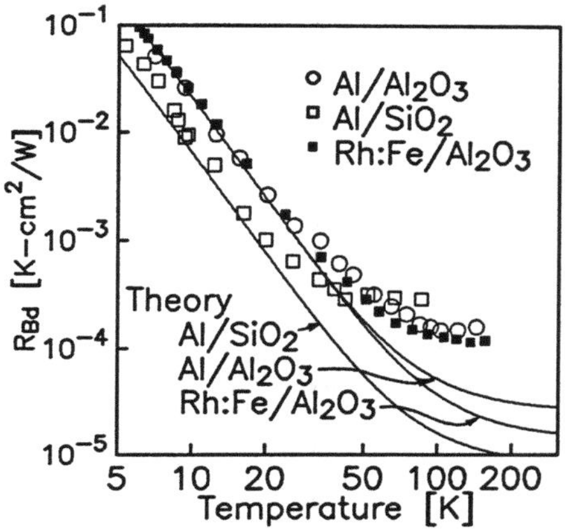

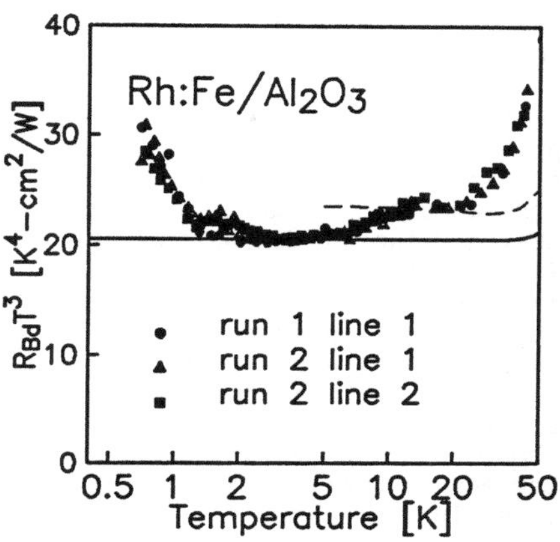

Fig. 1. Thermal boundary resistance of Al on sapphire, Al on quartz, and Rh:Fe on sapphire. The acoustic mismatch predictions are shown as the solid lines.

Fig. 2. Thermal boundary resistance of Rh:Fe on sapphire, multiplied by T^3 to remove the strong temperature dependence. The solid line is the acoustic mismatch prediction, and the dashed line shows the prediction of a simple model which assumes that all phonons are diffusively scattered at the interface.

pure metals, we can only measure the thermal boundary resistance above 5K. Figure 1 shows the thermal boundary resistance between the pairs Al on sapphire, Al on quartz, and Rh:Fe on sapphire. Below about 50K there is remarkable agreement between the acoustic mismatch model (the solid lines) and the data. Above 50K the measured boundary resistance ceases to follow the predicted temperature dependence of acoustic mismatch theory, and levels off. Even with our technique, at such high temperatures and with such small boundary resistances, we encounter the problem of temperature extrapolation to the boundary. A possible reason is that the thermal conductivity in the bulk within the first few hundred Å of the surface is lower than that deep in the bulk because of damage caused by, for example, polishing.

The data on the Rh:Fe/sapphire interface extend to below 1K. This is possible because of the temperature dependence of the resistivity of Rh:Fe, which looks like that of a pure metal at high temperatures, but below 10K the resistivity continues to drop monotonically with a slope high enough to be useful as a thermometer to well below 1K[3]. Thus the remarkable properties of Rh:Fe allow us to make the first measurement of the thermal boundary resistance that spans the entire temperature range from below 1K to over 100K. Figure 2 shows the thermal boundary resistance between Rh:Fe and sapphire, multiplied by the cube of the temperature to remove the strong temperature dependence. Plotted are three indistinguishable sets of data on two separate lines of Rh:Fe in experiments done several months apart. There are four features of interest. The upturn above 40K is discussed above. We believe the upturn below 2K is the result of the finiteness of the electron-phonon coupling, leading to a thermal resistance between the electrons and the Rh:Fe lattice, which varies as T^{-4}. In copper, this thermal resistance has been determined to be $5T^{-4}$ K^5cm^3/W[4]. The rise observed here corresponds to a value of $(1.5 \pm 0.5)T^{-4}$ K^5cm^3/W.

Between 2K and 5K the thermal boundary resistance between Rh:Fe and sapphire agrees precisely with the acoustic mismatch value. Above 5K there is a transition and the boundary resistance increases 15% relative to the acoustic mismatch value, and remains 15% higher than the acoustic mismatch value up to 25K. We attribute this transition to the onset of diffuse scattering at the interface. A preliminary calculation, which will be discussed in a future publication, shows that the small increase in the thermal boundary resistance in the 9K to 25K interval is quantitatively consistent with the assumption that all the phonons are scattered diffusively at the interface. To conclude, we have observed the transition from specular to diffuse scattering in measurements of the thermal boundary resistance. The effect of diffuse scattering at the interface is small, and this may be the reason why it has not been noticed in the past.

Acknowledgements
One of the authors (ETS) was supported through a fellowship from the Fannie and John Hertz Foundation. The work was supported by Semiconductor Research Corporation contract 82-11-001. Photolithography was performed at the National Research and Resource Facility for Submicron Structures, at Cornell.

References

1. See for example: Tom Klitsner and R.O. Pohl, these proceedings; D. Marx and W. Eisenmenger, Z. Phys. B 48, 277 (1982); J. Weber, W. Sandmann, W. Dietsche, and H. Kinder, Phys. Rev. Lett. 40, 1469 (1978); E.S. Sabisky and C.H. Anderson, Sol. St. Comm. 17, 1095 (1975); and references therein.

2. See for example: P. Herth and O. Weis, Z. Angew. Phys. 29, 101 (1969); J.D.N. Cheeke, B. Hebral, C. Martinon, J. Physique 10 Colloq. C4-57 (1972); and several other publications by these authors.

3. R.L. Rusby, Temperature, Its Measurement and Control in Science and Industry IV, 865 (1971).

4. M.L. Roukes, M.R. Freeman, R.S. Germain, R.C. Richardson, and M.B. Ketchen, Phys. Rev. Lett. 55, 422 (1985).

The Reversibility of Kapitza Resistance

Xu Yun-hui , Zheng Jia-qi, and Guan Wei-yan

Institute of Physics, Academia Sinica, Beijing, People's Republic of China

1. Introduction

Although research of the Kapitza resistance has continued for many years, there still is not a definite and quantitative result for the reversibility of the effect [1,2]. The reason lies in the technical difficulty of reversing the heat flow from liquid helium to solid (L→S) for the same interface under stationary conditions. This paper presents an apparatus in which the Kapitza resistance could be measured at the same interface in the same run in both directions of the heat flow from S→L or L→S and reports the results of R_k measurements for oxygen-free copper and single crystal $Gd_3Ga_5O_{12}$ with (111) plane.

2. Apparatus and Experiment

We have measured the Kapitza resistance R_k by the steady state method. The apparatus built to accommodate heat flow in both directions is shown in Fig.1. This is an entirely symmetrical arrangement. The sample S was glued to tubes t_1 and t_2. The t_1 and t_2 were thin wall stainless steel tubes fitted with bellows.

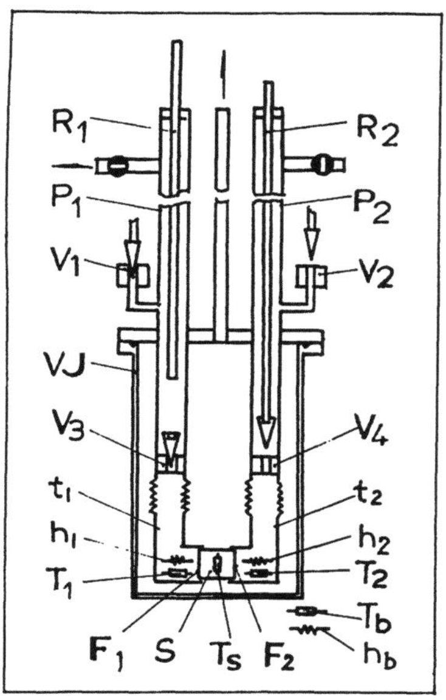

Fig.1 The apparatus for the Kapitza resistance measurement with the reversible heat flow

Each valve V_{1-4} used to fill liquid He into t_1 and t_2 from helium bath had a stainless steel needle with a 7.5° taper and a brass valve seat. Each valve must be tested carefully before use. The residual helium in the pipe P_1 (or P_2) could pumped out after valves V_1 and V_3 (or V_2 and V_4) were closed. Two screwdriver rods R_1 and R_2 could be raised and separated from their needles after operation. The R_1 and R_2 were also taken as good thermal sources for evaporating the residual helium in P_1 and P_2. Temperatures were measured by calibrated Allen-Bradley resistors T_1,T_2,T_s and T_b. The temperature of the bath was kept constant to an accuracy of $\pm 5\times 10^{-5}$ [K] by an electronic system. A vacuum jacket VJ was used to prevent the heat exchange with the main bath. All of samples were polished and degreased.

At the beginning, valves V_1,V_2,V_3 and V_4 were opened, tubes t_1 and t_2 could be filled with liquid helium from the main bath. To perform the R_k measurement, V_1 and V_3 were closed. While the P_1 was pumped out, the rod R_1 was raised up to a certain position to get a higher temperature, then it was pushed down to evaporate the residual helium above the closed valve V_1. By doing such an operation several times no helium existed in the pipe P_1 above V_1 and a vacuum was obtained. A steady heat flow $\dot{Q}$ was supplied by the heater h_1 in the liquid helium of the tube t_1. The $\dot{Q}$ went through interface F_1,sample S and interface F_2 to liquid helium in the tube t_2 connected with the main bath at T_b. The temperature difference ΔT_1 and ΔT_2 across two interfaces F_1 and F_2 determined by means of the sensitivity (dR/dT) of thermometers T_1 and T_s. The errors of T and ΔT were less than $\pm 5\times 10^{-3}$ [K] and $\pm 2\times 10^{-4}$ [K] respectively. The Kapitza resistances R_{k1}(L→S) and R_{k2}(S→L) at F_1 and F_2 were obtained directly from formula $R_k = A\Delta T/\dot{Q}$. Changing the bath temperature and repeating experimental procedures as described above, the temperature dependence of R_{k1}(L→S) and R_{k2}(S→L) could be obtained.

Second,to measure the heat leak from walls of tubes,electric leads and the residual gas escaping through valves,valves V_2 and V_4 had to be closed and the pipe P_2 was evacuated. A small heating power $\dot{q}$ was supplied to the liquid helium inside the tube t_1 in order to set up a temperature difference ΔT between the main bath and the insulated liquid helium contained in the tubes with sample and to keep same value as that of temperature dropping between the main bath and the heating bath with maximum heat flow $\dot{Q}_{max}$ during the Kapitza resistance measurement. Therefore $\dot{q}$ was equivalent to the total heat leak under the extreme condition. The value $\dot{q}/\dot{Q}$ was in the range of 2% at 1.93[K] to 4% at 1.39[K]. It was noted that the real heat leak while R_k was measuring must be less than this value $\dot{q}$, because one of tubes was connected with the main bath at T_b.

Third,opening V_1 and V_3 with the same experimental procedure, the Kapitza resistances R_{k2}(L→S) and R_{k1}(S→L) and their temperature dependence at the interfaces F_2 and F_1 could be measured in the same run on the same surface.

3. Results and Discussion

From these results shown in Fig.2,we might compare the values of the Kapitza resistance for an identical interface F_1(or F_2)

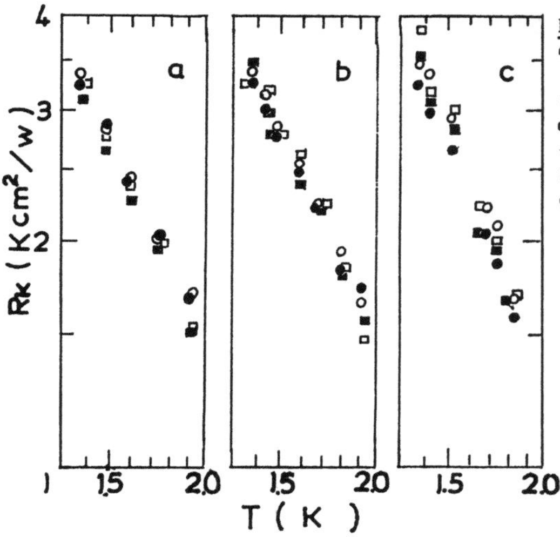

Fig.2 The experiment results of R_k. □ R_{k1}(L→S) ■ R_{k1}(S→L) ○ R_{k2}(L→S) ● R_{k2}(S→L) (a). Ø16.2×11.5[mm] oxygen-free copper. (b). Ø8.4×10[mm] oxygen-free copper. (c). Ø15.85×11 [mm] single crystal GGG with (111) plane

with different directions of the heat flow namely to study the reversibility of the Kapitza resistance. It indicated that for the same interface F_1(or F_2) the Kapitza resistance R_k(L→S) and R_k(S→L) agree within 5% in the temperature range of 1.39-1.93[K]. It means that the value of the kapitza resistance R_k is not affected by the reversed heat flow for an identical solid surface and is in agreement with the classical acoustic theory.

However it should be noted that the R_k values for the single crystal $Gd_3Ga_5O_{12}$ reported here are lower than that of our previous measurements [3]. It seems to be the different location of the heaters which is in the liquid helium in this paper and on the solid in ref.[3]. If one compares with the R_k data obtained from the steady state (dc) method with a heater on the solid,and from the second sound (ac) method with a heater in the liquid helium, there also appear similar differences [4-7]. The authors intend to complete this work and to discuss it in a subsequent communication.

Reference

1. Wey-yen Kuang (Guan Wei-yan), Sov.Phys. (JETP) 15, 635(1962)
2. A.F.G.Wyatt and G.J.Page, J.Phys.C: Solid St. Phys. 11, 4927 (1978)
3. Xu Yun-hui, Zheng Jia-qi and Guan Wei-yan, Proc. 17th Int. Conf. on Low Temp. Phys. Part.I Eds.V.Eckern et al. (North-Holland, 1984) P.687
4. F.Wagner, F.J.Kollarits, K.E.Wilkes and M.Yaqub, J. Low Temp Phys. 20, 181 (1975)
5. L.J.Challis and R.A.Sherlock, Phys.C: Solid St. Phys. 3, 1193 (1970)
6. J.A.Katerbery and A.C.Anderson, J. Low Temp. Phys. 42, 165 (1981)
7. N.J.Brow and D.V.Osborne, Phil. Mag. 3, 1463 (1958)

Study of Phonon-Surface Defect Coupling by Kapitza Transmission on "Real" Surfaces

H. Kinder[1], *A. de Ninno, D. Goodstein*[2], *G. Paternò, and F. Scaramuzzi*

ENEA, Frascati, Italy

S. Cunsolo

Univ. La Sapienza, Rome, Italy

On atomically clean surfaces, the anomalous Kapitza transmission is absent, but it can be restored by minute amounts of adatoms [1]. The adatoms supposedly form localized excitations ("two level systems") which absorb and emit phonons by direct coupling to the solid and to the helium [2]. The same mechanism may also work on ordinary, or "real" surfaces which have been held on air, or treated with organic solvents. A specific experiment for these ordinary surfaces was missing so far. In fact, the Kapitza transmission was generally found to be unspecific, and did not depend much on phonon polarization [3,4], frequency [5,6], angle of incidence [7], or on surface treatment [8].

However, these observations are expected from the defect model only in the limit of strong absorption, i.e. for transmission coefficients near unity. For small absorption coefficients, the transmission should be rather sensitive to all those parameters. Moreover, small transmission coefficients should be achievable also for ordinary surfaces, if the angle of incidence of the phonons is nearly normal and the frequencies are low.

Qualitatively, this can be easily understood. If the defects couple to the phonon strain field (i.e. by deformation potential) then their interaction will be strongly modified by the boundary condition of the surface. For phonons with normal wave vectors, the boundary condition imposes a node of the strain at the surface, and thus prevents them from coupling as long as the defect layer is sufficiently thin. For thicker layers or shorter wave lengths, the defects begin to feel the strain away from the node such that absorption or emission is more and more allowed. At a given frequency, this effect is actually stronger for transverse phonons, because they have shorter wavelengths than the longitudinal ones.

With finite angles to the normal, the node becomes incomplete, and a finite parallel strain remains even at the free surface. Therefore, the defect layer can now be stretched, and absorption/emission should increase with the angle, even for very thin layers. In experiments, the direction of the group velocity rather than the phase velocity is defined by the geometry. This usually leads to different angles, and thus to different interaction strengths for longitudinal and transverse phonons even in the limit of thin layers or long wavelengths.

No such dependence on angle is expected if the coupling were not by deformation potential (leading to complex elastic constants) but by the acceleration of the phonon, i.e. inertial coupling (leading to a complex density). Deformation potential coupling is known for tunneling centers in

[1] Permanent address: Techn. University of Munich, D-8046 Garching, FRG
[2] Permanent address: California Institute of Technology, Pasadena, USA

glasses, alkali halides, etc. while inertial coupling is known for mass defects.

For a simple experiment, we have chosen the (111) surface of silicon and have observed transmitted phonons propagating into the crystal along the normal direction (see inset in Fig. 2). The detector size was 1/3 of the crystal thickness. A transmission experiment is superior to the reflection geometry in the present context because the phonon direction is better defined. For this geometry, the longitudinal phonons have nearly normal wave vectors while the slow transverse phonons have several oblique modes with angles of 10° and 20° to the normal. Thus, the transverse phonons should couple more strongly because of both effects discussed above.

Quantitative calculations for this geometry were performed using the theory of KINDER and WEISS [9] assuming equal imaginary and real parts for the complex elastic constants of the defect layer. Fig. 1 shows the predicted transmission coefficients and the L/T-Ratio.

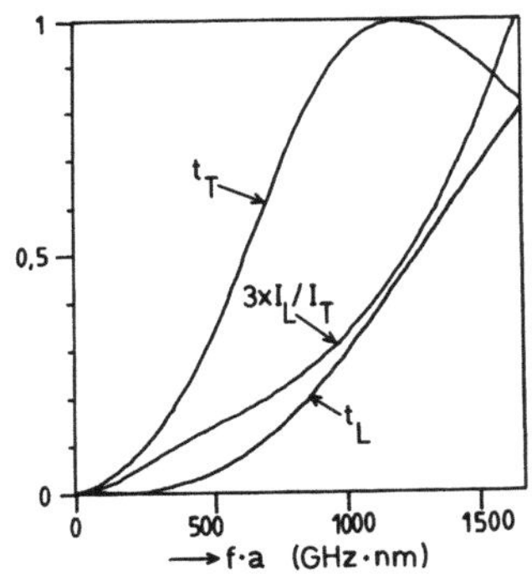

Fig. 1. Integrated energy flux transmission coefficients of longitudinal (L) and slow transverse (T) phonons as a function of the product of frequency and layer thickness. The transverse phonons increase more rapidly so that the L/T intensity ratio (I_L/I_T, scaled up by a factor of 3) is also increasing with frequency and layer thickness.

Experimental results [10] for the same geometry are presented in Figs. 2, 3, and 4. The effect of surface preparation is shown in Fig. 2.
The various surface conditions are not well characterized. But the very fact that the transmission coefficient can vary over more than an order of magnitude was not observed before. Even stronger than this is the variation of the L pulse. This can be seen more clearly on Fig 3.

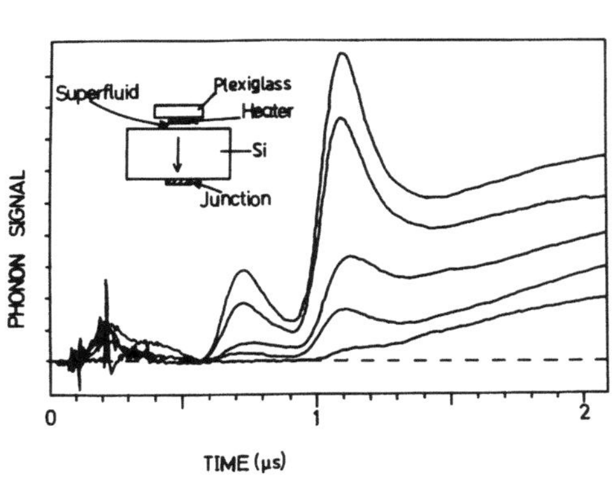

Fig. 2. Variation of phonon pulse signals with surface preparation. All pulses were taken with the same heater power (4W) and the identical Al junction as detector, and they are all plotted with the same scale. From bottom to top trace: (1) etched with HNO_3 and HF, (2) additional drop of isopropyl alcohol, (3) further exposure to pump oil, (4) re-etched with the junction saved, (5) wiped with lens tissue.

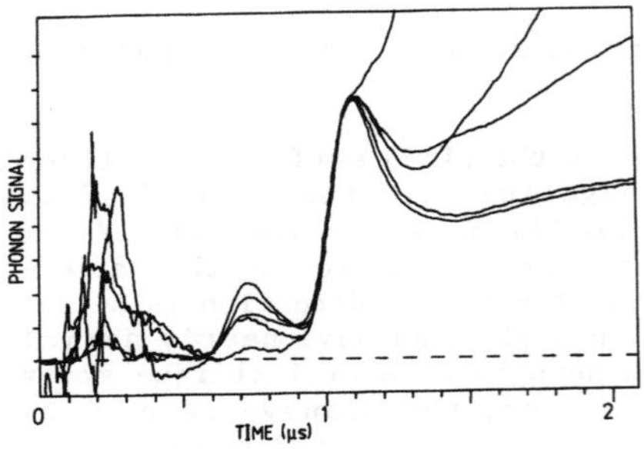

Fig. 3. Scales of traces of Fig. 2 adjusted to have the same T pulse heights. The relative heights of the L pulses vary in the same order as the absolute heights in Fig. 2. This is qualitatively in agreement with the L/T-ratio of Fig. 1.

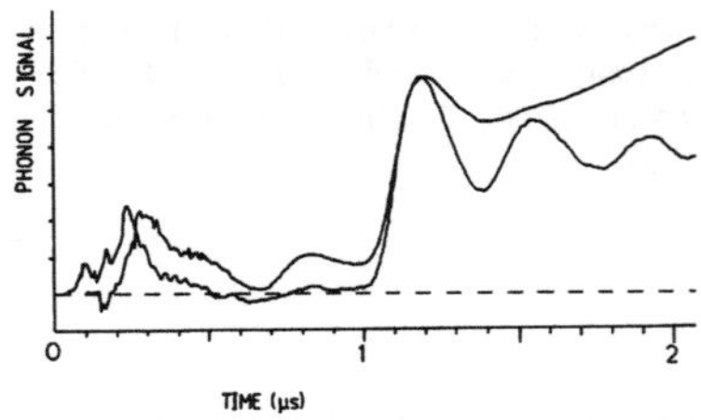

Fig 4. Phonon pulses of 4W and 0.08W transmitted through the same sample surface (No. 3 of Fig. 2), scaled to the same T peak height. Increasing frequency has the same effect as increasing contamination.

We have also studied the dependence on phonon pulse power for a given surface condition. With increasing power, the frequency increases. An example is shown in Fig. 4.

In conclusion, the experimental arrangement chosen here, using long wavelength phonons with nearly normal wave vectors, has in fact yielded the strong variations and the qualitative tendencies predicted by the theory. This implies that the surface defect layer was thinner than the wave length, and that the defect-phonon coupling is via the phonon strain (deformation potential) and not via the phonon acceleration (inertial). Further experiments are planned where we wish to measure the full angular distribution of the phonons transmitted into the solid.

References

1. H. C. Basso, W. Dietsche, and H. Kinder in: Proc. of the Seventeenth Int. Conf. on Low Temp. Phys. (LT17), ed. by U. Eckern, A. Schmid, W. Weber, and H. Wühl (North Holland 1984), p. 465; L. Köster, S. Wurdack, W. Dietsche, and H. Kinder, present conference.
2. H. Kinder, Physica (Amsterdam) 107B, 549 (1981)
3. C.-J. Guo and H. J. Maris, Phys. Rev. Lett. 29, 855 (1972)
4. J. Wolter and R. E. Horstmann, Phys. Lett. 61A, 238 (1977)
5. T. J. B. Swanenburg and J. Wolter, Phys. Rev. Lett. 33, 882 (1974)
6. W. Dietsche and H. Kinder, J. Low Temp. Phys. 23, 27 (1976)
7. A. R. Long, R. A. Sherlock, and A. F. G. Wyatt, J. Low Temp. Phys. 15, 523 (1974)
8. S. Burger, W. Eisenmenger, and K. Lassmann, in: Proc. of the Seventeenth Int. Conf. on Low Temp. Phys. (LT17), ed. by U. Eckern, A. Schmid, W. Weber, and H. Wühl (North Holland 1984), p. 659
9. H. Kinder and K. Weiss, present conference
10. H. Kinder, A. de Ninno, D. Goodstein, G.釋Paternò, F. Scaramuzzi, and S. Cunsolo, Phys. Rev. Lett. 55, 2441 (1985)

Interaction Between Phonons and ^{3}He-Quasiparticles in the ^{3}He-^{4}He Mixture Confined in Porous Media

K. Yakubo and T. Nakayama

Department of Applied Physics, Hokkaido University, Sapporo 060, Japan

1. Introduction

The problem of liquid or solid He in restricted geometry has recently been a subject of extensive studies both theoretically and experimentally[1]. It is expected that the dilute ^{3}He-^{4}He mixture confined in porous media should behave quite differently from the bulk mixture. This paper investigates the interaction between low-frequency phonons in porous media and ^{3}He-quasiparticles in the ^{3}He-^{4}He mixture. Porous media considered here consist of pores with different shapes and channels connecting these. This system is related to the problem of heat transfer between porous media and ^{3}He-^{4}He mixture[2,3]. It will be shown that ^{3}He-quasiparticles are regarded as ones localized in pores in a certain frequency and temperature region.

2. Localization of ^{3}He-quasiparticles

We consider the porous medium with the following characteristics: (1) the diameter of pores is $\sim$100A, (2) the configuration and shapes of pores are not uniform, (3) the pores are connected by channels. Now let us consider the case where the diameter of pores is D=100A, and the diameter and the length of channels are d=7A and ℓ=10A, respectively. The probability that a ^{3}He-quasiparticle moves to the adjacent pore through the narrow channel is calculated from the following effective potential picture. The zero-point energy E_0 of ^{3}He-quasiparticles in the channel is much higher than that in the pore, namely, E_0=1.41K for d=7A. Since the Fermi energy E_f of the 5% ^{3}He-^{4}He mixture becomes 0.34K, the barrier height between pores is 1.07K. Thus, ^{3}He-quasiparticles must tunnel through the barrier height of 1.07K to transfer into the adjacent pore. The tunneling probability is expressed by

$$\frac{1}{\tau} = \frac{4v_f k_1^2 k_2^2}{D[(k_1^2+k_2^2)^2 \sinh^2(\ell k_2)+4k_1^2k_2^2]} , \tag{1}$$

where $k_1=[2m_3^*(E_f+\delta)]^{1/2}/\hbar$ and $k_2=[2m_3^*(E_0-E_f-\delta)]^{1/2}/\hbar$. The symbols m_3^*, δ and v_f are the mass, level spacing and the Fermi velocity of ^{3}He-quasiparticle, respectively. From the typical values mentioned above, the inverse of lifetime becomes $1/\tau \sim 10^5 \sec^{-1}$. Therefore we see that the condition $\omega\tau > 1$, where the quantum picture is relevant, is valid for the frequency regime larger than 0.1MHz.

3. Energy spectrum of ^{3}He-quasiparticles

In the case of an isolated pore, the energy spectrum of ^{3}He-quasiparticles should be essentially the same as that for electrons in small metal particles, that is, the energy spectrum becomes discrete due to the quantum-size effect[4]. The mean-energy level spacing $\bar{\delta}$ is obtained by the density of state N(0) at the Fermi level ; $\bar{\delta}=1/N(0)V$, where V is the volume of a pore. The mean level-spacing $\bar{\delta}$ is estimated to be 0.83mK for the typical values for our system.

Pores are connected by channels and energy levels have the width ΔE whose magnitude depends on the degree of the overlapping of wave functions of ^{3}He-quasiparticles through the potential barrier mentioned in Section 2. The width is expressed as $\Delta E=\hbar/\tau_3$, where τ is the tunneling probability estimated in Eq.(1); i.e., $\Delta E \sim 10^{-3}$mK. This value is so small compared with $\bar{\delta}$ that the interpore connection due to channels does not change the discrete energy structure. Thus the system can be viewed as an assembly of isolated pores for the condition $\omega\tau >1$. In order to obtain the physical quantities for the actual system, the distribution function of the level spacing δ is needed for an assembly of pores with different shapes. We use the distribution function identical with that in the case of small metal particles[5]. The off-diagonal elements of the Hamiltonians for ^{3}He-quasiparticles take random values for every pores due to different shapes, where the distribution of the eigenvalues and the eigenvalue spacings are described by the orthogonal ensemble from the random matrix theory[6]. We adopt the Wigner distribution P(t) for the eigenvalue spacings: $P(t)=(\pi/2)t\exp(-t^2/4)$, where $t=\delta/\bar{\delta}$. Physical quantities must be averaged by this distribution function.

4. Interaction between phonons in porous media and ^{3}He-quasiparticles

Let us calculate the lifetime of phonons through the interaction with ^{3}He-quasiparticles. The wavelength of phonons considered here is much larger than the size of pores. These phonons change the pore volume; i.e., the Fermi-energy of ^{3}He-quasiparticles. This type of the interaction is known to be the deformation-potential coupling where the coupling constant is determined by the Fermi energy level of quasiparticles. The interaction Hamiltonian is written as

$$H_{int} = \sum_{\eta,q,\sigma} A_q a^+_{n+\hbar\omega,\sigma} a_{n,\sigma}(b_q+b^+_{-q}), \tag{2}$$

where

$$A_q = -\frac{2}{3} i E_f \left(\frac{\hbar q}{2\rho v_s}\right)^{1/2} .$$

Here q and v_s are the wave number and the velocity of phonons in a porous medium, respectively. ρ is the mass density of the medium. The symbols $a^+_{n,\sigma}$ and b^+_q are the creation operators of a ^{3}He-quasiparticle, whose energy level is specified by n, and of a phonon with the wave vector $\vec{q}$, respectively. We can calculate the lifetime of phonons from the relation

$$\frac{1}{\tau} = \frac{\pi}{2\hbar^2} [1-\exp(\hbar\omega\beta)] \sum_{m,n} \rho_n |x_{nm}|^2 \delta(\omega+\omega_{nm}) \quad , \tag{3}$$

where ρ_n is the Gibbs distribution function and x_{nm} is the matrix element for the operator $x=\sum A_q a^+_{n+\hbar\omega,\sigma} a_{n,\sigma}$ and $\omega_{mn}=(E_m-E_n)/\hbar$, respectively. The Gibbs distribution function is given by $\rho_n=\exp(-E_n\beta)/Z$ where Z is the partition function and $\beta=1/k_BT$. The partition function is expressed by, for low temperatures, $Z = z + 4\exp(-\bar{\delta}\beta)$, where z takes 1 for even number of ^{3}He-quasiparticles in a pore and 2 for odd number. We take n=0 in Eq.(3) because the excitation by a phonon with frequency $h\nu < \bar{\delta}$ occurs at the Fermi level. Thus the inverse of lifetime of phonons can be expressed by

$$\frac{1}{\tau} = \frac{4\pi E_f^{\,2} n\omega}{9\rho v_f^{\,2}} \sum_{m=1} \frac{1-\exp(-m\beta\delta)}{z+4\exp(-\beta\delta)} \, \delta(\hbar\omega - m\delta) \quad , \tag{4}$$

where n is the number of pores in unit volume. The inverse of the lifetime must be averaged out by the Wigner distribution function P(t). As a result, we have

$$\frac{1}{\tau} = g\,\omega^2 \sum \frac{1}{m^2} \exp\left[-\frac{\pi}{4}\left(\frac{\hbar\omega}{m\bar{\delta}}\right)^2\right] \frac{1-\exp(-\beta\hbar\omega)}{z+4\exp(-\beta\hbar\omega/m)} \quad , \tag{5}$$

where the definition is $g=(\pi^2\hbar^2 N^2 n)/8\rho v_s^2$, in which N is the number of ^{3}He-quasiparticles in a pore. We see that the inverse of the lifetime is proportional to ω^3 in the low-frequency region($\hbar\omega << \bar{\delta}$) and to ω in the high-frequency region ($\hbar\omega >> \bar{\delta}$). The latter case shows the same dependence as that for bulk ^{3}He-quasiparticles in the mixture. The high-frequency phonons diminish the quantum size effect. Finally we make the numerical estimation of τ for the typical experimental situation; T=1.5mK, 5% concentration of ^{3}He, pore diameter D=70A ($\bar{\delta}$=2.4mK), and phonon frequency ν=10MHz. For this situation, the lifetime of phonons becomes 10^{-3}sec.

References

1. For example, see Proc. 17th Inter. Conf. on Low Temp. Phys. Part I-III, ed. by U. Eckern et al., (North Holland, Amsterdam 1984).
2. T.Nakayama: In Phonon Scattering in Condensed Matter, ed. by W. Eisenmenger et al., Springer Ser. Solid-State Sci., Vol. 51 (Springer, Berlin 1984) p.155.
3. A.R. Rutherford, J.P. Harrison, M.J. Scott: J. Low Temp. Phys. 55, 157(1984).
4. For example, see a review by J.A.A.J. Perenboom, P. Wyder, F. Meier: Phys. Rep. 78, 173(1981).
5. L.P. Gor'kov, G.M. Eliashberg: Sov. Phys. JETP 21, 940(1965), and R. Denton, B. Muelschlegel, D.J. Scalapino: Phys. Rev. B7, 3589(1973).
6. T.A. Brody, J. Flores, J.B. French, P.A. Mello, A. Pandey, S.S.M. Wong: Rev. Mod. Phys. 53, 385(1981).

Part VI

Phonon Scattering in Insulators

Phonon Scattering by Distortion About a Point Defect

P.G. Klemens

Department of Physics and Institute of Materials Science,
University of Connecticut, Storrs, CT 06268, USA

1. Introduction

Point defects scatter phonons because of changes in mass and force constants at the defect site; a mass increase and an expansion both lower the local value of the phonon velocity and reinforce the perturbation. For point defects described by a mass difference ΔM, the relaxation rate is [1]

$$1/\tau = c\, a^3\, (\Delta M/M)^2\, (4\pi v^3)^{-1}\, \omega^4 \tag{1}$$

where M, a^3 and v are the normal mass, atomic volume and phonon velocity, c is the defect concentration and ω the angular frequency. To account for distortion, assume that the point defect occupies a volume $V+\Delta V$ instead of V; the effective value of $\Delta M/M$ then becomes [2,3]

$$\Delta M'/M = \Delta M/M + 2G\, \Delta V/V \tag{2}$$

where G is the Grüneisen constant.

While ΔM is known for solute atoms, ΔV is not. However, it may be estimated from the changes of the lattice spacing of the matrix with solute content, measured by X-ray diffraction. The distortion also has a strain energy which contributes to the energy of incorporation. Thus it is also possible to estimate ΔV from the limit of solubility of the solute atom. Both estimates are based on continuum elasticity, which may not be reliable for inclusions of atomic dimensions; furthermore they assume that elastic effects dominate both the lattice dilation and the energy of incorporation. The results derived here should serve as a guide of when distortion effects become important, rather than as a quantitative model.

2. The Model

In an isotopic elastic continuum, if a sphere of radius R $(1+\gamma)$ is fitted into a spherical hole of radius R, the displacement field is radial and has the form

$$u(r) = Ar + \beta R^3/r^2 \tag{3}$$

and consists of a uniform expansion of the matrix and a short-range non-dilatational shear strain field. Only the latter scatters phonons; the expansion A can be identified with a fractional change in the lattice spacing, i.e. $A=da/a$. The shear strain energy is composed of that due to the compression of the included sphere and the strain energy of the matrix. It becomes [4]

$$E = (9/2)cK_I(\gamma - A - \beta)^2 + (9/2)(1-c)KA^2 + 6c\mu\beta^2 \tag{4}$$

where c is the concentration of inclusions, μ and K are the shear and bulk molduli of the matrix, K_I the bulk modulus of the inclusion.

The parameters A and β can be expressed in terms of γ from the condition $\partial E/\partial A=0$ and $\partial E/\partial \beta=0$. One finds

$$A = c(1-c)^{-1}\beta(4\mu/3K) \tag{5}$$

$$\beta = \gamma[1 + 4\mu/3K_I + 4\mu c/3(1-c)K]^{-1}. \tag{6}$$

The overall dilation of the material due to the inclusions is $3A+3c\beta$.

3. Phonon Scattering

A substitutional solute atom is represented by a spherical inclusion. Phonons are scattered by the mass difference ΔM, and by the fractional volume change $\Delta V/V$. The latter is identified with 3β. Neither β nor γ are observable, but one knows the fractional change of the lattice spacing $da/a=A$, the linear expansion of the matrix. Thus, for small concentractions c,

$$\beta = (3K/4\mu)(1/c)da/a \tag{7}$$

so that

$$\Delta M'/M = \Delta M/M + 6G(3K/4\mu)(1/c)\ da/a. \tag{8}$$

The two terms reinforce the scattering if ΔM and da are of the same sign: if positive both reduce the local phonon velocity. If ΔM is large, M must be the larger of the masses of solute and solvent to minimize the perturbation.[3]

If one lacks data for da/a one can deduce the energy of incorporation from the solubility limit. If this energy is identified with the strain energy (4), E can be expressed in terms of β and the magnitude of β found. Depending on the sign of β chosen, one would then obtain two values of $\Delta M'/M$. Since there may be other contributions to the energy of incorporation, this is less reliable. One concludes that when the solubility range is limited, distortion must be important in phonon scattering.

Interstitials can be regarded as the case when $\gamma=1$; β can then be deduced from (6). The choice of K_I is uncertain. Since the inclusion is under considerable compression, it should be larger than K. Taking $K_I=2K, \beta=2/3$ and

$$\Delta M'/M = M_i/M+2G$$

where M_i is the mass of the interstitial. If the atom fits into a pre-existing cavity in the lattice, γ and β are smaller.

4. Comparing with Observations

In dilute alloys of Si in Ge, $\Delta M/M=-0.61$. From the difference in atomic volume $\gamma=-0.12$ and, from (6), $\beta=\gamma/2=-0.06$. Thus $\Delta M'/M=-0.97$, so that scattering should be roughly twice of what it would be from mass difference alone.

For the lattice thermal conductivity of copper alloys, where G=2, there are data for (1/c)da/a from X-ray measurements, as well as thermal conductivity data. [5] Values of $\Delta M'/M$ calculated from (8) are usually too large.

Thus for Sn in Cu, $(1/c)da/a=0.29$, while $\Delta M/M=0.46$, so that $(\Delta M'/M)^2=19.1$, while the observed value [5] is only 2.4. The effect of distortion in (8) appears to be overestimated by a factor 4, perhaps because part of the observed lattice expansion on alloying is due to an increased pressure of the electron gas. A correction based on a free electron gas model exceeds the required correction.

Cu-Al is interesting because the observed scattering is weak, making $\Delta M'/M$ about 0.2 and of either sign. Mass difference and distortion seem to almost cancel. Now $\Delta M/M=-0.57$ and since $(1/c)da/a=0.14$, the second term in (8) is 1.9. To obtain the observed value of $\Delta M'/M$, the distortion term would have to be reduced by a factor of about 2.5.

References

1. P.G. Klemens: Proc. Phys. Soc. A 68, 1113 (1955)
2. P. Carruthers: Rev. Mod. Phys. 33, 92 (1961)
3. M. W. Ackerman and P. G. Klemens: J. Appl. Phys. 42, 968 (1971)
4. P.G. Klemens: Int. J. Thermophysics 7, 197 (1986)
5. N. Sadanand: Doctoral Diss., Univ. of Connecticut (1979)

Phonon Scattering by Defects and Grain Boundaries at High Temperature in Polycrystalline Lanthanum Tellurides of Various Compositions

J.W. Vandersande[1], *C. Wood*[1], *and D. Whittenberger*[2]

[1]Jet Propulsion Laboratory, California Institute of Technology, Pasadena, CA 91109, USA
[2]NASA, Lewis Research Center, Cleveland, OH 44135, USA

1. Introduction

During the past ten years, both theoretical and experimental work on the thermal conductivity of SiGe alloys has shown the importance of grain boundary scattering at high temperatures [1,2]. A grain size of less than 5 m reduced the lattice thermal conductivity by around 35% below the single crystal value at 1000K in heavily-doped alloys [3]. This decrease is believed to be due to the scattering of low-frequency (long wavelength) acoustic phonons by the grain boundaries. These phonons carry a large fraction of the heat in these alloys at high temperatures because the high-frequency (short wavelength) phonons are scattered by the Si-Ge mass difference.

Low-frequency phonons are scattered not only by grain boundaries but also by any other extended defects such as precipitates, pores and second phase material. The effect of grain boundaries and a second phase on the high temperature thermal conductivity was recently observed in lanthanum sulfide alloys [4]. There it was found that when a second phase was present both inside the grains as well as at the grain boundaries, the lattice component of the measured thermal conductivity was reduced by about 40%. As reported here, other extended defects were found to reduce the high temperature thermal conductivity of lanthanum telluride alloys.

2. Experimental Details

Lanthanum telluride specimens ($LaTe_y$) in the single phase regime $1.33<y<1.50$ were prepared by vacuum hot pressing lanthanum telluride powder. The samples were n-type with donor concentrations varying from $4.6x10^{21} cm^{-3}$ to "zero" as the composition varied from $LaTe_{1.33}$ (a semimetal) to $LaTe_{1.50}$ (an insulator), respectively. Also, the number of lanthanum vacancies increases linearly from "zero" as the composition varies from $LaTe_{1.33}$ to $LaTe_{1.50}$. The $LaTe_{1.52}$ and $LaTe_{1.60}$, although having excess Te over La_2Te_3 were, surprisingly, found also to be n-type.

The thermal conductivity was obtained from the simultaneous measurement of diffusivity and specific heat between 300 and 1000 ^oC, using a flash diffusivity apparatus described elsewhere [4,5].

$LaTe_y$ specimens were materialographically prepared utilizing a procedure developed for ceramic materials. Light optical examination of the as polished surfaces revealed that all samples contained a dark gray second phase, identified as La-rich La_2O_2Te, distributed within a light gray matrix. Several specimens also contained metallic inclusions in the form of small Nb particles or extended Te nodules. Quantitative image analysis of all second phases was undertaken on polished samples, and the results are summarized in Table 1.

Table 1. Description of the Microstructure of $LaTe_y$

Composition	Vol. Frac. Percent		Approx. Dia. μm		Nearest Neighbor Distance, μm	Avg. Grain Size μm	Sample Number
	La-rich Phase	Metal. Phase	La-rich Phase	Metal. Phase	La-rich Phase		
$LaTe_{1.33}$	1.1	-	1.1	-	5.6	10	3048
$LaTe_{1.40}$	4.7	-	2.	-	4.2	10	3026
$LaTe_{1.42}$	3.6	5.0[1]	1.1	35.	2.4	14	3093
$LaTe_{1.45}$[2]	1.1	3.3[1]	1.3	31.	5.7	11	3096
$LaTe_{1.45}$	1.2	1.6[3]	1.2	5.	4.5	8	3039
$LaTe_{1.52}$	2.3	0.2[3]	1.7	-	6.0	8	3042

[1] Te nodules. [2] Annealed 4 h at 1400°C. [3] Nb particles.

3. Experimental Results and Discussion

The measured total thermal conductivities at 500°C for all the samples are plotted in Figure 1 as a function of composition.

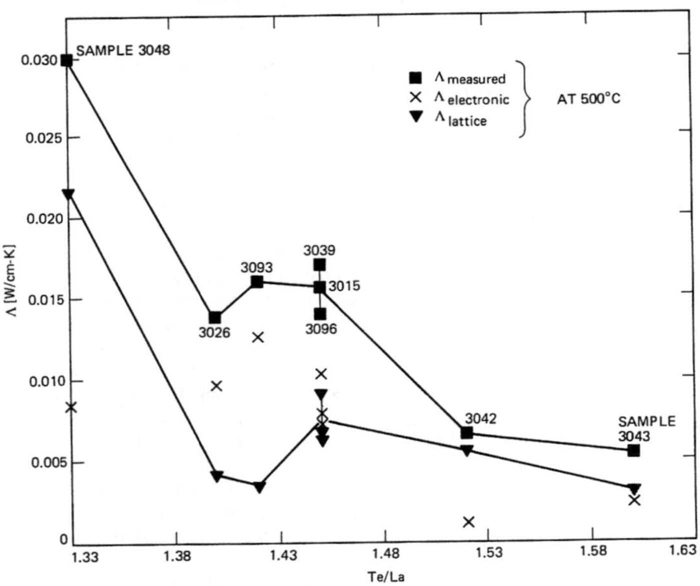

Figure 1. The Total, Electronic, and Lattice Thermal Conductivity at 500°C. of Various $LaTe_y$ Samples Versus Composition.

Also shown in the figure are the electronic components of the thermal conductivity which were calculated from the measured electrical conductivity. The degree of degeneracy for the Lorenz factors for this calculation were determined from the measured Seebeck coefficients. The lattice components of the thermal conductivity were found by subtracting the electronic component from the total conductivity.

It is important to note that both the total and lattice conductivities do not decrease monotonically with increasing Te composition as would be expected from the increase in the number of vacancies as the composition goes from $LaTe_{1.33}$ to $LaTe_{1.50}$. However, there is a dip in both curves in the $LaTe_{1.40}$ - $LaTe_{1.42}$ composition range. The lattice components at $LaTe_{1.40}$ and at $LaTe_{1.42}$ were actually less than that found at $LaTe_{1.52}$ which is known to have considerably more vacancies and has a smaller grain size. The usual high temperature phonon scattering mechanisms thus cannot explain the observed results. Additional phonon scattering mechanisms must be present and the most likely one appears to be the scattering of low-frequency phonons by extended defects. In this case, the defects are the lanthanum-rich second phase and the metallic inclusions. Note that from the microstructural analysis, the largest amount of second phase material and the smallest distance between these second-phase defects occurs in samples with the $LaTe_{1.40}$ - $LaTe_{1.42}$ compositions, which is exactly where the dip occurs in Figure 1.

The electronic component for the $LaTe_{1.33}$ sample appears to be unrealistically low (and thus the lattice component too high) and is very likely due to the fact that oxygen was absorbed between thermal and electrical conductivity measurements.

Acknowledgements

The work described in this paper was carried out at the Jet Propulsion Laboratory/ California Institute of Technology, under contract with the National Aeronautics and Space Administration.

References

1. H. R. Meddins and J. E. Parrott, J. Phys. C: Solid St. Phys.,9, 1263 (1976)
2. N. Savvides and H. J. Goldsmid, J. Phys. C.: Solid St. Phys. 13, 4657 (1980)
3. D. M. Rowe and V. S. Shukula, J. Appl. Phys. 52 (12), 7421 (1981)
4. J. W. Vandersande, C. Wood, A. Zoltan, and D. Whittenberger, to be published in the Proceedings of the 19th International Thermal Conductivity Conference, Cookeville, TN. (1985).
5. C. Wood and A. Zoltan, Rev. Sci. Instrum. 55, 235 (1984)

Phonon Absorption-Spectroscopy in the Presence of Strong Elastic Phonon Scattering

J. Mebert[1], *O. Koblinger*[2], *S. Döttinger*[1], *and W. Eisenmenger*[1]

[1]Physikalisches Institut, Universität Stuttgart, Pfaffenwaldring 57, D-7000 Stuttgart 80, Fed. Rep. of Germany
[2]IBM-Deutschland, GmbH, Schönaicherstr. 220, D-7030 Böblingen, Fed. Rep. of Germany

1. Introduction

Rare Earth doped crystals are known to show very sharp absorption lines in IR-absorption measurements due to 4f-intra-shell transitions of the Rare Earth dopants. The line sharpness is caused by screening the 4f-orbit by electrons occupying the spatialy extented 5s and 5p shells. Absorption lines have been found at 21.2 cm^{-1} and 32.8 cm^{-1} in Er^{3+} doped CaF_2 [1] and at 52 cm^{-1} in Er^{3+} doped LaF_3 [2].
In this energy range phonon spectroscopy is a good and useful method in absorption measurements. The predominant advantages are the very high frequency resolution of about 10 GHz (in IR typical 30 GHz) and the existence of other coupling mechanisms for the 4f-intra shell transitions. In IR-spectroscopy these transitions are characterized by a small oscillator strength because they are forbidden in first order by the parity selection rule. Phonon transitions are induced by orbit-lattice coupling allowing transitions without parity change.
In pulse measurements a strong elastic phonon scattering can be observed. Such elastic phonon-scattering was also published by different other authors [3/4] in recent times. In this work we show that in the presence of a strong phonon scattering background absorption structures can only be well resolved by reducing sample thickness to the phonon mean free path. This mean free path can be determined by analyzing the pulse shape of 285 GHz phonons. By reducing sample thickness to the appropriate value of 0.3 mm the 21.2 cm^{-1} crystalline field transition in $CaF_2:Er^{3+}$ could be evaluated with the very high resolution of 5 GHz. In experiments performed on a 1mm thick $LaF_3:Er^{3+}$ sample we observed an absoption line at 14.2 cm^{-1} not visible in FIR absorption measurements.

2. Phonon spectroscopy measurements at $CaF_2:Er^{3+}$

Phonon spectroscopy measurements with an Al and a Sn tunneling junction as voltage tunable phonon generator and detector respectively were applied to Er^{3+} doped CaF_2 using the well known modulation technique. The basic intention was to evaluate the crystalline field absorption lines reported above. Fig. 1 shows the corresponding phonon absorption spectrum. The sample thickness in this case was d = 2 mm. At 636 GHz (21.2 cm^{-1}) a weak, hardly resolved absorption line can be recognized. This structure can be ascribed to the 21.2 cm^{-1} absorption line known from FIR measurements.

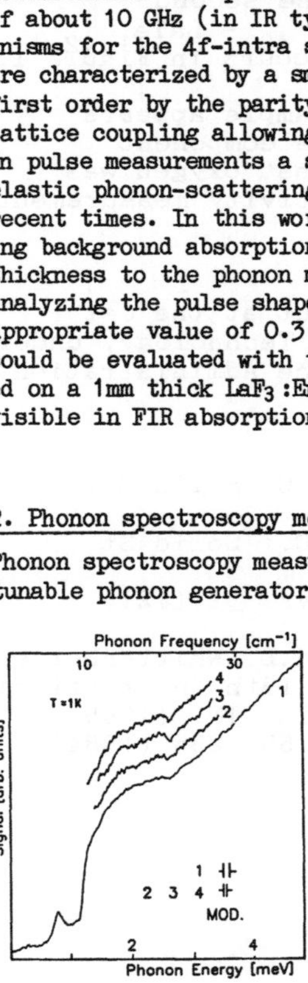

Fig. 1 Phonon absorption spectrum of $CaF_2:Er^{3+}$
d = 2 mm

3. Pulse measurements with superconductive Sn-tunnel junctions

In this technique superconductive Sn-tunnel junctions are used as well as phonon generator and phonon detector. The advantage of this arrangement is the possibility of investigating nonequilibrium phonon propagation at the fixed frequency of 285 GHz coupled with a high S/N ratio. Fig 2a, b, and c show the Sn/Sn pulse-measurements of $SrF_2:Eu^{2+}$, $LaF_3:Er^{3+}$ and $CaF_2:Er^{3+}$ respectively.

The generator pulse width was 0.2 µs in case a, 0.4 µs in case b and 0.3 µs in c. The detector signal on the other hand extends over a range of about 10 µs. This large extension of the detector signal indicates a strong elastic scattering of the 285 GHz phonons increasing both the pulse propagation time and the pulse width Such nonequilibrium phonon propagation determined by strong elastic scattering processes is called "diffusive" phonon propagation. The essential scattering mechanisms leading to diffusive phonon propagation are mass defect scattering and scattering processes at lattice imperfections [5].

4. The influence of strong elastic phonon scattering on phonon absorption measurements

In phonon absorption spectroscopy absorption structures are indicated by absorption and reemission of resonant phonons i.e. elastic scattering of the resonant phonons at the absorption centers. In the presence of a strong elastic scattering background the resonant scattering at absorption centers is a weak additional effect, hardly detectable. By reducing the sample thickness to the elastic scattering mean free path of the phonons the elastic scattering background can be reduced and the resonant phonon scattering at absorption centers becomes resolvable. In a first order approximation the mean free phonon path in the frequency range from 200 GHz to 1 THz can be estimated by the mean free path of 285 GHz phonons.
This value can be determined by fitting the pulse measurements (Fig. 2) to an analytic expression for the detector signal based on the diffusive equations well

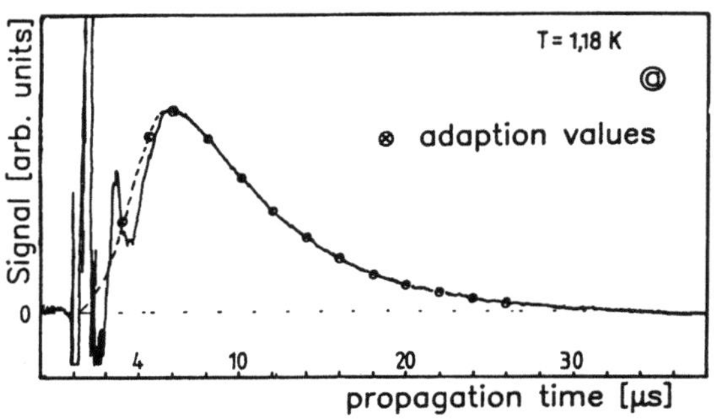

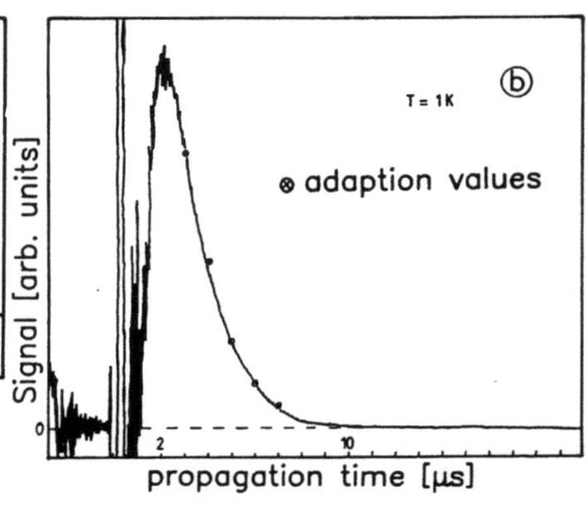

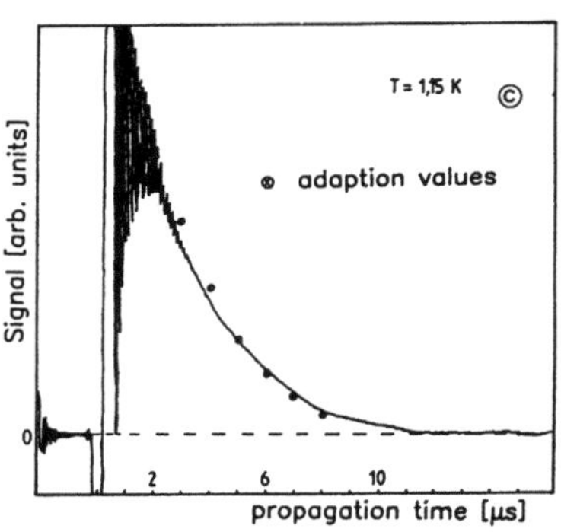

Fig. 2 Sn/Sn pulse measurements of a) $SrF_2:Eu^{2+}$
b) $LaF_3:Er^{3+}$
c) $CaF_2:Er^{3+}$

known from kinetic theory. The adaption values are shown in Fig. 2. The values obtained for λ(285 GHz) are

λ = 1.8 ± 0.6 mm in $SrF_2:Eu^{2+}$; λ = 0.7 - 1 mm in $LaF_3:Er^{3+}$

λ = 0.2 - 0.4 mm in $CaF_2:Er^{3+}$

5. Phonon spectroscopy measurements at thin samples (d=λ(285 GHz))

By reducing the sample thickness to the value of mean free path of 285 GHz phonons the resonance structure at 21.2 cm^{-1} hardly detectable at the thick $CaF_2:Er^{3+}$ sample (see 2.) could now be well resolved (Fig.3 a). Fig.3 b shows the high resolution phonon absorption spectrum of this resonance ($\Delta\nu$ = 5.2 GHz). An additional signal modification can be seen at 30 cm^{-1} (b) with less intensity. This absorption can be ascribed to respective Er^{3+} crystalline field transitions already known from FIR-measurements (see 1.).

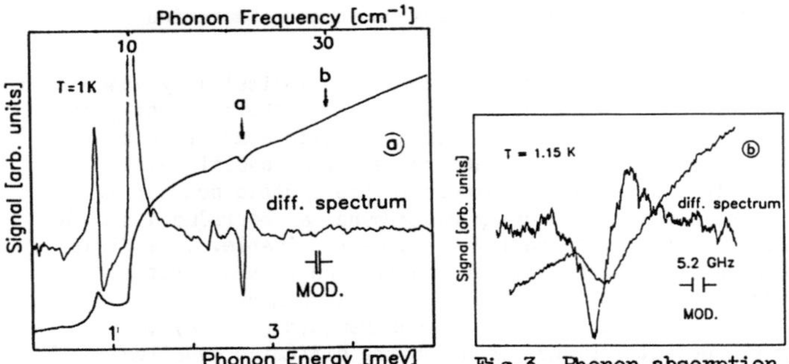

Fig.3 Phonon absorption spectrum of $CaF_2:Er^{3+}$; d=300 μm

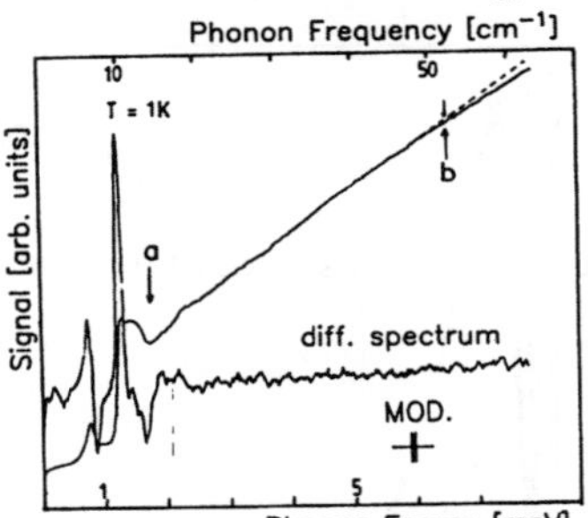

Fig. 4

Phonon absorption spectrum of $LaF_3:Er^{3+}$
d = 1 mm

Phonon absorption measurements were also applied to appropriate thin $LaF_3:Er^{3+}$ samples. Fig. 4 shows the related spectrum to a d=1 mm thick sample.

At 14.2 cm^{-1} a deep absorption line was found not visible in FIR-measurements (a). This line so far could not be identified. For frequencies higher than 52 cm^{-1} a signal reduction is observed (b). This can be explained by the frequency limit of the TA-branches in the phonon dispersion relation of LaF_3. Phonons with frequencies higher than this limit cannot propagate in the sample.

[1] Ward/Clayman J. Phys. C Solid State Phys. Vol.8, p 872 (1975)
[2] Hadni/Strimer Phys. Rev. B; 5,11, 4609 (1972)
[3] Bron, Keilmann Phys. Rev. B 12, 2: 2496 (1975)
[4] Eisfeld, Renk Int. Conf. in Phonon Scattering, Brown University Providence 1979
[5] Klemens Proc. Phys. Soc. 68, 12A, 73 (1955)

Spin-Wave Heat Conduction and Magnon-Phonon Coupling in EuS Ferromagnets

G.V. Lecomte[1], *H. v. Löhneysen*[+,2], *J. Wosnitza*[2], *and W. Zinn*[3]

[1]Labor für Tieftemperaturphysik, Universität/GH Duisburg, D-4100 Duisburg, Fed. Rep. of Germany
[2]Zweites Physikalisches Institut der RWTH Aachen, D-5100 Aachen, Fed. Rep. of Germany
[3]Institut für Festkörperforschung der KFA Jülich, D-5170 Fed. Rep. of Germany

Heat transport by magnons in a ferromagnet, although well established experimentally, remains sometimes elusive to observe in a thermal-conductivity experiment. Indeed, as Walton pointed out /1/, the magnitude of the coupling existing between phonon and magnon systems, γ, must lie in the right range related to the other scattering mechanisms and to the physical dimensions of the specimen. If γ is too large, one can only observe scattering between the strongly coupled phonons and magnons. On the other hand, if γ is weak to the point that the phonon mean free path for a collision with a magnon becomes larger than the specimen's length, heat cannot enter the spin-wave system in sufficient amount so that the magnon conductivity, although large, cannot be detected.

In a previous paper /2/, we showed a magnetic field effect on the heat transport reaching 70 % of the total thermal conductivity κ at 1.5 K in a single crystal of EuS. The aim of the present study is to observe at least qualitatively the influence of sample preparation by measuring κ in another single crystal of different quality, and in a thin film of EuS deposited on glass. We can also compare these experimental values to those found in a sintered sample /3/. At the same time, we wanted to confirm our values in view of a recent paper /4/ finding a different field dependence of κ in a EuS single crystal.

Both single crystal samples were prismatic rods of about 1 mm^2 cross-section and a length of 5 to 6 mm, oriented along a [100] direction. The thermal conductivity κ was measured using a steady-state method already described /5/. EuS films were prepared by high-temperature vapor deposition on a thin glass substrate (30 μm). The film thermal conductivity was obtained by subtracting from the total heat conductance the value observed for the bare substrate in the course of preliminary runs. This procedure neglects the effect of phonons crossing over from the film into the glass or conversely, but their contribution to the apparent heat conduction is expected to be small. The main source of error lay in the determination of the geometry factor with a film thickness known only to within 20 %, leading to an estimated overall accuracy of 30 % on the absolute film conductivity.

κ (T) with and without a large applied magnetic field is plotted in double logarithmic representation in Fig. 1. Data for sample 1 have already been presented elsewhere /2/. Sample 2, a specimen a little smaller than sample 1, exhibits a markedly lower κ at high T. Below 4 K, the temperature

---

[+] Physikalisches Institut der Universität, D-7500 Karlsruhe, FRG

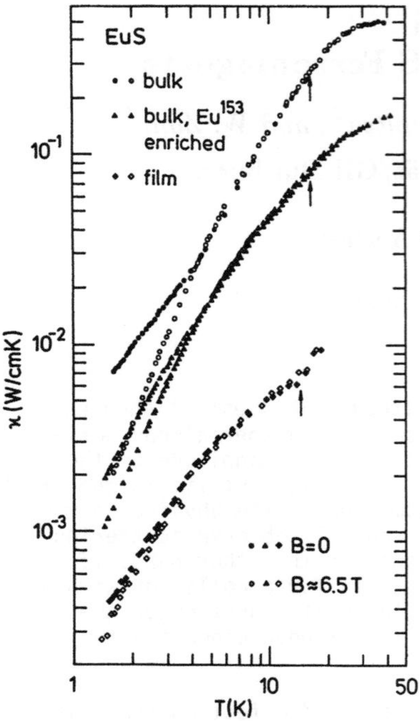

Fig. 1 Thermal conductivity κ vs. temperature T of EuS for B = 0 and in a magnetic field B ≃ 6.5 T. The enriched sample had been used for neutron scattering, the effect of isotope scattering can be neglected for the other samples. Arrows indicate Curie temperatures.

dependence of κ in zero field can be well approximated with a T^2 law. A large field (6.5 T) reduces κ for T < 4 K, this effect coming close to saturation only at our lowest temperature, 1.5 K, where the decrease amounts to 45 % of the zero-field value. Since the same field affects more strongly ($\Delta\kappa/\kappa \simeq 80\%$) the conductivity in sample 1, the high-field κ which is representative of phonon heat transport alone, differs only by a factor 1.5 between both bulk specimens at low temperature. From approximately 5 to 20 K, a magnetic field increases κ by a small, but significant amount of the order of 5 % for both samples. The EuS film measured here had a thickness of 1.5 µm. With respect to bulk EuS, κ is smaller by one order of magnitude. The pronounced dip in the vicinity of the Curie temperature T_c = 13.5 K for thin films /6/ is reproducible, and was also found in a thinner (600 nm) film during preliminary experiments. This dip is reduced in a large applied field. Although smaller than in the bulk samples, a reduction of κ in an applied field is clearly seen below 3 K.

For comparison, the data of McCollum et al. /3/ in the 1.5 to 4 K range overlap approximately the values found for sample 2, with a slightly steeper T-dependence, while the data of Arzoumanian et al. /4/ lie above those for sample 1 by a factor of roughly 2.5 at 5 K, with a well defined maximum at about 13 K, and an increase of κ with increasing B between at least 2 and 15 K.

In the simplest, non-interacting picture, magnon and phonon contributions add up independently to give the total conductivity: $\kappa = \kappa^m + \kappa^{ph}$ where κ^{ph} does not include phonon-magnon scattering and similarly for κ^m. Such a model cannot describe the experimental situation, because a minimum degree of coupling of the spin waves to the lattice is required to allow a heat flow through the magnon system. A non-infinite coupling between lattice and spins leads to different temperatures for magnons (T_m) and for phonons (T_{ph}) in the end regions of the sample /1/. Hence, it is essential to know the distance $1/\alpha$ required for T_m to come close to T_{ph} to properly analyze the outcome of experiments.

Assuming that interactions between different types of excitations occur less frequently than between similar heat carriers, one obtains relations for T_m and T_{ph} along the axis x of a rod of length 2L under steady state axial heat flow q. The ratio of the temperature increases at x is given by /1/:

$$\frac{\Delta T_m}{\Delta T_{ph}} = \frac{\alpha x \cosh\alpha L - \sinh\alpha x}{\alpha x \cosh\alpha L + (\kappa^m/\kappa^{ph}) \sinh\alpha x}$$

where the origin is chosen at the middle of the sample, and

$$\alpha = \left(\frac{\kappa^m \kappa^{ph}}{\gamma\ (\kappa^m + \kappa^{ph})}\right)^{-1/2}$$

where the phonon-magnon coupling γ is expressed as the heat flux per unit volume from phonons to magnons for a unit value of $T_m - T_{ph}$.

The magnon heat conductivity is calculated from the approximate quadratic dispersion curve /7/ and from the relaxation times given by Forney and Jäckle /8/. Quantitative description of the κ -increase in a field in the vicinity of 9 K is obtained with an inverse relaxation time for phonon-magnon scattering $(\tau^{ph}{}_m)^{-1} = 2 \cdot 10^6\ T^{3/2}\ K^{-3/2}\ s^{-1}$, or twice the value given in /8/. For sample 1, κ^m increases with decreasing T between 3 and 5 K in agreement with /8/. The maximum value of κ^m at 2.6 K corresponds to a point-defect concentration of 0.1 %, a reasonable value if taken to represent both chemical impurities and crystal defects. From the value of $1/\alpha$, κ^m is expected to decrease strongly only at very low temperatures (0.025 K). While the results of McCollum et al. /3/ can also be reasonably fitted in this picture, we are unable to account for the field dependence of κ observed by Arzoumanian et al. /4/.

In conclusion, we have presented thermal conductivity measurements for a range of different samples of EuS which can be quantitatively accounted for in terms of phonon and magnon heat transport.

This work was carried out within the research program of Sonderforschungsbereich 125 Aachen-Jülich-Köln.

REFERENCES

1. D. Walton: Proc. 1st Intl. Conf. on Phonon Scattering in Solids, ed. by H.J. Albany (CEN-Saclay 1972) p. 295;
   D.J. Sanders, D. Walton, Phys. Rev. B 15, 1489 (1977)
2. G.V. Lecomte, H.v.Löhneysen, W. Zinn: Proc. 4th Intl. Conf. on Phonon Scattering in Solids, ed. by W. Eisenmenger, K. Lassmann, S. Döttinger (Springer, Berlin 1984) p. 466
3. D.C. McCollum, R.L. Wild, J. Callaway: Phys.Rev. 136, A 426 (1964)
4. C. Arzoumanian, A.M. de Goer, B. Salce, F. Holtzberg: Proc. 17th Intl. Conf. on Low Temp. Physics, ed. by U. Eckern, A. Schmid, W. Weber, H. Wühl (North-Holland, Amsterdam 1984) p. 169
5. G.V. Lecomte, H.v. Löhneysen, W. Zinn: J. Magn.Magn. Mat. 38, 235 (1983)
6. U. Köbler: private communication; see also: J. Köhne, G. Mair, W. Rasula, B. Saftic, W. Zinn: J. de Phys. Colloque 41, C - 127 (1980)
7. H.G. Bohn, W. Zinn, B. Dorner, A. Kollmar: Phys. Rev. B 22, 5447 (1980)
8. J.-J. Forney, J. Jäckle: Phys. Kond.Mat. 16, 147 (1973)

Effects of Elastic Anisotropy on Phonon Scattering in Non-Metallic Materials

H.H. Sample[1], *K.A. McCarthy*[1], *M.B. Koss*[1], *and A.K. McCurdy*[2]

[1]Physics Department, Tufts University, Medford, MA 02155, USA
[2]Department of Electrical Engineering, Worcester Polytechnic Institute, Worcester, MA 01609, USA

1. Introduction

It has long been known that certain elastically anisotropic materials also show anisotropies in thermal conductivity, at both low and high temperatures. In this paper, we report thermal conductivity measurements on oriented single crystals of two such materials: tellurium dioxide (TeO_2) and mercurous chloride (Hg_2Cl_2), in the 1.6-100 K temperature range. In the Umklapp scattering region, we find that elastic anisotropy results in substantial thermal conductivity anisotropy in Hg_2Cl_2, but not in TeO_2. Some of these results have been reported previously [1].

2. Tellurium Dioxide

TeO_2 has a structure with tetragonal symmetry, and has unusual acoustic properties. For sound waves propagating along <100> directions, the fast shear wave has a higher velocity than the longitudinal wave. On the other hand, a shear wave propagating in a [110] direction has a velocity approximately one-seventh that of the above longitudinal wave. Highly anisotropic ballistic phonon propagation has been observed at 1.6 K using phonon imaging techniques [2].

Figure 1 shows the thermal conductivity as a function of temperature for three high-purity crystals of TeO_2, with their long axes (the heat-flow directions) oriented along [100], [110], and [001]. All three samples were from the same Crystal Technology Inc. growth melt, and the [100] and [110] samples were cut from neighboring regions of the same piece (the phonon-imaging sample mentioned above also came from this same piece). The samples all had approximately square cross-sections with $D = (\text{Area})^{1/2}$ between 3.3 and 3.8 mm, overall sample lengths L of about 23 mm, and ratios D/L of about 0.15. All had rough surfaces. The dashed line in Fig. 1 shows the results for the as received [001] sample with optically polished surfaces and L = 27 mm.

Between 1.6 and 10 K, the effects of elastic anisotropy on the thermal conductivity are clearly evident. At 1.75 K, the experimental K[110]/K[100] is about 2.2, in fair agreement with the value 1.6 predicted by phonon focusing calculations appropriate for the boundary scattering regime. However, these same calculations predict that the [001] conductivity should be comparable to that of the [100] sample; experimentally it is nearly equal to that of the [110] sample. The discrepancies between experiment and theory are perhaps due to the fact that our measurements do not extend to low enough temperatures, and because the calculations were performed using room temperature elastic constants. In the Umklapp scattering range, the [100] and [110] conductivities are identical to within our experimental precision; the [001] sample has only a slightly larger (10 to 20%) conductivity at the highest temperatures.

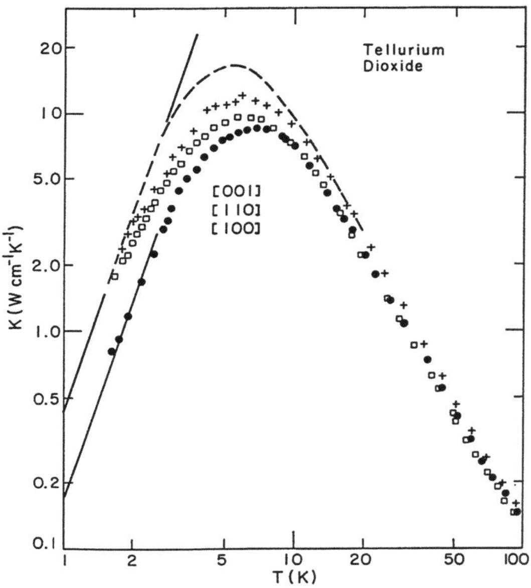

Fig. 1. Thermal conductivity of oriented single crystals of TeO_2. Long axis of specimen (heat flow direction) along [001] (crosses), [110] (open squares), and [100] (closed circles). All samples from same Crystal Technology Inc. growth melt, and had rough surfaces. Dashed line is [001] sample with optically polished surface and longer length. The solid lines indicate T^3 behavior.

3. Mercurous Chloride

At room temperature, Hg_2Cl_2 has a structure consisting of strongly-bound linear chains of Cl-Hg-Hg-Cl molecules connected end-to-end; the chains are weakly coupled together into a body-centered tetragonal lattice, with the c-axis along the chain [3]. The material is highly anisotropic elastically: the longitudinal velocity along the [001] direction is 3.34 km/s, whereas the slowest transverse velocity (along [110] directions) is only 0.347 km/s, comparable to the speed of sound in air. At 185 K, the material undergoes a second-order ferroelastic transition. The chain dimensions remain essentially unchanged, but the basal square elongates slightly along one of the [110] directions and contracts along the other, so that these directions are no longer equivalent. The low temperature phase has orthorhombic symmetry, and is slightly less elastically anisotropic than the room temperature structure [1]. For convenience, we continue to refer crystal directions to the tetragonal cell.

Figure 2 shows thermal conductivity results for a set of Hg_2Cl_2 samples with various orientations. At low temperatures, the measured conductivities are at least a factor of 15 lower than the predictions of phonon focusing. Further, we have observed decreases (never increases) in the low temperature conductivity of some samples, induced by demounting, cutting, and remounting them. And for one sample, a decrease was induced merely by warming it to room temperature (see the solid and open circles in Fig. 2). We therefore believe [1] that the low temperature conductivity of Hg_2Cl_2 samples is limited by systems of planar defects, which are visible in most specimens, oriented primarily perpendicular to [110] directions. The planar defects evidently scatter phonons in a manner similar to crystal boundaries, as indicated by the T^3 behavior of all conductivities at low temperature.

The high temperature conductivities of Hg_2Cl_2 specimens are intrinsic properties of the specimens, and are not affected by warming to room tempera-

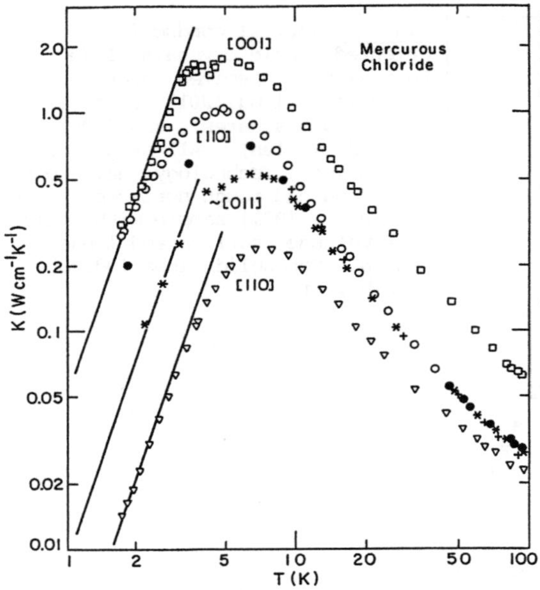

Fig. 2. Thermal conductivity of oriented single crystals of Hg_2Cl_2. Long axis of specimen (heat flow direction) along [ØØ1] (open squares), [11Ø] (open circles), about 1Ø° from [Ø11] (asterisks and crosses), and [11Ø] (inverted triangles). All samples have rough surfaces. [ØØ1] and lower [11Ø] samples are a matched pair cut from the same boule, and have nearly the same cross sectional dimension D of about 3.3 mm, smaller than for the other samples. The solid circles are for the upper [11Ø] sample after cycling to room temperature. The solid lines indicate T^3 behavior.

ture or by handling. In contrast to the situation for TeO_2, the Umklapp region conductivities for different orientations of Hg_2Cl_2 vary markedly. For example, the conductivity of the [ØØ1] sample (along the chains) is 2.3 to 2.7 times larger than that of the other orientations measured. In this respect, our results are similar to those of de Göer, et al. for alpha-HgI_2 [4].

The [ØØ1] and lowest conductivity [11Ø] samples shown in Fig. 2 were a "matched" pair, having been cut from the same high purity boule, with nearly the same cross sectional dimension D of about 3.3 mm. The other three samples are believed to be slightly less pure, and have dimensions D as large as 1Ø mm. We do not believe that the higher conductivity of the second [11Ø] sample (the circles in Fig. 2) is related to its larger D, although this remains to be verified. Rather we speculate that the difference between the two [11Ø] samples at high temperatures is related to different amounts of the two non-equivalent [11Ø] directions being oriented along the heat flow axis of the samples. We note that at 123 K, the calculated ratio of the two [11Ø] longitudinal group velocities is 1.3, about the same as the ratio of the two [11Ø] sample conductivities at the highest temperatures.

Acknowledgement

We acknowledge support from the Tufts University Faculty Research Fund.

References

1. K.A. McCarthy, H.H. Sample and A.K. McCurdy, Proc. 19th Int. Therm. Cond. Conf., Cookeville, TN, October, 1985, in press
2. D.C. Hurley, J.P. Wolfe and K.A. McCarthy, Phys. Rev. B33, 4189 (1986)
3. C. Barta, Kristall und Technik 5, 541 (197Ø)
4. A.M. de Göer, M. Locatelli, I.F. Nicolau, J. Phys. Chem. Sol. 43, 311 (1982)

Low-Temperature Thermal Conductivity of Single Crystal Diacetylene

D.T. Morelli and J.P. Heremans

Physics Department, General Motors Research Laboratories, Warren, MI 48090–9055, USA

Since the discovery in 1977 [1] that doped polyacetylene possesses a metallic conductivity, much activity has centered on the possibility of using this and other conjugated polymers as quasi-one-dimensional conductors. In the case of polyacetylene, comparison between experimental results and theory is complicated by its complex morphology and instability in air. Polydiacetylene (PDA), on the other hand, forms stable and quite large single crystals. This system in its pristine (undoped) state can serve as an important testing ground for comparing the intrinsic properties of conjugated polymers with theory. To this end we have carried out detailed measurements of the low-temperature thermal conductivity of diacetylene bis(p-toluene sulfonate) of 2,4 hexadiyne-1,6-diol. (In what follows we will refer to this crystal in its monomer form as TS and in its polymer form as PTS.)

Figure 1 shows the results of our thermal conductivity measurements on three TS samples and three PTS samples. The dashed line indicates the results of WYBOURNE et al. [2], measurements taken on a PTS sample in the chain direction. Looking first at the TS samples, we see that K(T) is essentially isotropic and reminiscent of the thermal conductivity of a typical crystalline dielectric. The thermal conductivity of the polymerized samples, on the other hand, is quite different in both anisotropy and temperature dependence. Along the chains K(T) exhibits an extremely broad maximum from 10-20 K, while across the chains the thermal conductivity flattens above about 10 K. Thus in comparison to monomer samples, in PTS phonon-phonon U-processes appear to be extremely weak or totally absent. An even more interesting result is that the conduction of heat across the carbon chains in PTS is some 30-50 times less than conduction along the chain, an anisotropy not observed in unpolymerized samples.

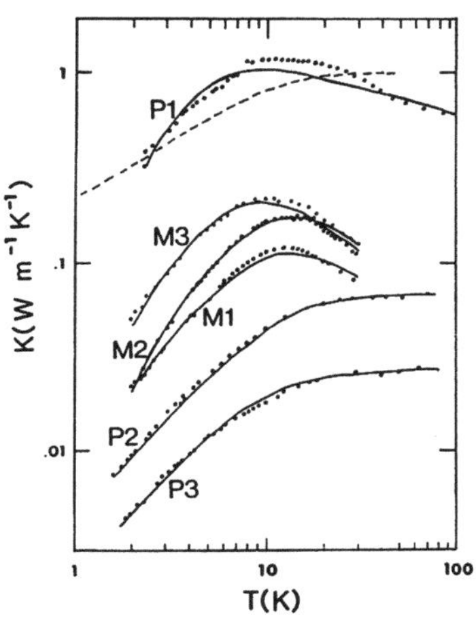

Fig. 1. Thermal conductivity of single crystal diacetylene. M1: monomer, ∥ chain; M2 and M3: monomer, ⊥ chain; P1: polymer, ∥ chain; P2 and P3: polymer, ⊥ chain. The dashed line represents the results of WYBOURNE, et. al. (2); the solid lines are fits discussed in the text

In analyzing thermal conductivity data the standard approach [3] is to assume that the phonon spectrum is isotropic in three-dimensional $\vec{q}$-space, i.e., that the allowed phonon wavevectors are contained in a sphere of maximum radius equal to the Debye wavevector q_D. For many solids this postulate is quite reasonable since their first Brillouin zones are not too different from spherical in shape. The thermal conductivity can be written according to the kinetic formula

$$K(T) = \sum_{s} [v(s)]^2 \int_{\text{all } \omega} h\omega \, \tau(\omega,s) \, g(\omega,s) \frac{df}{dT} (\omega,T) \, d\omega \tag{1}$$

where v is the sound velocity and τ the relaxation time of mode s with frequency ω, and f is the phonon distribution function and $g(\omega,s)$ the density of states. Assumptions regarding the phonon spectrum determine the form of $g(\omega,s)$.

In a material with a highly anisotropic crystal structure (and therefore anisotropic Brillouin zone), evidently the foremost assumption that the phonon wavevectors are distributed uniformly in $\vec{q}$-space cannot be retained. In the case of PTS, the allowed phonon wavevectors in the chain direction ($q_\parallel$) are about three times larger than perpendicular to the chain ($q_\perp$). For $q < q_\perp$ the phonon wavevectors are distributed uniformly and the material is essentially three-dimensionally isotropic. For $q_\perp < q < q_\parallel$, however, wavevectors are allowed to exist only along the chain direction. In this regime the phonons act one-dimensional and $g(\omega,s) = \omega^o/v$. Thus we see that for these phonons $K \sim v$: in one dimension the <u>fast</u> phonons dominate heat conduction. In PTS, longitudinal phonons travel in the chain direction about 6-7 times faster than longitudinal phonons traveling across the chains [4]. This anisotropy of fast phonon modes will produce an anisotropy in the thermal conductivity <u>if</u> the conditions of one-dimensionality of the phonon spectrum are met. Furthermore, according to PEIERLS [5], in one dimension U-processes cannot exist if there is any dispersion in $q(\omega)$ near the band edge. The result is that the quasi-one-dimensionality of the PTS structure can qualitatively account for both the anisotropy and temperature dependence of K(T) observed in Fig. 1. The one-dimensional nature of the phonon spectrum, however, disappears exponentially below temperatures corresponding to $q_\perp$. In PTS, heat capacity studies [6] indicate that $T_\perp \approx 50$ K and $T_\parallel \approx 2400$ K. Thus below 50K PTS should behave three-dimensionally, and the large anisotropy observed at lower temperatures cannot be attributed to the phonon spectrum. This implies that the origin of the anisotropy is in the scattering of the phonons, i.e., in $\tau(\omega,s)$. One is thus led to consider in detail the possible scattering processes which can occur in TS and PTS.

TEM studies of TS single crystals [7] have shown that the monomer contains both edge and screw dislocations distributed essentially isotropically throughout the crystal. YOUNG, et al. [8], however, have shown that in PTS crystals stacking faults of spacing on the order of a micron are formed parallel to the chain direction and we confirmed this observation on the samples we used. YOUNG and PETERMANN [9], also using TEM, estimate the dislocation density in a PTS single crystal as $\lesssim 10^{13}$ m^{-2}. This is about two orders of magnitude higher than dislocation densities in other systems [10-12] whose thermal conductivity has been shown to be limited by dislocation scattering. We thus believe that our data can be understood on the basis of phonon-dislocation scattering at low T and phonon-phonon scattering at high T.

For TS, we assume that phonons scatter off of the strain fields of isotropically distributed dislocations. In PTS, phonons propagating parallel to the chains will scatter off of dislocations which are <u>not</u> aligned in the chain direction; phonons propagating perpendicular to the chains will scatter off of stacking faults. The results of this fitting procedure are indicated by solid lines in Fig. 1 and are quite good for all samples. From the stacking fault-phonon scattering rate we find stacking fault spacings of 0.5 μm and 2.5 μm for samples P3 and P2, respectively, in good agreement with TEM observations. For the TS samples, M1, M2, and M3 we find dislocation densities of 3.9, 3.1, and 1.4 x 10^{19} m^{-2}, respectively. As mentioned above, there has been no independent determination of dislocation densities in TS. For the PTS sample P1 we find $N_D \approx 2 \times 10^{17}$ m^{-2}, about 10^4 times greater than that reported from TEM studies of these crystals. Such a large discrepancy between dislocation densities inferred from thermal conductivity data and observed densities is not unusual for dislocation-dominated systems [10-12]. The difference is usually attributed to the association or mobility of dislocations in the lattice, both of which will increase the scattering rate significantly. No other scattering mechanism, in particular the optical-acoustic phonon scattering process put forth by WYBOURNE et al. [2] can account, even qualitatively, for the observed temperature dependence of all our samples.

One last rather fundamental result must be pointed out: the Umklapp scattering rate for PTS samples is at least two orders of magnitude weaker than in TS samples. We believe that this is a manifestation of the quasi-one-dimensionality of the PTS phonon spectrum, which becomes more and more pronounced above 50 K, and severely restricts anharmonic phonon-phonon interactions.

ACKNOWLEDGMENTS

The authors would like to thank Dr. M. Sakamoto for preparation of the monomer crystals, and acknowledge useful discussions with Professors A. C. Anderson, M. S. Dresselhaus, P. G. Klemens, and C. Uher.

REFERENCES

1. H. Shirakawa, E. G. Louis, A. G. MacDiarmid, C. K. Chiang, and A. G. Heeger: J. Chem. Soc. Chem. Commun., 578 (1977).
2. M. N. Wybourne, B. J. Kiff, and D. N. Batchelder: Phys. Rev. Lett. 53; 580 (1984).
3. R. Berman: Thermal Conduction in Solids, (Clarendon Press, Oxford, 1976).
4. W. Rehwald, A. Vonlanthen, and W. Meyer: Phys. Stat. Sol. (a) 75, 219 (1983).
5. R. Peierls: Ann. Phys. Lpz. 3, 1055 (1929).
6. I. Engeln and M. Meissner: J. Polymer Science: Polymer Phys. Ed. 18, 2227 (1980).
7. M. Dudley, J. N. Sherwood, D. J. Ando, and D. Bloor: Mol. Cryst. Liq. Cryst. 93, 223 (1983).
8. R. J. Young, R. T. Rend, and J. Petermann: J. Mat. Sci. 16, 1835 (1981).
9. R. J. Young and J. Petermann: J. Polymer Science: Polymer Phys. Ed. 20, 961 (1982).
10. R. L. Sproull, M. Moss, and H. Weinstock: J. Appl. Phys. 30, 334 (1959).
11. A. C. Anderson and S. C. Smith: J. Phys. Chem. Solids, 34, 111 (1973).
12. S. G. O'Hare and A. C. Anderson: Phys. Rev. B 9, 3730 (1974).

Dislocation-Phonon Interaction Under Hydrostatic Pressure

Y. Hiki, T. Kosugi, and Y. Kogure

Faculty of Science, Tokyo Institute of Technology,
Oh-okayama, Meguro-ku, Tokyo 152, Japan

1. Introduction

Thermal phonons produce a prominant resistive force against a moving dislocation in a crystal, especially at higher temperatures, and the force F is proportional to the dislocation velocity v, namely F = Bv. The coefficient B is called the dislocation damping constant, which represents the strength of the dislocation-phonon interaction. The present study is intended to investigate the mechanisms of the damping of moving dislocation by measureing the pressure dependence of damping constant in alkali halide and aluminum crystals through an acoustic method. The effect of contained impurities on the pressure dependence is also studied.

2. Dislocation Damping in Alkali Halide Crystals

The pressure dependence of ultrasonic attenuation α in LiF and NaCl single crystals have been measured at sound frequencies of f = 27 - 99 [MHz]. The details of the experimental method have already been described [1 - 2]. The measurements were made for the two states of the specimen: the deformed state by compression to introduce dislocations, and the irradiated state by γ rays to suppress the dislocation motion. By taking the difference of the sound attenuation between the two states, the pressure dependence of the decrement (internal friction) due to dislocations was determined.

According to the theory of overdamped resonance of dislocations [3], the decrement Δ (= α[dB/μsec]/8.686f[MHz]) can be expressed as

$$\Delta = D\omega\tau/(1 + \omega^2\tau^2), \tag{1}$$

where $\omega = 2\pi f$, and D and τ are parameters containing various quantities related to the dislocations. By differentiating (1) by pressure P,

$$\frac{1}{\Delta_o}\frac{d\Delta}{dP} = [\frac{1}{D_o}\frac{dD}{dP}] + \frac{1 - \omega^2\tau_o^2}{1 + \omega^2\tau_o^2}[\frac{1}{\tau_o}\frac{d\tau}{dP}], \quad \Delta_o = \Delta(P = 0) \text{ etc.} \tag{2}$$

Values of $(1/D_o)(dD/dP)$ and $(1/\tau_o)(d\tau/dP)$ can be determined by fitting (2) to the experimental $(1/\Delta_o)(d\Delta/dP)$-vs-ω. The pressure dependence of damping constant can be evaluated by inserting these values into the formula

$$\frac{1}{B_o}\frac{dB}{dP} = \frac{1}{\tau_o}\frac{d\tau}{dP} - \frac{1}{D_o}\frac{dD}{dP} + \frac{1}{G_o}\frac{dG}{dP}, \tag{3}$$

which is obtained from the relation [3] $B \propto \tau/D$. The last term in the RHS of (3) is the pressure dependence of shear modulus calculatable from the 2nd- and the 3rd-order elastic constants of the material.

Table I. Pressure dependence of damping constant in units of 10^{-11} [cm^2/dyn]

Crystal	Specification	$(1/B_o)(dB/dP)$
LiF	Specimen I (pure, Harshaw)	-3.17
	Specimen II (impure, IAO)	+20.0
	theory	-2.0
NaCl	Specimen I (pure, Harshaw)	-0.18
	theory	+0.30
Al	State I (annealed and deformed)	+0.13
	State II (irradiated once)	+7.59
	State III (irradiated twice)	+5.71
	State IV (slowly cooled)	-14.2
	State V (rapidly cooled)	-2.43
	theory	+0.11

The determined values of $(1/B_o)(dB/dP)$ for LiF and NaCl crystals are shown in Table I. LiF-I and NaCl were supplied by Harshaw, and LiF-II by the Institute of Applied Optics (IAO). A large difference exists between LiF-I and LiF-II. By a mass spectroscopic analysis, LiF-II was found to contain much more impurities than LiF-I. Now we mention here that a number of theoretical studies have been made for the dislocation damping constant B by considering various types of dislocation-phonon interactions. The theories can be divided into two categories: the nonlinearity and the fluttering mechanisms. We have calculated the pressure dependence of the damping constant along the formulation of these theories [4], and evaluated results (nonlinearity mechanism) are also shown in the table. The experimental results for LiF-I (pure) and NaCl are agreeable with the theoretical values. However, the result for LiF-II (impure) is very different from any theoretical conclusions based on simple dislocation-phonon interactions. The effect of impurities thus appears to be very important.

3. Dislocation Damping in Aluminum - Impurity Effect

The effect of impurities on the pressure dependence of damping constant has been investigated systematically by using an aluminum specimen. A zone-refined Al single crystal (99.9999 %) was used, and the specimen was successively treated to change the concentration and the state of impurities. State 0 (reference state): The specimen was well annealed by repeatedly increasing and decreasing temperature between 380 [°C] and 630 [°C]. The dislocation density was considered to be 10^4 - 10^5 [cm^{-2}]. The dominant impurity in the specimen was 0.6 [ppm] Si.

State I: The specimen was axially compressed by 0.1 [%]. The density of dislocations was estimated to be 10^7 - 10^8 [cm^{-2}].
State II: The specimen was irradiated in a reactor (KUR) to introduce Si solute atoms through the following nuclear reaction: $^{27}Al(n, \gamma)^{28}Al$(2.27 min) $\rightarrow$ $^{28}Si + \beta^- + \gamma$. The concentration of silicon impurities was estimated to be 3.0 [ppm]. The specimen was then deformed by 0.01 [%].
State III: The specimen was further irradiated and the silicon concentration was evaluated to be 5.0 [ppm]. It was then deformed by 0.06 [%].
State IV: The specimen was annealed at 600 [°C], and slowly cooled to room temperature (30°C/hour). Then, the specimen was deformed by 0.06 [%].

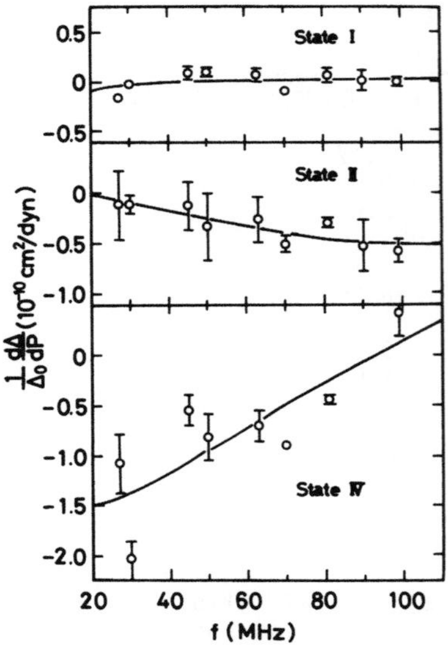

Fig. 1. Examples of the pressure dependence of the decrement due to dislocations on the sound frequency. To obtain the dislocation contribution, the pressure dependent decrement in the state 0 was subtracted from those in the States I - V. The measurements were made by using various sound frequencies. Thus the data shown were obtained. The solid lines are results of the fit of the data to (2).

State V: The specimen was annealed at 600 [°C], and rapidly cooled (600°C → 100°C/15 min). The specimen was then deformed by 0.05 [%].
The experimental values of the pressure dependence of damping constant for the State I - V are summarized in Table I.

In the State I, the concentration of the impurities is small and they are distributed uniformly in the specimen. The experimentally determined value of $(1/B_o)(dB/dP)$ is close to the theoretical value based on the dislocation-phonon interactions. However, the pressure dependence is strongly altered through the specimen treatments. The results can be qualitatively explained as follows [5]. As the concentration of Si atoms increases by the irradiation, they apt to make small clusters such as dimers (State II and III). By the annealing and the slow cooling, large clusters of Si atoms may be formed (State IV). The size of the clusters decreases by the annealing and the rapid cooling (State V). The shape of the clusters may be non-spherical, and they can rotate according to the dislocation vibration. The cluster rotation delays from the dislocation motion, resulting a dislocation damping. The response time for the cluster reorientation is considered to be sensitive to the pressure, and to be dependent on the cluster size. The remarkable correlation of $(1/B_o)(dB/dP)$ on the state of impurities could be understood on this basis.

References

1. Y. Hiki and T. Kosugi: Jpn. J. Appl. Phys. 21, Suppl. 23-1, 23 (1982)
2. T. Kosugi, Y. Kogure and Y. Hiki: J. Phys. Soc. Jpn. 54, 2565 (1985)
3. A. V. Granato and K. Lücke: J. Appl. Phys. 27, 583; 789 (1956)
4. Y. Kogure, T. Kosugi and Y. Hiki: J. Phys. Soc. Jpn. 54, 4592 (1985)
5. T. Kosugi, Y. Kogure, Y. Hiki and T. Kino: J. Phys. Soc. Jpn. 55, 1203 (1986)

Phonon Scattering by Dislocations in LiF and Si

T. Suzuki

Institute of Industrial Science, University of Tokyo,
Roppongi, Minato-ku, Tokyo 106, Japan

1. Introduction

Two different mechanisms have been discussed about the phonon scattering by dislocations. One is static scattering by the strain field and/or the topological disorder surrounding dislocations, and the other is dynamic scattering by fluttering mechanism. By many workers thermal conductivity measurements have been utilized to investigate the phonon scattering mechanism, as reviewed by ANDERSON [1]. Usually the dynamic scattering is stronger than the static scattering by more than one order of magnitude. In crystals, however, there are various causes which hinder the free fluttering of dislocations. Pinning by point obstacles or mutual cutting of dislocations shortens the free length of dislocations. Interaction with other parallel dislocations or segregation of impurities to dislocations acts as potential trough. The most important and only intrinsic hindrance for free fluttering is Peierls potential. The depth of Peierls potential is greatly different among the crystal kinds. The phonon scattering, thus the thermal conductivity in a certain crystal containing dislocations is closely related to the magnitude of the Peierls stress τ_P. In this paper the emphasis is on the influence of Peierls potential on the phonon scattering.

2. Theoretical Background

The dislocation-phonon interaction by fluttering mechanism has been discussed in detail by NINOMIYA [2], and the scattering width of phonons is calculated. When dislocations flutter freely without any hindrance, the relaxation rate τ_D^{-1} is calculated by averaging the scattering width given by him with respect to the incident angle. The result is [3]

$$\tau_D^{-1} \simeq 2\pi^2 v^2 N / \omega[\ln(2\omega^2/3\omega_D^2)]^2, \qquad (1)$$

where ω is the phonon frequency, ω_D the Debye frequency, v the sound velocity and N the dislocation density. When τ_D of (1) is the dominant term of total relaxation rate, the thermal conductivity κ is $T^{3.5-4}$.

When a dislocation lies in a potential trough, the low-frequency fluttering is restricted and the scattering of low-frequency phonons is not so strong as in the case of free fluttering. If the potential is approximated by $\frac{1}{2}Px^2$, the relaxation rate of phonons is given by [2]

$$\tau_D^{-1} \simeq \frac{1}{8} v^2 N \omega^3 / [(\omega_P^2 - \omega^2)^2 + \beta^2\omega^4] \qquad \text{for edge dislocations}, \qquad (2)$$

where $\omega_P=\sqrt{P/m}$, $m \sim \rho b^2$ being the mass per unit length of a dislocation, ρ the density, b the Burgers vector and β is of the order of 0.1. For screw dislocations τ_D^{-1} is much larger than (2) for edge dislocations: 80 times at $\omega >> \omega_P$ and 9 times at $\omega << \omega_P$. The temperature dependence of κ is then,

$$\kappa \sim T^4 \qquad \text{for } T >> T_P , \qquad (3)$$

$$\sim T_P^4 \text{ (independent of T)} \qquad \text{for } T << T_P , \qquad (4)$$

where $T_P \simeq \hbar\omega_P/5k$. If we consider the Peierls potential which has the Peierls stress τ_P, then $P \simeq 4\tau_P$, and T_P is given by

$$T_P \simeq \frac{2\hbar}{5kb}\sqrt{\tau_P/\rho} \ . \qquad (5)$$

As a rough estimate, $T_P \sim 10K$ for τ_P=100MPa. If T_P is far below the measurement region of T, κ is determined by (1). On the contrary, when T_P is sufficiently large, κ should be determined by the static scattering mechanism.

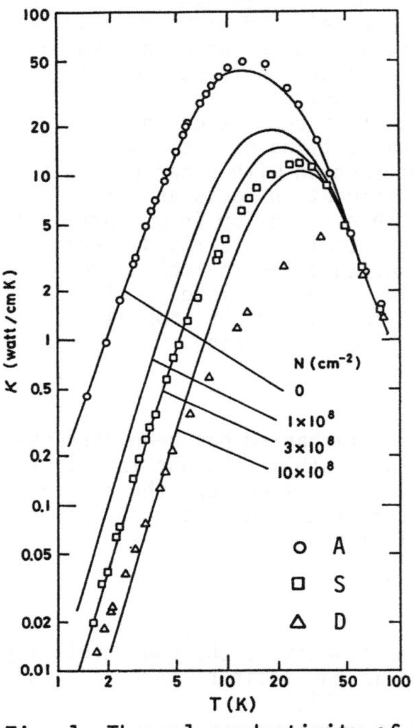

Fig. 1. Thermal conductivity of LiF. A—undeformed, etch pit density n=5×10^5cm^{-2}, S—deformed 6% in single slip, n=1.3×10^8cm^{-2}, D—deformed 13% in double slip, n=3.5×10^8cm^{-2}. Deformation was made by compression along <100> at the room temperature.

The solid lines are calculated based on the theory of free fluttering mechanism (1) for various values of dislocations density N.

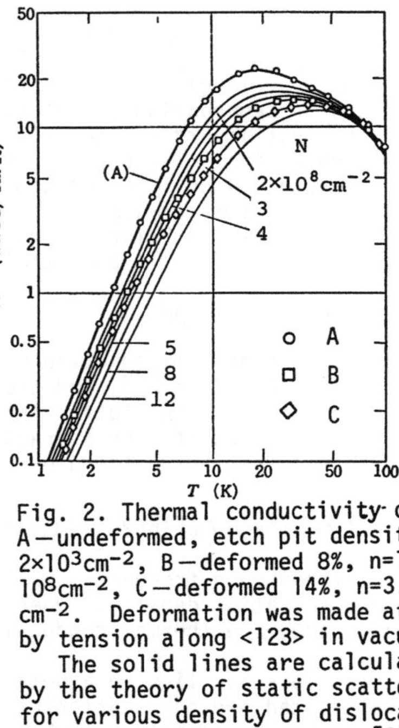

Fig. 2. Thermal conductivity of Si. A—undeformed, etch pit density n=2×10^3cm^{-2}, B—deformed 8%, n=1.7×10^8cm^{-2}, C—deformed 14%, n=3.5×10^8 cm^{-2}. Deformation was made at 900°C by tension along <123> in vacuum.

The solid lines are calculated by the theory of static scattering for various density of dislocations N, The relaxation rate is [8]

$$\tau_D^{-1} \simeq 0.33 \cdot \gamma^2 b^2 N \omega$$

Where γ is the Grüneisen constant. The present crystals contain 5×10^{17} cm^{-3} of oxigens. For the calculation the resonance scattering by the oxigens was taken into account.

3. Thermal Conductivity of LiF and Si

In Fig. 1 the thermal conductivity of plastically deformed LiF [3] is compared with the theory of free fluttering (1). For the sample S, which was deformed in single slip, the relaxation rate of (1) fits the data quite well. A small disagreement around the peak may be due to the point defects introduced by the deformation. The $\kappa(T)$ curve of the sample D which was deformed in double slip shows weaker T-dependence than the theory predicts. This should be attributed to the shortening of pinning length, because during the deformation in double slip dislocations cut with each other very frequently. Edge dislocations in LiF has τ_P of 20MPa [4], thus from (5) $T_P \sim 1K$. However, the data below 1K measured by ANDERSON et al.[5] show no evidence for the change of the slope of $\kappa(T)$ around 1K. The strong temperature dependence of κ ($\sim T^3$) below 1K may be due to the scattering by screw dislocations which have smaller τ_P than edge dislocations.

The thermal conductivity of Si is insensitive to dislocations (Fig. 2). The similar result for Ge was reported by SATO et al.[6]. Scattering by free fluttering mechanism (1) is too strong to describe the observed data. We must consider the effect of Peierls potential. Analysis of the data using (2) was attempted, but it was impossible to reproduce the observed $\kappa(T)$ curves throughout the temperature range, and it was found that τ_P should be larger than 1.2GPa. CASTAING et al.[7] made deformation experiments of Si under hydrostatic pressure and estimated τ_P to be of the order of shear modulus. We can conclude that the Peierls potential of Si is too large to observe the dynamic scattering. In Fig. 2 the data of Si are compared with theory of static scattering. The agreement is good, except that the value of N to fit the data at $T > 10K$ is about twice the value for $T < 5K$. If about a half of total dislocations are dipoles with the separation of $2\text{-}3\times10^{-6}$cm, such trend of $\kappa(T)$ curve is understood.

References

1. A.C. Anderson: in Dislocations in Solids ed. by F.R.N. Nabarro, Vol.6 (North Holland, New York, 1983) p.235
2. T. Ninomiya: Nat. Bur. Stand. (U.S,), Spec. Publ. 317, 315 (1970)
3. T. Suzuki and H. Suzuki: J. Phys. Soc. Japan 32, 164 (1972)
4. T. Suzuki and H. Kim: J. Phys. Soc. Japan 39. 1566 (1975)
5. A.C. Anderson and M.E. Malinowski: Phys. Rev. B5, 3199 (1972)
6. M. Sato and K. Sumino: J. Phys. Soc. Japan 36, 1075 (1974)
7. J. Castaing, P. Veissiere, L.P. Kubin and J. Rabier: Phil. Mag. A44, 1407 (1981)
8. P.G. Klemens: Proc. Phys. Soc.,London A68, 1113 (1955)

Phonon Scattering by a Grain Boundary in Silicon

R.A. Brown

School of Mathematics and Physics, Macquarie University,
North Ryde, NSW 2113, Australia

1. Introduction

As discussed in a recent review [1] thermal conductivity [2] and heat pulse [3] experiments have shown the interaction of phonons with grain boundaries to be weak and probably due to anharmonic coupling with the distortion field of the (static) boundary. Theoretical discussions of this mechanism have been based on formulations [4,5] developed before data on third--order elastic constants (TOECs) were available.The aim of the present paper is to analyze the heat pulse experiments of [3] using a theoretical framework[6-8]which, although lacking the attractive simplicity of Klemens'approach[4,5],offers an exact description of the anharmonic interaction in terms of TOECs.

2. Theory

The scattering of phonons propagating ballistically along the positive z-axis is described by a depletion rate [8]

$$\tau_K^{-1} \equiv -\frac{1}{N_K}\frac{dN_K}{dt} = \frac{(8\pi\hbar)^{-2}}{L_1L_2L_3}\sum_{j'}\int_{k_3'<0} dS' \; |h_{K'K}|^2/|\nabla_{\underset{\sim}{k}'}\omega_{j'}(\underset{\sim}{k}')| \quad , \qquad (1)$$

in which N_K is the occupation number of mode $j\underset{\sim}{k} \equiv K$ of energy $\hbar\omega_j(\underset{\sim}{k})$ in a crystal of volume $L_1L_2L_3$ and the integral is over the back-scattered half[8]of the k-space surface $\omega_{j'}(\underset{\sim}{k}')= \omega_j(\underset{\sim}{k})$. The matrix elements $h_{K'K}$ as given by (4.6) of [9] depend on the wavevectors,polarizations and frequencies of the indicated phonons,the TOECs and Fourier transforms(FTs)of the boundary's distortion fields.

Following[10]the 10° tilt boundary on (001)in Si is modeled as an array of edge dislocations parallel to[010]and alternately having Burgers vectors b_1 = [101]a/2 and b_2 = [$\bar{1}$01]a/2. For misorientation Θ the density of *each* type is [11] $\rho = (2/a)\sin(\Theta/2)$,leading to a periodic structure with period $D \equiv 1/\rho$ along [100]. Denoting a typical component of the boundary's distortion field by $\Phi(\underset{\sim}{r})$ we have

$$\Phi(\underset{\sim}{r}) \equiv \Phi(x,z) = \sum_m \phi(x - mD,z) \qquad (m = -\infty \text{ to } \infty) \qquad (2)$$

where $\phi(x,z)$is the corresponding distortion component due to a single pair of dislocations. Defining the 3-dimensional FT $\tilde{\Phi}(\underset{\sim}{\kappa})$ of $\Phi(\underset{\sim}{r})$ as in [9] we find

$$\tilde{\Phi}(\underset{\sim}{\kappa}) = 4\pi^2\rho\,\delta(\kappa_2)\,\tilde{\phi}(\kappa_1,\kappa_3)\sum_m \delta(\kappa_1 - 2m\pi\rho) \qquad (3)$$

where $\tilde{\phi}$ is the 2-dimensional FT

$$\tilde{\Phi}(\kappa_1,\kappa_3) = \int\int_{-\infty}^{\infty} \exp[i(\kappa_1 x + \kappa_3 z)]\, \phi(x,z)\, dx\, dz \quad . \tag{4}$$

The index m in (3) can be interpreted as labelling the diffraction orders from the periodic array of dislocation pairs.

From (4.6) of [9] and (3) above each $h_{K'K}$ can be written

$$h_{K'K} \equiv h_{j'j}(\underset{\sim}{k}',\underset{\sim}{k}) = 4\pi^2 \rho\, g_{j'j}(\underset{\sim}{k}',\underset{\sim}{k}) \sum_m \delta(k_1 - k_1' - 2m\pi\rho) \tag{5}$$

in which each g is a linear combination of FTs [of type (4)] of the distortion components appropriate to a single pair. We introduce spherical polar coordinates by

$$k_1' = k'\sin\xi'\cos\mu',\quad k_2' = k'\cos\xi',\quad k_3' = k'\sin\xi'\sin\mu' \tag{6}$$

and adopt the Debye approximation $k' = \omega/v'$ with v' constant for each polarization branch, independent of propagation direction. Then (5) in (1) with $\delta(\kappa_1) = (\pi\kappa_1)^{-1}\sin(\tfrac{1}{2}L_1\kappa_1)$, etc. (in the limit as $L_1 \to \infty$) yields

$$\tau_{j\underset{\sim}{k}}^{-1} = (4\omega\rho^2/\hbar^2 L_3)\sum_{j'm} v_{j'}^{-2}[\omega^2 v_{j'}^{-2} - (k_1 - 2m\pi\rho)^2 - k_2^2]^{-\frac{1}{2}} |g_{j'j}(\underset{\sim}{k}'_{j'm}, \underset{\sim}{k})|^2 \tag{7}$$

where $\underset{\sim}{k}'_{j'm}$ has components

$$k_1' = k_1 - 2m\pi\rho,\quad k_2' = k_2,\quad k_3' = -[\omega^2 v_{j'}^{-2} - (k_1 - 2m\pi\rho)^2 - k_2^2]^{\frac{1}{2}} \tag{8}$$

and the sum is over polarizations and diffraction orders for which (7) is real.

For the principal modes studied in [3] $k_1 = k_2 = 0$. Taking $v_{j'} = 4.69\ \mathrm{kms}^{-1}$ (appropriate to the quasitransverse mode propagating along [101]) as the smallest possible [12] value of $v_{j'}$, the dominant phonon approximation ($\nu = 10^{11}\,T$) shows that *only* m = 0, corresponding to pure back-reflection with $k_1' = k_2' = 0$, $k_3' = -\omega/v_{j'}$, contributes to (7) for the temperatures T<15K probed in[3]. Accordingly (7) yields

$$\tau_j^{-1} = (4\rho^2/\hbar^2 L_3)\sum_{j'} |g_{j'j}(-\omega\hat{\underset{\sim}{z}}/v_{j'}, \omega\hat{\underset{\sim}{z}}/v_j)|^2/v_{j'} \quad . \tag{9}$$

Enormous simplifications accrue in (4.6) of [9] due to the high symmetry and we find

$$\left.\begin{aligned} \tau_q^{-1} &= (\omega^2\rho^2/4L_3)[v_T^{-1}|\alpha\tilde{n}_{11} + \beta\tilde{n}_{33}|^2_{(0,2k)} + v_L^{-1}|\gamma\tilde{n}_{13} + \delta\tilde{\omega}_{13}|^2_{(0,k+v_T k/v_L)}] \\ \tau_p^{-1} &= (\omega^2\rho^2/4L_3)\ v_T^{-1}|\zeta\tilde{n}_{11} + \eta\tilde{n}_{33}|^2_{(0,2k)} \end{aligned}\right\} \tag{10}$$

where the $\tilde{n}_{ij}$ and $\tilde{\omega}_{ij}$ denote the 2-dimensional FTs of the static strains and rotations associated with a single dislocation pair, v_L and v_T denote the speeds of the longitudinal and (degenerate) quasi (q) and pure(p) transverse modes . The arguments κ_1, κ_3 appropriate to the FTs appear as parenthetic subscripts. The elastic constants enter (10) through

$$\alpha = (c_{12} + c_{166})/c_{44},\ \beta = (c_{11} + c_{166})/c_{44},\ \gamma = (c_{11} + c_{12} + 2c_{44} + 2c_{166})/(c_{11}c_{44})^{\frac{1}{2}}$$
$$\delta = (c_{11} - c_{12} - 2c_{44})/(c_{11}c_{44})^{\frac{1}{2}},\ \zeta = (c_{12} + c_{144})/c_{44},\ \eta = (c_{11} + c_{166})/c_{44} \ .$$

For simplicity we evaluate (10) using the dislocation distortion fields appropriate to an *isotropic* body. Standard components refer to a dislocation along Oz and slip plane normal to Oy. On transforming these to refer to the crystal axes we find for a *symmetric* boundary with dislocations $\underset{\sim}{b}_1$ and $\underset{\sim}{b}_2$ located at $(-\tfrac{1}{4}D,0)$ and $(\tfrac{1}{4}D,0)$ respectively,

$$-\tilde{n}_{11}(0,\kappa) = \tilde{n}_{33}(0,\kappa) = ib/[(1-\nu)\kappa\sqrt{2}],\ \tilde{n}_{13}(0,\kappa) = 0,\ \tilde{\omega}_{13}(0,\kappa) = 2ib/\kappa\sqrt{2} \tag{11}$$

where $b = a/\sqrt{2}$ is the Burgers vector and ν is Poisson's ratio. On using (11) in (10) and defining [3] reflection coefficients by $R = 1 - \exp(-L_3/\tau v) \approx L_3/\tau v$, we find

$$\left.\begin{aligned} R_q &= L_3/v_T\tau_q = (b^2/32D^2)\left\{[(\alpha-\beta)/(1-\nu)]^2 + [4\delta/(1+v_T/v_L)]^2 v_T/v_L\right\} \\ R_p &= L_3/v_T\tau_p = (b^2/32D^2)\,[(\zeta-\eta)/(1-\nu)]^2 \end{aligned}\right\} \quad (12)$$

independent of frequency in $T \lesssim 15$K. Taking elastic constants from [13] we obtain $v_T = 5.8$ kms^{-1}, $v_L = 8.4$ kms^{-1}, $\alpha-\beta = -1.3$, $\delta = -0.50$, $\zeta-\eta = 2.8$ and with $\nu = 0.22$ [14], a = 5.4 A and D = 30 A for $\theta = 10.3°$ [10] equations (12) yield $R_q = 0.0019$, $R_p = 0.0063$.

3. Summary and Discussion

We calculated reflection coefficients for transverse modes incident normally on a symmetric 10° tilt boundary on (001) in Si. Within the dominant phonon approximation these are temperature independent in T<15K; diffraction effects are trivial within this range. The boundary structure was treated exactly within the dislocation model, both strain and rotation components of its distortion field being included. The anharmonic coupling was described exactly using the measured TOECs. The principle uncertainties arise from using the Debye model and distortions appropriate to an isotropic body.

Our results are consistent with experimental upper bounds [3] of 1-2% and indicate that a modest increase in sensitivity (or a slightly greater misorientation angle, e.g. $\theta \sim 13°$ yields $R_p \sim 0.01$) would enable detection. Rotations account for none of our R_p and only 26% of R_q whereas the broadly comparable R = 0.3% quoted in [3] is based on a model [4,5] in which *only* the effects of lattice rotation are retained.

References

1. A.C. Anderson: in *Dislocations in Solids,* ed.F.R.N.Nabarro (North Holland, New York 1983) Vol.6, p.235
2. E.P. Roth and A.C. Anderson: Phys.Rev. B<u>17</u>, 3356 (1978)
3. E.P. Roth and A.C. Anderson: Phys.Stat.Sol. B<u>93</u>, 261 (1979)
4. P.G. Klemens: Proc.Phys.Soc. A <u>68</u>, 1113 (1955)
5. P.G. Klemens: Solid State Physics <u>7</u>, 1 (1958)
6. Y. Kogure and Y. Hiki: J.Phys.Soc. Japan <u>36</u>, 1597 (1974)
7. D. Eckhardt and W. Wasserbäch: Phil.Mag. A<u>37</u>, 621 (1978)
8. R.A. Brown: Solid State Commun. <u>51</u>, 89 (1984)
9. R.A. Brown: J.Phys.C: Solid State Phys. <u>16</u>, 1009 (1983)
10. K. Hubner and W. Shockley: Proc.Int.Conf.Semicond.,ed. A.C. Stickland (Inst.Phys. and Phys.Soc., London 1962) p.157
11. S. Amelinckx and W. Dekeyser: Solid State Physics <u>8</u>, 325 (1959)
12. B.A. Auld: *Acoustic Fields and Waves in Solids* (Wiley, New York 1973) Vol.I, Fig. 7.2
13. H.J. McSkimmin and P. Andreatch; J.Appl.Phys. <u>35</u>, 3312 (1964)
14. O.L. Anderson: in *Physical Acoustics,* ed. W.P. Mason (Academic Press, New York 1965) Vol.III B, p.43 (Appendix II).

Nonlocal Thermal Conduction in Insulators

F.W. Sheard, G.A. Toombs, and S.R. Williams

Department of Physics, University of Nottingham, University Park, Nottingham, NG7 2RD, UK

The thermalisation of nonequilibrium phonons in an insulating crystal at low temperatures, where phonon-phonon interactions are negligible, is not well understood [1]. Experiments on the temporal decay of phonon pulses [2] have suggested a mechanism involving inelastic scattering at surface defects. The equilibration of a phonon distribution may also be studied using steady-state heat conduction by observing the spectral redistribution of phonons which occurs when there is a spatial change in the phonon scattering probability [1,3]. This results in a nonlinear variation in temperature which occurs over a distance comparable with the inelastic mean free path (mfp). However to treat effects which occur on the same scale as the mfp a nonlocal treatment of thermal conduction is required.

A suitable model is defined by the Boltzmann equation for the phonon distribution function $N_{\tilde{q}}(x)$, in which the scattering terms are represented by means of relaxation times:

$$\mu v \frac{\partial N_{\tilde{q}}}{\partial x} = - \frac{N_{\tilde{q}} - N_q^0\{T(x)\}}{\tau_i} - \frac{N_{\tilde{q}} - \overline{N}_q}{\tau_e(x)} , \quad (\mu = q_x/q) .$$

The inelastic processes are taken to have a constant mfp $\ell_i = v\tau_i$ (v = phonon velocity) and relax $N_q(x)$ to the Bose-Einstein distribution $N_q^0\{T(x)\}$ at the local temperature $T(x)$. But the elastic scattering processes can only relax $N_q(x)$ to the angle-averaged distribution $\overline{N}_q(x) = \frac{1}{2}\int_{-1}^{1} N_q(x)d\mu$. The spectral redistribution arises from the spatial variation of the elastic mfp $\ell_e(x) = v\tau_e(x)$. Since $\overline{N}$ is not an equilibrium distribution the mode temperature $T_q(x)$, defined by the equality $\overline{N}_q(x) = N_q^0\{T_q(x)\}$, is frequency dependent.

In our model the ends x = 0 and d of the crystal are terminated by black bodies at slightly different temperatures $T_0 + \Delta T_0$ and T_0 respectively ($\Delta T_0 << T_0$). Considering the temperature deviations $\Delta T_q(x)$ and $\Delta T(x)$ from T_0, we obtain the integral equation [4]

$$\Delta T_q(x) = \tfrac{1}{2} E_2\{g(x,0)\}\Delta T_0 + \tfrac{1}{2} \int_0^d E_1\{g(x,x')\mathrm{sgn}(x-x')\}D_q(x')dx' ,$$

where $E_n(u) = \int_0^1 \exp(-u/\mu)\mu^{n-2}d\mu$, $\mathrm{sgn}(x) = 1$ $(x > 0)$ or -1 $(x < 0)$,

$$D_q(x) = \frac{\Delta T_q(x)}{\ell_e(x)} + \frac{\Delta T(x)}{\ell_i} , \quad \frac{1}{\ell(x)} = \frac{1}{\ell_e(x)} + \frac{1}{\ell_i} , \quad g(x,x') = \int_{x'}^{x} \frac{ds}{\ell(s)} .$$

If ℓ_i is independent of phonon frequency ω_q, we find that $\Delta T(x)$ is a weighted average of the mode temperatures

$$\Delta T(x) = \int \Delta T_q(x) C(\omega_q) d\omega_q / \int C(\omega_q) d\omega_q \ ,$$

where $C(\omega_q)$ is the heat capacity per unit frequency range. We emphasise that in a spatially uniform crystal (ℓ_e independent of x), $T_q(x) = T(x)$ for all modes and the temperature gradient dT/dx is constant except near the black-body boundaries where some spectral and angular redistribution of the phonon modes occurs.

Numerical calculations have been made for a crystal containing different resonant elastic scatterers in the two halves, taking $\ell_e^{-1}(\omega_q) = A/\{(\omega_q - \omega_r)^2 + \Gamma^2\}$, with $\omega_r = \omega_1$ in region 1 ($x < \frac{1}{2}d$) and ω_2 in region 2 ($x > \frac{1}{2}d$). For heat flow from region 1 to 2, modes near ω_1 are strongly scattered in region 1 and therefore suffer a larger temperature drop than average. For modes near ω_2 the temperature drop is principally in region 2. Hence the mode temperature is depressed near ω_1 and raised near ω_2. This behaviour is confined to within a distance $\lesssim \ell_i$ from the interface at $x = \frac{1}{2}d$ as shown by the results in Fig. 1(a) for $\ell_i = 0.2d$. The associated temperature nonlinearity can be seen in Fig. 1(b) as an increase in the magnitude of the temperature gradient. The change in gradient near x = 0 and x = d is an end effect due to the black-body boundaries. When $\omega_1 = \omega_2$ the two halves of the crystal are identical and there is no nonlinear behaviour near the interface. The change in temperature gradient upon tuning the resonance frequencies has been observed

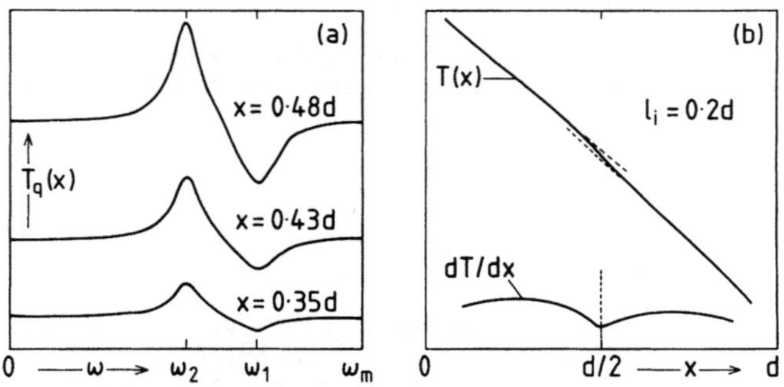

Fig. 1(a) Frequency variation of mode temperatures $T_q(x)$ at different points in crystal. (b) Spatial variation of local temperature T(x) and gradient dT/dx

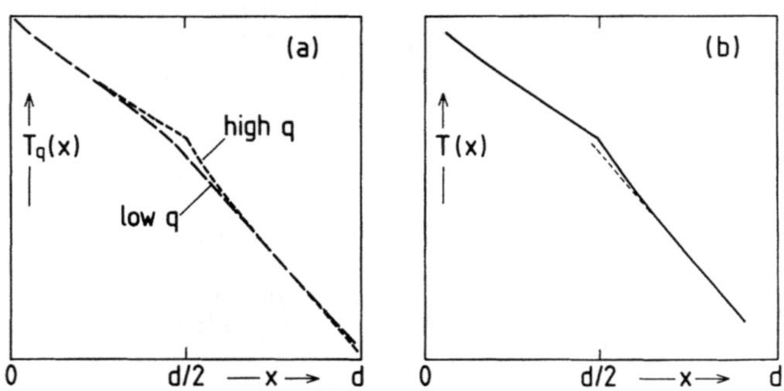

Fig. 2 Spatial variation of (a) mode temperature $T_q(x)$ and (b) local temperature T(x) for crystal-disordered solid system

experimentally [3]. However, because the numerical calculations are restricted to a finite number of spatial points (28) and frequencies (28) we have not been able to precisely simulate the experimental conditions. The maximum frequency ω_m was chosen such that $\hbar\omega_m/k_BT \simeq 7$.

We have also considered the case when region 1 is a crystal with $\ell_e = \ell_i = 0.2d$ and region 2 is a disordered material in which the high-frequency phonons are more strongly scattered ($\ell_e = 0.05d$, $\omega > \frac{1}{4}\omega_m$) than low-frequency phonons ($\ell_e = 0.2d$, $\omega < \frac{1}{4}\omega_m$). The splitting of the mode temperatures near the interface and the consequent temperature nonlinearity in region 2 are shown in Figs. 2(a) and (b). These results are in qualitative agreement with those of Jäckle [5] who treated the problem using an approximate diffusion equation for $T_q(x)$.

1. L.J. Challis: In Phonon Scattering in Condensed Matter, ed. by W. Eisenmenger, K. Lassmann and S. Döttinger, Springer Ser. Solid-State Sci., Vol.51 (Springer, Berlin, Heidelberg, 1983), p.2.
2. H.J. Trumpp and W. Eisenmenger: Z. Phys. B28, 159 (1977).
3. L.J. Challis, A.A. Ghazi and M.N. Wybourne: Phys. Rev. Lett. 48, 759 (1982).
4. F.W. Sheard, G.A. Toombs and S.R. Williams: In Phonon Physics, ed. by J. Kollar, N. Kroo, N. Menyard and T. Siklos (World Scientific, Singapore, 1985) p.435.
5. J. Jäckle: Z. Phys. B52, 133 (1983).

Low-Temperature Phonon Scattering in "Pure" and Li-Doped $KTaO_3$

B. Salce[1] and L.A. Boatner[2]

[1]Département de Recherche Fondamentale, SBT/LCP, C.E.N. Grenoble, 85 X F-38401 Grenoble Cedex, France
[2]Solid State Division, Oak Ridge National Laboratory, Oak Ridge, TN 37830, USA

The cubic perovskite $KTaO_3$ is a so-called incipient ferroelectric : it does not undergo a ferroelectric transition, in spite of the presence of a soft large wavelength transverse optic phonon, as this zone center mode is stabilized by the zero-point thermal fluctuations. Raman and neutron scattering experiments have shown that its energy lies below 30 cm^{-1} at $T < 40$ K, so it was expected that $KTaO_3$ should be a good candidate to study the interaction between acoustic and optic phonons. Previous thermal conductivity K(T) measurements have indicated that no phonon scattering occurred below 1 K in a "pure" (P) $KTaO_3$, as the calculated Casimir limit was reached, but the existence of a true minimum near 7 K was evidenced [1]. This minimum was not significantly affected by adding paramagnetic impurities (Co, Ni, Ag, Cu), by annealing (oxygen, hydrogen, vacuum) or by applying a magnetic field up to 7 Tesla [2]. More recently, a large static electric field E effect was reported, and it was assumed that the scattering around 7 K was an intrinsic effect [3].

On the other hand, it seems to be established that replacing Ta by Nb, or K by Li and Na, produces materials undergoing phase transitions of different kinds (ferroelectric, glass-like) [4] [5]. This makes them a very attractive family of compounds as the transition temperature T_C can be increased from 0 K in a continuous way by changing the concentration x of the substitutional atoms. However, despite a substantial amount of work, the nature and the role of residual impurities in phase nucleations is not clarified. We report K(T) measurements carried out using the steady state heat flow method in the temperature range 70 mK - 150 K, on one new "ultra pure" (UP) $KTaO_3$ and other mixed compounds, in order to assist in answering these questions. The samples were grown by a flux technique using a high temperature reaction of Ta_2O_5, K_2CO_3 and Nb_2O_5 (Li_2CO_3, Na_2CO_3).

"Ultra-pure" $KTaO_3$

In figure 1, the data carried out on the UP sample are compared to results obtained on one P crystal representative of several samples grown from different melts. The two curves merge at high temperature and reach their respective Casimir limit below 1 K. The most striking feature for the UP sample is the shape of the K(T) curve in the range 1 K to 100 K, compared with the P one : the minimum disappears and the scattering decreases by nearly one order of magnitude around $T \sim 10$ K. However, applying a field $E = 6$ kV/cm increases the conductivity, as in the P sample. It is worthwhile to notice that K(T) measurements have previously been reported by Steigmeier [6] on 3 "pure" $KTaO_3$, but the data were different in form and magnitude from sample to sample, the "highest" comparing favourably with our UP crystal. Two questions are now arising : 1) Is the scattering still present in the UP sample either due to remaining impurities or to the nature of $KTaO_3$ lattice ? The presence of the E induced change suggests an

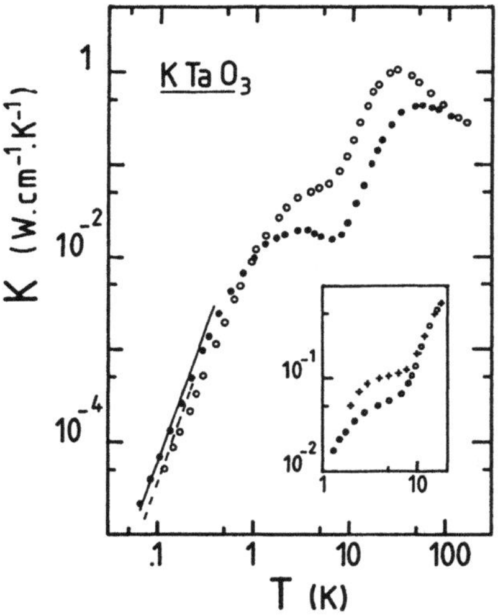

Figure 1

● P ; o UP
+ 6 kV/cm

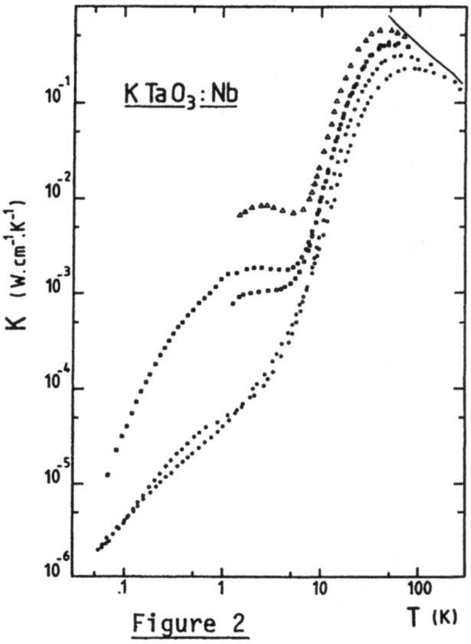

Figure 2

▲ .15 % ; ■ .3 % ; □ .4 %
o .8 % ; ● 3 %

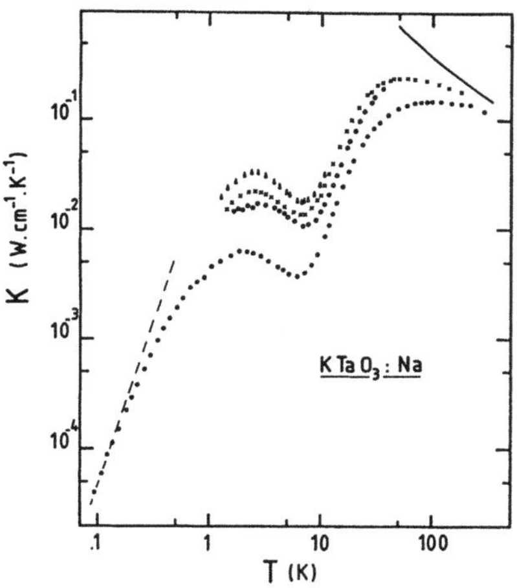

Figure 3

o .4 % ; ● .8 %
▲ 1.6 % ; ✕ 2,4 %

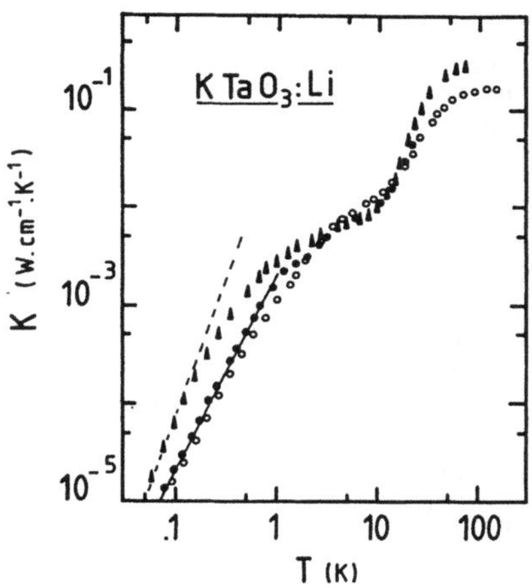

Figure 4

▲ .2 % ; ● .6 % ; o .8 %
—— T^2 – – – T^3

intrinsic soft mode effect, but we cannot rule out the existence of electric field sensitive impurities. 2) What is the source of the minimum detected on P samples ?

Mixed crystals

To try to identify this center, we report in figures 2,3 and 4 a set of K(T) measurements obtained on Nb, Na and Li-doped $KTaO_3$.

For $KTaO_3$:Nb, and by comparison with the P $KTaO_3$ conductivity, we must introduce a large amount of Nb (1500 ppm) to decrease significantly K(T).

Increasing x enhances the scattering down to the lowest temperatures. As it is known the soft mode energy falls to 0 with x, it seems likely to assign this effect to an interaction between the acoustic and optic phonons.

The data obtained on $KTaO_3$:Na are more puzzling : the minimum is strongly enhanced but the calculated T^3 limit is reached below 300 mK. However, the magnitude of K(T) does not correlate with x, what could mean either Na is not reponsible for the minimum or the calculated x value is not the true concentration.

Three Li-doped $KTaO_3$ have been studied and no minimum was detected. The less doped sample displays the T^3 behaviour at low temperatures but the two other crystals exhibit a $\sim T^2$ law below 1 K. Raman scattering study has shown that increasing x does not decrease the soft mode energy [7]. So, it seems unlikely that the remaining phonon scattering down to 100 mK can be accounted for by arguing an "acoustic-optic" phonon interaction. An alternative explanation could be the existence of a glass-like orientational disorder of the off-center Li^+ ions, which was evidenced by dielectric measurements [4]. In that case, the T^2 law could be the signature for phonon scattering on tunneling states existing in the glassy polarization phase [2] [8]. On the other hand, it can be expected that decreasing the Li concentration removes the collective effects. Then, from the K(T) data at T<1 K, we suggest that the concentration limit to get a glass-like dipolar phase is $x \simeq 0.4$ %.

To conclude, it is clear that the thermal conductivity of $KTaO_3$ is not really understood at the present time : a scattering due to the soft mode should be present but extrinsic impurities have a large influence. We have suggested that CO_3^{2-} ions could be the source of scattering as the presence of carbon was detected by ion channeling in the crystals [9]. However, we failed to correlate the C concentration with the scattering intensity. The data reported here support a possible role of Li or Na, as they are always present in samples at low concentrations, but more precise analysis is required.

References

1. A.M. De Goër, B. Salce and L.A. Boatner :"3rd I.C. on Phonon Scattering in Condensed Matter" (Providence U.S.), Plenum Press, N.Y.(1980) 243
2. B. Salce, A.M. De Goër and L.A. Boatner, Journal de Physique, Colloque C6, Sup. n° 12, 42 (1981) 424
3. B. Salce and L.A. Boatner, "2d I.C. on Phonon Physics" (Budapest - Hungary) (1985) in Press
4. U.T. Hochli and L.A. Boatner, Phys. Rev. B, 20 n° 1 (1979) p. 266
5. U.T. Hochli and D. Baeriswyl, J. Phys. C : Solid State Phys. 17 (1984) 311
6. E.F. Steigmeier, Phys. Rev. 168 n° 2 (1968) 523
7. R.L. Prater, L.L. Chase and L.A. Boatner, Phys. Rev. B, 23 n° 11 (1981) 5904
8. J.J. De Yoreo, R.O. Pohl and G. Burns, Phys. Rev. B, 32 n° 9 (1985) 5780
9. M. Dubus, B. Daudin, B. Salce and L.A. Boatner, Solid State Comm. 55 n° 8 (1985) 759

Ultrasonic Velocity and Modified Critical Behaviour in the Random-Field Ising System $DyAs_xV_{1-x}O_4$

J.H. Page [1,3] , *M.C. Maliepaard* [2,3] , *and D.R. Taylor* [3]

[1]Physics Department, University of Manitoba, Winnipeg R3T 2N2, Canada
[2]Cavendish Laboratory, Cambridge, CB3 OHE, UK
[3]Physics Department, Queen's University, Kingston K7L 3N6, Canada

Recently there has been considerable interest in the effects of random fields on Ising phase transitions [1-14]. In their pioneering paper, IMRY and MA [1] were the first to suggest that random fields destroy long-range order in Ising systems when the dimensionality d is less than $d_\ell = 2$ and that the phase transition which occurs for $d > d_\ell$ is characterized by drastically different critical behaviour. Subsequent analyses of the random-field Ising model (RFIM) have resulted in many debates on both of these issues [2]. While it is now generally agreed that the lower critical dimenionality d_ℓ is indeed equal to 2 [3], there is still controversy concerning the correct description of the critical properties. It has been suggested [4-6] that the modified critical behaviour of random-field systems can be explained in terms of a reduction in the effective dimensionality, whereby the random-field exponents in d dimensions are equal to those for zero random field in $\bar{d} = d - \delta$ dimensions. Evidence for a dimensionality shift δ of 1 has come from numerical simulations [7] of the RFIM and from experiments [8] on randomly-diluted antiferromagnets in a uniform field, a system believed to be equivalent to a pure ferromagnet in a random field [9]. However, the critical exponents determined in these studies have been shown to violate the Schwartz-Soffer inequality [10], leading to suggestions that the phase transition may be first order [7] or that these results do not reflect true equilibrium behaviour [10]. An alternative explanation has been suggested by more recent Monte Carlo simulations [11] which confirmed that the Schwartz-Soffer inequality is satisfied in strong random fields but found that in weaker fields, similar to those of the earlier simulations and experiments, the apparent exponents were influenced by crossover effects from random to non-random critical behaviour. These recent simulations are also consistent with new predictions for the dynamic scaling of random-field systems [12-14] in which the dimensional reduction hypothesis has been called into question.

In this paper, we present the results of ultrasonic velocity experiments that investigate the critical behaviour of a new random-field system formed from the mixed Jahn-Teller (JT) compounds $DyAs_xV_{1-x}O_4$. As discussed in Ref. 15, mixed JT compounds may be excellent physical realizations of the RFIM: the interactions between rare-earth ions that drive the structural phase transitions in these materials are well described by a pseudospin Ising Hamiltonian, and they involve couplings to lattice strains which enable large static random fields to be generated by crystal defects (e.g. by the As/V mismatch). In our previous electric susceptibility and order parameter measurements in $DyAs_xV_{1-x}O_4$ [15] we have found evidence of time- and history-dependent ordering below the phase transition temperature T_D characteristic of the metastability inherent to random-field systems. However, because of the resulting inaccessibility of the true equilibrium phase for $T < T_D$, we were unable to study the critical properties (e.g. of the order parameter). An important advantage of the present ultrasonic experiments is that critical behaviour above T_D can be investigated where equilibrium properties are expected.

In pure crystals of both $DyVO_4$ and $DyAsO_4$ and hence also in the mixed system, the lattice distortion is of B_{1g} symmetry [16], so that near the phase transition, the divergence of the pseudospin susceptibility χ is inversely proportional to the softening of the elastic constant $\frac{1}{2}(C_{11}-C_{12})$. This can be measured from the ultrasonic velocity in either of two ways [17]: (i) for transverse waves propagating in the [110] direction and polarized along the [1$\bar{1}$0], $\frac{1}{2}(C_{11}-C_{12}) = \rho v_T^2$ is measured directly, or (ii) for longitudinal waves propagating in the [100] direction, the measured elastic constant $C_{11} = \rho v_L^2 = \frac{1}{2}(C_{11}+C_{12}) + \frac{1}{2}(C_{11}-C_{12})$ contains an additional term $\frac{1}{2}(C_{11}+C_{12})$ which, however, remains constant at low temperatures, enabling the variation of $\frac{1}{2}(C_{11}-C_{12})$ to be determined. Since the ultrasonic attenuation is too large for transverse waves to be used for measurements close to the phase transition, the second method was used in this work. It is worth noting that longitudinal waves have been used successfully in the past to investigate critical behaviour near the JT phase transition in $TmVO_4$ [17].

The temperature dependence of the ultrasonic velocity is shown in Fig. 1 for [100] longitudinal waves in a sample of concentration $x \sim 0.15$. As the temperature is lowered, the velocity decreases continuously until it reaches a minimum value at $T_D = 7.9K$, indicative of a second order phase transition at T_D. The velocity minimum corresponds to a value of the elastic constant $\rho v^2 = 1.57 \times 10^{11}$ N/m²; this is in good agreement with the value of $\frac{1}{2}(C_{11}+C_{12})$ measured in other rare earth zircons [18], consistent with the expected result that $\frac{1}{2}(C_{11}-C_{12}) = 0$ at a B_{1g} second-order phase transition. We use this fact below to determine $\frac{1}{2}(C_{11}-C_{12})$ from our velocity data. In the low temperature phase, the velocity recovers towards its unperturbed value; superimposed on this is a small dip at about 3.0K where an antiferromagnetic transition, of no further interest here, is expected to occur. Note that the JT transition temperature is reduced by almost a factor of 2 with respect to that of pure $DyVO_4$ ($T_D = 14.6K$ [16]), showing that this mixed crystal is subject to strong random fields.

Figure 2 shows a log-log plot of $\frac{1}{2}(C_{11}-C_{12})$, obtained from $\rho[v(T)^2-v(T_D)^2]$, versus the reduced temperature $t = (T-T_D)/T_D$ with $T_D = 7.90K$.

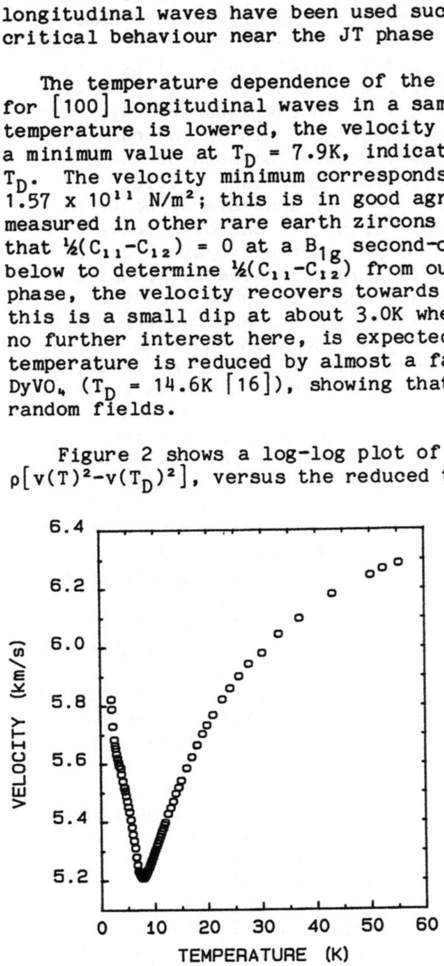

Fig. 1 Temperature dependence of the velocity of [100] longitudinal waves at 37.5 MHz in $DyAs_{.15}V_{.85}O_4$.

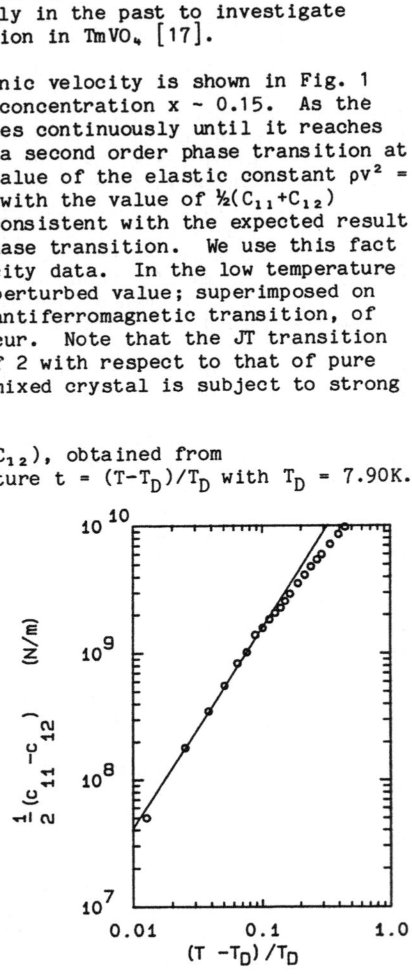

Fig. 2 Elastic constant $\frac{1}{2}(C_{11}-C_{12})$ vs. reduced temperature. The straight line has slope $\gamma = 1.6$.

(The density $\rho = 5.79 \times 10^3 kg/m^3$ was calculated from X-ray measurements of the lattice constants.) For $t \lesssim 0.1$, the elastic constant can be fitted to the power law $\frac{1}{2}(C_{11}-C_{12}) \propto \chi^{-1} \propto t^{\gamma}$ with $\gamma = 1.6 \pm 0.3$. While γ for this sample appears to be significantly larger than the pure 3-d Ising value 1.25, this result has to be viewed with some caution due to uncertainties in the fitting procedure and the possible effects of concentration gradients and departures from equilibrium. It is nonetheless interesting that our value of γ is quite close to $\gamma = 1.75$ as predicted for a dimensional reduction of unity and as found in simulations [7] and experiments [8] in weak random fields. Our result also overlaps with the value $\gamma = 2.0 \pm .5$ obtained in recent large-scale simulations by OGIELSKI and HUSE [11], whose estimates of the exponents in strong random fields were found to satisfy the theoretically derived inequalities. Future work will attempt to improve our experimental accuracy with the aim of discriminating between the dimensional reduction hypothesis [4-6] and more recent theoretical ideas on the critical behaviour of random field Ising systems [12-14].

Acknowledgements

Research support by NSERC of Canada is gratefully acknowledged.

References

1. Y. Imry and S.-k. Ma: Phys. Rev. Lett. 35, 1399 (1975)
2. For a short review, see A. Aharony: J. Magn. Magn. Mat. 54-57, 27 (1986)
3. J.Z. Imbrie: Phys. Rev. Lett. 53, 1747 (1984)
4. A. Aharony, Y. Imry and S.-k. Ma: Phys. Rev. Lett. 37, 1364 (1976)
5. M. Schwartz: J. Phys. C 18, 135 (1985)
6. Y. Shapir: Phys. Rev. Lett. 54, 154 (1985)
7. A.P. Young and M. Nauenberg: Phys. Rev. Lett. 54, 2429 (1985)
8. D.P. Belanger, A.R. King and V. Jaccarino: Phys. Rev. B 31, 4538 (1985)
9. S. Fishman and A. Aharony: J. Phys. C 12, L729 (1979).
10. M. Schwartz and A. Soffer: Phys. Rev. Lett. 55, 2499 (1985)
11. A.T. Ogielski and D.A. Huse: Phys. Rev. Lett. 56, 1298 (1986)
12. A.J. Bray and M.A. Moore: J. Phys. C 18, L927 (1985)
13. J. Villain: J. Physique 46, 1843 (1985)
14. D.S. Fisher: Phys. Rev. Lett. 56, 416 (1986)
15. D.R. Taylor, E. Zwartz, J.H. Page and B.E. Watts: J. Magn. Magn. Mat. 54-57, 57 (1986)
16. J.H. Page, D.R. Taylor and S.R.P. Smith: J. Phys. C 17, 51 (1984) and references therein
17. J.H. Page and H.M. Rosenberg: J. Phys. C 10, 1817 (1977)
18. R.L. Melcher: In Physical Acoustics Vol. 12, ed. by W.P. Mason and R.N. Thurston (Academic, New York 1976)

Quasi Resonant Transport Behaviour of Nonequilibrium Phonons in Insulating Crystals

R. Ruckh and E. Sigmund

Institut für Theoretische Physik, Universität Stuttgart, D-7000 Stuttgart 80, Fed. Rep. of Germany

1. Introduction and Microscopic Model

A quantum mechanical model is developed in order to investigate the dynamics of nonequilibrium acoustic phonons in insulating crystals, like $Al_2O_3(Cr^{3+})$, $CaF_2(Eu^{2+})$. The defect ions are described as electronic two level centers disturbed statistically on lattice sites $\vec{m}$ of the host crystal. The model Hamiltonian of the electron phonon system reads ($\hbar=1$)

$$H^O = \sum_m \Delta\, \sigma_3^{\vec{m}} + \sum_\varkappa \omega_\varkappa\, b_\varkappa^+ b_\varkappa \tag{1}$$

$$H^I = \sum_{\vec{m}} \sum_\varkappa \alpha_\varkappa\, \sigma_1^{\vec{m}}\, (b_\varkappa\, e^{i\varkappa m} + \text{h.c.})$$

Δ is the levelsplitting of the two level centers, σ_i^m are the electronic operators of a SU(2) algebra, $b_\varkappa^+(b_\varkappa)$ creates a phonon of frequency $\omega_\varkappa$ and $\alpha_\varkappa$ represents the electron phonon coupling. Applying a unitary transformation $U=e^S$ where $S \sim \sigma(\sigma^+ b - \sigma^- b^+)$ the Hamiltonian in eq. (1) and expanding it up to 3rd order in $\alpha_\varkappa$ the resonant interaction in (1) can be eliminated yielding weakly interacting new quasiparticles ("dressed phonons and electrons"). Neglecting elastic scattering and phonon trapping effects the interaction

$$H^I = \sum_{\varkappa_1} \sum_{\varkappa_2} \sum_{\varkappa_3} \sum_m \frac{4\,\Delta\, \alpha_{\varkappa 1}\, \alpha_{\varkappa 2}\, \alpha_{\varkappa 3}}{(\Delta^2-\omega_{\varkappa 1}^2)(\Delta^2-\omega_{\varkappa 2}^2)} \left(\omega_\varkappa\, \sigma_2^m + i\Delta\sigma_1^m \right) \cdot e^{im(\varkappa_1 + \varkappa_2 - \varkappa_3)}\, b_{\varkappa 3}^+\, b_{\varkappa 2}\, b_{\varkappa 1} + \text{h. c.} \tag{2}$$

These "anharmonic phonon decay" processes are due to the electron phonon coupling (Fig. 1). In contrast to the "conventional" anharmonic phonon decay, have the electronic energy of a two level center is simultaneously changed by $\pm\Delta$, (Fig. 1).

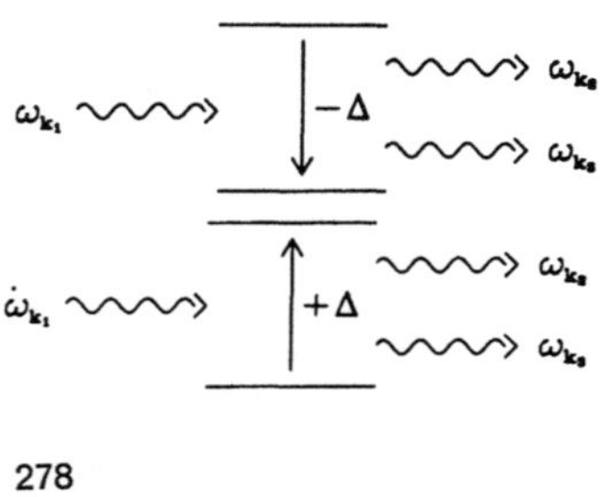

Fig. 1: "Anharmonic" phonon decay due to the electron phonon interaction.

2. Phonon Transport

In order to calculate the spatial, spectral and time evolution of a nonequilibrium phonon distribution a Boltzmann equation for one phonon Wigner functions $\rho_p(\varkappa,r,t)$, is derived where the collision term bears the 3-Phonon decay processes of eq. (1). The electronic dynamics are neglected, which implies that all electronic processes are much faster than the changes in the phonon distribution. The phonon Boltzmann equations are solved numerically by replacing the quasicontinous phonon spectrum by five discrete modes (2). The scattering between these five modes contains both up- and down- conversion processes, resulting from the energy conservation of the interaction processes in eq. (2)

$$(3) \qquad \omega_{\varkappa 1} \pm \Delta \rightarrow \omega_{\varkappa 2} + \omega_{\varkappa 3}$$

Therefore, due to the "spin flip" of $\pm\Delta$ the phonons $\varkappa_2,\varkappa_3$ which are created in the decay process may be higher (up conversion) or lower (down conversion) in frequency than the $\varkappa_1$-phonon.

The up- and down conversion as well as the resonant character of the electron phonon coupling (3) stabilize phonons which are nearly in resonance with the electronic two-level centers whereas the nonresonant phonons decay rapidly. This result is shown in (Fig. 2), where at t=0 5 phonon modes are created at mode temperature T_n(n=1,... 5)

$$(4) \qquad T_n(r,\ t) = \frac{\omega_n}{\varkappa_B}\ (1 + 1/\rho_p(\varkappa,r,t)$$

k_B=Boltzmann constant. Fig. 2 shows the numerical solution of the Boltzmann equations 100 nsec after generation of the pulse.

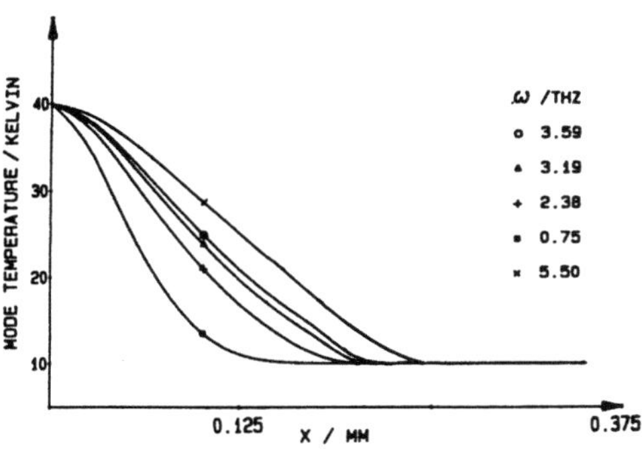

Fig. 2: Numerical solution of the Boltzmann equation for five phonon modes (Δ= 5 THz).

3. Conclusions

The model calculations presented here show that the dynamical behaviour of nonequilibrium phonons interacting with electronic two-level systems can lead to a quite different ("quasiresonant") behaviour than phonons which undergo e.g. decay processes due to lattice anharmonicities ("quasidiffusive phonon transport") (3).

In addition we note that for high two-level concentrations the model presented here may be used to describe coherent motion of electronic or phonon solitary waves. These solitary excitations can be stabilized by interaction processes similar to those in eq. (2).

References

1) R. Ruckh, E. Sigmund: to be published in Physica B
2) R. Ruckh, E. Sigmund: J. Phys. C: in print
3) W. Bron, Y. Levinson, J. O'Connor: Phys. Rev. Lett. **49**, 209 (1982)

Thermal Conductivity of LaB_6: High Magnetic Field Study

Y. Peysson, B. Salce, and A.M. de Goër

Département de Recherche Fondamentale, Service des Basses Températures, Laboratoire de Cryophysique, Centre d'Etudes Nucléaires, 85 X, F-38041 Grenoble Cedex, France

The properties of rare-earth hexaborides have been extensively studied during the last years, especially the "dense Kondo" system CeB_6 [1]. An interesting feature of this compound is the orbital degeneracy of the ground state Γ_8 of the Ce^{3+} ions, which can be subjected to a Jahn-Teller effect [2][3]. In such a case, the lattice contribution to the heat transport could be strongly affected. Previous thermal conductivity measurements [4][5] have suggested that the phonon (K_{ph}) and electronic (K_e) components (supposed to be additive : $K = K_{ph} + K_e$) are of the same order of magnitude in the Kondo regime so that the separation of the two contributions is not easy. On the other hand, the thermal conductivity of the isostructural non magnetic metal LaB_6 displays the classical shape of a normal metal, with $K_e \gg K_{ph}$ [4][5]. In order to determine the phonon contribution for a normal hexaboride metal, we have investigated the heat transport in LaB_6 from 1.5 K to 15 K using the Corbino method where a large magnetic field is applied perpendicularly to the heat flow in order to reduce the electronic component [6]. In that case, K_e can be divided into two contributions $K_{e\perp} + K_{e/\!/}$ respectively perpendicular and parallel to the magnetic field. For a normal metal, it has been shown that the electronic thermal conductivity displays the same behaviour as the electrical conductivity. In particular, $K_{e\perp}$ decreases as H^{-2} in the high field limit for all directions such that closed orbits alone occur for the path of the electrons. With the Corbino geometry, the heat flow is radial, $K_{e/\!/} = 0$ and it should be possible for high enough magnetic fields to determine K_{ph} which is supposed to be field independent.

For this experiment, we have used a thin disk-shaped single crystal of LaB_6 prepared by the floating zone method (diameter 8 mm, thickness 0.5 mm). A small orifice has been opened at the center of the disk by spark erosion, in order to introduce a thin wire fastened to the main heater. The heat flows from the center of the disk towards the outer edge. The sample holder is shown in Fig. 1. The Allen Bradley thermometers have been calibrated for various magnetic fields. The thermal conductivity has been measured by the usual steady state heat flow method using a classical ^4He cryostat. The magnetic field was applied perpendicularly to the disk and was parallel to the <111> axis of the crystal.

The K(T) results for different values of the applied field are given in Fig.2. In zero field, there is a satisfactory agreement with the curve measured using the standard geometry [3], in view of the larger uncertainty on K in the Corbino experiment, for such a highly conducting sample [5]. By applying the field, K(T) is strongly reduced by a factor of 12 for H = 7 T. For each value of the field we have fitted our data by a polynomial function, in order to calculate the variation of K with the magnetic field at a given temperature. K(H) is plotted versus H^{-2} in Fig.3 for T = 2,4,6,8 K. For high magnetic fields (H > 4 T), K(H) is rapidly decreasing with the field. However it can be seen that the H^{-2} regime is still not achieved.

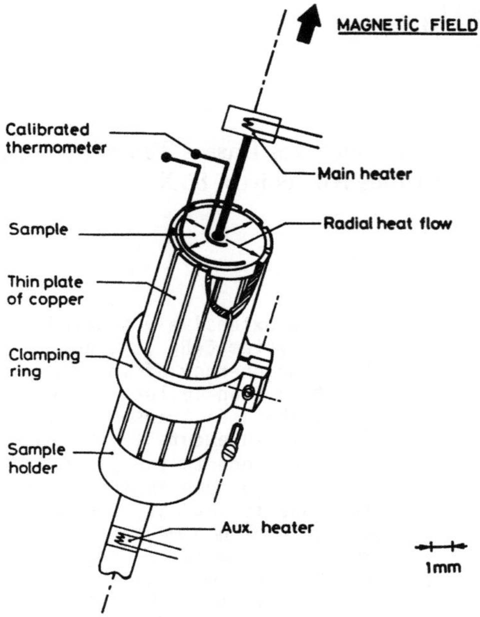

Fig.1 - Schematic view of device for measuring heat transport properties with the Corbino Geometry

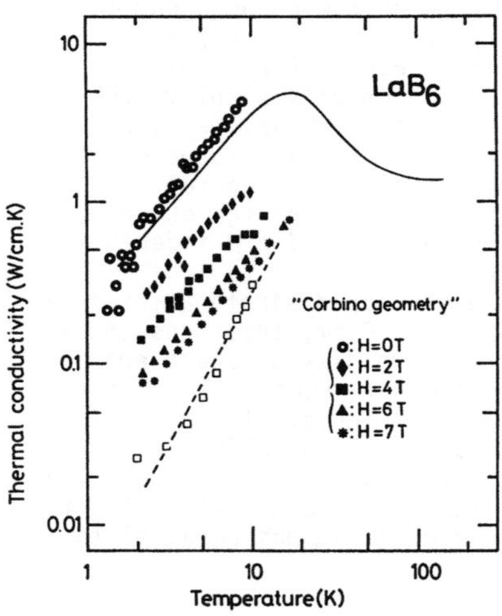

Fig. 2 - Thermal conductivity of LaB_6 for various magnetic fields. The dashed line represents the phonon contribution to the heat transport extrapolated from the $K(H) = f(H^{-2})$ plot (see fig. 3)

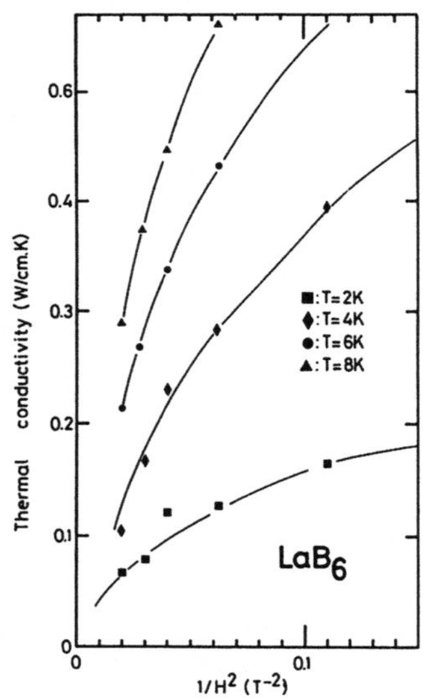

Fig. 3 - Plot of the thermal conductivity of LaB_6 versus H^{-2} for various selected temperatures

This is consistent with the magnetoresistance data performed along various high symmetry directions of the LaB_6 structure [7]. Hence it is only possible to get an upper limit for K_{ph} by extrapolation to infinite field using a mean square fit. This estimation of K_{ph} is reported in Fig.2. K_{ph} has roughly a T^2 variation with the temperature, as expected for phonon scattering by conduction electrons [8]. But in view of the limited range of magnetic field explored in this experiment, this behaviour might be somewhat fortuitous. Moreover, K_{ph} which is estimated to be 25 times smaller than the electronic component from our analysis seems to be a rather large value, as for usual metal $K_{ph}/K_e \approx 10^{-3}$. Further experiments to higher fields is then necessary to improve the validity of our K_{ph} extrapolated value. Nevertheless we must emphasize that the thermal conductivity of CeB_6 is, in the same temperature range, one order of magnitude smaller than $K_{ph}(T)$ [2][3] so that strong phonon scattering is certainly present in the Kondo regime.

Acknowledgments

The authors are thankful to Dr. S.Kunii for providing the sample and the technical assistance of Mr. J.Blanchard and Mr. J.A.Favre is gratefully acknowledged.

References

1. T. Komatsubara, T. Suzuki, M. Kawakami, S. Kunii, T. Fujita, Y. Isikawa, A. Takase, K. Kojima, M.Suzuki, Y. Aoki, K. Takegahara, T. Kasuya : J. Magn. Magn. Mat. 15-18 (1980) 963
2. E. Zirngiebl, B. Hillebrands, S. Blumenröder, G. Guntherodt, M. Loëwenhaupt, J.M. Carpenter, K. Winzer, Z. Fisk : Phys. Rev. B30, (1984) 4052
3. Y. Peysson, C. Ayache, J. Rossat-Mignod, S. Kunii, T. Kasuya : J. de Physique 47 (1986) 113
4. Y. Peysson, C. Ayache, B. Salce, S. Kunii, T. Kasuya : J. Magn. Magn. Mat. (1986)
5. Y. Peysson : Thesis Grenoble (1986)
6. M. Hubers, J.F.M. Klein, H. Van Kempen, H.N. Delang, J.S. Lass, A.R. Miedema, P. Wyder : Proc. Int. Conf. on Phonon Scattering in Solids, ed. H.J. Albany (Paris 1972) p.169
7. A.J. Arko, G. Crabtree, D. Karmi, F.M. Mueller, L.R. Windmiller, J.B. Ketterson, Z. Fisk : Phys. Rev. B13 (1976) 5240
8. J.M. Ziman : Electrons and Phonons (Clarendon Press, Oxford 1960)

Diagrammatic Picture of Pure Dephasing Processes in Phonon-Phonon Interactions

B. Perrin

D.R.P., CNRS, UA. 71, Université Paris VI, Tour 22, 4, place Jussieu, F-75252 Paris Cedex 05, France

1 INTRODUCTION

One of the standard theoretical ways to study phonon-phonon interactions in pure crystals is the temperature Green's function technique which leads to the expressions of frequency shift and width induced by interactions. Usually lowest-order diagrams satisfactorily explain phonon lifetimes and shifts[1] in simple crystals for which all vibrational modes have the same status. However, since a few years a number of results have been obtained on optical modes ($\vec{K}\simeq\vec{0}$) by high-resolution spontaneous Raman scattering[2], CARS[3] or picosecond CARS[4] in molecular crystals, for which internal vibrational modes play a particular role. These modes may be few dispersed and the discussion of results about them mainly concerns the distinction between the contributions of depopulation processes (T_1) and dephasing processes (T_2) to internal phonons (vibrons) lifetimes. Depopulation processes relax the energy of a vibrational mode towards other modes, and dephasing processes may be defined as the other contributions to linewidth. In fact this distinction depends on the point of view adopted to describe the internal vibrations :

- from a liquid state point of view a CARS experiment corresponds to the preparation of a state where the different contributing oscillators illuminated in a given volume are excited with a spatial coherence defined by $\vec{K}$. Depopulation effects are related to energy escape from these oscillators and thus, from the whole vibrational branch.
- from a solid state point of view the spatially coherent states discussed above are a natural basis and correspond to delocalized modes. Then any process which varies the occupation number of such modes may be considered as a depopulation process.

The whole discussion about depopulation and dephasing contributions to phonons and vibrons lifetime may be led within the scheme of thermal Green functions, and we will now describe diagrams which correspond to the different processes.

2 DEPOPULATION AND SPATIAL COHERENCE LOSS PROCESSES

Diagram 1b clearly gives a depopulation process while diagram 1a only gives a frequency shift, and we have to consider higher-order diagrams to account for coherence loss processes. Among fourth-order diagrams, diagram 2a gives again a depopulation contribution but diagram 2b may be considered either as a depopulation process for the delocalized mode (j, $\vec{K}$) or a dephasing process for the vibrational state j of the molecule. Diagrams 2b spread over the whole Brillouin zone delocalized states well defined in phase space and may contribute to localization effects against the intermolecular coupling responsible for internal modes dispersion and delocalization.

Fig. 1 - Lowest order diagrams

Beside the depopulation and spatial coherence loss processes described above, there is another contribution to vibrational linewidth of molecules in dense media considered as a pure dephasing and which corresponds to a frequency random modulation. This process has been studied following different ways such as stochastic theory[5], density matrix formalism[6],.... In papers devoted to phonon-phonon interactions in molecular crystals[2,7,8] which use the thermal Green function techniques, random modulation is always discussed outside the scheme of this formalism and is not treated on the same footing with the other contributions to phonon or vibron lifetimes.

In the next section we show that it is possible to account for random modulation of an internal or external mode λ_o $(j_o,\vec{K})$ with a diagrammatic expansion of thermal Green function by summing an infinite series of particular diagrams.

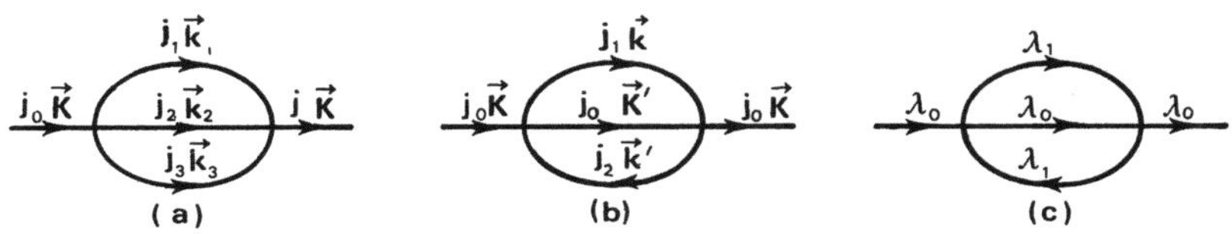

Fig. 2 - Fourth order diagrams

3 FREQUENCY RANDOM POPULATION PROCESSES

Diagram 1a which contributes to frequency shift means that the frequency ω_o of mode λ_o is shifted by $\delta\ (\bar{n}_1+\frac{1}{2})$ when the crystal is strained by populating of mode λ_1 $(j_1,\vec{k})$; however this implicitly assumes that phonon population in mode λ_1 is instantaneously set to its thermal value. In fact when these two modes are isolated from the thermal bath and coupled by the operator $\delta\ (a_o^\dagger a_o+\frac{1}{2})\ (a_1^\dagger a_1+\frac{1}{2})$, the resonance ω_o of mode λ_o is split in a series of resonances at $(\omega_o'+n_1\delta)$ with relative intensity $\exp\ (-\beta\hbar n_1\omega_1)$ where $\omega_o'=\omega_o+\delta/2$. The pole of the phonon propagator G_{λ_o} only dressed with diagram 1a is $(\omega_o'+\bar{n}_1\delta)$, barycentre of this multiplicity of resonances.

We consider now diagrams 3 only built with phonon lines λ_o and λ_1 linked through vertices $V\ (\bar{\lambda}_o\lambda_o\bar{\lambda}_1\lambda_1)$ $(=\hbar\delta/24)$. A diagram with a q bubbles insertion contributes to the self-energy of phonons λ_o through a resonant term $\delta^{p+1}(\omega-\omega_o)^{-p}$ and cannot be neglected in comparison with diagrams 1a (q=o) or 2c (q=1). The summation of this infinite series of diagrams with the appropriate pairing scheme numbers gives rise to a new pole $\omega_o+\delta+O(\bar{n}_1)$ in the propagator of mode λ_o besides the pole $\omega_o+O(\bar{n}_1)$. In fact the summation shown on Fig. 4 may be viewed in terms of a Dyson equation for the two-phonon Green function G_{λ_o,λ_1} which has a pole $\omega_o+\omega_1+\delta(1+\bar{n}_o+\bar{n}_1)$; then summation of diagrams 3 only displays the interaction of G_{λ_o,λ_1} with the propagator G_{λ_1} leading to the pole $\omega_o+\delta+O(\bar{n}_1)$.

Fig. 3 - Diagram contributing to frequency random modulation

Fig. 4 - Dyson equation for the two-phonon Green function G_{λ_o,λ_1}

Now we have only to dress phonon λ_1 with a life-time $\tau=1/2\gamma$ to account for the random modulation of frequency ω_o by fluctuations of the occupation number in mode λ_1. The summation of diagrams 3 with dressed phonons λ_1 and phonons λ_o renormalized by insertion of the diagram 1a in phonon lines λ_o (this corresponds to the shift of ω_o towards $\omega_o'+\delta\bar{n}_1$) gives for the self-energy of phonon λ_o

$$S_{\lambda_o}(\omega) = -\beta\hbar\delta(\bar{n}_1+\tfrac{1}{2}) - \beta\hbar\delta^2\bar{n}_1(\bar{n}_1+1)(\omega-\bar{\omega}_o-\delta+2i\gamma)^{-1} \tag{1}$$

where $\bar{\omega}_o=\omega_o'+\bar{n}_1\delta$. Moreover we have assumed that $\bar{n}_o<<1$ and $\bar{n}_1<<1$. Thus the spectral density of phonon λ_o is given by

$\omega > o$

$$A_{\lambda_o}(\omega) = \frac{2}{\beta\hbar}\;\frac{2\delta^2\,\bar{n}_1\gamma}{((\omega-\bar{\omega}_o)(\omega-\bar{\omega}_o-\delta)-\delta^2\bar{n}_1)^2+4\gamma^2(\omega-\bar{\omega}_o)^2} \tag{2}$$

For large lifetimes τ the spectral density has two lines around ω_o' and $\omega_o'+\delta$. For a faster modulation these two lines merge into a single one for which it is possible to define effective position and width given by

$$\left\{\begin{aligned} \omega_e &= \omega_o+\frac{\delta}{2}+\frac{\delta}{1+\delta^2\tau^2}\,\bar{n}_1 \\ \gamma_e &= \frac{\delta^2\tau}{1+\delta^2\tau^2}\,\bar{n}_1 \end{aligned}\right. \tag{3}$$

which are the classical expressions obtained by other ways [6,9]. These results show that frequency random modulation may be taken into account through the thermal Green function formalism and a diagrammatic expansion. The technique of causal Green function determined by the equation of motion method seems also to be a way to discuss this problem[10] .

Finally it must be noticed that interference effects between contributions to phonon lifetimes of diagrams 1,2 and 3 may be studied by inserting any of these three diagrams on phonons lines λ_o to calculate the other ones.

REFERENCES

1-A.A. Maradudin and A.E. Fein, Phys.Rev. 128 (1962) 2589
2-P. Ranson, R. Ouillon and S. Califano, Chem. Phys. 86 (1984) 115
3-P.L. Decola, R.M. Hochstrasser and H.P. Trommsdorff, Chem. Phys. Let.72 (1980) 1.
4-B.H. Hesp and D.A. Wiersma, Chem. Phys. Letts 75 (1980) 423
5-R.Kubo, in Fluctuations, Relaxation and resonance in Magnetic systems,ed by D. Ter Harr (Plenum, New York, 1962) 23.
6-R.M. Shelby, C.B. Harris, and P.A.Cornelius, J.Chem.Phys. 70 (1979) 34
7-L.A. Hess and P.N. Prasad, J. Chem. Phys. 72 (1980) 573
8-J. Kalus, J. de Chimie Physique 82 (1985) 137
9-R.J. Abott and D. Oxtoby, J. Chem. Phys. 70 (1979) 4703
10-M.A. Ivanov, L.B. Kvashnina and M.A. Krivoglaz, Sov. Phys-Solid State 7 (1966) 1652.

Part VII

Phonon Imaging

Propagation of Large-Wave-Vector Phonons

*S. Tamura**

Department of Physics, University of Illinois at Urbana-Champaign, Urbana, IL 61801, USA

1. Introduction

It is well known that the dispersion curves of acoustic phonons in solids are approximated to be linear in the magnitude of the wave vector $q=|\vec{q}|$ for small $\vec{q}$. However, as $\vec{q}$ approaches the Brillouin zone boundaries, they appreciably deviate from the linearity in q. (A typical example for the phonon dispersion curves in GaAs is shown in Fig. 1 [1,2]). The phonons possessing such large wave vectors are excited thermally in solids as lattice vibrations and play important roles in characterizing the thermal and other physical properties of each solid.

Various novel techniques for generating and detecting high-frequency, nonequilibrium phonons being developed recently [3] have made it possible to study experimentally the propagation characteristics of large-wave-vector phonons. In this paper, we shall review mainly theoretical aspects of the current studies on the large-wave-vector phonons in nonmetallic solids. Especially, the dispersive effects on the isotope scattering and anharmonic deay, which are important ingredients determining the phonon liftimes or mean-free-paths, and focusing effects governing the ballistic phonon propagations will be discussed.

2. Isotope scattering

A number of naturally occurring elements are composed of several isotopes. The phonons propagating in elemental and compound materials consisting of such elements are scattered by fluctuations in mass distribution caused by randomly distributing isotopic atoms as well as by other lattice imper-

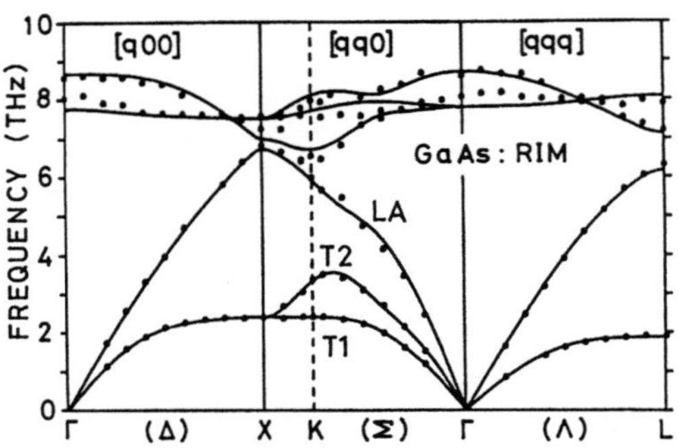

Fig. 1 Phonon dispersion curves in GaAs. Experimental data are from ref. 1. Solid curves are calculations based on RIM [2]. T1 and T2 denote the lower and upper TA branches, respectively.

* Permanent address: Department of Engineering Science, Hokkaido University, Sapporo 060, Japan

fections [4,5]. The scattering probability of the phonons by the isotopes increases with increasing the frequency because short-wavelength phonons are more sensitive to the mass fluctuations due to isotopic atoms. This is an elastic scattering and yields the mixing of mode polarizations. The scattering rate of a phonon $\lambda=(\vec{q},j)$ (where j denotes a mode) in cubic, multiatomic lattices is given by [6]

$$\tau_I^{-1}(\lambda) = \frac{\pi^2}{3N}\,\nu_\lambda^2 \sum_\sigma g_\sigma\, |\vec{e}(\sigma|\lambda)|^2 \sum_{\lambda'} \delta(\nu_\lambda-\nu_{\lambda'})\, |\vec{e}(\sigma|\lambda')|^2, \tag{1}$$

where N is the number of unit cells in a crystal, ν is the phonon frequency, $\vec{e}$ is the unit polarization vector, and g_σ is the factor representing the magnitude of the mass fluctuation due to σ atom in a unit cell. ($g_\sigma=0$ if σ atom does not have isotopes.) In this equation, both the squared amplitude $|\vec{e}(\sigma|\lambda)|^2$ associated with the initial phonon and the one-phonon density of states weighted by the squared amplitudes of final phonons λ', the second sum of Eq. (1), are frequency-dependent. As the frequency increases the ordinary one-phonon density of states defined by $d_1(\nu)=\Sigma_\lambda\,\delta(\nu-\nu_\lambda)$ increases more rapidly than ν^2 predicted by Debye model, because in most crystals acoustic branches exhibit normal dispersion, that is, the dispersion curves are convex upwards. This property of lattice vibrational spectrums of acoustic modes is reproduced commonly by published models of the lattice dynamics. However, the wave-vector dependence of the squared amplitudes is more complicated and highly model dependent.

Now, we shall concentrate our discussion on the isotope scattering of the phonons in GaAs [6,7]. Because the element As does not possess isotopes, only the lattice vibrations of Ga atoms scatter the phonons. In Fig. 2 the theoretical wave-vector dependences of the squared amplitudes of Ga atoms $|\vec{e}(\mathrm{Ga}|\lambda)|^2$ have been plotted. The models employed here are the extended rigid-ion model (RIM) [2] and valence overlap-shell model (OSM) [8]. As the wave vector approaches the zone-boundaries of the [q,q,0] and [q,q,q] directions, $|\vec{e}(\mathrm{Ga}|\lambda)|^2$ for TA branches are predicted to decrease in RIM but increase in OSM [9]. This implies that the dispersive effects on the isotope scattering of the phonons in GaAs will be enhanced in OSM but suppressed in RIM. Here, we note that in the low-frequency limit Eq. (1) for GaAs is reduced to the simple Klemens' formula [4], $\tau_{I,K}^{-1}=7.38\times10^{-42}\times\nu^4$ (sec^{-1}) which exhibits no anisotropy and mode dependence. The calculated relaxation times versus frequency up to the lowest zone-boundary

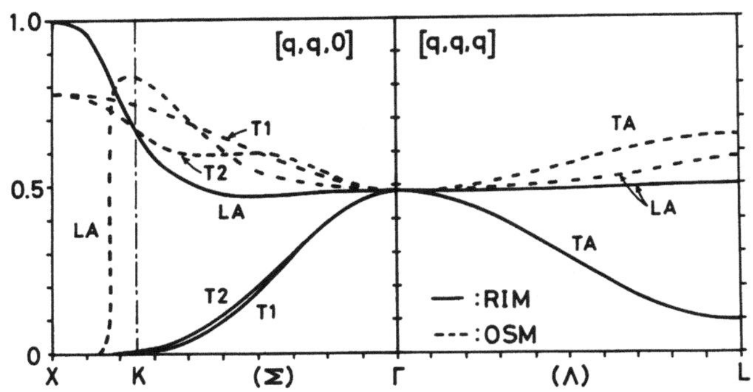

Fig. 2 Wave-vector dependences of squared amplitude of Ga atom. Solid lines and dashed lines are predicted by RIM and OSM, respectively

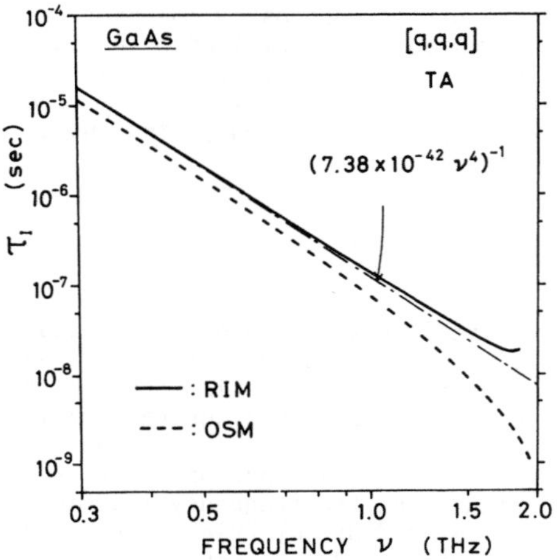

Fig. 3 Relaxation times of TA phonons by the isotope scattering. Solid and dashed lines represent those predicted by RIM and OSM, respectively. Dot-dash line is the relaxation time $\tau_{I,K}$ predicted by the Klemens' formula

frequency 1.9 THz in the [q,q,q] direction have been shown in Fig. 3. Indeed, we find that at finite frequencies τ_I for the TA phonons calculated by RIM becomes rather longer than $\tau_{I,K}$ predicted by the Klemens' formula being valid for nondispersive frequencies, whereas OSM predicts τ_I much shorter than $\tau_{I,K}$. However, this figure indicates that dispersive effects on the isotope scattering are rather small at frequencies lower than 1 THz, though they become appreciable and highly model-dependent at frequencies close to the zone-boundaries. Approximating τ_I by $\tau_{I,K}$ and multiplying it by phonon group velocity v including dispersion, the mean-free-paths of [q,0,0]-propagating TA phonons limited by isotope scattering are estimated to be 1.7 mm ($v=3.05\times10^5$ cm/sec) at 0.7 THz and 0.39 mm ($v=2.84\times10^5$ cm/sec) at 1.0 THz.

3. Anharmonic deay

High-frequency phonons satisfying $\nu \gg k_BT/h$ (T being the ambient temperature) are down-converted via three-phonon processes into a pair of low-frequency phonons. The decay rate is given by [10]

$$\tau_A^{-1}(\lambda) = \frac{\pi\hbar}{8N\nu_\lambda} \sum_{\lambda',\lambda''} \frac{|\Phi(\lambda,\lambda',\lambda'')|^2}{\nu_{\lambda'}\nu_{\lambda''}} \Delta(\vec{q}-\vec{q}\,'-\vec{q}\,'')\,\delta(\nu_\lambda-\nu_{\lambda'}-\nu_{\lambda''}), \tag{2}$$

where Φ is the Fourier transform of the third-order atomic force constants of anharmonic lattices, and $\Delta(\vec{q})=1$ if $\vec{q}$ is zero vector or a reciprocal lattice vector and $\Delta(\vec{q})=0$ otherwise. The energy and crystal momentum conservations for the processes described in Eq. (2) impose the selection rule that a phonon cannot decay into a set of phonons of higher phase velocities [11]. This kinematical constraint severely restricts the decay of high-frequency TA phonons especially in the presence of normal upwards frequency dispersion [7,12]. This can be seen by studying the weighted two-phonon density of final states $d_2(\lambda)= D_2(\nu=\nu_\lambda, \vec{q})$ available for a decaying phonon, where D_2 is defined by

$$D_2(\nu,\vec{q}) = \sum_{\lambda',\lambda''} \frac{\Delta(\vec{q}-\vec{q}\,'-\vec{q}\,'')}{\nu_{\lambda'}\nu_{\lambda''}}\,\delta(\nu-\nu_{\lambda'}-\nu_{\lambda''}). \tag{3}$$

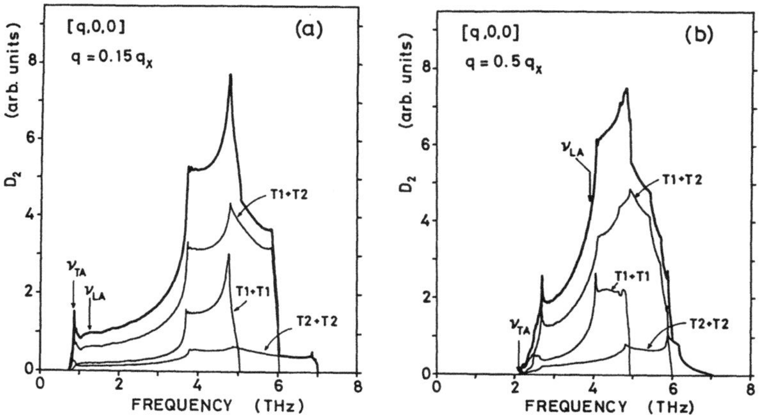

Fig. 4 Two-phonon densities of states D_2 in the [q,0,0] direction, (a) q= $0.15q_X$, (b) q=$0.5q_X$

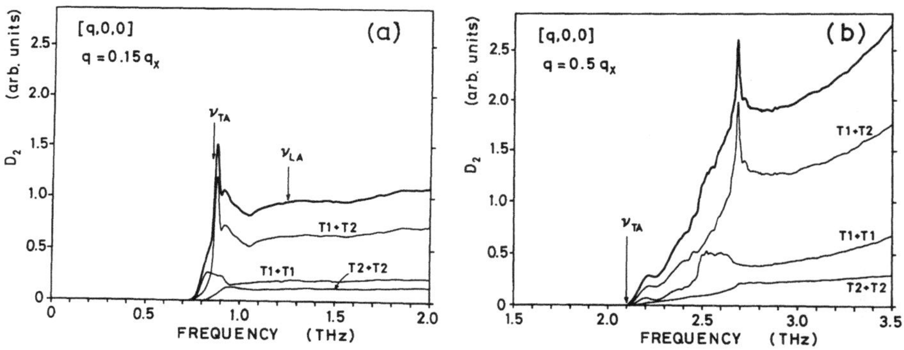

Fig. 5 Magnifications of Fig. 4

This quantity expresses the extent of the phase space to which the spontaneous decay of a phonon is possible. In Eq. (3) the sum over mode polarizations will be restricted to the TA modes because the relevant final states for the decay of a TA phonon are those consisting of the TA phonons. It can be shown that the contribution of the final states consisting of one LA phonon and one TA phonon is in general nonvanishing but can practically be neglected.

In Figs. 4 and 5 we have plotted D_2 calculated by RIM [2] for phonons of q=$0.15q_X$ and $0.5q_X$ in the [q,0,0] direction of GaAs, where $q_X = 2\pi/a$ is the magnitude of the wave vector at X point of the Brillouin-zone boundary and a is the lattice constant. (Note that D_2 is proportional to ν^0 in the low-frequency limit.) The physically meaningful densities of states $d_2(\lambda)$'s are those of D_2's at frequencies which satisfy the on-frequency shell condition $\nu = \nu_\lambda$. These physical frequencies ν_λ's are indicated by ν_{TA} and ν_{LA}, and their values are ν_{TA} =0.85 THz and ν_{LA}=1.24 THz for q=$0.15q_X$,and ν_{TA} =2.10 THz and ν_{LA} =3.92 THz for q =$0.5q_X$. From these figures we can understand that the TA phonons of the frequency 0.85 THz have an appreciable magnitude of the density of final states which is the same order of magnitude as that of 1.24-THz LA phonons propagating in the same direction. (We note that d_2 for the decay channel LA→LA+TA is much smaller than d_2 for LA→TA+TA [13].) However, at much higher frequencies the phase space of the final two-phonon states for the TA phonon decay decreases rapidly due to the effects of dispersion. We observe that the TA phonons of 2.1-THz frequency no more decay spontaneously into a pair of lower frequency phonons.

Unfortunately, we can not predict the absolute magnitudes of the decay rates of the acoustic phonons at dispersive frequencies since we do not know the anharmonic part of the interatomic potential of GaAs. For the estimation we show in Fig. 6 the decay rates of 1-THz phonons in GaAs calculated by the continuum elasticity model. Note that this model predicts the decay rates proportional to ν^5 [14,15]. The results suggest that for phonons of frequencies lower than 1 THz the anharmonic decay rates are much smaller than the scattering rates by isotopic atoms. It is for near-zone-boundary LA phonons that the spontaneous decay dominates the isotope scattering.

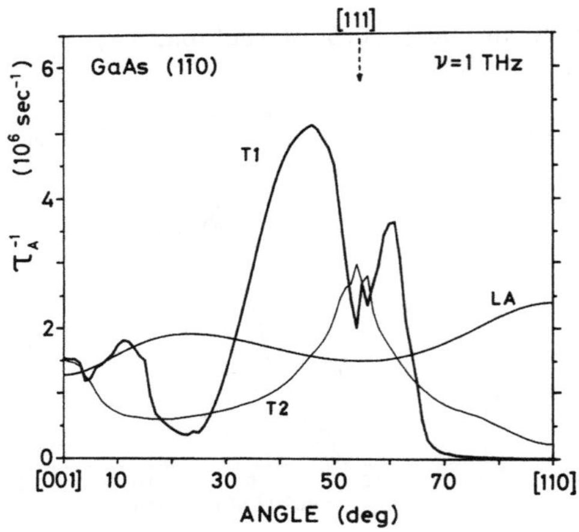

Fig. 6 Spontaneous decay rates of 1-THz phonons in GaAs obtained by long-wavelength approximation

4. Focusing effects

In crystalline solids the constant-frequency (slowness) surfaces of the phonons spanned in the $\vec{q}$-space are not spherical in general owing to the presence of the anisotropy of underlying lattices. Then, even if the phonons are emitted from a source isotropically in the $\vec{q}$-space, the distribution of the group-velocity vectors, whose directions are given by outward normals of the slowness surfaces, is anisotropic and the phonon energy flux concentrates along certain directions and deconcentrates along others [16-18]. This anisotropy is called phonon focusing [19]. In fact, it turns out that in some directions the phonon intensity emitted from a point source becomes infinite [20]. These directions called caustics are originated from zero-Gaussian curvature, parabolic points on the slowness surfaces and separate distinctively the focusing directions from the defocusing directions of the phonons. Hence, the focusing patterns of the phonons are expected to be very sensitive to the distortion of the slowness surfaces due to frequency dispersion.

The first experimental observation of the dispersive effects on the phonon focusing was made by DIETSCHE et al. [21] with a thin Ge sample.

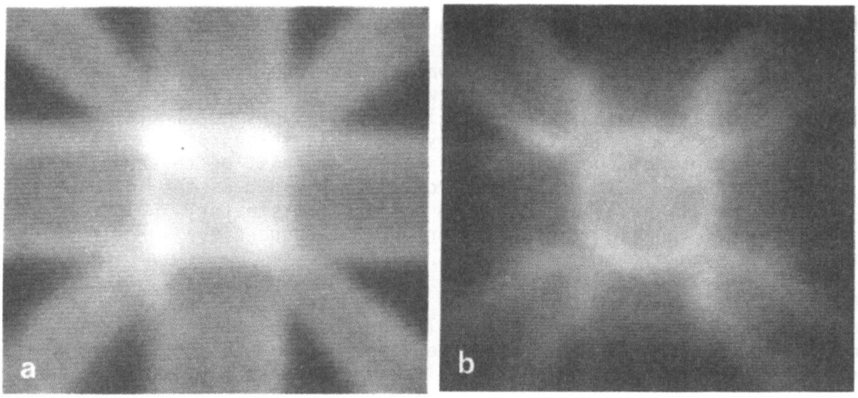

Fig. 7 Experimental phonon images detected by, (a) Aℓ bolometer and, (b) PbBi tunnel junction

They used Pb-O-Pb tunneling junction as the phonon detector to select out the phonons of frequencies higher than 0.7 THz. Computer simulations of the focusing of dispersive phonons in Ge were then published [22,23]. More recently the focusing of large-wave-vector phonons have been studied for GaAs [24-26], mainly to examine the possibility of ballistic transport of highly dispersive TA phonons over macroscopic distances [27]. Figures 7(a) and 7(b) show the experimental phonon images in the (001) plane of a GaAs sample of 0.4-mm thickness detected by Aℓ bolometer and PbBi tunneling junction, respectively [28]. Here we note that the Aℓ bolometer acts as a detector of low-frequency (∿0.3 THz) phonons, whereas the PbBi tunneling junction works as a detector of the phonons with frequencies higher than 0.85 THz. The sharp features of the phonon focusing patterns in the latter image is due to the isotope scattering, which prohibits the ballistic propagation over sample thickness of the phonons of frequencies higher than 1 THz (see Sec.2). Evidently the marked differences between these images are due to the dispersion.

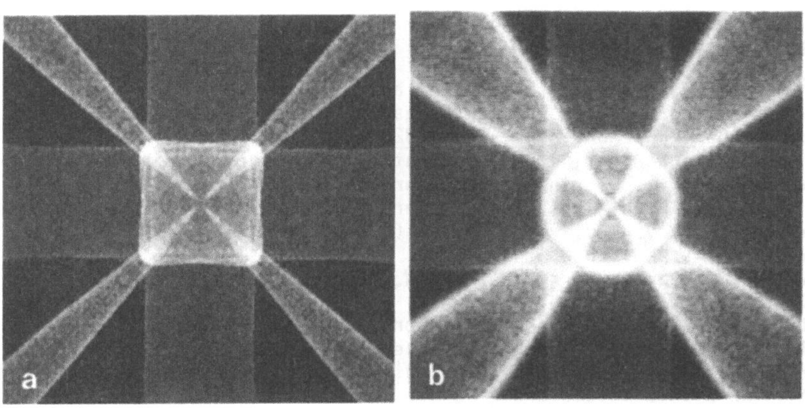

Fig. 8 Theoretical phonon images, (a) $\nu=0$ THz, (b) $\nu>0.85$ THz

The corresponding theoretical images for phonon frequencies of 0 THz and higher than 0.85 THz are given in Figs. 8(a) and 8(b), respectively. The latter image is constructed by the 14-parameters- shell model [1] including the effects of the isotope scattering [29]. The coincidence of the experimental images with the theoretical images is very good. This implies that the shell model predicts correctly the whole shapes of the constant-frequency surfaces of the TA phonons in GaAs at least up to about 1 THz. The other current models of the lattice dynamics such as the dipole and bond charge models have been shown to be less satisfactory in predicting the experimental phonon images [26] or phonon frequencies in directions other than the high symmetry directions.

Thus we can expect that the phonon imaging technique applied to the dispersive phonons provides a powerful means of determining the lattice dynamics. Recently, it has also been applied to InSb and succeeded in evaluating the published models of the lattice dynamics [30].

5. Summary

In this paper we have reviewed the propagation characteristics of large-wave-vector phonons mainly in GaAs. So far a number of experiments on the large-wave-vector phonons were performed in various samples other than GaAs [3]. However, except for a few cases, there is little evidence that the experimetal results should be attributed to the propagation or scattering of fully dispersed large-wave-vector phonons. To study experimentally the phonon dynamics being characteristic to dispersive regime, it would be necessary to choose more dispersive solids.

On the theoretical side, there also exist some difficulties associated with the calculations of transport properties of large-wave-vector phonons. This is caused by the fact that we should employ lattice-dynamics models. In particular, the explicit calculations of scattering rates require very long computer times. Moreover, for many crystals we do not even have plausible lattice-dynamics models which can be usable to calculate the properties of the phonons beyond the harmonic approximation. In order to understand the dynamics of large-wave-vector, nonequilibrium phonons further efforts need to be expended both theoretically and experimentally.

References

1. Experimental data are from G. Dolling and J. L. T. Waugh: in Lattice Dynamics, edited by R. F. Wallis (Pergamon, London, 1965), p. 19.
2. S. Tamura and T. Harada: Phys. Rev. B32, 5245 (1985).
3. See, Phonon Scattering in Condensed Matter, edited by W. Eisenmenger, K. Lassmann, and S. Dottinger (Springer-Verlag, New York, 1984).
4. P. G. Klemmens: Proc. Phys. Soc. London. Sect. A68, 1113 (1955).
5. P. Carruthers: Rev. Mod. Phys. 33, 92 (1961).
6. S. Tamura: Phys. Rev. B30, 849 (1984).
7. M. Lax, V. Narayanamurti, P. Hu, and W. Weber: J. Phys. (Paris) Colloq. 42, C6-161 (1981).
8. K. Kunc and H. Bilz: in Proc. Intl. Conf. Neutron Scattering, edited by R. M. Moon, Gatlinburg, ORNL, Tennessee, p. 195.
9. Corresponding behaviors of $|\vec{e}(\mathrm{Ga}|\lambda)|^2$ in the bond-charge and deformation-dipole models are given in ref. 7 and ref. 6, respectively.
10. A. A. Maradudin and A. E. Fein: Phys. Rev. 128, 2589 (1962).
11. M. Lax, P. Hu, and V. Narayanamurti: Phys. Rev. B23, 3095 (1981).

12. H. J. Maris: Phys. Lett. 17, 228 (1965).
13. K. Okubo and S. Tamura: Phys. Rev. B28, 4847 (1983).
14. R. Orbach and L. A. Vredevoe: Physics 1, 91 (1964).
15. S. Tamura and H. J. Maris: Phys. Rev. B31, 2595 (1985).
16. F. Rosch and O. Weis: Z. Phys. B27, 33 (1977).
17. J. C. Hensel and R. C. Dynes: Phys. Rev. Lett. 43, 1033 (1979).
18. G. A. Northrop and J. P. Wolfe: Phys. Rev. Lett. 43, 1424 (1979); Phys. Rev. B22, 6196 (1980).
19. B. Taylor, H. J. Maris, and C. Elbaum: Phys. Rev. Lett. 23, 416 (1969); Phys. Rev. B3, 1462 (1971).
20. P. Taborek and D. Goodstein: Solid State Commun. 33, 1191 (1980).
21. W. Dietsche, G. A. Northrop, and J. P. Wolfe: Phys. Rev. Lett. 47, 660 (1981).
22. S. Tamura: Phys. Rev. B25, 1415 (1982) and ibid. B28, 897 (1983).
23. G. A. Northrop: Phys. Rev. B26, 903 (1982).
24. J. P. Wolfe and G. A. Northrop: in Phonon Scattering in Condensed Matter, Ref. 3, p. 100.
25. G. A. Northrop and J. P. Wolfe: in Nonequilibrium Phonon Dynamics, edited by W. E. Bron (Plenum, New York, 1985), p. 665.
26. G. A. Northrop, S. E. Hebboul, and J. P. Wolfe: Phys. Rev. Lett. 55, 95 (1985).
27. R. G. Ulbrich, V. Narayanamurti, and M. A. Chin: Phys. Rev. Lett. 45, 1432 (1980).
28. D. C. Hurley, S. E. Hebboul, and J. P. Wolfe: unpublished.
29. S. E. Hebboul and J. P. Wolfe: unpublished.
30. S. E. Hebboul and J. P. Wolfe: "Imaging of dispersive phonons in InSb" ; this volume.

Phonon Focusing in Piezoelectric Crystals

A.K. McCurdy

Worcester Polytechnic Institute, 100 Institute Road,
Worcester, MA 01609, USA

1. Introduction

Phonon focusing in dielectric crystals depends upon the second-order elastic constants [1]. These constants determine the shape of the phase and group-velocity surfaces for each of the three modes and thus the phonon focusing properties of the crystal. For certain ratios between these constants, the transversely polarized group-velocity surfaces can exhibit cuspidal edges [1, 2]. It is along directions coincident with these cuspidal edges that the most intense phonon focusing can be observed [3]. Energy flow is highly enhanced along such strongly focused directions.

In piezoelectric crystals, however, the second-order elastic constants can be stiffened due to the coupling between the stress field and the accompanying electric field. The amount of elastic stiffening depends not only upon the piezoelectric and permittivity constants but also upon the direction of the wave vector for the propagating elastic wave [4]. By changing the ratios between the elastic constants, elastic stiffening modifies the phase and group-velocity surfaces and thus the phonon focusing properties of the crystal.

Phonon focusing calculations until recently have neglected effects of elastic stiffening in piezoelectric crystals. Since these effects have recently been observed in heat-pulse experiments [5], these effects will now be briefly examined for cubic and hexagonal crystals. More detailed results for these and other crystal symmetries will be presented elsewhere.

2. Results

Polar plots of the group-velocity surfaces are shown in Figs. 1 and 2 for piezoelectric CdS and InSb, respectively. Results used the elastic constants, piezoelectric constants, permittivity constants and densities at room temperature tabulated in the appendix of reference 4. Calculations performed for CdS and ZnO (both of hexagonal 6mm symmetry) show that piezoelectric stiffening increases the group velocity of the longitudinal mode along the c axis by 1.2% and 4.5%, respectively. Perpendicular to the c axis the velocity of the slower transverse mode in CdS and ZnO is increased by 1.8% and 3.5%, respectively. Calculations for both crystals show cusps about non-symmetry collinear axes (along which the group velocity is parallel to the wave vector) [6]. This collinear axis is 43.4 and 40.7 degrees to the c axis for CdS and ZnO, respectively. The cusp width for CdS and ZnO is 3.6 and 4.3 degrees, respectively. With piezoelectric stiffening ignored, however, the collinear axis for CdS and ZnO shifts to 45.3 and 45.1 degrees, respectively. Furthermore, the cusp width in CdS narrows to 1.8 degrees whereas the cuspidal feature in ZnO vanishes.

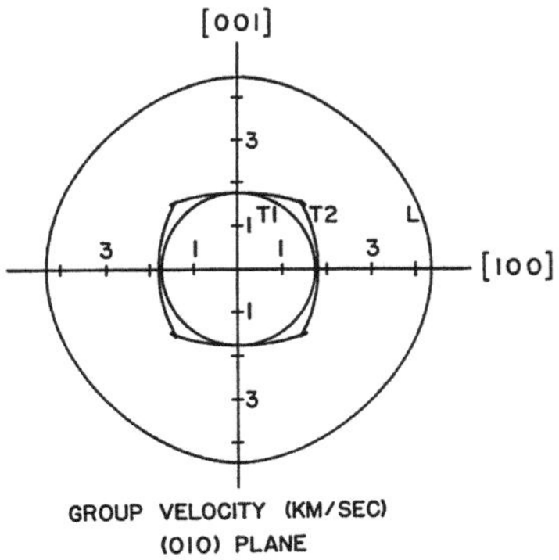

Fig. 1 Polar plots of the group-velocity surfaces (km/s) for piezoelectric CdS (hexagonal 6mm) showing the narrow cusps in the faster (T2) transverse mode. The velocity surfaces have rotational symmetry about the [001] axis

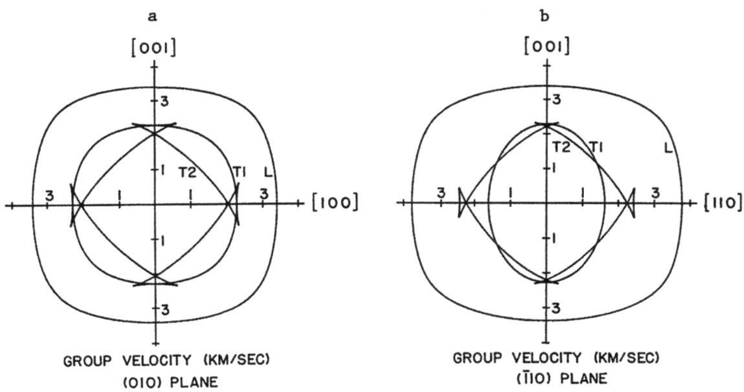

Fig. 2a,b. Polar plots of the group-velocity surfaces (km/s) for piezoelectric InSb (cubic $\bar{4}3m$) for (a) {100} planes and (b) {110} planes

Piezoelectric stiffening in InSb (cubic $\bar{4}3m$) increases the group velocity of the faster transverse mode by 5.8% along the <110> axes, but only increases the group velocity of the longitudinal mode by 2.7% along the <111> axes. Cuspidal features in piezoelectric InSb occur about the <100> and <110> axes. In {100} planes the cusp width about the <100> axis is 29.3 degrees, whereas in {110} planes the cusp width about the <100> and <110> axes, respectively, is 15.8 and 19.6 degrees. With piezoelectric stiffening ignored, however, the cusp width about the <100> and <110> axes in the {110} planes changes to 19.0 and 10.3 degrees, respectively. Piezoelectric stiffening does not affect the cuspidal features about the <100> axes, however, when wave vectors are restricted to the {100} planes.

3. Discussion

Piezoelectric stiffening in most crystals changes the phase velocity by less than several percent. As a result, effects of elastic stiffening can best be examined along directions coincident with cuspidal edges. Along such directions phonon focusing is intense, so that minimal stiffening of the elastic constants gives significant changes in the phonon intensity.

For CdS piezoelectric stiffening not only shifts the cusp location, but also doubles its width. For ZnO, however, the cusp is present only when piezoelectric effects are included. For InSb the cusps centered about the <100> axes in the {100} plane are not affected by the piezoelectric effect. However, because of piezoelectric stiffening of the faster transverse mode along the <110> axes, cusps in the {110} planes are widened about the <110> axes, but narrowed about the <100> axes. Note that the pair of cuspidal edges about the <110> axes are bisected by the {100} planes and extend for a wide angle toward the <100> directions. This cuspidal feature (for the non-piezoelectric case) is a general property of cubic crystals having similar elastic anisotropies [7].

Acknowledgement

Appreciation is due the College Computation Center at Worcester Polytechnic Institute for the use of their computing facilities, and to Mr. Jon D. Nelson for programming the computer plotting routines used in the figures.

References

1. H.J. Maris: J. Acoust, Soc. Am. 50, 812 (1971)
2. G.C. Winternheimer and A.K. McCurdy: Phys. Rev. B18, 6576 (1978)
3. G.A. Northrop and J.P. Wolfe: Phys. Rev. B22, 6196 (1980)
4. B.A. Auld: Acoustic Fields and Waves in Solids, Vol. 1 (J. Wiley & Sons, New York 1973)
5. G.L. Koos and J.P. Wolfe: Phys. Rev. B29, 6015 (1984)
6. A.K. McCurdy: Phys. Rev. B9, 466 (1974)
7. A.K. McCurdy: Phys. Rev. B26, 6971 (1982)

Phonon Focusing Near a Conic Point

D.C. Hurley, M.T. Ramsbey, and J.P. Wolfe

Physics Department and Materials Research Laboratory,
University of Illinois at Urbana-Champaign, 104 S. Goodwin Ave.
Urbana, IL 61081, USA

Phonon focusing in crystals is an intriguing result of elastic anisotropy. The sharp caustics formed by the channeling of the phonon wavevectors create a wide variety of focusing patterns -- a different one in each material. Since the focusing pattern depends solely on the ratios of the elastic constants, cubic crystals may be mapped into a two-dimensional "parameter space" according to their values of the ratios $a=C_{11}/C_{44}$ and $b=C_{12}/C_{44}$ [1,2]. With this classsification, the focusing patterns of individual materials are seen as a logical evolution of a few general caustic patterns. The patterns become more complex as the elastic anisotropy, characterized by the parameter $\Delta=a-b-2$, increases. Figure 1 shows the location in this parameter space of various materials and their values for a, b, and Δ.

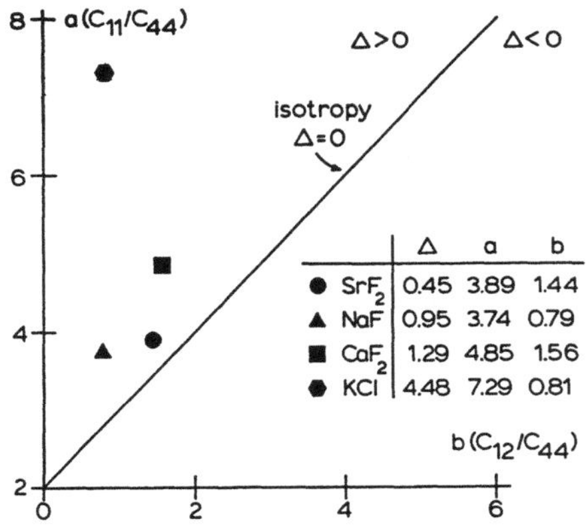

Fig. 1. Diagram of elastic-parameter space for cubic crystals. Materials are plotted according to the ratios of (low-temperature) elastic constants $a=C_{11}/C_{44}$ and $b=C_{12}/C_{44}$ Points for which the anisotropy factor $\Delta=a-b-2=0$ completely isotropic

For materials with positive values of Δ, the most intricate and interesting caustic structures are formed near the <111> crystalline directions. Phonon-imaging techniques [3] provide a means of detecting and studying large portions of the focusing pattern at one time. Figure 2 shows experimental phonon images for several different materials with positive Δ. The slow transverse (ST) three-cusp pattern gradually grows as the anisotropy increases. For small anisotropy ($\Delta = 0.45$, SrF_2), the fast transverse (FT) pattern contains no caustics. For a slightly larger anisotropy ($\Delta = 0.95$, NaF), a pattern of FT caustics is present, as indicated in the figure. For the more anisotropic crystals (CaF_2 and KCl), these structures become more extended.

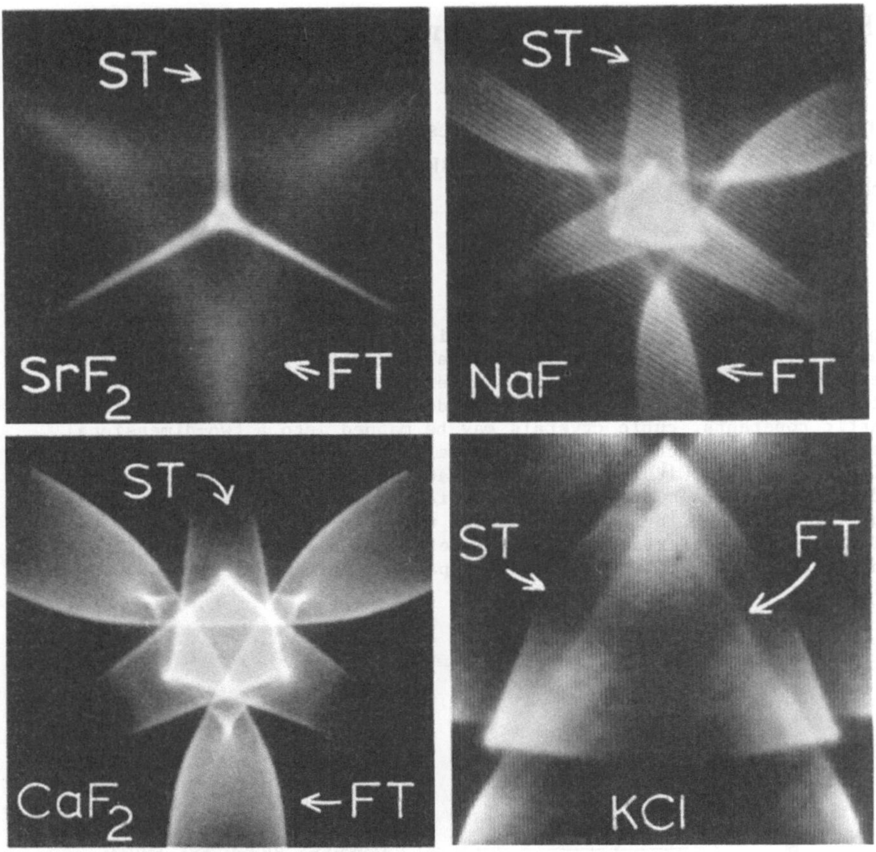

Fig. 2. Experimental phonon heat-pulse images of several cubic crystals for which $\Delta>0$. Phonons created by laser excitation of a copper film on one side of the crystal are detected on the opposite side by a superconducting bolometer biased at its transition temperature. These phonon images were recorded using a broad time gate, so that both transverse phonon modes were included. The images are centered on the [111] direction and span ±35° right to left

The <111> directions in cubic crystals have a special importance: they mark points of conical degeneracy for the two transverse modes, giving rise to an "internal conical refraction" [4]. Transverse phonons with k-vectors infinitesimally close to <111> are "refracted" into a well-defined cone of V-vectors. In addition, the ST and FT caustics in the immediate vicinity of <111> arise from a single fold line that oscillates from one transverse sheet to the other several times as it revolves around <111>. The group velocity (ray) surfaces in Fig. 3 illustrate this behavior. The circle on each sheet represents the propagation directions of phonons with k-vectors infinitesimally close to <111>. The two transverse wave sheets join together along this circle of conical refraction. The heavy lines show the fold line intersecting the conic circle at the dots.

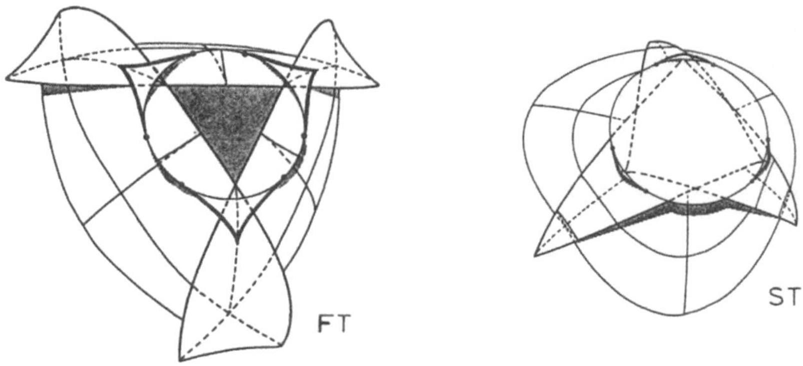

Fig. 3. Representation of the ray surface near the [111] direction for the fast transverse (FT) and slow transverse (ST) sheets in CaF_2. Thick lines represent fold lines. No phonons are focused into the central shaded region on the FT sheet

The focusing patterns can be quantitatively classified using only a few parameters [2]. For instance, the basic features of the ST mode structure can be described by specifying the opening angle α between the <111> direction and one of the three cusp structures. We have measured α in the experimental phonon images and compare it with values calculated from the low-temperature elastic constants in Table 1. A detailed numerical characterization of the caustic patterns in cubic crystal for both positive and negative Δ may be found in Ref. 2. Given an arbitrary set of elastic constants, the focusing pattern can be empirically determined without detailed calculation.

Table 1. Theoretical and experimental values for α, the angle between the ST cusp tip and the <111> direction (indicated in line drawing)

	α (Theoretical)	α (Experimental)
SrF_2	3.2°	—
NaF	8.8°	9.8° ± 1.5°
CaF_2	12.2°	13.0° ± 1.5°
KCl	32.9°	33.1° ± 1.5°

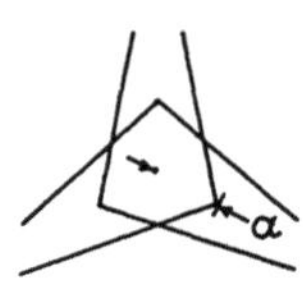

This work was supported by the National Science Foundation under the MRL Grant DMR 83-16981 and equipment support from NSF DMR 80-24000.

References

1. A.G. Every: Phys. Rev. B 24, 3456 (1981).
2. D.C. Hurley and J.P. Wolfe: Phys. Rev. B 32, 2568 (1985).
3. G.A. Northrop and J.P. Wolfe: In Nonequilibrium Phonon Dynamics, edited by W.E. Bron (Plenum, New York, 1985), p. 165.
4. A.G. Every: Phys. Rev. B (to be published).

A Model for Calculating Pseudosurface Wave Structures in Phonon Imaging

A.G. Every

Department of Physics, University of the Witwatersrand, Johannesburg 2001, South Africa

The intensity variation in phonon images derives principally from two sources: phonon focusing in the bulk of the crystal, and the directional emissivity of the surfaces across which the phonons enter and leave the crystal. The role that phonon focusing plays has been explored in numerous publications [1]. Recently surface directivity effects have been reported in highly polished sapphire [2,3] and diamond [4] single crystals. Pronounced halos of enhanced phonon intensity are observed close to the critical cones for mode conversion of transverse to longitudinal waves at the crystal surface, and these have been interpreted as the signature of pseudosurface waves (PSW's) in the quasi-free crystal surface. In this paper we describe a simple model for treating the directional dependence of the phonon emissivity and absorptivity of the crystal surface.

The phonon transmission probability across an interface between a crystal and metal heater or detector overlayer depends in a complicated way on the phonon polarization and direction, the elastic properties of both crystal and overlayer, on the nature of the boundary conditions and on the condition of the crystal's surfaces. In the experiments on sapphire and diamond the crystal surfaces were flat on the scale of the dominant phonon wavelength and the metal overlayers were either of low acoustic impedance compared to the crystal or weakly bonded to the crystal surfaces [3]. Under these conditions the crystal surface vibrates almost as if free, with the overlayer having merely a slight damping effect on this motion. The precise nature of the boundary conditions and the specific properties of the overlayer have, in these circumstances, little influence on the directional dependence of the phonon transmission. These conclusions are derived from the extensive calculations of EVERY et al.[3] on isotropic media, and also on their experimental observation that varying the composition of the metal heater film has no discernible effect on the channeling structures.

Our model draws from these ideas, while also taking into account the elastic anisotropy of the crystal. The pivotal assumption of our model is that a phonon of a bulk wave of the crystal has a transmission probability across an interface which is proportional to the square of the amplitude of vibration caused by that (normalized) mode impinging on that free crystal surface. This amplitude depends on the nature of the incoming wave as well as the three outgoing waves which are generated. All four waves are phase matched in the surface, i.e. have the same value of $S_{//}$, the component of slowness parallel to the crystal surface. The slowness vectors also satisfy the Christoffel characteristic equation. The amplitudes of the outgoing waves are obtained from a set of 3 linear equations representing the free surface boundary conditions.

Our image construction process consists in firstly generating a large number of randomly distributed values of $S_{//}$, and solving Christoffel's equation for the perpendicular components $S_{\perp}$. The ray vector for each

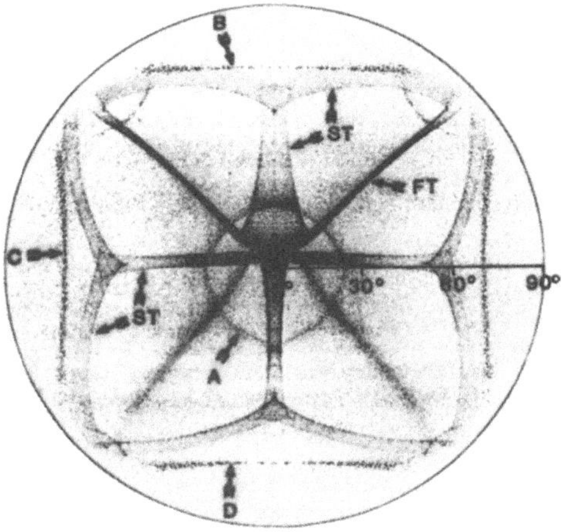

Fig.1: Polar diagram of ST and FT phonon intensity for the (58°/30°) surface of sapphire. Intensity is measured by the darkness of the greyscale.

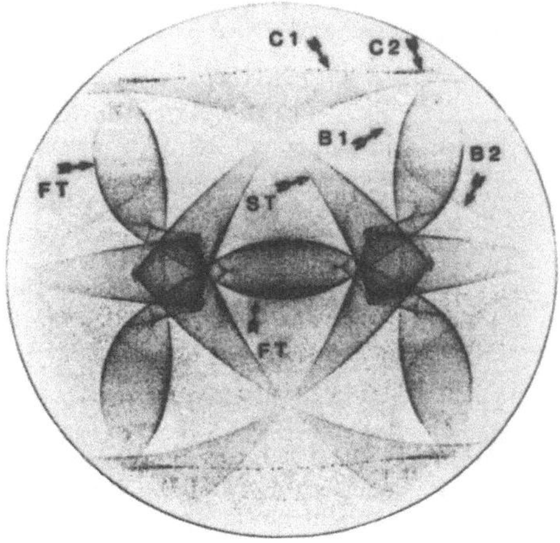

Fig.2: Phonon intensity diagram for the (110) surface of CsCℓ

incoming wave is then calculated and given a weighting proportional to the squared amplitude of vibration it brings about in the surface. Finally these weighted ray vectors are sorted in direction to form a polar diagram of phonon intensity. Details of the procedure are provided elsewhere [5,6].

Figure 1 shows a phonon intensity diagram for sapphire calculated in this way. The sharp structures labelled ST and FT are slow transverse and fast transverse phonon focusing caustics. The structures A, B, C and D are surface-derived features associated with sharp resonances in surface vibration for certain bulk wave directions. These resonances arise from deep minima in the determinant of the boundary condition equations. LIM and FARNELL [7] have attached the term "pseudosurface wave" to such minima, an extension of the concept of a genuine surface wave or Rayleigh wave for which the boundary condition determinant vanishes. A PSW consists predominantly of an evanescent partial wave component whose amplitude falls off exponentially with distance from the surface. In addition however, the PSW has a small bulk wave component through which it gradually radiates its energy into the bulk.

The focusing caustics and halo-like feature A in Fig. 1 are in very good agreement with the experimental images of KOOS et al. [2] and EVERY et al. [3] for sapphire. The halo is generated by modes which lie very close to the ST → L mode conversion critical cone, and is associated with a PSW which is longitudinal in character. A similar but much sharper halo is predicted for the diamond (100) surface, again in very good agreement with experiment [4]. This L PSW also features in the interpretation of surface Brillouin scattering [8]. The structure BCD was out of view in the original experiments. The PSW it results from is not related to any particular critical cone. It is of a similar nature to a prominent PSW that occurs on the (100) surface of crystals such as Si and Cu, and which has been the subject of investigation using a variety of ultrasonics, light scattering and other techniques [9-12].

Figure 2 shows the phonon intensity diagram for the (110) surface of CsCℓ. The focusing caustics are labelled ST and FT. The L PSW is fairly broad and obscured by its overlap with the FT caustics occurring in the central region of the diagram. The structure B1B2 occurs close to the ST → FT critical cone while C1C2 belongs to the most prominent PSW on this surface [9].

We have calculated phonon intensity plots for a variety of materials and surface orientations, and they all show at least some presence of PSW structures [5,6]. In an actual situation where the overlayers do not strictly represent negligible loading of the crystal surface, the PSW structures can be expected to be somewhat broader and fainter than shown here. Roughening of the crystal surface has a similar effect [3].

In conclusion, phonon imaging provides a complete picture of all the PSW's existing in a crystal surface and also conveys important information about the surface conditions.

This work was supported by the Foundation for Research Development.

References

1. For a recent review of phonon imaging see G.A. Northrop and J.P. Wolfe: In Nonequilibrium Phonon Dynamics, ed. by W.E. Bron (Plenum, New York 1985) p.165
2. G.L. Koos, A.G. Every, G.A. Northrop, and J.P. Wolfe: Phys. Rev. Lett. 51, 276 (1983)
3. A.G. Every, G.L. Koos, and J.P. Wolfe: Phys. Rev. B29, 2190 (1984)
4. D.C. Hurley, A.G. Every and J.P. Wolfe: J. Phys C 17, 3157 (1984)
5. A.G. Every: Solid State Commun. 57, 691 (1986)
6. A.G. Every: Phys. Rev. B33, 2719 (1986)
7. T.C. Lim and G.W. Farnell: J. Acoust. Soc. Am. 45, 845 (1969)
8. R.E. Camley and F. Nizzoli: J. Phys. C 18, 4795 (1985)
9. G.W. Farnell: In Physical Acoustics, ed. by W.P. Mason and R.N. Thurston (Academic, New york 1970), Vol.6, p.109
10. R.G. Pratt and T.C. Lim: Appl. Phys. Lett. 15, 403 (1969)
11. F.R. Rollins, T.C. Lim, and G.W. Farnell: Appl. Phys. Lett. 12, 236 (1968)
12. J.R. Sandercock: Solid State Commun. 26, 547 (1978)

Imaging of Spatial Structures with Ballistic Phonons*

R.P. Huebener, E. Held, W. Klein, and W. Metzger

Physikalisches Institut II, Universität Tübingen, Morgenstelle 14, D-7400 Tübingen, Fed. Rep. of Germany

The generation of ballistic phonons by scanning the surface of a single-crystalline specimen at liquid-He temperatures with a laser or electron beam has been used recently for investigating the phonon focusing effect, i.e., the anisotropic ballistic phonon propagation /1/. The same scanning principle can also serve for imaging structural inhomogeneities in a crystal with ballistic phonons /2/. An object placed between phonon source and detector disturbs the ballistic phonon propagation and influences the detector signal for a distinct coordinate of the scanning beam (see Fig. 1). If two phonon detectors are operated simultaneously, three-dimensional imaging with ballistic phonons becomes possible.

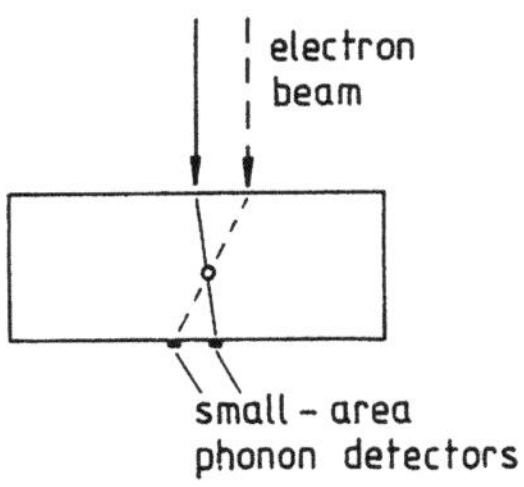

Fig. 1. Principle of acoustic tomography with ballistic phonons

We have realized this principle of three-dimensional acoustic tomography using a scanning electron microscope equipped with a liquid-He stage /3/. In our apparatus the top of the sample can be scanned with the electron beam, whereas the bottom of the sample carrying the phonon detectors is in direct contact with the liquid-He bath. In recent experiments performed with Ge samples we have shown that the region, acting as the source of ballistic phonons, is heated by the electron beam typically to about 10 K, corresponding to a dominant phonon frequency of about 600 GHz /2/. Therefore, the acoustic imaging principle can simply be discussed in terms of geometric optics. Each of the phonon detectors just observes the "shadow" generated by the object.

As a typical example, we show in Fig. 2 the acoustic image of two holes, drilled sideways into a sapphire single

*Supported by a grant of the Deutsche Forschungsgemeinschaft

Fig.2. Acoustic image (left) and optical image (right) of two laser-drilled holes in z-oriented single-crystalline sapphire. Beam parameters: 26 kV; 0.5 µA; 20 kHz modulation; lock-in detection. The background features in the acoustic image are due to phonon focusing.

crystal using a laser technique. (crystal thickness = 2 mm; bath temperature = 2.06 K). A superconducting thin-film bolometer fabricated from O_2-doped Al with an effective area of 10 µm x 10 µm has been used for phonon detection. Comparison with the optical image, which is also shown, indicates that nearly all details are well reproduced by the acoustic image based on ballistic phonons.

The spatial or angular resolution of acoustic imaging by means of ballistic phonons is determined primarily by the area of the phonon source and detector. In principle, the detector area can be made as small as only a few µm in diameter using standard microfabrication techniques. Hence, we must concentrate on the phonon source. If the beam intensity is modulated at the angular frequency ω, the radius of the heated region showing the corresponding temperature modulation is approximately given by

$$\eta = (2K/\omega)^{1/2} \quad , \qquad (1)$$

where K is the thermal diffusion constant of the sample material. If the modulated signal is detected, the effective source diameter shrinks with increasing frequency ω according to Eq. (1). Recently, we have demonstrated this effect shown by the ballistic phonon signal /2,4/. An accurate calculation of the source diameter from (1) is difficult, since the local temperature profile of the source and the temperature dependence of K must be taken into account. Further, the effective source diameter can be reduced by a proper overlay film placed on the sample to be investigated. Therefore, an experimental

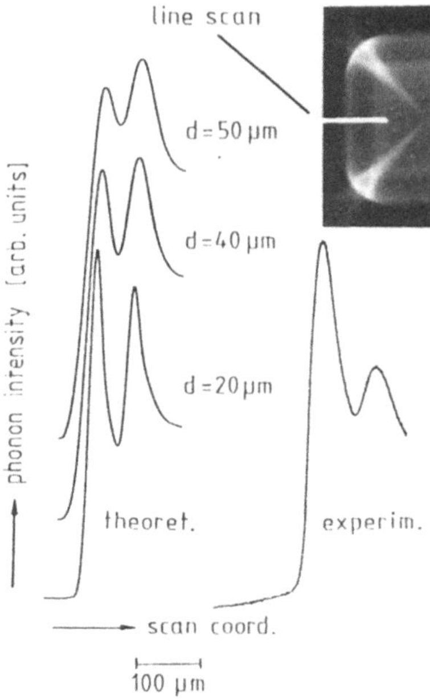

Fig.3. Ballistic phonon intensity of [001] oriented Ge for the line scan shown on the inset. Left: theoretical curves obtained for the different values of the source diameter indicated. Right: experimental curve (crystal thickness = 2 mm; bath temperature = 1.96 K; detector area = 10 μm x 10 μm; beam parameters: 26 kV, 1 μA, 30 ns pulses, 250 kHz, boxcar detection with 10 ns time window).

measurement of the effective source diameter and the corresponding resolution limit becomes highly important. In principle, such an experiment can be performed using a simple model structure such as shown in Fig. 2. However, we have used phonon focusing for determining the effective source diameter as described in the following.

We have calculated /5/ the anisotropic intensity of the ballistic phonons for the case of [001] oriented Ge, assuming a phonon source of variable diameter and a point-like phonon detector. A Gaussian distribution was assumed for the intensity of the phonon source as a function of the radial distance from the center. Twice the radius at which the intensity has decreased to 1/e of the value at the center was taken as the source diameter. In Fig. 3 we show a series of theoretical curves with the diameter d of the phonon source as parameter. For comparison, an experimental curve is also shown. All results given in Fig. 3 refer to the ST-mode and a line scan, the position of which is shown in the inset. By comparing the first peak on the left of the experimental curve with the theoretical results, we find for the diameter of the phonon source d = 40 μm. This value has been obtained without high-frequency beam modulation. According to Eq. (1) the effective source diameter can be reduced considerably by means of high-frequency beam modulation and by detection of the modulated signal.

References:

1. W.Eisenmenger, K.Lassmann, S.Döttinger: In Phonon Scattering in Condensed Matter, Springer Verlag, Berlin, 1984
2. R.P. Huebener, W.Metzger: Scanning Electron Microscopy, 1985/II, p. 617

3. H.Seifert: Cryogenics 22, 657 (1982)
4. W.Metzger, R.P.Huebener: In Phonon Physics, ed. by J.Kollar, N.Kroo, N.Menyhard, T.Siklar, World Scientific Publishing Co., Philadelphia, 1985, p. 959
5. G.A.Northrop: Computer Physics Communications 28, 103 (1982)

Dispersive Phonon Imaging in InSb

S.E. Hebboul, D.J. van Harlingen, and J.P. Wolfe

Physics Department and Materials Research Laboratory,
University of Illinois at Urbana-Champaign,
104 S. Goodwin Ave., Urbana, IL 61801, USA

It has been recognized for several years that the highly anisotropic intensity patterns associated with phonon focusing should undergo radical changes as the phonon wave vectors approach the Brillouin zone boundary [1-4]. This is simply a consequence of the distortion of the acoustic slowness surfaces in the dispersive regime. Phonon focusing patterns of Large-$\vec{k}$ phonons must contain detailed information about the curvatures of the slowness surfaces, and thus they provide a potentially new tool for gauging the validity of lattice-dynamics models.

Despite this promise, experimental successes have not come easily. Basically, one must image the heat flux with spatial and frequency resolution. The key requirements are ballistic propagation and monochromatic detection. The ballistic propagation of short wavelength phonons is limited, in most crystals, by mass-defect (or isotope) scattering, which increases roughly as ν^4. This difficult constraint is something of a blessing in disguise. When the low-pass filter effect associated with isotope scattering in the bulk of the crystal is combined with a detector having a sharp frequency threshold, a remarkably monochromatic detection can be achieved.

These ideas are made more quantitative by the Monte Carlo simulations in Fig. 1. Curve (1) is a 10 K Planck distribution as produced by a heated metal film on the crystal surface. Curve (2) is the transmission probability for ballistic phonons through a 2.0-mm-thick crystal of InSb, using the calculated isotope scattering rate of TAMURA [5]. Curve (3) is the actual transmitted distribution of ballistic phonons, as would be detected by a broadband detector. In contrast, curve (4) shows the frequency distribution of ballistic phonons traversing a 0.5-mm-thick crystal and detected by a detector with a sharp onset (Curve (5), corresponding to a Pb tunnel junction with $2\Delta = 650$ GHz). Note that the width of the distribution is only ~10% of the center frequency. The predicted phonon-focusing pattern displays sharp caustic structures quite different from the sharp lower-frequency structures at the left.

The data in Fig. 2 are a confirmation of these predictions from actual phonon-imaging experiments on InSb. At left is the image for a 2-mm-thick crystal with a broadband Aℓ bolometer. At right is the image of a 0.5-mm-thick crystal with a Pb tunnel junction detector.

Salient results which will be described in a longer paper [6] include the following: (1) The theoretical images in Fig. 1 employ a BCM lattice dynamics model previously fit to neutron scattering data. (2) Several other lattices dynamics models predict qualitatively different patterns at 700 GHz [7]. (3) The phonon wavevectors in the dispersive image of Fig. 2 have a magnitude of about 30% of the zone boundary. (4) Contrary to

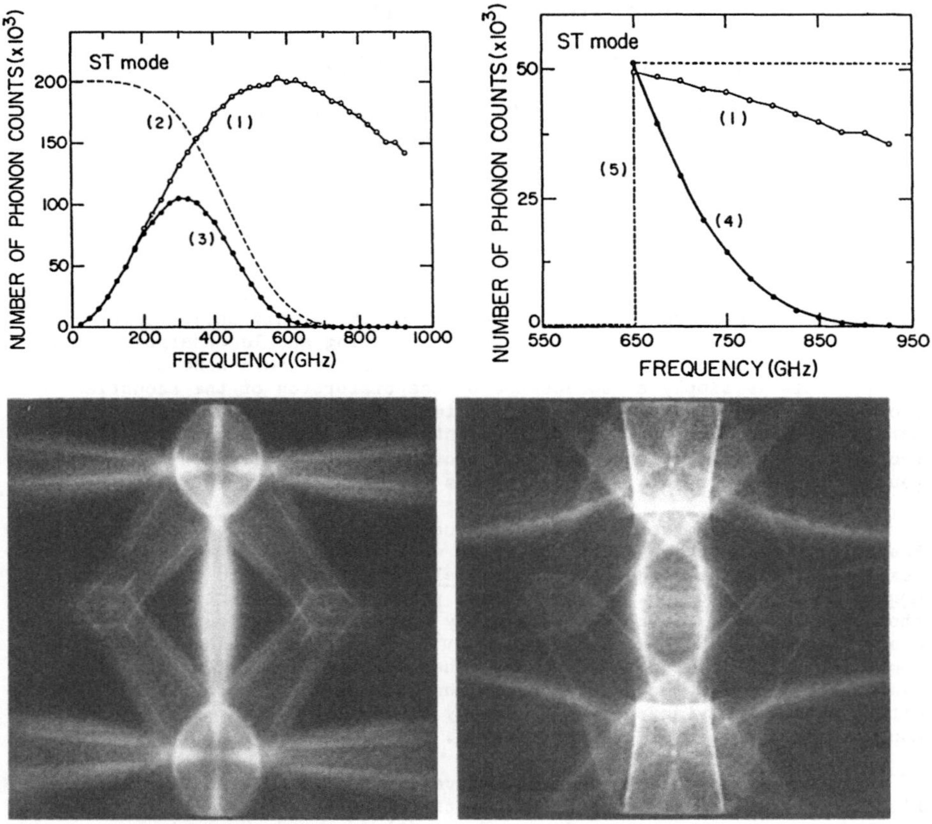

Fig. 1. Monte Carlo Calculations of phonon distributions

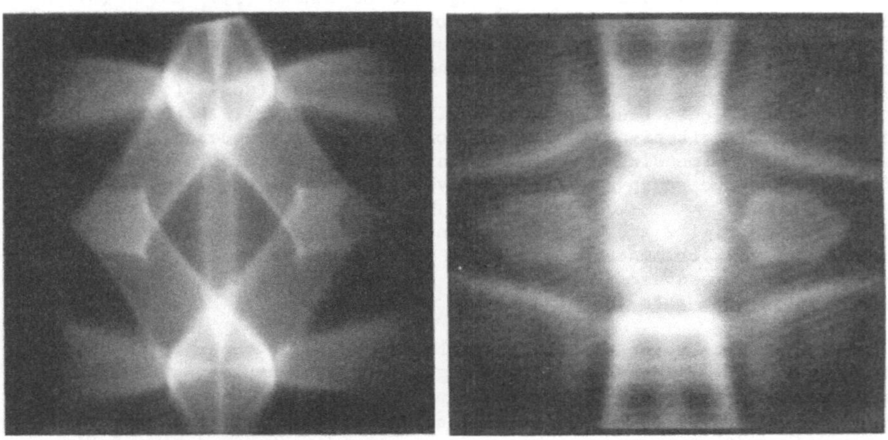

Fig. 2. Experimental phonon images

previous beliefs, direct photoexcitation of the crystal holds no advantage to heating a metal film on the crystal. This observation, understandable from the calculations of Fig. 1, opens the door to dispersive imaging of a wide variety of large-gap insulating crystals.

Dispersive phonon imaging is a fact. The future lies in pushing the observations further into the Brillouin zone, measuring the dispersive effects in different crystals, and developing methods for quantitative input to lattice dynamics models.

We thank E. Swiggard of the Naval Research Laboratory for providing the InSb crystal. This research is supported in part by the National Science Foundation under a Materials Research Laboratory grant NSF DMR 83-16981.

References

1. G.A. Northrop and J.P. Wolfe: Phys. Rev. B22, 6196 (1980)
2. W. Dietsche, G.A. Northrop, and J.P. Wolfe: Phys. Rev. Lett. 47, 660 (1981)
3. S. Tamura: Phys. Rev. B28, 897 (1983)
4. G.A. Northrop: Phys. Rev. B26, 903 (1982)
5. S. Tamura: Phys. Rev. B30, 849 (1984)
6. S.E. Hebboul and J.P. Wolfe: Phys. Rev. B, accepted for publication
7. This was also shown for GaAs: G.A. Northrop, S.E. Hebboul, and J.P. Wolfe: Phys. Rev. Lett. 55, 95 (1985)

Extraordinary Caustics in the Phonon Focusing Patterns of Crystals

A.G. Every

Department of Physics, University of the Witwatersrand, Johannesburg 2001, South Africa

Catastrophe theory provides a convenient framework for classifying the phonon focusing caustics of crystals [1-4]. There are however many prominent focusing singularities which, because they arise in the vicinity of degeneracy between different branches of the phonon dispersion relation, or because they are consequent upon the special properties of the Christoffel wave equation, are not readily accommodated in elementary catastrophe theory. The "extraordinary" caustics described here have been obtained from computer generated phonon images. The search for such structures is not complete, and a mathematical analysis of the possibilities would be very useful.

For a specific crystal, and in the absence of dispersion, the catastrophe generating function depends on two control parameters, the polar coordinates Θ and Φ of the direction of the phonon flux. There are, however, a large number of crystals in existence, whose elastic moduli are distributed over a wide range of values. Moreover, these moduli can, to some extent, be continuously varied by manipulation of temperature, pressure and stoichiometry etc. Thus in a real sense the elastic moduli may be regarded as additional control parameters. In this expanded control parameter space one deals with families of phonon images in which can be observed the unfolding of higher dimensional catastrophes such as the butterfly and hyperbolic umbilic [3]. In this expanded space one also witnesses the unfolding of the extraordinary structures which are the subject of this paper.

The degeneracies that occur in the solution of Christoffel's equation give rise to some intriguing phonon focusing phenomena. This is particularly true of trigonal crystals slightly removed from hexagonal symmetry. Figure 1 shows meridian sections of the slowness surfaces of some representative hexagonal media. These surfaces have full rotational symmetry about the hexad axis along the z-direction. Whatever the values of the elastic constants, the pure transverse and quasi transverse sheets of the slowness surface have a tangential degeneracy in the z-direction (point 7). A lowering of the symmetry to trigonal, which causes the elastic constant C_{14} to be finite, transforms this tangential degeneracy into a set of four conical degeneracies. One of these conical points lies in the z-direction, the other three are located symmetrically around it. Since the outer of the two slowness sheets meeting at a conical point must have at least one principal curvature negative, while elsewhere this sheet must possess convex regions, this implies the existence of lines of zero Gaussian curvature (parabolic lines). These lines map onto phonon focusing caustics. Thus the slightest deviation from hexagonal symmetry results in the formation of caustics. The three-bladed structure B in Fig.2 is the pattern of caustics formed in this way for a fairly wide range of elastic constants.

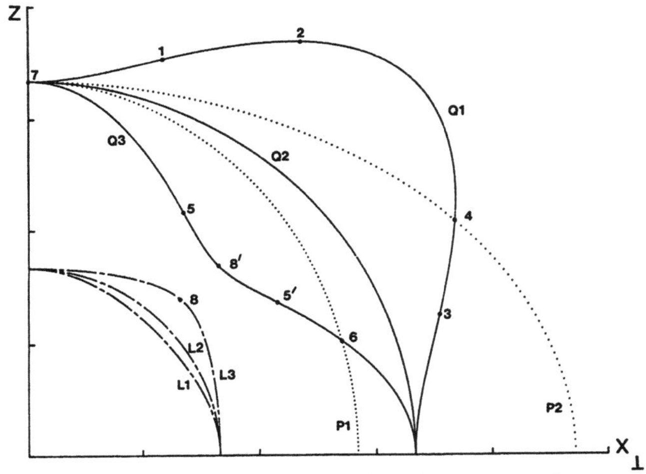

Fig.1: Meridian sections of the slowness surfaces of representative hexagonal media. Expressed in units of C_{44}, $C_{11} = 4$ and $C_{33} = 4$ for all the curves. The remaining elastic constants are as follows: For the quasi T and quasi L branches which do not depend on C_{12}, $C_{13} = 3$ (for Q1 and L1), $C_{13} = 2$ (for Q2 and L2) and $C_{13} = 0$ (for Q3 and L3). For the pure T branch which does not depend on C_{13}, $C_{12} = 1.25$ (for P1) and $C_{13} = 3.0$ (for P2).

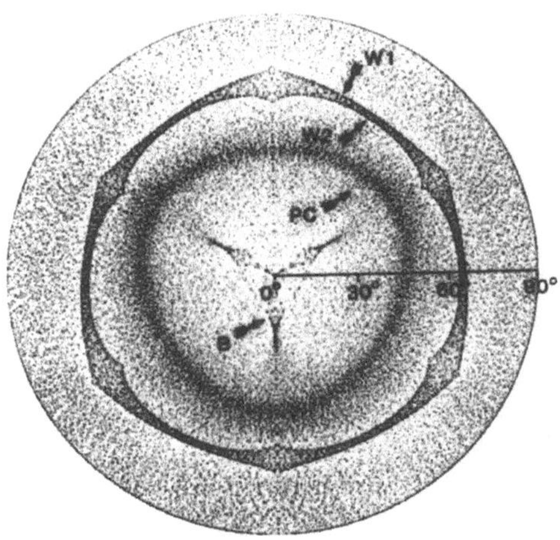

Fig.2: Polar plot of the ST and FT phonon flux for a trigonal medium with $C_{11}=4$, $C_{33}=4$, $C_{12}=1.8$, $C_{13}=1.6$ and $C_{14}=0.025$, all in units of C_{44}.

The intersection of curves P1 and Q3 at point 6 (and similarly the intersection at point 4) in Fig.1 signifies a circular line of degeneracy on which the pure T and quasi T sheets cross through each other. Immediately C_{14} deviates from zero this degeneracy is completely lifted except at six regularly spaced conical points. This requires the reclassification of the T branches into slow and fast T. Each of these two sheets consists of segments of the original pure and quasi T sheets joined together by narrow transitional strips. For the ST branch the strips are negatively curved, and the bounding parabolic lines are responsible for the caustics labelled W1 and W2 in Fig.2. They materialise over their entire lengths immediately the crystal symmetry is lowered to trigonal. Another instance of this kind in which caustics form globally is found in cubic media. For $C_{12} = -C_{44}$ the slowness surface consists of three intersecting oblate spheroids, and there are no caustics whatsoever. Immediately C_{12} deviates from $-C_{44}$ the lines of degeneracy are lifted, leaving only point tangential and conical degeneracies. This is accompanied by the formation of a complex and extensive pattern of caustics.

The feature labelled PC in Fig.2 is a band of large but finite phonon intensity. A suitable change in the elastic constants causes this band to evolve into a pair of adjacent circular caustics. These arise from parabolic lines passing through points 5 and 5' in Fig.1. In Hexagonal symmetry this transformation occurs globally.

The Gaussian curvature at a conical point diverges, resulting in a circle or ellipse of zero phonon intensity, i.e. an anticaustic. Lines of zero Gaussian curvature under certain conditions pass through a conical point, thus giving rise to a collision between a caustic and anticaustic. The effect on the caustic is to cause a gradual fading to zero of the integrated intensity of the caustic as it approaches the anticaustic.

More details on these and related phenomena are provided elsewhere [5].

To conclude, phonon focusing, particularly in the non dispersive regime of the Christoffel equation, provides some interesting challenges to catastrophe theory as regards the prediction and classification of caustics arising in the vicinity of degeneracy, and also the effects of symmetry lowering distortions. The fairly dramatic changes in the focusing patterns that such distortions can produce renders phonon imaging a potentially valuable technique for studying phase transitions and for the measurement of certain hard-to-obtain elastic constants.

This work was supported by the Foundation for Research Development.

References

1. For a recent review on phonon imaging see G.A. Northrop and J.P. Wolfe: In Nonequilibrium Phonon Dynamics, ed. by W.E. Bron (Plenum, New York 1985) p. 165
2. P. Taborek and D. Goodstein: Phys. Rev. B22, 1550 (1980)
3. A.G. Every: Phys. Rev. B24, 3456 (1981)
4. D. Armbruster and G. Dangelmayr: Z. Phys. B52, 87 (1983)
5. A.G. Every: Phys. Rev. B, to be published

Micro Phonography of Buried Doping Structures in Si

H. Schreyer, W. Dietsche, and H. Kinder

Physik Department E 10, Technische Universität München,
D-8046 Garching, Fed. Rep. of Germany

Recently, we had shown that a superconducting tunnel junction can be used as a spatially resolving phonon detector [1]. In the original paper the detector was used to observe the phonon focusing properties in Si. Later, "phonographs" of the phonon emission of a 2DEG have been made [2].

In this contribution, we report on the observation of the doping structures in commercial Si wafers. The experimental set-up is shown as inset in Fig. 1. The wafers with 0.5 mm thickness were doped on one side with 10^{15} cm^{-2} B by ion implantation. The doping structure was "illuminated" from the other side of the crystal with phonons generated in a NiCr heater film. In the doped regions, the phonons were backscattered causing a "phonon shadow". The spatial distribution of the transmitted phonons was measured with the new detector. This way a "contact print" of the doping distribution was obtained.

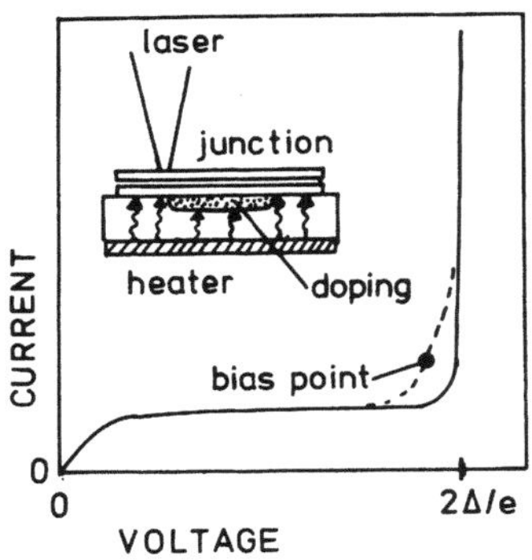

Fig. 1. Current-voltage characteristic of the Al tunnel junction. Inset: Experimental set-up

In order to get a tunnel-junction detector in the spatially resolving mode it must be locally excited by a focused laser beam. In the focal area the energy gap of the superconductor is reduced due to the creation of excess quasiparticles. Thus the current-voltage characteristic (Fig. 1) changes from the solid to the dashed line. The operating point is chosen along the dotted load line. Therefore, the tunnel current flows through the illuminated spot only. This spot can be raster-scanned across the junction area. If this spot coincides with an area of high phonon flux the local gap is further reduced. Thus, the modulation of the junction voltage with position is a measure of the phonon-flux distribution. It is remarkable that only a fraction of 5×10^{-4} of the junction area is laser excited in a typical experiment. In the earlier work, a spatial resolution of about 100 μm was reached. With an improved optics shown in Fig.2 we got an

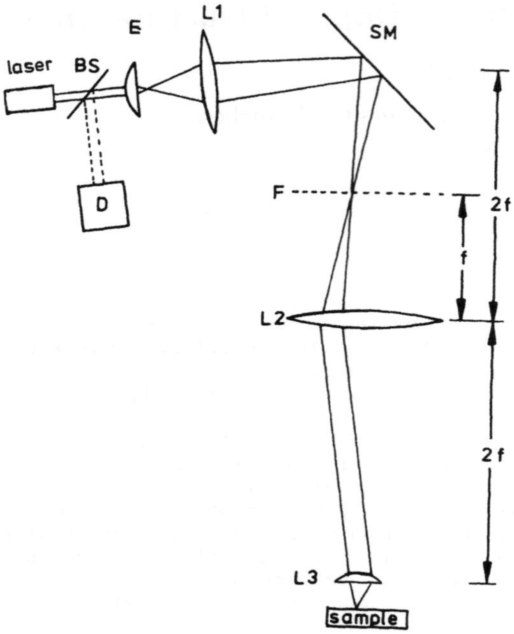

Fig. 2. The HeNe laser beam was first expanded (E) and then focused by lens L1. The focal plane of L1 (F) was halfway between the swiveling mirrors (SM) and the lens L2. The focal length of L2 equalled the distance F-L2. This way, the mirrors were imaged onto the focusing lens L3. The light beam at L3 was always parallel and passed through the central area of the lens minimizing aberrations. A control of the focus size was possible by monitoring the light which was reflected back from the sample with the photo detector D near the beam splitter (BS). With the detector the optical resolution was controlled

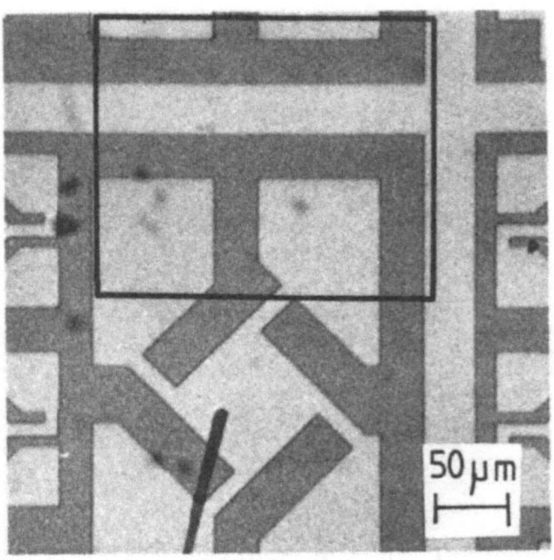

Fig. 3. The darker areas were undoped and had a width of 36 μm. Also outlined (solid line) is the size and the position of the Al tunnel junction.
The picture in Fig. 4 corresponds to this square.

optical resolution of 3 μm. An area of 0.5 x 0.5 mm^2 on the sample could be scanned.

A conventional photograph of the doping structure studied by us is shown in Fig. 3 [3]. The heater film on the opposite surface was much larger (1 x 1 mm^2) to ensure a homogeneous phonon illumination, unaffected by phonon focusing. The power of the laser at the sample was about 10 μW over 10^{-5} mm^2. The heater power was about 3mW.

A microphonograph of the structure of Fig. 3 is shown in Fig. 4. The phonon shadows cast by the doping structures can be seen very clearly as

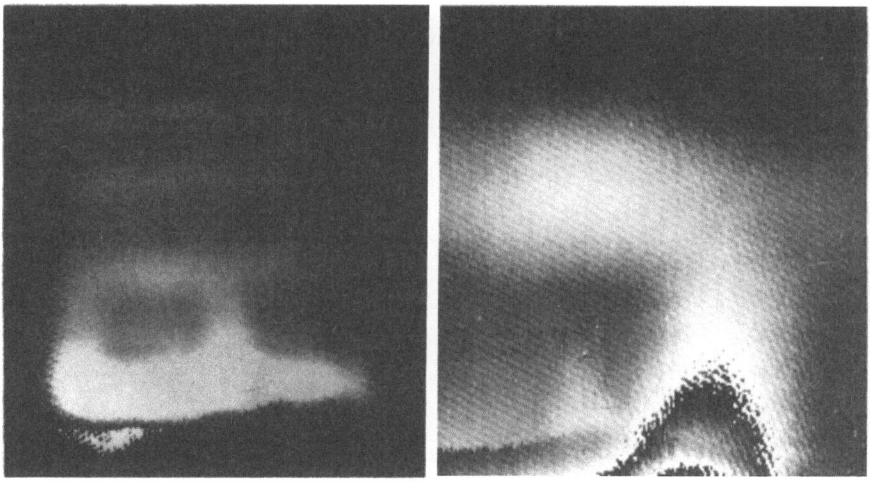

Fig. 4 Microphonograph of the wafer of Fig.3. The picture on the right-hand side is taken at a larger magnification.

the dark areas demonstrating the feasibility of this technique. The intensities of the structures varied over the detector area. We suspect that this was caused by inhomogeneities in the Al films forming the junction. We estimate that the spatial resolution is less than 10 μm, i.e. close to laser-focus size.

In conclusion, we have shown that laser scanned Al junctions can be used to take "phonographs". We expect that this technique will be very useful in the future. An important improvement may be the combination with monochromatic phonon sources. This could lead to an enhancement of the contrast, if frequency dependent scattering is exploited.

1. H. Schreyer, W. Dietsche, and H. Kinder: in LT17-Contributed papers, ed. by U. Eckern et al. (North-Holland, 1984) pg. 665. A better focusing phonograph than published in [1] is reproduced in H. Kinder: Nonequilibrium Phonon Dynamics ed. by W.E. Bron, Nato Advanced Study Institute (Plenum, 1985) pg. 129
2. W. Dietsche: New Developments in Tunnel-Junction Phonon Spectroscopy, this volume
3. The sample was supplied by Fa. Siemens, München

Part VIII

Large-Wavevector Phonons and Optical Techniques

High-Frequency Phonon Injection

J.I. Dijkhuis

Fysisch Laboratorium, Rijksuniversiteit Ultrecht, P.O. Box 80.000, NL-3508 TA Utrecht, The Netherlands

1. Introduction

In the past few years there has been a significant progress in the study of the dynamics of high-frequency acoustic phonons [1-3]. A rich variety of phenomena could be disclosed through the application of new experimental techniques to inject high-frequency phonons. The aim of this paper is to review, though not exhaustively, the various methods for high-frequency phonon injection and the main results obtained with them in a number of crystal types. The emphasis will be on incoherent phonon generation throughout the entire Brillouin zone by optically pumping impurity centers dispersed in the crystal and on surface generation of the phonons by thin-film heating. Most of the work discussed in this paper relies on methods for frequency-selective phonon detection in order to unravel the phonon dynamics [3]. In section 2 the volume generation is discussed of monochromatic phonons by direct transitions, including the dynamics of the resonant phonons, stimulated emission and the spectral width of the produced phonons. In section 3, the volume generation of broadband phonons is examined. Surface generation of high-frequency phonons is discussed in connection with thin-film heaters in section 4 and point-contact phonon generators in section 5.

2. Monochromatic phonon generation

One-phonon emission processes dominate at low temperatures for transitions between electronic impurity levels that are sufficiently close together. This direct process can be used for the optical generation of monochromatic phonons. In optically pumped ruby, 29-cm^{-1} phonons have been produced in the $2\bar{A}(^2E) \rightarrow \bar{E}(^2E)$ transition of Cr^{3+} [4]. Also in Al_2O_3, but doped with V^{4+}, direct conversion of 52.6-cm^{-1} and 28.8-cm^{-1} far-infrared radiation to phonons of that energy has been demonstrated [1,2,3]. For sufficiently narrow one-phonon transitions, strong resonant scattering of the phonons may occur, and many collisions are suffered by a phonon during its lifetime (for a review see, e.g., Ref.5 and references therein). In ruby, the phonon transport to the temperature bath is bottlenecked, and may involve spectral diffusion by inelastic scattering [4,6], spatial diffusion [6], or a combination of both [3], quasidiffusion [2], a spectral wipe-out from resonance by energy transfer [7] or by weak exchange interactions between Cr^{3+} ions [8]. Other systems in which the dynamics of resonant phonons have been examined include the Zeeman levels of $\bar{E}(^2E)$ of Cr^{3+} in Al_2O_3 in a magnetic field [9] and the $2\bar{A}(^2E)$ and $\bar{E}(^2E)$ levels of other Cr^{3+}-doped laser materials such as alexandrite [10] and YAG [11,12]. For the case of doping with rare-earth ions in crystals such as CaF_2 and SrF_2 [13,14], and LaF_3 [15], anharmonic decay appears to limit the phonon lifetime and has been determined to go with the fifth power of the frequency, in accord with theoretical considerations [3,13]. This decay is exclusively due to longitudinal acoustic phonons. The transverse mode is infinitely stable [16], unless mode conversion occurs by scattering at impurities.

The strong resonant scattering and long lifetimes observed for high-frequency phonons in these systems, suggest the use of such one-phonon transitions for the production of intense monochromatic phonon beams by amplification. To achieve this, a population inversion must be maintained sufficiently long. There are, however, very few experiments yet. Stimulated emission of 24.7-cm^{-1} phonons has been demonstrated following pulsed pumping of V^{4+} in Al_2O_3 by far-infrared radiation at $52.6\ cm^{-1}$ [1]. Selective optical excitation of the $2\bar{A}(^2E)$ level of Cr^{3+} in Al_2O_3 at 1.6 K by a pulsed dye laser has been shown to result in a directional beam of 29-cm^{-1} phonons within a cone of ~ 6° around the c axis, possibly due to phonon focusing effects [17]. A three order of magnitude speed up of spin-phonon relaxation due to stimulated phonon emission has been observed also in ruby but in a strong magnetic field after population inversion of the Zeeman transition of $\bar{E}(^2E)$ by pulsed dye-laser excitation [18]. In the latter experiment, the high phonon-occupation numbers reached appear to lead to nearly collinear three-phonon processes. The generation of monochromatic and possibly coherent beams of high-frequency phonons by stimulated or perhaps superradiant [9] emission remains a challenge.

The spectral width of the phonon packets produced in a direct process is largely determined by the intrinsic width of the electronic transition involved. Here, we present preliminary results of two methods applied to alexandrite at 1.5 K to determine this spectral width [19]. The first measurement exploits the $40\ cm^{-1}$ phonons produced in the transition $2\bar{A}(^2E) \rightarrow \bar{E}(^2E)$ as a probe [4,8,10]. The second scheme takes advantage of high-resolution optical spectroscopy [20,21]. In Fig.1 the R_2-fluorescent intensity, which emanates from $2\bar{A}(^2E)$, as well as the lifetime of $2\bar{A}(^2E)$ are plotted as a function of a magnetic field. These parameters are a direct measure for the degree of bottlenecking of the $40\ cm^{-1}$ phonons. The magnetic field lifts the degeneracy of the Kramers levels $2\bar{A}(^2E)$ and $\bar{E}(^2E)$ [4,10], and at sufficiently high fields, four 40-cm^{-1} phonon packets are produced instead of one, each of which only communicates with exclusively its own transition. This results in a reduction of the bottlenecking, as observed through the decrease of the R_2-fluorescent intensity and the lifetime of $2\bar{A}(^2E)$. Knowing the relevant g-factors, the spectral width of the phonons can be estimated. Here, we find for alexan-

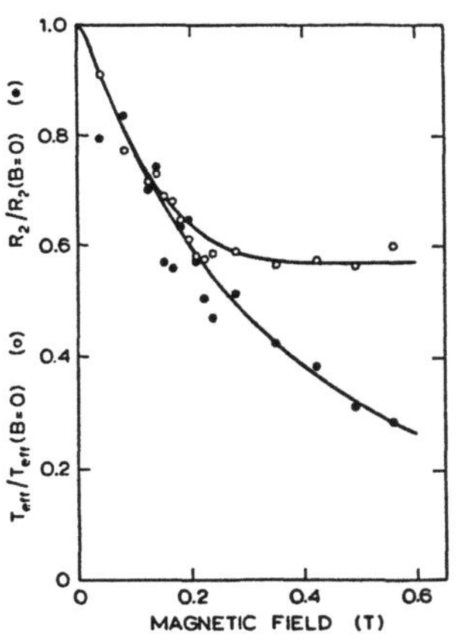

Fig. 1. Dependence on the magnetic field of the $2\bar{A}(^2E)$ relaxation time and the R_2 intensity in alexandrite at 1.5 K. Field is parallel to the c-axis. From the relevant g-factors, the width of the bottlenecked $40\ cm^{-1}$ phonons is deduced to be $0.3\ cm^{-1}$.

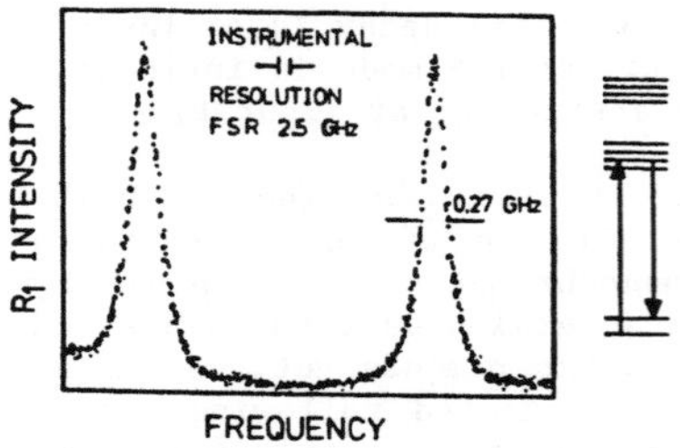

Fig. 2a. Fluorescent spectrum of the transition $\bar{E}(^2E)$ → $^4A_2(\pm1/2)$ in alexandrite at 1.5 K upon selective excitation of $^4A_2(\pm3/2)$ → $\bar{E}(^2E)$ with a single-frequency laser. Free-spectral range of the Fabry-Perot interferometer is 2.5 GHz and its finesse ~ 30.

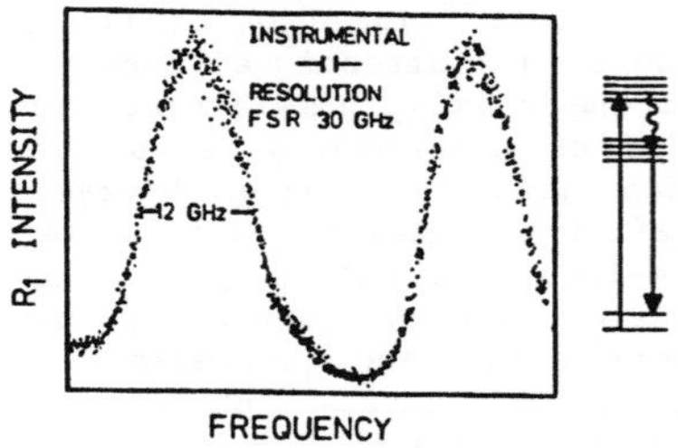

Fig. 2b. Same as Fig. 1a, but now upon selective excitation of $^4A_2(\pm3/2)$ → $2\bar{A}(^2E)$ and with a F.S.R. = 30 GHz.

drite 0.3 cm^{-1}. In order to measure the width of the $2\bar{A}(^2E)$ → $\bar{E}(^2E)$ directly by high-resolution optical spectroscopy in the same sample at 1.5 K, we have selectively excited the $\bar{E}(^2E)$ level from the $^4A_2$ ground-state with a 1-MHz-linewidth single-frequency laser. Consequently, we only excite the line over a frequency bandwidth corresponding to the homogeneous linewidth of the individual Cr^{3+} ions. The homogeneous line-width is accessible for measurements through the observation of the narrowed fluorescence from $\bar{E}(^2E)$. This is accomplished by means of a Fabry-Perot interferometer. From Fig.2a a value of $\Delta\nu$ = 0.27 GHz is found. This value is comparable to the one measured in ruby, and may be due to interactions of Cr^{3+} with the surrounding Al nuclei [20]. Upon exciting $2\bar{A}(^2E)$, however, we find a much broader line of 12 GHz or 0.36 cm^{-1} (Fig. 2b). This additional broadening corresponds to the inhomogeneous broadening of the $2\bar{A}(^2E)$ → $\bar{E}(^2E)$ transition. We have verified this in an experiment at 10 K, in which we have observed a much narrower fluorescence from $2\bar{A}(^2E)$ under selective excitation of that level. The conclusion is that the values found by the phonon spectroscopy method coincide with those obtained by high-resolution optical spectroscopy. The study of spectral phonon dynamics could take advantage of high-resolution optical techniques to generate and detect phonons with optical precision.

3. Broadband phonon generation

The very fast nonradiative decay following optical excitation of the broad vibronic levels of impurity ions leads to the injection of many phonons in the crystal. In ruby, for example, a direct measurement of the relaxation $^4T_2$ to ^2E, about 5000 cm^{-1} lower in energy, yields an upper limit of only 7 ps [22]. This means that very locally and in an extremely short time-span, very-high frequency phonon modes will be populated to a significant degree. Of these, the TA modes may be long lived and detectable at least at low temperatures and in absence of mode conversion [16]. This has been demonstrated in TlCl by probing the resonant far-infrared absorption involved in the transition TA + hν → LA, following optical excitation of impurities [3]. On the other hand, these optically produced high-frequency

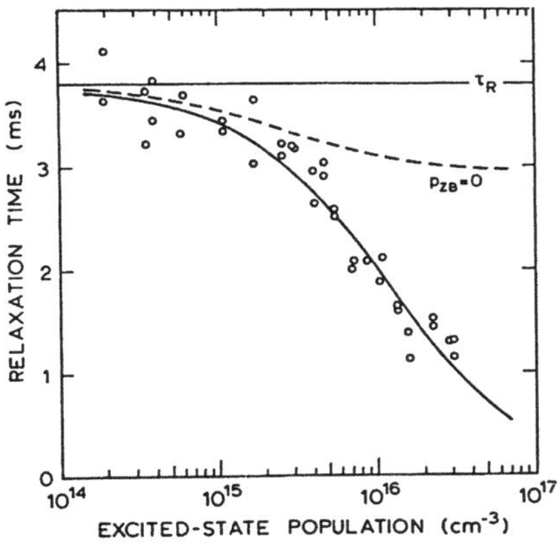

Fig. 3. Effective relaxation T_{eff} in 130 at ppm ruby vs the excited-state population at 1.5 K, as measured through the recovery of the fluorescent intensity of the upper Zeeman level of $\bar{E}(^2E)$ to $^4A_2$ following removal of microwave excitation. Active zone is a cylinder of diameter 1.5 mm and length of 4 mm, pumped with an Argon laser. Magnetic field is parallel to the c-axis and splits the $\bar{E}(^2E)$ states by 32.7 GHz (see Ref. 24).

TA phonons can induce Raman processes between metastable levels. This has been suggested as the explanation of the discrepancy between time-resolved and stationary experiments on the bottlenecking of 29-cm^{-1} phonons in ruby [5] and worked out in more detail in a set of quasistationary experiments confirming the longevity of near-zone-boundary phonons [23]. Also in YAG doped with Cr^{3+}, such Raman processes are observed to connect the $\bar{E}(^2E)$ and $2\bar{A}(^2E)$ levels [11,12]. In another type of measurement in ruby at 1.5 K which is shown in Fig.3, the spin-lattice-relaxation time of the Kramers doublet $\bar{E}(^2E)$ is given as a function of the metastable population N*, maintained by stationary optical pumping in $^4T_2$ [24]. The observed acceleration by a factor of over 3 relative to the radiative lifetime τ_R, is attributed to two-phonon processes induced by large-wave-vector TA phonons optically generated in the decay $^4T_2 \rightarrow {}^2E$, and allows for a rough estimate of the occupation numbers and lifetimes of these phonons [24]. We approximate the induced Raman rate by $2p_{ZB}/T_{RAM}$, in which p_{ZB} is the effective occupation number of the TA phonons present in the crystal and T_{RAM} represents the Raman-relaxation time for the case of phonon-occupation numbers p equal to unity. Here we have neglected the wave-vector dependence of p with the justification that the matrix elements strongly favor the large wave-vectors [23,24]. T_{RAM} is calculated adopting a realistic density of states of TA phonons deduced from inelastic-neutron-scattering data in Al_2O_3 and amounts to about 0.5×10^{-10} s [24]. In this calculation, the contributions are included of both the Raman processes connecting the Zeeman levels directly and those via $2\bar{A}(^2E)$. Our measurement displayed in Fig.3 shows a speedup from 3.8 ms to 1 ms yielding $p_{ZB} \sim 2 \times 10^{-8}$. From this value we can estimate the effective lifetime, τ, by examination of the feeding into the phonon modes. Due to the local character of the phonon generation in the decay from $^4T_2 \rightarrow E$, 5500 cm^{-1} below, we may adopt a stationary feeding per mode, ϕ, that is independent of the wave vector and, of course, linear in the optical pump intensity. At $N^* = 4 \times 10^{16}$ cm^{-3} a value for ϕ is computed of 8×10^{-3} s^{-1} [23,24]. Equating $p_{ZB} = \phi.\tau$, we thus find $\tau \sim 2 \times 10^{-6}$ s. This is a remarkably long lifetime, but, of course, subject to some uncertainty because of the approximations made in the calculation of T_{RAM} and ϕ. These results nevertheless suggest only weak elastic scattering and negligible mode conversion of large-wave-vector TA phonons. This corroborates another finding in ruby, that a roughly linear dependence ot p_{ZB} has been observed on the dimensions of the excited zone [23]. This fact points to ballistic flow or at least only small-angle scattering of the phonons. In Section 5 we will present experiments carried out with a point-contact phonon generator, that shows ballistic flow of very- high-frequency phonons over distances as large as 0.5 mm.

4. Thin film heaters

A major break-through in the study of the transport of high-frequency phonons was achieved in 1964, when the first heat-pulse experiment, in quartz and sapphire were performed [25]. In this type of experiments the heat pulse is usually detected with a bolometer, which is essentially not frequency selective. The first measurements with frequency-selective heat-pulse detection have been carried out in 1971 in ruby [26]. Here, 29-cm^{-1} phonons present in the heat pulse induce transitions from $\bar{E}(^2E) \rightarrow 2\bar{A}(^2E)$ upon arrival at optically-excited Cr^{3+} ions and an enhanced fluorescence from the $2\bar{A}(^2E)$ level can be recorded. Using this technique, detailed information on the dynamics of 29-cm^{-1} phonons in ruby has been obtained [5,6] such as trapping time, anisotropy of the electron-phonon interaction, phonon focusing and phonon-phonon interactions, and Raman processes connecting $\bar{E}(^2E)$ and $2\bar{A}(^2E)$ [27]. Vibronic-sideband spectroscopy [1,2] and stress tuning [13,14] of electronic transitions of rare-earth ions in SrF_2 and CaF_2 allowed for a time- and spectrally-resolved study of the dynamics of phonons up to frequencies of more than 3 THz. A fourth-power dependence of the elastic phonon scattering in $CaF_2:Eu^{2+}$ on frequency has been observed at least upto about 1 THz, with absolute values in good agreement with point-defect and isotope scattering. At lower frequencies ballistic transport over several mm accompanied with phonon focusing has been found [28]. Evidence for quasidiffusive transport of phonons with a broad spectral distribution has been obtained in $SrF_2:Eu^{2+}$ [2]. In Cr^{3+}-doped YAG, a thermal mode of propagation of heat pulses at high power densities produced turned out to satisfactorily describe the experimental facts including a linear dependence of the signal on distance [29]. The latter experiments point to a predominance of anharmonic processes in YAG over elastic scattering of the phonons in the temperature interval from 2-20 K. In solid solutions of aluminum-based garnets, however, the pulse propagation has been shown to be predominantly quasidiffusive [30].

5. Point-contact phonon generator

A very small and efficient device to inject very high-frequency phonons into a crystal at low temperatures is a metallic point contact [31,32]. These contacts are made by pressing a metal whisker on a thin (~ 500 Å) metal film evaporated on the surface of the crystal. The linear dimensions of the contact thus obtained can be much smaller than the mean free path of electrons in the metal. The application of a voltage V across the contact thus allows for a free acceleration and a gain in energy amounting to eV of the electrons. This excess energy is readily transferred to the phonons, resulting in a non-thermal mode population upto an energy equal to eV, but, of course, limited to the maximum phonon energy in the metal [31]. The edge of the emitted phonon spectrum at eV, in principle allows for quasimonochromatic very-high-frequency-phonon spectroscopy. This promising feature of the device has, however, not found experimental confirmation yet. The first experiments on high-frequency phonon injection in ruby using a Au-Au point contact, have been described very satisfactorily when complete thermalization of the phonons was assumed [31,32]. Here we give a very rough estimate of the dimension of the hot surface around the point contact. We resort to our observation in ruby at

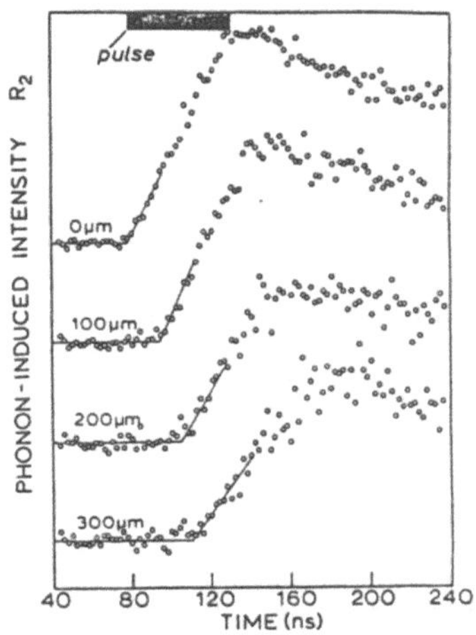

Fig. 4. Optical detection of phonons generated by a Au-Au point contact vs time in ruby at 1.5 K at various distances from the contact. Volume of detection $(50\ \mu m)^3$ is prepared by a 200 mW Argon laser modulated with 1 kHz. Point contact is fired every 1.2 µs in a pulse of 50 ns and a power of 10 mW every 1.2 µs during the light-off period. Slow decay (~ 1 µs) is due to bottlenecking of 29 cm^{-1} phonons.

1.5 K [33] of softening [34] of the Au-Au point contact at a dissipated power of about 10 mW. At the softening temperature (for Au: 400 K) the contact area increases and consequently a decrease of the resistance is observed. From temperature measurements of thermal phonon generators on the other hand, we infer for the configuration of gold on sapphire a radiation temperature of 400 K at a power density of $10^4\ W/mm^2$ [35]. This means that the typical diameter of the phonon radiator is about 1 µm. In order to examine to what extent the phonons penetrate into the crystal at 1.5 K, we have plotted in Fig.4 the R_2 fluorescence from a detection volume of about $(50\ \mu m)^3$ positioned at several distances from the point contact. Ballistic phonon flow is observed with a velocity very close to the transverse acoustic velocity. The leading edge corresponds to the length of the electrical pulse applied to the point contact (50 ns) and consequently indicates ballistic motion of the phonons through the excited zone as well. This observation excludes the predominance of effects of the resonant 29 cm^{-1} phonons, because they suffer strong diffusive motion in the excited zone. The signal of Fig.4 therefore must be due to Raman scattering of high-frequency phonons. In this connection it is noted that the device has a very short elecrical response (< 1 ns). This enables, together with its small size, an examination of ballistic-phonon transport over very small distances. In order to check the consistency, we perform a rough calculation of the temperature reached at the point contact from the measured signal height. The typical signal height reached after a ballistic phonon pulse of 50 ns duration at a power of ~ 10 mW at a distance between generator and detector of about 50 µm corresponds to roughly exciting up to 0.5% of Cr^{3+} from $\bar{E}(^2E)$ to $2\bar{A}(^2E)$ with a rate of $10^5\ s^{-1}$. Identifying this rate with p_{ZB}/T_{RAM}, we find $p_{ZB} \sim 10^{-5}$. ($T_{RAM} = 0.1$ ns for Raman processes connecting $\bar{E}(^2E)$ and $2\bar{A}(^2E)$ [23]). This would mean that an effective population number of large-wave vector phonons of about 10^{-5} is arriving at the detector about 50 µm away from the generator. This extrapolates to a mode population at the generator of approximately a factor of 2×10^4 higher, when due account is taken of the ratio of the phonon flux at the detector and the generator. For an average energy of TA phonons in ruby of, say, 200 cm^{-1} this value corresponds to an effective temperature of 160 K. This value is, of course, subject to uncertainties in the calculation of T_{RAM} and in the experimental parameters, but the result is in fair agreement with the above finding based on the observation of softening of the point contact. For a more detailed analysis, we refer to Ref. 31 in which the power dependence is examined of the phonon-induced fluorescent intensity. We conclude that large-wave-vector TA phonons are efficiently produced by the point contact and may travel over macroscopic distances. The latter result is in concord with the observations described in section 3. The thin-film-point-contact phonon generator is a very promising device for the injection of very-high-frequency phonons in an otherwise cold crystal by virtue of its small size, its fast electrical response, and its efficiency in reaching high temperatures with very small pulse powers only.

6. Conclusions

Optical pumping of impurity centers dispersed in a crystal is an efficient method for the generation of high-frequency acoustic phonons. Due to the local character of the generation process, phonons throughout the entire Brillouin zone can be produced. The intrinsic width of direct phonon transitions at impurities can be very small, allowing for the production of monochromatic phonons. The fast nonradiative decay of the broad vibronic levels of the impurities allows for the production of broad-band acoustic phonons of wave vectors up to the zone boundary. Electronic Raman transitions induced by such phonons can be measured, among others, by excited-state ESR experiments. In ruby a longevity and ballistic flow, or at least small-angle scattering, is found for large-wave-vector TA phonons. These phonons can also be produced by a thin-film-point-contact phonon generator. High-frequency phonons injected by a point contact into a ruby crystal appear to travel ballistically to the detector located up to 0.5 mm away. The device apparently produces a hot area of ~ 1 μm diameter with a temperature of ~ 400 K already at a power level of ~ 10 mW. The application of a point-contact phonon generator to crystals other than ruby seems interesting to gain new information on the dynamics of high-frequency phonons.

Acknowledgement

It is a pleasure to acknowledge discussions with H.W. de Wijn, M.J. van Dort and K.Z. Troost.

References

1. W.E. Bron: Rep. Progr. Phys. 43, 301 (1980)
2. W.E. Bron: Nonequilibrium Phonon Dynamics, NATO ASI B124 p. 1 (1985)
3. K.F. Renk: International School of Physics "Enrico Fermi", (1985)
4. J.I. Dijkhuis, A. van der Pol, H.W. de Wijn: Phys. Rev. Lett. 37, 1554 (1976)
5. A.A. Kaplyanskii, S.A. Basun, V.L. Shekhtman: Journ. de Phys. C6 439 (1981)
6. S.A. Basun, A.A. Kaplyanskii, V.L. Shekhtman: Sov. Phys. JETP 55 1119 (1982)
7. R.S. Meltzer, J.E. Rives, W.C. Egbert: Phys. Rev. B25 3026 (1982)
8. R.J.G. Goossens, J.I. Dijkhuis, H.W. de Wijn: Phys.Rev.B32 7065 (1985)
9. J.I. Dijkhuis, K. Huibregtse, H.W. de Wijn: Phys. Rev. B20 1835 (1979)
10. R.J.G. Goossens, J.I. Dijkhuis, H.W. de Wijn: Phonon Scattering in Condensed Matter (Springer) p. 112 (1984)
11. A.P. Abramov, I.N. Abramova, I.Ya. Gerlovin, K. Razumora: Sov. Phys. Solid State 27 10 (1985)

12. R.J.G. Goossens, E.E. Koldenhof, J.I. Dijkhuis, H.W. de Wijn: Journ. de Phys. C7 259 (1985)
13. R. Baumgartner,M. Engelhardt, K.F. Renk: Phys.Rev.Lett.47 1403 (1981)
14. A.V. Akimov, A.A. Kaplyanskii, A.L. Syrkin: JETP Lett. 33 393 (1981)
15. R.S. Meltzer, J.E. Rives: Phys. Rev. B28 4786 (1983)
16. M. Lax, P. Hu, V. Narayanamurti: Phys. Rev. B23 3095 (1981)
17. P. Hu: Phys. Rev. Lett. 44 417 (1980)
18. J.G.M. van Miltenburg, G.J. Jongerden, J.I. Dijkhuis, H.W. de Wijn: Phonon Scattering in Condensed Matter (Springer) p. 130 (1984); and to be published
19. M.J. van Dort, J.I. Dijkhuis, H.W. de Wijn, to be published
20. A. Szabo: Phys. Rev. Lett. 25 924 (1970)
21. D.J. Sox, S. Majetich, J.E. Rives, R.S. Meltzer: Journ. de Phys. C7 493 (1985)
22. S.K. Gayem, W.B. Wang, V. Petricevic, R. Dorsinville, R.R. Alfano, Appl. Phys. Lett. 47 455 (1985)
23. R.J.G. Goossens, J.I. Dijkhuis, H.W. de Wijn: Phys.Rev.B32 5163 (1985)
24. J.G.M. van Miltenburg, J.I. Dijkhuis, H.W. de Wijn: Phonon Scattering in Condensed Matter (Springer) p. 119 (1984); and to be published
25. R.J. von Gutfeld: Physical Acoustics 5 (Academic), p. 233 (1968)
26. K.F. Renk, J. Deisenhofer: Phys. Rev. Lett. 26, 764 (1971)
27. A.P. Abramov, I.N. Abramova, I. Ya. Gerlovin, I.K. Razumova, Sov. Phys. JETP 52, 659 (1980)
28. A.P. Abramov, I.N. Abramova, I. Ya. Gerlovin, I.K. Razumova: Sov. Phys. Solid State 22 556 1980; ibid 24 37 (1982); Opt. Spectr. USSR 51 6 (1981).
29. A.P. Abramov, I.N. Abramova, I. Ya. Gerlovin, I.K. Razumova, Sov. Phys. Solid State 22 1354 (1980)
30. S.N. Ivanov, E.N. Khazanov: Sov. Phys. JETP 61 172 (1985)
31. R.J.G. Goossens, J.I. Dijkhuis, H.W. de Wijn, A.G.M. Jansen, P. Wyder: Phonon Scattering in Condensed Matter (Springer) p. 46 (1984); Physica 127B 422 (1984)
32. R.J.G. Goossens, J.I. Dijkhuis, H.W. de Wijn: Journ. Lumin. 34 19 (1985)
33. K.Z. Troost, J.I. Dijkhuis, H.W. de Wijn: to be published
34. R. Holm: Electric Contacts (Gebers, Stockholm) (1946)
35. P. Herth, O. Weis: Z. angew. Physik 29 101 (1970)

Nonequilibrium Electron-Phonon Dynamics†

W.E. Bron

Department of Physics*, Indiana University,
Bloomington, IN 47405, USA

1. INTRODUCTION

I have been requested to present a brief review of recent advances in our knowledge of the dynamics of electrons, phonons and their mutual interactions. To keep within the time and space restrictions of the review, I have chosen to illustrate progress in this field by reviewing three recent investigations carried out on polar semiconductors and which involve measurements, for the most part, directly in the time domain.

A number of sign-posts exist in our knowledge of these dynamics. For example, it has been known for some time that highly excited free carriers, i.e., electron-hole pairs, in polar semiconductors decay into lower energy states via the emission of low wavevector, longitudinal optical (LO) phonons primarily through the so-called "Frohlich Interaction" [1]. The decay proceeds through a cascade of carrier-phonon interactions terminating in the creation of an electron-hole plasma (gas or liquid) at some minimum of the conduction band [2]. Fig. 1 is an illustration of the excitation and decay in a hypothetical direct gap semiconductor with simple parabolic bands.

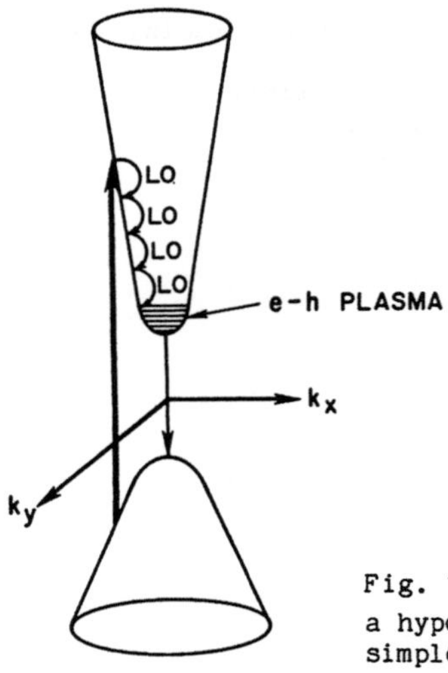

Fig. 1 Excitation and decay processes for a hypothetical semiconductor with a simple band structure

†Supported through ARO DAAG 29-83-K-0091

*Address after July 1, 1986 is Department of Physics, University of California, Irvine, CA 92717.

There are a number of auxiliary features to note in connection with the dynamics of the decay process. For typical polar semiconductors the electron effective mass is approximately $m^* \approx 0.25\ m_e$ [3]. This means that the curvature of a simple parabolic conduction band is such that the LO phonon which can interact with an electron in the cascade must possess a wavevector, $\vec{q}$, of the order of $10^5\ cm^{-1}$ and that this value needs to change as the cascade proceeds. This conclusion holds true for electrons with energies of up to 20 LO phonon energies above the band minimum. Accordingly, measurements of the decay by time-resolved incoherent Raman scattering, or by two-, or three-wave mixing experiments, such as the CARS method, all need to be performed in the backward scattering geometry if phonons with such relatively high wavevectors are to be detected. The effects of more complicated band structures has also been considered [2].

It has also been predicted [1] that the average time interval between successive steps in the cascade is of the order of 100 to 200 femtoseconds which is shorter than the carrier-carrier scattering time in semiconductors with carrier concentrations of less than $10^{17}\ cm^{-3}$ [1].

The interaction of phonons with the electron-hole plasma, on the other hand, is not as well understood. It has been found [4] experimentally in GaP that the LO phonon frequency increases as the plasma concentration exceeds about $0.5 \times 10^{18}\ cm^{-3}$, but that little effect is observed below that concentration and that, curiously, no change in the phonon lifetime is observed up to concentrations of about $5 \times 10^{18}\ cm^{-3}$. The observed frequency shifts have been attributed to coupled plasmon-phonon modes.

More recently it has been discovered [5] that in polar semiconductors, such as GaP, the LO phonons themselves decay into acoustic phonons in times of the order of picosecond, and that the dominant decay mechanism, at least at low temperatures, are three-phonon anharmonic processes. Finally, a recent investigation [6] on systems which contain an electron-hole plasma with concentration less than $0.5 \times 10^{18}\ cm^{-3}$, indicates an additional effect is present, namely, that the strength of nonlinear phonon generation, such as coherent Raman excitation (CRE), is effected by the presence of the plasma.

It should already be apparent to the reader that the dynamics of electrons and phonons in these materials exhibits many interesting and multifaceted properties. In the following sections, I shall address a number of these facets in more detail.

2. EXPERIMENTAL RESULTS AND DISCUSSION

2.1 Free-Carrier Lifetime and Phonon Decay

The first experiment which I wish to cite was performed in GaAs by VON DER LINDE, et al. [7] who used time-resolved Raman scattering to study the dynamics of nonequilibrium incoherent optical phonons and their interaction with photoexcited hot electrons and holes.

In the experiment, photoexcited electron-hole pairs (e-h pairs) were generated with light from pulses from dye-lasers with autocorrelation duration of 2.5 ps. The electron-hole pair excess energy was 17 times the LO-phonon energy of 36.5 meV. Subsequent decay of the e-h pairs, as illustrated in Fig. 1, was monitored by a probe pulse (of the same laser frequency) which preceded or followed the pump pulse. The Raman (backward) scattering off of the LO phonon distribution, as measured by time delayed

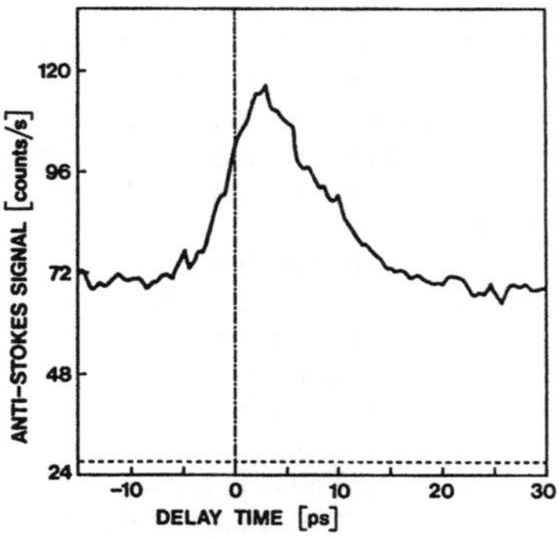

Fig. 2 Time-resolved anti-Stokes Raman scattering intensity

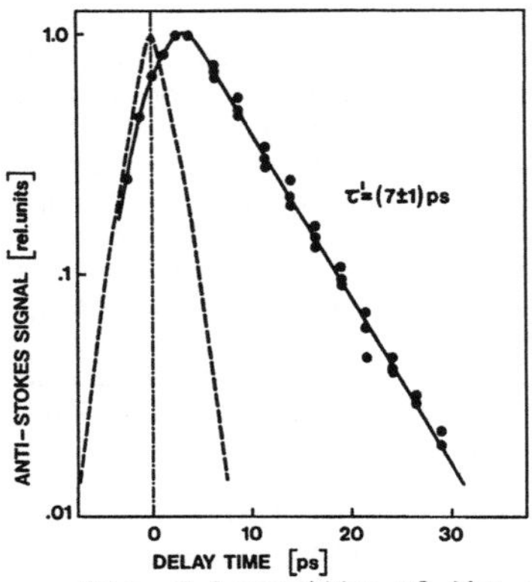

Fig. 3 Logarithm of the anti-Stokes Raman scattering intensity and laser autocorrelation signal.

probe pulse is, accordingly, a function of the temporal evolution of the LO phonons (with $\vec{q} \sim 10^5$ cm^{-1}). The experimentally measured anti-Stokes Raman signal is illustrated in Fig. 2. The zero of the delay time is fixed at the maximum of the pump pulse. The time-independent background of approximately 70 counts per second is the result of carrier excitation by the probe pulse, and the weak signal at about 28 counts per second results from residual stray laser light and detector dark current.

The rise of the signal above ~ 70 counts per second indicates the growth of the LO phonon through the decay of the photoexcited electron-hole pairs. The rise time of the signal of approximately 6 ps is related to the decay time of the e-h pairs into the LO phonons. But, since the temporal resolution in this experiment was only 2.5 ps, no further attempt was made to use this information to determine the electron-phonon interaction time.

The decrease in the anti-Stokes signal is a function of the decay of the LO phonon into acoustic phonons. As shown in Fig. 3, the decay follows an exponential function, exp $(-\Delta t/\tau')$, with $\tau' = 7 \pm 1$ ps when the sample is held at 77°K. The solid curve in Fig. 3 is a convolution of the temporal evolution of the number of phonons, and the temporal evolution of the probe pulse, whereas the dashed curve is the laser autocorrelation signal. Thus, in this case, the phonon decay is easily resolved.

The experiment, described above, was repeated by KASH, et al. [8] through incoherent Raman scattering measurements in the spectral domain, but carried out with a probe pulse after various delays relative to the pump pulse. In this experiment the laser pulse duration was only 0.6 ps (600 fs). The pump and probe pulses were polarized perpendicularly to each other as was the LO phonon Raman (backward) scattering excited by the pump and the probe pulses. However, the pump pulse was observed to also produce a spectrally broad, but unpolarized emission band. This emission, plus the signal from stray laser light and detector dark current, was compensated for by taking difference measurements between spectra taken with the probe delay time, Δt, fixed at various values less that obtained when $\Delta t \ll 0$ ps.

The results are illustrated in Fig. 4 which shows the intensity of the anti-Stokes Raman signal as a function of Raman shift for various values of

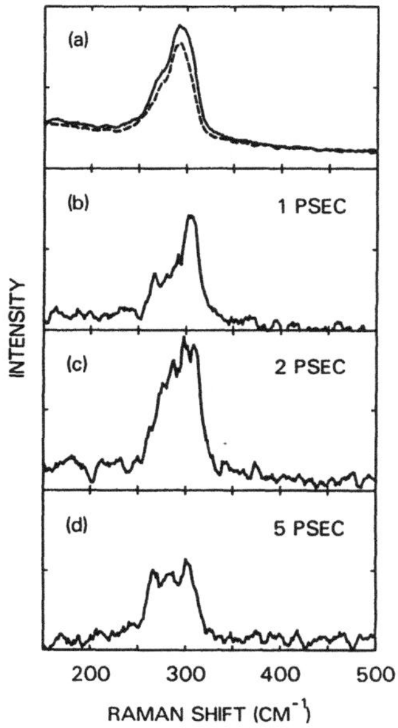

Fig. 4 Time-resolved anti-Stokes Raman difference signal for Δt = 1, 2 and 5 ps.

Δt for a molecular-beam epitaxically grown GaAs sample held at room temperature. In Fig. 4a the solid line is observed for Δt = 1 ps, whereas the dashed line is observed for Δt = -20 ps. Similar differences for Δt = 1, 2 and 5 ps are indicated in Fig. 4b, c, d. Note that the peak and integrated intensities of the anti-Stokes Raman line increases and then decreases in a manner similar to that shown in Figs. 2 and 3. However, due to the better temporal resolution obtained by KASH, et al. compared to VON DER LINDE, et al. it becomes possible to resolve the electron cascade in more detail. It is found that the total cascade consists of 12 cascade steps each of approximately 165 fs long; in very good agreement with the predictions [1]. The phonon decay time near 80K was found to be 7.5 ps, in good agreement with the earlier results of VON DER LINDE, et al. [7].

The observed decay time, τ, is related to the fundamental phonon lifetime only in a remote way. As should be apparent from Fig. 1, the e-h pair decay produces an ensemble of optical phonons with a spread of $\vec{q}$. The decay of such an ensemble may well differ from the decay of a well-defined coherent phonon state, as recently reported by KUHL and BRON [5,9].

2.2 Effects of Electron-Hole Plasma

We now turn our attention to the interaction of phonons with an electron-hole plasma, i.e. to the situation which prevails once the "hot" electrons have decayed to some minimum of the band structure. The lifetime of e-h pairs against recombination varies from many nanoseconds to microseconds, thus the dynamics of the interaction should easily be observable with modern picosecond laser spectroscopy.

We have recently reported [10] detailed CARS measurements of the dephasing time of coherent optical phonons in GaP containing very low carrier concentrations. The observed phonon dephasing time is attributable solely to decay due to anharmonic phonon-phonon interactions. The interaction, at least in GaP, results only in the creation of longitudinal acoustic (LA) phonons of half the energy of the LO phonons. These experiments have been repeated with the addition of an electron-hole plasma produced by absorption of additional laser light in the same manner as illustrated in Fig. 1. Since the effective wavevector of the LO phonon produced by our CARS technique is $\geq 3 \times 10^3$ cm^{-1}, and since the temporal resolution of the CARS measurements was limited to ~ 1.3 ps, no observation could be made on the interaction during the free carrier cascade. Similarly, since the concentration of the e-h plasma produced by the cascade was less than ~ 10^{17} cm^{-3}, no shift in the LO phonon frequency or change in dephasing time could be expected. None was in fact observed. However, we were surprised to find that a strong (approximately 10-20%) decrease was observed in the magnitude of the CARS signal associated with the coherent phonon state. Although the plasma does not in this limit form a coupled state with the phonons, the results imply that it does apparently interfere with the production of the coherent phonons.

The last mentioned phenomenon has been recently investigated in more detail by RHEE and BRON [6]. In this experiment coherent phonon generation proceeds through nonlinear interaction in GaP of pulse trains emitted from two dye lasers driven simultaneously by 5 ns duration pulses from a N_2-laser. One dye laser is operated at a frequency ω_ℓ = 18,725 cm^{-1}, which together with another laser at ω_s = 18,322 cm^{-1}, makes up the difference frequency $\omega_{LO} = \omega_\ell - \omega_s$ = 403 cm^{-1} required to excite coherent LO phonons. The maximum energy per laser pulse is ~ 60 µJ resulting in an initial LO phonon concentration $n_{LO} \approx 5 \times 10^{16}$ cm^{-3}, located within the overlap volume of the two laser beams.

The laser operating at ω_ℓ also serves to excite the luminescence from the bound exciton in GaP [11] (zero phonon line at 18,688 cm^{-1}) with which the vibronic sidebands are associated. In Fig. 5 we present the observed dependence of the anti-Stokes vibronic sideband intensity, as measured at the LO and the LA peak frequencies, as a function of the power in the s-laser when the power of the ℓ-laser is fixed at 135 MW/cm^2.

It can be shown [10] that the intensity of the anti-Stokes vibronic sideband should be proportional to the product of the power of the two lasers (ℓ and s lasers) which drive the coherent phonon generation. The

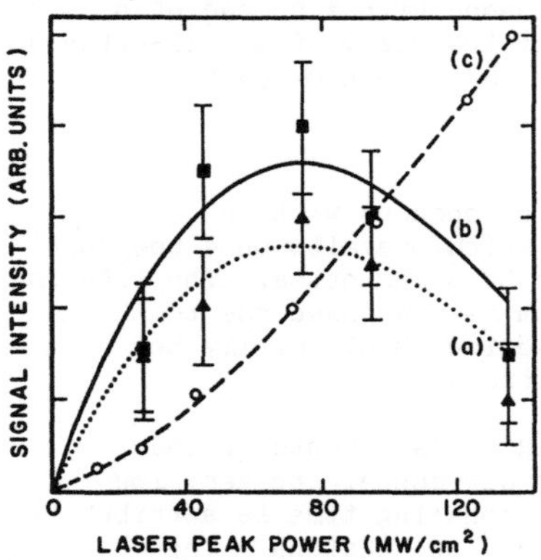

Fig. 5 Anti-Stokes vibronic intensity as a function of the s-laser peak power for (a) the LO peak, and (b) the LA peak frequency. (The data marked c is not discussed in this paper). Symbols are experimental points, solid and dotted lines are theoretical fits to the data. The ℓ-laser peak power is fixed at 135 MW/cm^3.

results shown in Fig. 5 obviously disagree with this prediction. The origin of the discrepancy arises from the presence of competing forces within the driving force, F_{LO}, through which the coherent LO phonons are generated. The pertinent expression for the driving force is

$$F_{LO} \propto [R + \frac{8\pi e^* |\chi^{(2)}|}{\varepsilon_\infty} \{1 - \frac{24\pi \chi^{(3)}}{0.017} (|\vec{E}_\ell|^2 + |\vec{E}_s|^2)\}]E_\ell E_s, \quad (1)$$

where we have made the simplification $\chi^{\to(3)} = \chi^{(3)}_{1111} = \chi^{(3)}_{1122} = \chi^{(3)}_{1221}$ for the elements of the third order nonlinear electronic susceptibility, $(c/\omega)^2[(2\vec{k}_\ell - \vec{k}_s)^2 - \varepsilon(\omega)] \approx -0.017$ with $\vec{k}_i$ being the corresponding wavevector of the laser light, $\varepsilon(\omega)$ is the dielectric constant at the frequency ω, C is the speed of light, and R is the effective Raman tensor element, $\chi^{(2)}$ the effective second order nonlinear electrons susceptibility tensor element, and E_ℓ and E_s are the field intensities of the ℓ- and s-lasers. In the event of high enough values of the laser power it becomes possible for the higher order terms to lead to a net decrease in the F_{LO}, which in turn results in a decrease in the production of coherent LO phonons, and hence to a decrease in the anti-Stokes intensity.

The most significant contribution comes from $\chi^{(3)}$ which contains the response of bound electrons, $\chi_b^{(3)}$, and that from the carriers in the plasma, $\chi_f^{(3)}$. The major contribution comes from terms in $\chi_f^{(3)}$, i. e.,

$$\chi_f^{(3)} \propto N^*(\sigma_\ell|E_\ell|^2 + \sigma_s|E_s|^2), \quad (2)$$

in which N* is the concentration of bound excitons which is itself proportional to the laser power, and σ_ℓ and σ_s are, respectively, the cross section for the excitation of a bound exciton to form a carrier through a further absorption of ℓ- or s-laser light. Thus, the nonlinearity in the last term of Eq. 1 (produced by the nonlinearity as expressed in Eq. 2) can account for the observed variation in the anti-Stokes sideband intensity, as is shown by the calculated dependence on the s-laser power indicated by the solid and dotted lines of Fig. 5.

Thus, although the effect of an electron-hole plasma on phonon lifetimes and frequency can be observed only at high plasma densities, the interference of lower density plasma with the formation of optical phonons can be readily observed and explained.

3. Conclusion

Investigations on the dynamics of electronic carriers and of phonon, plus their interaction, has received a recent revival with the advent of pico- and femto-second lasers. A number of facets of these dynamics is now clear. In polar semiconductors, hot carrier decay proceeds via a cascade of emission of LO phonons with each step of the cascade lasting of the order of 150 fs. The LO phonons produced possess wavevectors of the order of 10^5 cm^{-1} and decay in GaAs at 77K in about 7 ps [7,8] and in GaP at 5K in about 26 ps [9]. The interaction of an electron-hole plasma on phonon frequencies is not readily observed unless the plasma concentration exceeds 10^{17} cm^{-3}, but the presence of an even lower concentration plasma has striking effects on coherent phonon generation when the excitation laser power exceeds the order of 100 MW/cm^2.

1. E.M. Conwell and M.O. Vassel: IEEE Trans. Electron. Devices 13, 22 (1966).

2. C.L. Collins and P.Y. Yu: Phys. Rev. B30, 4501 (1984).
3. E.M. Conwell: High Field Transport in Semiconductors (Springer, Berlin, Heidelberg 1967).
4. J.E. Kardontchik and E. Cohen: Phys. Rev. Lett. 42, 669 (1979).
5. For a review, see W.E. Bron: in Nonequilibrium Phonon Dynamics, ed. by W.E. Bron (Plenum, New York and London, 1985).
6. B.K. Rhee and W.E. Bron: to be submitted for publication elsewhere.
7. D. von der Linde, J. Kuhl and H. Klingenberg: Phys. Rev. Lett. 44, 1505 (1980).
8. J.A. Kash, J.C. Tsang and J.M. Hvam: Phys. Rev. Lett. 54, 2151 (1985).
9. J. Kuhl and W.E. Bron: Sol. State Commun. 49, 935 (1984).
10. W.E. Bron, J. Kuhl and B.K. Rhee: Submitted to Phys. Rev. B.
11. D.G. Thomas, M. Gershenson, and J.J. Hopfield: Phys. Rev. 131, 2397 (1963).

Monte Carlo Calculations of Phonon Transport

M. Lax [a], *V. Narayanamurti, R.C. Fulton, and N. Holzwarth* [b]
AT & T Bell Laboratories, Murray Hill, NJ 07974, USA

1. Introduction : To understand recent experiments [1] on propagation of high frequency TA phonons in GaAs we need numerical/analytic procedures for handling transport across layers not much larger than the mean free path. When mode conversion: TA → LA (by isotope scattering) and down conversion: LA → TA + TA and LA → TA + LA (by anharmonic coupling) are included analytic solutions appear hopeless, and Monte Carlo calculations have been performed.

To check the Monte Carlo program, even crude analytic approximations are useful. To our surprise, a modified diffusion procedure has been found to work remarkably well when down-conversion is neglected, and the thickness is say 10 mean free paths. When down-conversion is present, we only attempt to predict the flux of detected phonons that have not been down converted. Even here, the modified diffusion procedure with loss works remarkably well. For shorter slabs, the finite propagation time across the slab must be considered, and this can be accomplished by replacing the diffusion equation equation by a telegrapher's equation.

2. Previous work: Improvements on previous work have been made (a) by using a sharper one-sided smoothing algorithm that prevents phonons from appearing to arrive faster than the speed of sound. (b) by taking a Monte Carlo step in time followed by a branch to a possible outcome, rather than vice-versa. (c) by using the Tamura [3] anharmonic coefficients based on experimental higher-order elastic coefficients rather than the much larger coefficient used by Guseinov and Levinson [4].

3. Test Cases and Diffusion Theory: The simplest test case is that in which TA phonons scatter only elastically without the possibility of conversion to LA phonons, and hence without the possibility of down-conversion. Unfortunately, even for this simple case, no analytical solution exists. However, a diffusion approximation based on the concept of extrapolation length [5], used in steady state neutron transport theory is found to work surprisingly well even for our pulse case. See Fig. 1 for the case $\sigma = 1/3$ for which isotope scattering is 1/3 of the theoretical result [2].

The extrapolation length is a concept obtained from the solution of the Milne problem [5] for light transport in a stellar atmosphere, treated as a semi-infinite medium. At distances far from the (plane) boundary, the solution of the steady state transport equation is found to approach asymptotically the solution of a diffusion equation for which the density vanishes not at the real boundary, but outside the medium a distance away of $z_0 = 0.7106\lambda$, the "extrapolation length", where λ is the mean free path. We have extended this boundary condition by applying it to both sides of a slab. If the slab extends from $z = 0$ to w the boundary conditions used are:

$$n(-z_0, t) = n(w + z_0, t) = 0 \tag{1}$$

Note that the condition (1) has been applied for all time to a time dependent problem, although z_0 was calculated from the steady state solution only.

The detected current is calculated using Fick's law at the real boundary:

$$J_z = -\left[D\partial n/\partial z \right] / 1.2306 \tag{2}$$

The factor 1.2306 is needed because it is not appropriate to assume that the diffusion result is valid close to, or at the surface. Since the usual Fick's law flux is too large by a factor 1.2306 compared to the exact solution of the Milne problem, we apply precisely the same reduction factor to the slab problem.

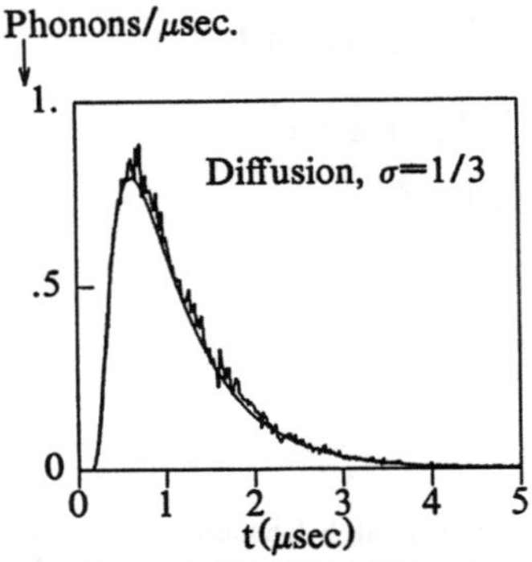

Fig. 1 Comparison between a Monte Carlo calculation containing only TA → TA scattering and a modified diffusion calculation for the case $\sigma = 1/3$ corresponding to a ratio of thickness to mean free path slightly less than ten. Vertical scale is normalized to one incident phonon.

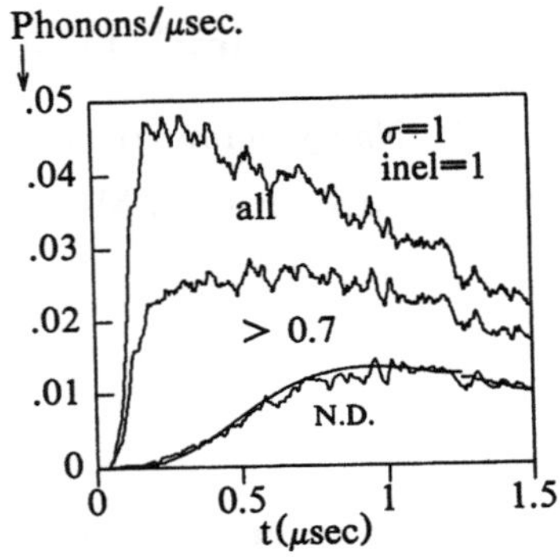

Fig. 2 A comparison between Monte Carlo calculations with down conversion with the standard ($\sigma = 1$ and inelastic $= 1$) rates of isotope scattering and down-conversion using Tamura's anharmonic coefficients. N.D. represents phonons that were never down-converted.

The diffusion approximation works when the thickness w is large compared to the mean free path λ. Moreover, we learn that the detected flux varies not as

$$\exp(-w/\lambda) \quad \text{but as} \quad (1.2306)^{-1}\,[\,1+(w/2z_0)\,] \tag{3}$$

The diffusion approximation breaks down when the thickness is not much larger than the mean free path. The most notable error, that the diffusion solution permits phonon arrival in zero time can be overcome by using the telegrapher's equation in place of the diffusion equation. When transverse phonons can scatter into longitudinal phonons (and vice-versa), we extend our theory by using coupled diffusion equations with a single extrapolation length. The latter is a weighted average of the corresponding extrapolation lengths for the transverse and longitudinal cases weighted by the fraction of time spent as transverse/longitudinal.

When down-conversion takes place, there is no available approximate theory. However, we can use our generalized diffusion theory by regarding down-conversion as absorption. The phonon flux at the detector then represents the fraction of phonons that get to the detector without down-conversion. This information can also be extracted from the Monte Carlo data. A sample comparison of our Monte Carlo results for the non-down-converted phonons is compared with the prediction of diffusion with absorption in Fig. 2.

4. Monte Carlo Results: Having verified that our Monte Carlo code agrees well with an approximate diffusion theory, even when down conversion is included in the form of absorption, we can now display one of our many Monte Carlo runs. Here we deal with the case in which the initial frequency is 1.5 Thz corresponding to TA phonons fairly close to the zone boundary in GaAs. We display the case in which $\sigma = 0.1$. The results are displayed for detectors at varying positions along the detector wall corresponding to distances of 0.2mm and 0.4mm. In Fig. 3 we display the pulse received by these detectors. The reason for our choice of short distances and weak isotope scattering is evident in this figure which shows considerable structure which disappears after many scatterings (as shown in Fig. 1). In Fig. 4 we display the arrival time of the edges found in the pulse. These edges appear at times dictated by the LA and TA speeds respectively. Thus the edges show a linear time versus distance behavior. The peak arrival times describe a more complex behavior because there are two peaks, and a shift occurs in which one dominates as one passes along the detector series.

5. Summary: In the absence of multiple down-conversion, the theory of Guseinov-Levinson [4] can not be applied to our experimental results obtained with a 0.7Thz cutoff. For moderate to large ratios of thickness to mean free paths we have obtained an approximate theory in good agreement with Monte Carlo results, providing us with a good understanding of phonon transport in thin slabs.

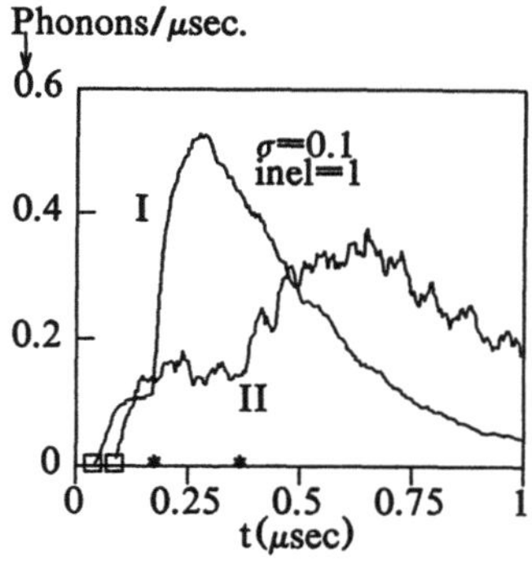

Fig. 3 A plot of the phonons detected at the first and second detectors against time. These curves show sharp kinks at the arrival times for LA and TA phonons. The second detector curve has been multiplied by 10 for visibility.

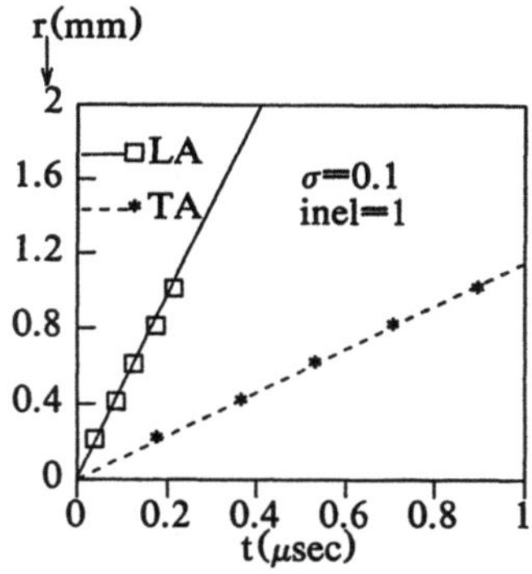

Fig. 4 A composite plot of arrival times versus distance to the five detectors. The straight lines correspond to the velocities 4.86mm/μsec and 1.15mm/μsec of LA and TA phonon frequencies of ω=1.5Thz

[a] Also Physics Dept. City College of New York, NY 10031. Work at City College supported in part by ARO and DOE.

[b] Present address: Physics Dept., Wake Forest Univ., Winston-Salem, NC 27109

[1] R.G. Ulbrich, V. Narayanamurti and M.A. Chin Phys. Rev. Lett. 45, 1432 (1980)

[2] M. Lax, V. Narayanamurti, R. Ulbrich and N. Holzwarth, in Proc. 4th Int'l Conference on Phonon Scattering in Condensed Matter, 1984, pp. 133-135.

[3] S. Tamura, Phys. Rev. 897 (1983)

[4] N.M. Guseinov and Y.B. Levinson, Sov. Phys. JETP 452,459 (1983)

[5] B. Davison and J.B. Sykes, *Neutron Transport Theory* Oxford, 1957

Dynamics of 38 cm^{-1} Phonons in Alexandrite

S. Majetich, J.E. Rives, and R.S. Meltzner

Department of Physics and Astronomy, University of Georgia, Athens, GA 30602, USA

The dynamics of 29 cm^{-1} phonons resonantly trapped by excited Cr^{3+} ions in ruby exhibits a dependence on N^*, the excited Cr^{3+} ion density, which is quite complicated. The phonons exhibit pure spatial diffusion at low N^* /1/, spectral diffusive behavior at intermediate N^*/2,3/, and at large N^*, spectral wipeout, whereby the phonons are spectrally shifted far into the wings of the resonance by either resonant phonon assisted energy transfer /4/ and/or inelastic Raman scattering from exchange coupled Cr^{3+} pairs /5/. Inelastic scattering from Cr^{2+} centers has also been suggested as a mechanism for the phonon loss /6/.

In order to aid the interpretation of phonon dynamics in ruby it is useful to examine other similar Cr^{3+} doped systems which might exhibit simpler behavior. GOOSSENS et al./7/ studied alexandrite under quasi-cw conditions, pumping the broad bands of Cr^{3+} and concluded that the phonon lifetime is proportional to $N^{*1/2}$ and that the phonon loss is controlled by diffusive transport rather than anharmonic decay. In order to better understand alexandrite we report on experiments with nanosecond pulsed excitation pumping directly the $^2E(2)$ state of the Cr^{3+} mirror sites/8/ over a wide range of N^* ($5\times10^{16}cm^{-3}<N^*<5\times10^{18}cm^{-3}$) and excited spot radii (0.25mm<2R<1.5mm).

Phonon generation is similar to that used in ruby/4/. A 5ns pulse of light of bandwidth 0.3cm^{-1} containing up to 4mJ of energy excites the $^2E(2)$ level in the E//b polarization. Rapid relaxation from the excited $^2E(2)$ level to the metastable $^2E(1)$ level (spontaneous relaxation time $T_1\approx10^{-10}$s) generates monochromatic 38cm^{-1} phonons. The phonon bandwidth is 13GHz, based on non-resonant fluorescence line-narrowing measurements/9/ yielding an approximate value of $\Sigma\simeq2\times10^{18}cm^{-3}$ for the number of phonon modes within the Cr^{3+} resonance. The phonons are detected from the $^2E(2)$ hot luminescence induced by the resonantly trapped phonons. To protect the photomultiplier tube from scattered laser light at the detection wavelength, the tube is gated off until 100ns after the excitation pulse.

Examples of the temporal profile of the $^2E(2)$ population, displayed as N_2/N^*, are shown in Fig. 1. The initial value of N_2 immediately after excitation scales with P^2 where P is the laser power, as expected for phonon-induced hot luminescence. The decay curves are non-exponential as observed previously in the quasi-cw experiments/7/. At the highest powers this may be due, in part, to large deviations from equilibrium, but at low powers this must be characteristic of the phonon loss. We consider only the initial decay time of N_2, represented by τ_2, obtained from the slope of the dashed lines of Fig. 1. From the amplitude of the temporal behavior of the luminescence from $^2E(1)$ and the value of N^* we find $\Sigma\simeq2\times10^{18}cm^{-3}$, consistent with our estimate based on the resonance width. Hence for alexandrite, unlike ruby, ($\Sigma_{ruby}\simeq3\times10^{16}cm^{-3}$), $N^*/\Sigma\equiv b<1$ over most of the experimental range of N^*.

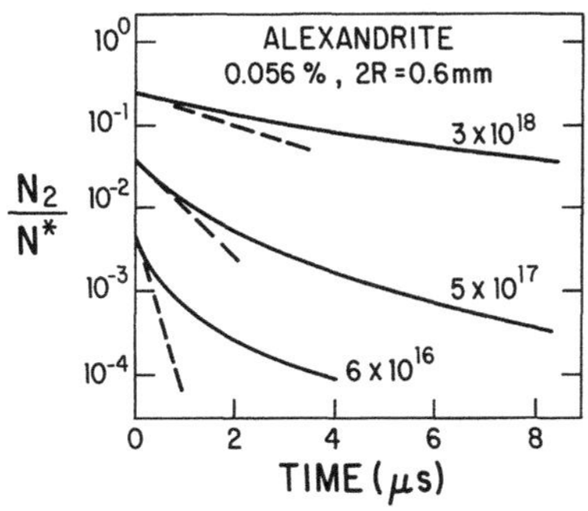

Fig. 1 Time dependence of fractional $^2E(2)$ population after pulsed excitation to the $^2E(2)$ level for three values of N*. Dashed lines show the initial slope from which τ_2 is obtained

A summary of the dependence of τ_2 on N^* is shown for three samples in Fig. 2. The data exhibit $\tau_2 \propto N^{*q}$, with q=0.6 for $N^*<2\times10^{18}cm^{-3}$. There is an indication that τ_2 saturates at the highest N^*, but data at still larger N^* are required for confirmation. For comparison, results on 0.054% ruby are also shown in Fig. 1. The behavior is similar but with q=0.5, as observed previously/3/. The dependence in alexandrite of τ_2 on spot radius R is shown in Fig. 3. The data for two samples indicate that $\tau_2 \propto R^v$ with v=1.1 at $N^*=10^{17}cm^{-3}$ and v=0.9 at $N^*=5\times10^{17}cm^{-3}$.

For pure anharmonic decay we expect $\tau_2=\tau_{AH}(1+b)$, which should be independent of R and $N^*(b<1)$. The dependence of τ_2 on N^* and R eliminates this as the dominant mode of decay and we conclude that $\tau_{AH}>1\mu s$ for the LA $38cm^{-1}$ phonons. The dependence of τ_2 on R at all N^* indicates that spatial transport plays a dominant role in the phonon loss. Ballistic transport is eliminated because τ_2 exceeds the ballistic time for either TA or LA phonons for all N^* and R. For pure spatial diffusion, with phonon velocity c,

$$\tau_2 = (R/cT_1)^2 T_1 b(1+b) \qquad (1)$$

which yields a much stronger dependence of τ_2 on N^* and R than observed. Therefore spatial diffusion, controlled by spectral diffusion, must be operative whereby the phonon frequency is transferred to the wings of the resonance.

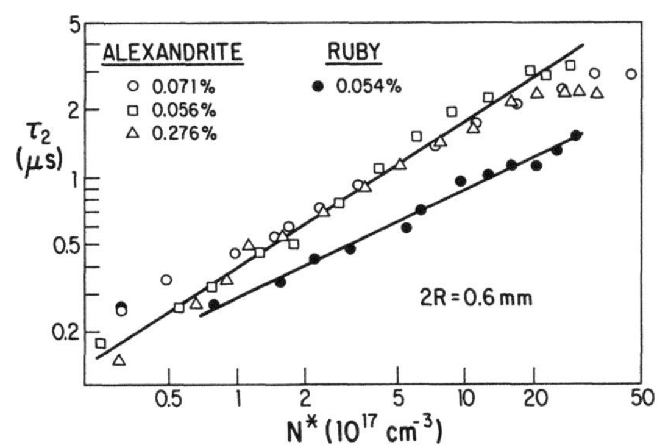

Fig. 2 Dependence of τ_2 on N* for alexandrite and ruby for a 0.6mm spot diameter. The solid lines are best fits to the data

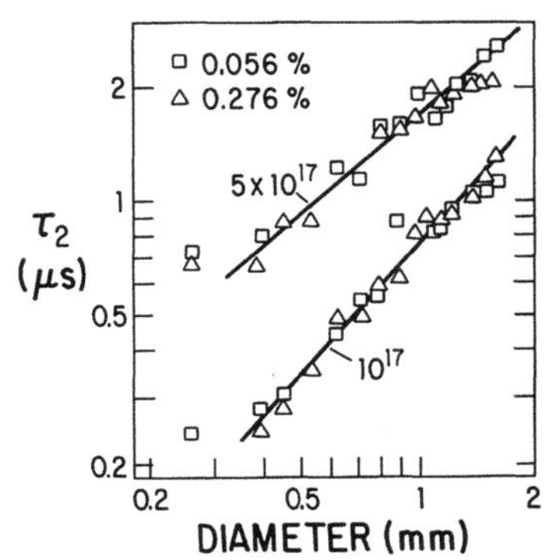

Fig. 3 Dependence of τ_2 on spot diameter in alexandrite for two values of N*. Solid lines are best fits to the data

When there is rapid communication between the center and the wings of the resonance, and where b>>1, we expect $\tau_2 \simeq N^{*q}R^v$ where q=v=0.5 for a Lorentzian resonance and q=v=1 for a Gaussian resonance/10/. However, for most of this data b<1, where models of spectral diffusion yield an even weaker dependence of τ_2 on N^*. Thus the data are consistent with a Gaussian-like resonance, which is probably appropriate for alexandrite ($\Delta\nu \simeq 13$GHz), although it is not known whether rapid communication between the center and wings of the resonance is a correct assumption.

The indication of possible saturation of τ_2 at the highest available N^* may be evidence for phonon loss due to exchange-coupled pairs, as was seen in ruby. However, because of the much broader resonance in alexandrite, one expects these processes will be much weaker than in ruby since a much smaller fraction of Cr^{3+} pairs would be capable of spectral wipeout. On the other hand, since b<1 for alexandrite, phonon loss due to pairs can be much more effective at a given N^* since the phonons propagate for a large fraction of the time, $(1+b)^{-1}$, in contrast to the situation in ruby where they spend most of the time locked up as electronic excitations.

Alexandrite and ruby exhibit a similar dependence of τ_2 on N^* and R in the range $5\times10^{16}cm^{-3}<N^*<2\times10^{18}cm^{-3}$. For alexandrite $\tau_2 \propto N^{*0.6}R^{1.1}$ which is consistent with spectral-spatial diffusion for a Gaussian-like resonance with b<1. The study of the behavior at large N^* needs to be extended. A proper understanding of the dynamics will probably require a detailed model analyzed with computer simulations for the spectral-spatial diffusion.

We acknowledge support from the U.S. Army Research Office, Grant DAAG29-84-G-0021. We thank Allied Corp., Synthetic Crystals Div., for the alexandrite samples.

1. J.I. Dijkhuis and H.W. de Wijn, Sol. St. Commun. 31, 39 (1979).
2. J.I. Dijkhuis and H.W. de Wijn, Phys. Rev. B20, 1844 (1979).
3. G. Pauli, K.F. Renk, G. Klimke and H.J. Kreuzer, Phys. Stat. Sol. (b)95, 503 (1979).
4. R.S. Meltzer, J.E. Rives and W.C. Egbert, Phys. Rev. B25, 3026 (1982).
5. R.J.G. Goossens, J.I. Dijkhuis and H.W. de Wijn, Phys. Rev. B32, 7065 (1985).
6. U. Happek, T. Holstein and K.F. Renk, Phys. Rev. Lett. 54, 2091 (1985).
7. R.J.G. Goossens, J.I. Dijkhuis and H.W. de Wijn, Phonon Scattering in Condensed Matter, 1983, eds. W. Eisenmenger, K. Lassmann and S. Dottinger (Springer, Berlin 1984) p. 112.
8. R.C. Powell, L. Xi, X. Gang and G.J. Quarles, Phys. Rev. B32, 2788 (1985).
9. D.J. Sox, S. Majetich, J.E. Rives and R.S. Meltzer, Journal de Physique 46, C7-493 (1985).
10. J.A. Giordmaine and F.R. Nash, Phys. Rev. 138, A1510 (1965).

Three-Phonon Zone-Boundary Processes and Melting of Solids

B.H. Armstrong

Department of Materials Science and Engineering, University of Washington, Seattle, WA 98195, USA

1. Introduction

Vibrational theories of melting [1,2], which presuppose a critical amplitude of vibration, have enjoyed considerable empirical success. But the critical amplitude determined empirically for a vibrational catastrophe is surprisingly small and a fundamental definition of such a catastrophe is lacking. An hypothesis is advanced herein addressing these issues. Lindemann's law is obtained from the 3-phonon transition rate without reference to a critical amplitude upon the assumption that zone-boundary (ZB) phonons cease to be valid excitations at the melting point. Corollary to this assumption is the onset of single-particle random movement on the time scale $(2\omega_D)^{-1}$ where ω_D is the Debye angular frequency. Results obtained for the Lindemann constant and for diffusion constants compare favorably with experiment for alkali metals and halides.

2. Theory

Breakdown of ZB phonon validity at the melting point has been suggested earlier and investigated experimentally to a limited extent [3,4,5]. We pursue this idea further in the context of the Debye approximation, along with a test for onset of partial decoupling of ion motion from the lattice. The validity condition of the phonon description is expressed as the minimum Heisenberg uncertainty product $\omega\tau(\omega) \geq 1/2$ where ω is phonon angular frequency and $1/\tau$ is the total 3-phonon transition rate. We assume this rate determines the linewidth because the 4-phonon rate remains smaller to the melting point [6]. Further assuming that $1/\tau$ monotonically increases with ω yields

$$1/\tau\,(\omega_D) \leq 2\omega_D. \tag{1}$$

We apply (1) to the Roufosse-Klemens (RK) high-temperature 3-phonon rate for simple cubic (sc) crystals [7]. Combining their expressions for U- and N-processes and evaluating the result at $q_D = (6\pi^2)^{1/3}/a = \omega_D/v$ yields

$$1/\tau\,(\omega_D) \cong (27)2^{1/2}\pi^3(\gamma^2 kT)/(Ma^2\omega_D) \tag{2}$$

The Grüneisen constant is labelled γ, M and a^3 are unit cell mass and volume, v is Debye mean velocity, q_D wave number at the ZB, and T is absolute temperature. Approximate equality is indicated in view of the approximations involved in the derivation. Neglect of dispersion suggests this result will overestimate the true rate at the ZB. Also, this sc result will be applied to other structures. Hence, we multiply (2) by 4/9, chosen to produce agreement with experimental results discussed below. This factor can be calculated in principle. With this adjustment, (1) and (2) yield

$$kT \leq Ma^2\omega_D{}^2/(6\cdot 2^{1/2}\pi^3\gamma^2). \tag{3}$$

Identification of this maximum T with the melting temperature T_m yields

$$kT_m = T_D^2 V_o^{2/3} A/C^2 \qquad (4)$$

which is Lindemann's melting law. The Debye temperature is denoted T_D, V_o is atomic volume, and A is atomic weight. The Lindemann constant C is defined by

$$C = (72)^{1/4}\pi^{3/2}\hbar\gamma L^{1/2}/k = 96.15\gamma\ [\mathrm{sec\ K^{-1}}] \qquad (5)$$

where L is Avagadro's number. The ratio of vibrational amplitude to interparticle spacing does not appear as a free parameter as in the Lindemann theory. Instead, this ratio is now determined to the precision of the uncertainty product. This evaluation of C is reminiscent of some provided earlier, but the physics is different. The Landau-Rumer transition rate, which retains validity toward high T [8], reduces to the same form as (2), but with a smaller numerical coefficient. Comparison with experiment is shown in Table 1, where γ is approximated by the thermal expansion value γ_{th}. The alkali halides are approximated as ideal Debye solids with each ion as a fictitious unit cell. The "atomic weight" ascribed to each is half the molecular weight. Thus, the Lindemann constant for these compounds is given by $C = 2^{1/2}(96.15\gamma) = 136.0\gamma$. The 4 halides shown are in order of increasing relative ion mass difference. C calculated according to (5) is labelled C(TH) whereas C(EXP) is calculated from (4). The Debye temperatures T_D are room-temperature elastic constant values. Results for the 16 Li, Na, K, and Rb halides for which data were found show a trend (expected from the derivation [7]) toward increasing error with increasing relative mass difference. This error was $\leq$ 10% for all but the two of greatest mass difference.

Table 1. Comparison of theory/experiment for Lindemann C

	T_D[K]	T_m[K]	th	C(TH)	C(EXP)
Li	312[a]	453	0.93[c]	89.4	91.6
Na	132[a]	371	1.15[c]	111	95.2
K	92[a]	336	1.15[c]	111	112
Rb	55[b]	312	1.13[c]	109	109
Cs	40[b]	302	1.04[c]	100	110
RbBr	130[a]	953	1.47[d]	200	200
KCl	226[a]	1043	1.49[e]	203	203
NaF	473[a]	1253	1.51[e]	205	214
RbI	102[a]	913	1.51[d]	205	194

[a]Ref.[9], [b]Ref.[10], [c]Ref.[11], [d]Ref.[12], [e]Ref.[13]

If the equality holds for (1) at T_m, then random ion motion on the time scale $\tau \sim (2\omega_D)^{-1}$ should appear at T_m as ions partially decouple from the lattice. This prediction can be tested by computation of self-diffusion constants for the liquid at $T = T_m$. For large wave number, the scattering law for a liquid reduces to that of a perfect gas. Hence, we use the expression $D = kT_m\tau/M_I$, where M_I is ion mass, for the diffusion constant D. With $\tau = 1/(2\omega_D)$ the results labelled D(TH) in Table 2 are obtained and compared to experimental liquid values D(EXP). For the halides, M_I is taken as the reduced mass and D is the sum of the + and - ion contributions. The units of D are cm^2/s. This comparison clearly supports the hypothesis of onset of single-particle motion on the time scale $(2\omega_D)^{-1}$ at the melting point (cf., Table 3.1 of [15]).

Table 2. Comparison of theory/experiment for diffusion constants

	Li	Na	K	In	NaCl	RbCl	CsCl	NaI
T_D[K]	312[a]	132[a]	92[a]	90[a]	303[a]	162[a]	170[b]	153[a]
D(TH)x10^5	6.7	3.8	3.0	1.3	8.1	7.7	6.1	10
D(EXP)x10^5	6.1[c]	4.2[c]	3.7[c]	1.6[c]	15.1[d]	8.7[d]	7.0[d]	10.2[d]

[a]Ref.[9], [b]Ref.[14], [c]Ref.[15], [d]Ref.[16]

Obviously, (2) is not precise on a mode-by-mode basis because γ and ω_D are gross averages. The accuracy of the results suggests that this averaging is important to the description of the melting process, which will be discussed in detail elsewhere. It is likely that longitudinal ZB mode validity must fail on average over all directions for decoupling with random motion to occur, with attendant volume increase (for simple crystals). Sudden onset of decoupling when the average ZB mode loses validity would be consistent with observed melting behavior. Initial failure for longitudinal modes would permit the required sudden volume change. This in turn causes vanishing of one of the shear moduli [17]. Thus, ion wandering could begin to complete the transition to the liquid state. Finally, (3) can be reduced to $v \geq 2.4\ v_{th}$, where v_{th} is the particle thermal velocity. This suggests that near equality of phonon and particle velocities facilitates decoupling.

References

1. L.L. Boyer: Phase Transitions 5, 1 (1985)
2. A.R. Ubbelohde: The Molten State of Matter (Wiley, New York 1978)
3. P.A. Fleury: in Anharmonic Lattices, Structural Transitions and Melting, ed. by T. Riste (Noordhoff, Leiden 1974)
4. J. Skalyo, Jr., Y. Endoh, and G. Shirane: Phys. Rev. B9, 1797 (1974)
5. B. Hennion and M. Schott: J. Phys. Lett. (France) 45, L621 (1984)
6. D.J. Ecsedy and P.G. Klemens: Phys. Rev. B15, 5957 (1977)
7. M. Roufosse and P.G. Klemens: Phys. Rev. B7, 5379 (1973)
8. B.H. Armstrong: Phys. Rev B23, 883 (1981); Phys. Rev. B32, 3381 (1985)
9. O.L. Anderson: in Physical Acoustics, ed. by W.P. Mason (Academic, New York 1965)
10. K.A. Gschneidner, Jr.: in Solid State Physics Vol. 16, ed. by F. Seitz and D. Turnbull (Academic, New York 1964)
11. V.G. Vaks, E.V. Zarochentsev, S.P. Kravchuk, V.P. Safronov, and A.V. Trefilov: Phys. Stat. Sol. (b) 85, 749 (1978)
12. Z.P. Chang and G.R. Barsch: J. Phys. Chem. Sol. 32, 27 (1971)
13. K.D. McLean and C.S. Smith: J. Phys. Chem. Sol. 33, 279 (1972)
14. A.A.Z. Ahmad, H.G. Smith, N. Wakabayashi, and M.K. Wilkinson: Phys. Rev, B6, 3956 (1972)
15. T.E. Faber: Introduction to the Theory of Liquid Metals (Cambridge, London 1972)
16. R.E. Young and J.P. O'Connell: Ind. Eng. Chem. Fundam. 10, 418 (1971)
17. J.H.C. Thompson: Phil. Mag. 44, 131 (1953)

Quasicrystalline and Crystalline Phonon Density of States: Neutron Scattering Measurement on Al_4Mn

P.F. Miceli[1], *S.E. Youngquist*[1], *D.A. Neumann*[1], *H. Zabel*[1], *J.J. Rush*[2], *and J.M. Rowe*[2]

[1]Department of Physics and Materials Research Laboratory, University of Illinois at Urbana-Champaign, Urbana, IL 61801, USA
[2]National Bureau of Standards, Gaithersburg, MD 20899, USA

The recent discovery[1,2] of rapidly quenched AlMn alloys exhibiting diffraction patterns with five-fold symmetry has stimulated intense experimental and theoretical research interest. The icosahedral point group symmetry associated with these novel materials is not consistent with lattice translations and therefore provides a new challange to our understanding of the physical properties of solids. While detailed information on the structure is not yet complete, calculations of the electronic and phonon density of states for quasiperiodic structures have been carried out[3-5]. Characteristic of these results are the existence of localized states as well as the self-similarity exhibited in any slice of reciprocal space.

To experimentally address these ideas, we have used inelastic neutron scattering to perform the first measurement of the vibrational density of states in a sample with icosahedral symmetry. Because of the powder nature of these samples, we obtain an orientational average of the vibrational modes rather than a phonon dispersion. The latter would be favorable if sufficiently large single (quasi)crystals were available. In order to determine the role of structure, we have performed the same experiment on a crystalline sample of the same compostion. Previous x-ray mesurements[6] on the quenched $Al_{1-x}Mn_x$ alloys have shown that the largest fraction of icosahedral phase is present for $x = .20$ with a negligible amount of coexisting crystalline Al. This fact, along with the ability to produce a crystalline phase at the same composition[7] (crystalline Al_4Mn is obtained by annealing the icosahedral phase), made $Al_{.80}Mn_{.20}$ the ideal composition to study.

We obtained the vibrational density of states from our inelastic neutron scattering data through the incoherent approximation[8], where the coherent dynamic scattering function is replaced by its incoherent counterpart. The resulting $\bar{g}(E)$ represents the vibrational density of states weighted by the coherent neutron scattering lengths and the vibrational amplitudes of Al and Mn. The measurements were done at the National Bureau of Standards research reactor on two triple-axis spectrometers, one of which was also equipped with a Be-graphite-Be filter for purposes of energy analysis. The resolutions in $\bar{g}(E)$ ranged from ~.5 meV at low energy transfers (0-15 meV) to ~3 meV at higher energy transfers. Further technical details are published elsewhere[9].

The measured, neutron weighted vibrational density of states $\bar{g}(E)$ of icosahedral and crystalline $Al_{0.80}Mn_{0.20}$ are shown in figures 1a and 1b, respectively, and are compared in figure 1c. In both samples, $\bar{g}(E)$ exhibits the usual E^2 dependence expected for a three dimensional system at low energies. In fact, $\bar{g}(E)$ is nearly identical for the two samples in this region, indicating that they must be elastically similar. Between

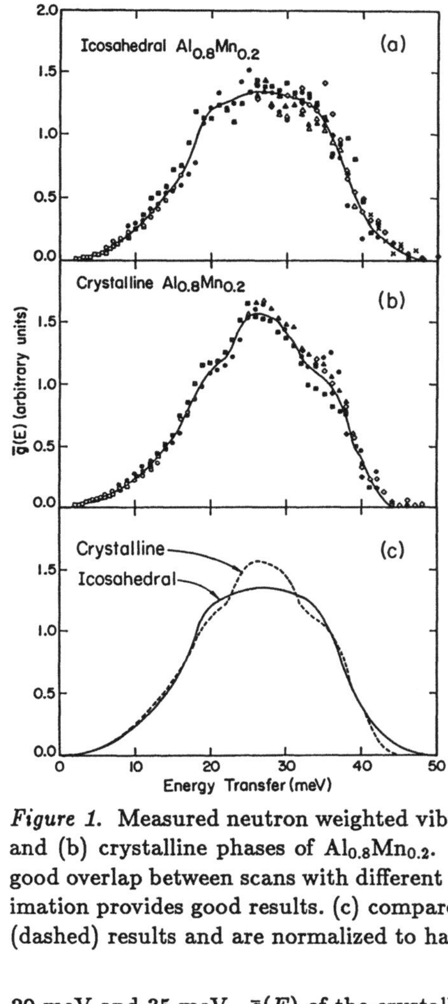

Figure 1. Measured neutron weighted vibrational density of states for (a) icosahedral and (b) crystalline phases of $Al_{0.8}Mn_{0.2}$. The solid lines are drawn as a guide. The good overlap between scans with different Q demonstrates that the incoherent approximation provides good results. (c) compares the icosahedral (solid) and the crystalline (dashed) results and are normalized to have the same area.

20 meV and 35 meV, $\bar{g}(E)$ of the crystalline sample shows some structure, including a pronounced maximum at 27 meV, while the spectra of the icosahedral sample is essentially featureless. The icosahedral density of states also displays a distinct high energy tail compared to that of the crystalline material. The smoothing of features in $\bar{g}(E)$ for the icosahedral phase relative to the crystalline phase might be attributed to the non-propagating nature of the higher energy vibrational states characteristic of systems lacking translational symmetry.

There is little dependence on the scattering vector, Q, in any of our measurements, indicating that the incoherent approximation is indeed satisfied inspite the fact that Al is a completely coherent scatterer and Mn has only a small incoherent scattering cross section. The lack of Q dependence is not suprising since the size of the Brillouin zone of the crystalline sample is less than 0.04 $Å^{-1}$, while for a material without translational symmetry it is essentially the inverse of the crystallite size.

In conclusion, crystalline and icosahedral $Al_{.80}Mn_{.20}$ exhibit pronounced differences in $\bar{g}(E)$ above ~18 meV, while appearing identical below this energy. This suggests that the interatomic atomic forces are similar on length scales larger than the interatomic distances, however, shorter wavelength vibrations are altered due to the detailed structural differences between these materials. We have not observed any features in $\bar{g}(E)$ which could clearly be identified as localized states, although one may speculate that the increased $\bar{g}(E)$ around 45 meV is an indication of localized excitations. The absence of sharp features due to these modes is not too surprising since the measurment takes an orientational average of all modes and convolutes it with the instrumental resolution. The results presented here do, however, provide a direct test for models of the interatomic forces and dynamics of the quasicrystalline state.

We thank H. L. Fraser for providing us with the quenched AlMn ribbons. This work was supported in part by the National Science foundation under grant no. DMR83-04890 and by the U. S. Department of Energy Division of Materials Science under contract no. DE- AC02-76ER01198. Two of us (S. E. Y. and D. A. N.) gratefully acknowledge fellowships from AT&T Bell Laboratories.

References

1. D. Shechtman, I. Blech, D. Gratias and J. W. Cahn, Phys. Rev. Lett. **53**, 1951 (1984).
2. R. Field and H. L. Fraser, Mater. Sci. Eng. **68**, L17 (1984).
3. F. Nori and J. P. Rodiguez, Private communication.
4. T. Odagaki and D. Nguyen, Phys. Rev. B **33**, 2184 (1986).
5. J. P. Lu, T. Odagaki, and J. L. Birman, Phys. Rev. B **33**, 4809 (1986).
6. S. E. Youngquist, P. F. Miceli, D. G. Wiesler, H. Zabel and H. L. Fraser, to be published.
7. M. A. Taylor, Acta Met. **8**, 256 (1980).
8. V. S. Oskotskii, Sov. Phys. Solid State **9**, 420 (1967).
9. P. F. Miceli, S. E. Youngquist, D. A. Neumann, H. Zabel, J. J. Rush, and J. M. Rowe, to be published.

Transient Frequency Distributions of Anharmonically Decaying Acoustic Phonons

U. Happek[1], *K.F. Renk*[1], *Y. Ayant*[2], *and R. Buisson*[2]

[1]Institut für Angewandte Physik, Universität Regensburg, D-8400 Regensburg, Fed. Rep. of Germany

[2]Laboratoire de Spectrométrie Physique, Université de Grenoble, Saint-Martin-D'Hères, France

We have studied transient frequency distributions of spontaneously decaying phonons in CaF_2 at low temperature. After pulsed phonon excitation by phonon-combination absorption of infrared radiation, we observed a decay cascade of acoustic phonons. We present a theoretical description of the cascade by a scaling function, that is characteristic for spontaneous phonon decay in a crystal containing mode-mixing elastic scattering centers.

We generated phonons by excitation with CO_2 laser radiation at a frequency of 28 THz [1]. By multiphonon absorption mainly optical phonons are produced. The optical phonons decay very fast into high-frequency acoustic phonons and these in turn decay further. Because of elastic scattering at impurities and isotopes, acoustic phonons at fixed frequency but different modes are in an equilibrium. We have detected phonons by vibronic sideband spectroscopy [2]; for this purpose the crystal was weakly doped with Eu^{2+} ions (0.01 mol%). The crystal (size 6x5x3 mm^3), immersed in liquid helium at a temperature of 1.6 K, was illuminated with pulsed radiation from a Q-switched CO_2 laser (pulse width 50 ns, energy 10^{-4}J, repetition rate 5000 cps). Due to the low absorption coefficient at the laser frequency (~2 cm^{-1}), and by choosing a large laser beam diameter, phonons were generated homogeneously in the crystal volume.

Fig. 1 shows experimental signal curves for different detection frequencies ν. At $\nu \simeq 2.0$ THz the signal increases during the laser pulse and then decreases exponentially, which we attribute to spontaneous phonon decay. At $\nu \simeq 1.0$ THz the rise time of the signal is longer than the duration of the laser pulse, phonons are still created after the initial excitation, eventually by decay of phonons around 2 THz. The decay is slower than for phonons at 2 THz; for high frequencies ($\nu > 1$ THz) the phonon lifetime τ decreases as $\tau^{-1} = \lambda\nu^5$, where $\lambda \simeq 1 \cdot 10^{-55}$ s^4. This behavior is consistent with theoretical studies [3] and earlier experimental results [4].

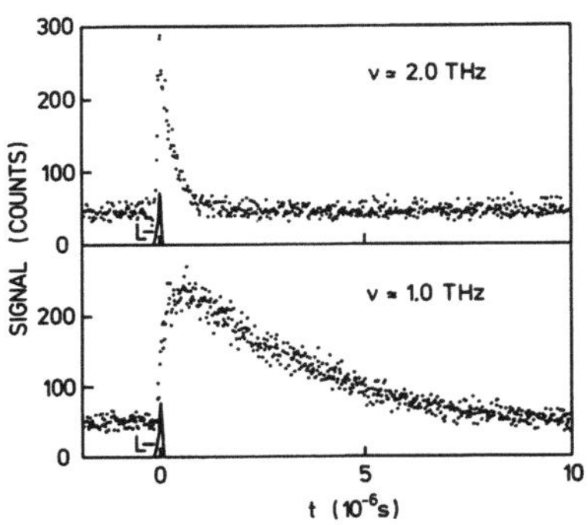

Fig. 1. Phonon-induced fluorescence signals and CO_2 laser pulse (L)

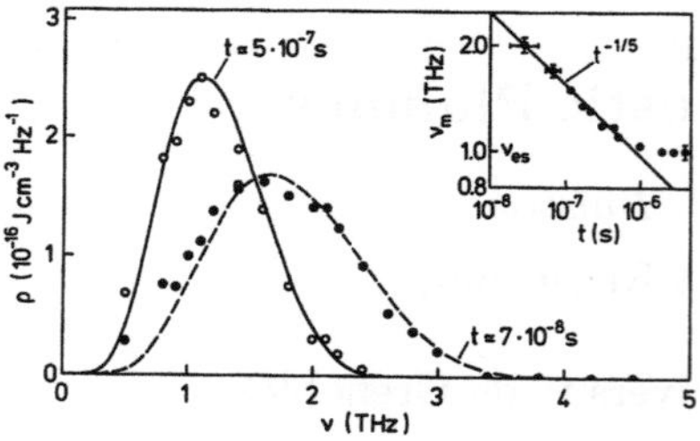

Fig. 2. Spectral energy density of phonons according to experiment (points) and theory (curves); inset, frequency of maximum energy density

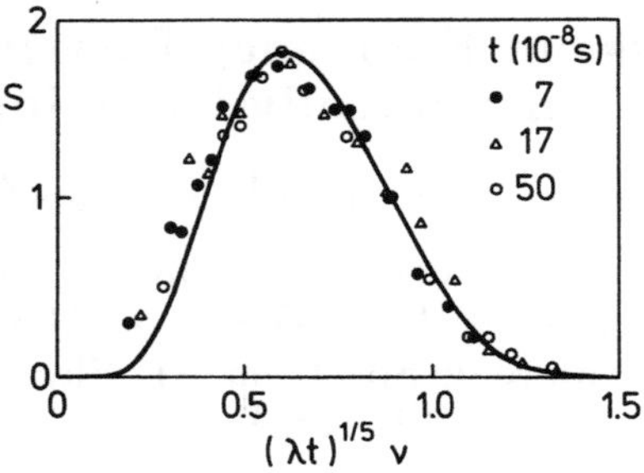

Fig. 3. Scaling function S according to theory (solid line) and experiment (points)

From the signal curves we have determined the spectral energy density $\rho(\nu,t) = p(\nu,t)Z(\nu)h\nu$ for fixed times t after excitation (Fig. 2). $p(\nu)$ is the phonon occupation number, determined from the ratio of anti-Stokes and Stokes fluorescence intensities at the detection frequency, $Z(\nu) = 12\pi\nu^2/c^3$ the density of states and c ($\simeq 4\cdot10^5$ cm/s) an average velocity of sound. Shortly after excitation ($t \simeq 7\cdot10^{-8}$s) the distribution is broad and shows a sharp cutoff at high frequencies. At a later time the distribution has narrowed and the cutoff has shifted to lower frequencies. The total energy density remains almost constant, indicating that the phonon redistribution occurs without energy loss. The frequency of maximum energy density, ν_m, shifts at small time as $t^{-1/5}$. For large times ($t > 5\cdot10^{-7}$s) ν_m remains constant, and the total energy decreases because of phonon loss to the helium bath.

We introduce a dimensionless variable $u = (\lambda t)^{1/5}\nu$ and a dimensionless spectral-energy density function $S(u) = W_0^{-1}(\lambda t)^{-1/5}\rho(\nu,t)$ where W_0 is the total energy density. We find that the u-dependence of S is almost the same for different times (Fig. 3), i.e. S(u) is a scaling function for the decay cascade. A phonon of frequency ν' has a probability per unit time of $2\lambda\nu'^4 g(\nu/\nu')d\nu$ to deliver simultaneously phonons at frequencies ν and $\nu' - \nu$. This leads to the normalization $\int_0^1 g(\nu/\nu')d(\nu/\nu') = 1$. We assume that at time $t = 0$ phonons at frequency ν_0 of an energy density W_0 are generated. The decay cascade can then be described by the integro-differential equation

$$\frac{\partial f}{\partial t} = -\lambda\nu^5 f + 2\lambda\int_\nu^{\nu_0}\nu'^4 g\left(\frac{\nu}{\nu'}\right)f(\nu')d\nu' + 2\lambda\nu_0^4 g\left(\frac{\nu}{\nu_0}\right)\frac{W_0}{h\nu_0}\exp(-\lambda\nu_0^5 t) \qquad (1)$$

where $f(\nu,t) = (h\nu)^{-1}\rho(\nu,t)$ is the frequency distribution. Introducing $f_0 = h\,W_0^{-1}(\lambda t)^{-2/5}f = u^{-1}S$, the cascade can be described after few steps ($u^5 >> u_0^5$) by

$$\frac{df_0}{du} + \frac{2}{u}f_0 = -5u^4 f_0 + \frac{10}{u}\int_u^\infty f_0(u')g\left(\frac{u}{u'}\right)u'^4 du'. \qquad (2)$$

Writing the decay characteristic in the form $g = N_\kappa\cdot(\nu/\nu')^\kappa(1-\nu/\nu')^\kappa$, where N_κ is a normalization constant and κ (≥ 1) a parameter, we have numerically solved (2) for different values of κ. For large frequencies ($u > 1$) the asymptotic solution is $S \simeq u^{-1}\exp(-u^5)$, at small frequencies ($u<<1$), S increases as $u^{(\kappa+1)}$. For $\kappa = 8$ we obtain a theoretical function S (solid line in Fig. 3) which is in good agreement with the experiment. In Fig. 2 we show the energy distribution curves obtained from the theoretical function. At small frequencies we observed more phonons than predicted by the theory, this may be due to splitting of phonons into three phonons.

The decay characteristic for $\kappa = 8$ corresponds to the decay of longitudinal phonons mainly. For the decay of transverse phonons a broader characteristic is expected. Using a Monte Carlo technique, Wolfe and Northrop [5] have calculated a decay cascade for an isotropic dispersionless medium with two phonon branches of slightly different velocities and found a scaling function which has a similar form as we obtain for $\kappa = 2$. This scaling function has its maximum at a larger value than in the experiment; therefore, our analysis suggests that longitudinal phonon decay is at least as important as transverse phonon decay. We note that Kazakovtsev and Levinson [6] have analyzed (1) by a different mathematical procedure, and proofed the existence of a scaling law and derived the asymptotic behavior (for $u \ll 1$ and $u \gg 1$).

In our experiment the observed phonons are far from thermal equilibrium and have small occupation numbers. Therefore anharmonic phonon recombination [7] can be neglected. According to our laser pulse energy, thermalization is expected at a time $t \simeq 10^{-3}$s, but after few μs almost all phonons have escaped into the helium bath.

In summary, we observed a decay cascade of high-frequency phonons and show that the cascade can be described by a scaling law.

The work was supported by the Deutsche Forschungsgemeinschaft.

References

1. U. Happek, R. Baumgartner, and K.F. Renk: in Phonon Scattering in Condensed Matter, ed. by W. Eisenmenger, K. Laßmann, and S. Döttinger (Springer, Berlin 1984), p. 37.
2. For a survey, see Nonequilibrium Phonon Dynamics, ed. by W.E. Bron (Plenum Press, New York 1985), NATO ASI Series B: Physics, Vol. 124.
3. A. Berke, A.P. Mayer, and R.K. Wehner: Solid State Commun. 54, 395 (1985); P.F. Tua and G.D. Mahan: Phys.Rev. B 26, 2208 (1982); S. Tamura and H.J. Maris: Phys. Rev. B 30, 2595 (1985); S. Tamura, this volume.
4. R. Baumgartner, M. Engelhardt, and K.F. Renk: Phys. Rev. Lett. 46, 1210 (1962).
5. J.P. Wolfe and G.A. Northrop: in Ref. 1, p. 100.
6. L. Kazakovtsev and Y.B. Levinson: phys. stat. sol. b 96, 117 (1979).
7. W.L. Schaich, Sol. State Commun. 50, 3 (1984).

Optical Detection of Large-Wave-Vector Phonons

R.J.G. Goossens, J.I. Dijkhuis, and H.W. de Wijn

Fysisch Laboratorium, Rijksuniversiteit Utrecht, P.O. Box 80.000, NL-3508 TA Utrecht, The Netherlands

During the last few years there has been considerable debate on the lifetime of near-zone-boundary phonons [1-5]. By quite general theoretical arguments near-zone-boundary phonons of the lowest acoustic branch cannot decay by anharmonic processes [4], although in real crystals mode conversion induced by impurities, defects, and isotopes presumably limits the intrinsic lifetime.

In this paper, we examine the dynamics of near-zone-boundary acoustic phonons taking advantage of optically pumped Cr^{3+} centers in diluted ruby. The phonons are generated by the breakup of the energy associated with the nonradiative decay following optical pumping, and detected by the enhancement of the R_2 luminescence due to Raman transitions invoked by these phonons (Fig.1) [6]. The rate of these transitions is in fact directly accessible to experimental examination by measuring the total feeding, $\phi_{2\bar{A}}^{tot}$, in $2\bar{A}(^2E)$ via a modulated pumping scheme [7]. We have

$$\phi_{2\bar{A}}^{tot} = \phi_{2\bar{A}} + p_{ZB}/T_{RAM} , \tag{1}$$

where $\phi_{2\bar{A}}$ is the direct optical feeding into $2\bar{A}(^2E)$ ($\sim 20\ s^{-1}$), p_{ZB} is the occupation number of the near-zone-boundary phonons averaged over the

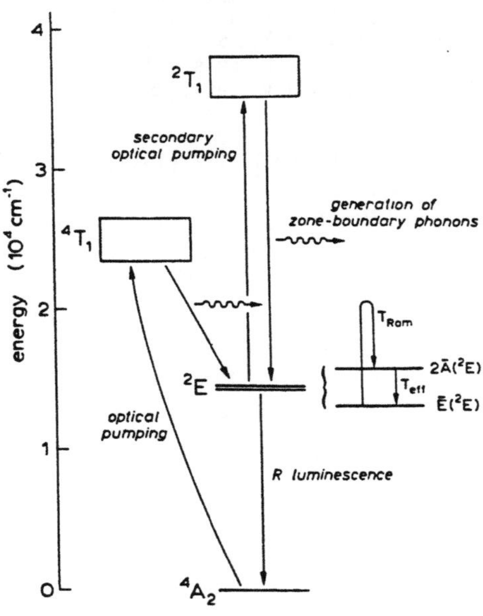

Fig.1 Schematic energy-level diagram of $Cr^{3+}:Al_2O_3$. Near-zone-boundary phonons are generated in the non-radiative decays following primary and secondary optical pumping. These phonons induce additional feeding into $2\bar{A}(^2E)$ by Raman transitions starting from $\bar{E}(^2E)$. Diagram applies to pumping with the 458 nm argon line; at 514 nm, pumping is primarily to the $^4T_2$ band located around 18000 cm^{-1}, while no efficient secondary pumping occurs.

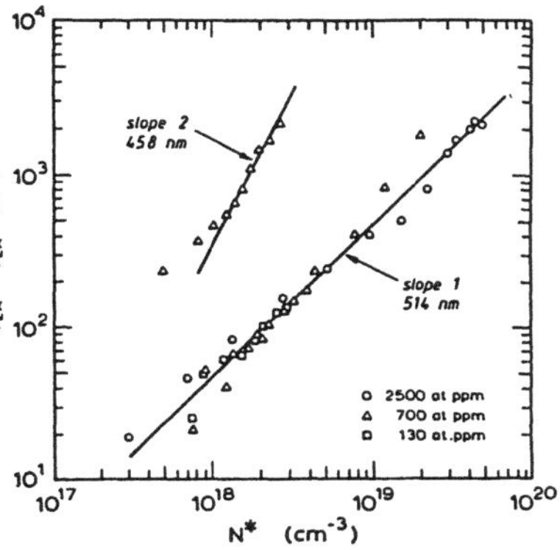

Fig.2 Raman part of the feeding into $2\bar{A}(^2E)$ in 130, 700, and 2500 at.ppm ruby at 1.5 K vs N* upon optical pumping at 514 or 458 nm. The excited zone has diameter 60 μm. Data for 700 and 2500 at.ppm are corrected for reabsorption. Slopes 1 and 2 point to generation of near-zone-boundary phonons in the primary and secondary optical pumping cycles, respectively.

modes active in Raman transitions, and T_{RAM} is the effective time constant of the Raman processes [7]. Here, the longitudinal acoustic phonons have such short lifetimes ($\sim 10^{-11}$ s) as to be ineffective in Raman transitions. It is further noted that near-zone-boundary phonons are favored in effecting Raman transitions by the transition probability of the process, which approximately goes with $k^2k'^2/\omega\omega'$, where ω is the angular frequency and the primed quantities refer to the outgoing phonon ($\hbar\omega-\hbar\omega'$ = 29 cm^{-1}).

For 130, 700 and 2500 at.ppm ruby at 1.5 K, the measured feeding is, after subtraction of $\phi_{2\bar{A}}$, presented in Fig.2 vs the excited state population N*. For experimental details and the calibration of the N* scale we refer to [7]. For the case of pumping with the 514 nm line of an argon laser, we thus find a linear dependence of p_{ZB}/T_{RAM} on N*, i.e., the optically produced near-zone boundary phonon occupation is proportional to the pumping intensity. For the case of pumping with 458 nm, however, a quadratic dependence is found. We attribute this to excited-state absorption of the laser pump light to still higher states. The resultant secondary pumping provides an additional generation of near-zone-boundary phonons, proportional to N^{*2} rather than N*, provided a significant fraction of the Cr^{3+} ions is in the metastable states. A quantitative analysis of these data, including a calculation of T_{RAM}, permits an estimate of the lifetimes of the near-zone boundary phonons. This is done in [7] and [8].

The size dependence of p_{ZB}/T_{RAM} in 700 at.ppm ruby is presented in Fig.3 for a selection of diameters of the pumped cylinder (R = 30–280 μm). For each diameter these measurements have been carried out as a function of the power of the laser operating in the all-lines mode to determine the effects of the secondary optical pumping cycle with increasing N*. Indeed, at small R an increase of ϕ with laser power faster than linear is observed, while at large R the increase is strictly linear. To extract the size dependence of p_{ZB}, we recall that N* is proportional to the laser power density $P/\pi R^2$. We further adopt at this point a power-law dependence of the form $p_{ZB} \propto L^\gamma$, where the size dependence is contained in the exponent γ. The quantity L is the typical dimension of the zone. Accordingly, we have

$$\phi_{2\bar{A}}^{tot} - \phi_{2\bar{A}} = \alpha[P/\pi R^2 + \beta(P/\pi R^2)^2]R^\gamma \,, \qquad (2)$$

where L is identified with the cylinder radius R. The solid curves in Fig.3 represent a fit of Eq.(2) to all data at the same time, with α, β,

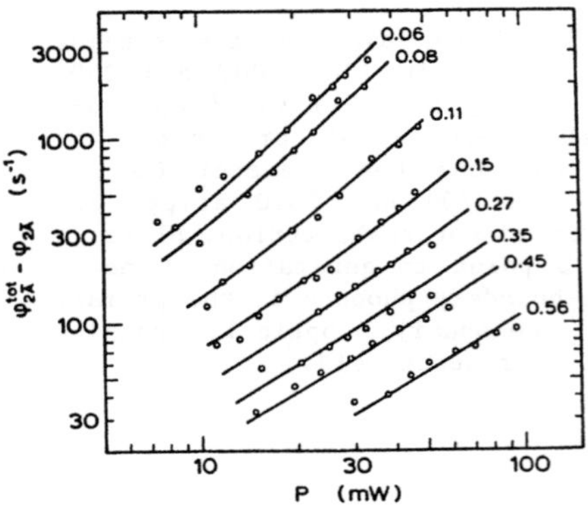

Fig.3 Raman part of the feeding into $2\bar{A}(^2E)$ in 700 at.ppm ruby at 1.5 K vs the laser power for various beam diameters, as indicated (in mm). The laser is operating in the all-lines mode. Solid lines represent a fit of Eq.(2) to the data, demonstrating P_{ZB} to scale approximately with R.

and γ as adjustable parameters. The quantity β is a measure of the contribution of the secondary pumping cycle. The fitted values for α and β coincide within the uncertainties with independent estimates (see [7]). The most important point is that $\gamma = 0.8 \pm 0.2$. The latter result, within errors equal to unity, indicates that the predominant loss mechanism for near-zone-boundary phonons is removal out of the excited zone by ballistic flight, or for that matter small-angle scattering, at least in the case of zone diameters up to about 0.6 mm.

In conclusion, with a technique of modulated pumping it has been demonstrated that the feeding into $2\bar{A}(^2E)$ is substantially larger than the direct optical feeding. The additional feeding per metastable Cr^{3+} is linear in N*, but becomes quadratic in case the laser wavelength allows for a secondary pumping cycle. The experiments further show the additional feeding to scale approximately with the excited-zone dimensions. These phenomena are explained by a Raman process connecting the $\bar{E}(^2E)$ and $2\bar{A}(^2E)$ populations. The process is induced by near-zone-boundary acoustic phonons produced in the breakup of energy released in the nonradiative decays following the optical excitations.

References

1. W. Grill, O. Weiss: Phys. Rev. Lett. 35, 588 (1975)
2. P. Hu, V. Narayanamurti, M.A. Chin: Phys. Rev. Lett. 46, 192 (1981)
3. R.G. Ulbrich, V. Narayanamurti, M.A. Chin: Phys. Rev. Lett. 45, 1432 (1980)
4. M. Lax, P. Hu, V. Narayanamurti: Phys. Rev. B 23, 3095 (1981)
5. H. Lengfellner, K.F. Renk: Phys. Rev. Lett. 46, 1210 (1981)
6. S.A. Basun, A.A. Kaplyanskii, V.L. Shekhtman: Fiz. Tverd. Tela 24, 1913 (1982) [Sov. Phys. Solid State 24, 1093 (1982)]
7. R.J.G. Goossens, J.I. Dijkhuis, H.W. de Wijn: Phys. Rev. B 32, 5163 (1985)
8. J.G.M. van Miltenburg, J.I. Dijkhuis, H.W. de Wijn: In Phonon Scattering in Condensed Matter (Springer, 1984) p. 119; and to be published

Spectral Wipeout of Imprisoned 29-cm^{-1} Phonons in Ruby

R.J.G. Goossens, J.I. Dijkhuis, and H.W. de Wijn

Fysisch Laboratorium, Rijksuniversiteit Utrecht, P.O. Box 80.000, NL-3508 TA Utrecht, The Netherlands

The relaxation of 29-cm^{-1} phonons resonant with the $\bar{E}(^2E)$-$2\bar{A}(^2E)$ transition of optically excited Cr^{3+} in Al_2O_3 (ruby) has yielded a remarkable enigma: While at low concentrations N^* of excited Cr^{3+} in a narrow-pumped zone the phonon relaxation time τ increases with N^*, such as expected for boundary-limited decay of trapped phonons, τ declines with N^* at high N^*. This is straightforwardly derived [1] from the effective decay time T_{eff} of the $2\bar{A}(^2E)$ level. Under the conditions met in the experiments, T_{eff} scales with $N^*\tau$ and tends to saturate at high N^* [2]. Apparently, at high N^* additional scattering processes that are able to bring the phonon excitation out of range become operative.

This study [3] contributes to solving as to what is the origin of the additional phonon relaxation in three ways: (i) New experimental data of T_{eff} indicate that the relevant quantity of the problem is the product N^*R, in which R is the radius of the laser-excited pencil, rather than N^* by itself (Fig.1). The experimental method employs modulated excitation via the broad bands with an argon laser, and time-resolved observation of the decay of the R_2 luminescent intensity emanating from $2\bar{A}(^2E)$ following removal of the optical pumping. (ii) A model is constructed which accounts for the measured T_{eff} for realistic values of the parameters involved. It

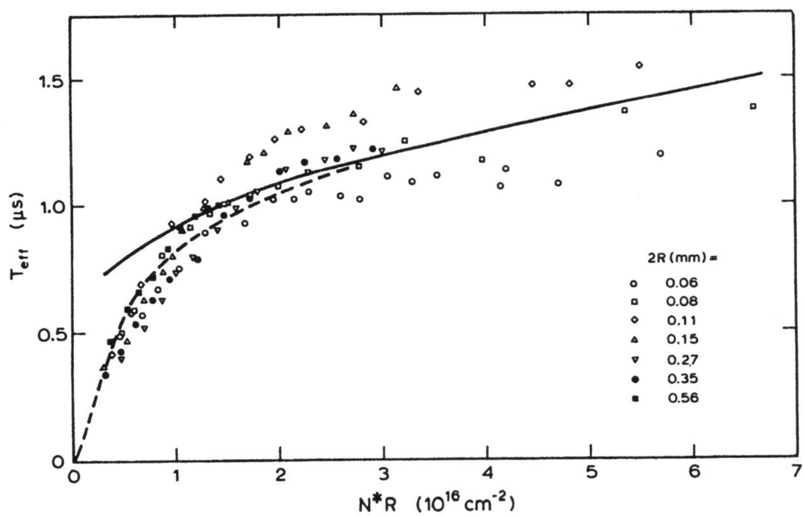

Fig.1 Effective decay time T_{eff} of $2\bar{A}(^2E)$ vs N^*R in 700 at.ppm ruby

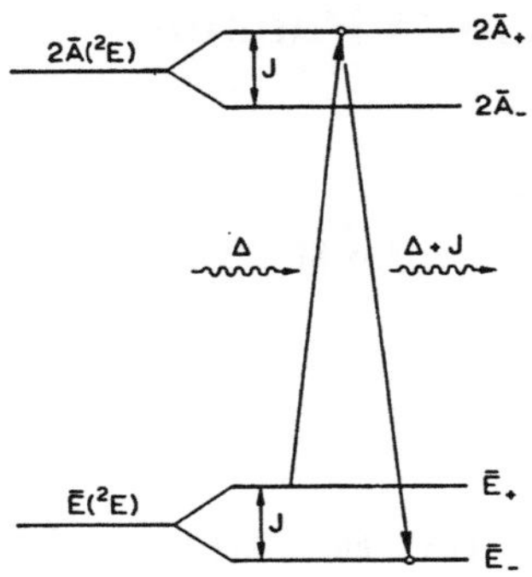

Fig.2 Frequency shifting mechanism involving one-site Orbach processes

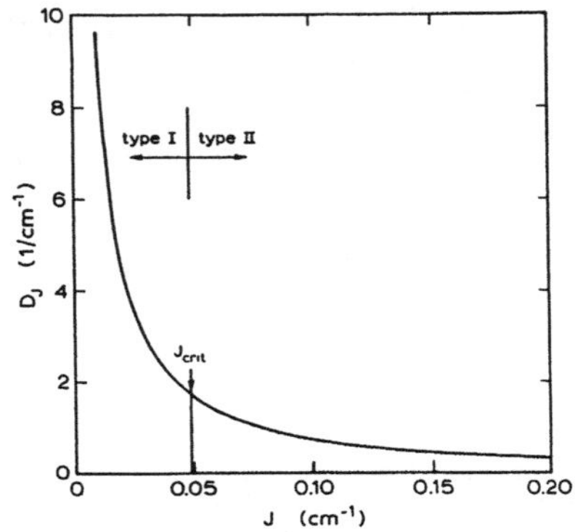

Fig.3 D_J vs J for 2500 at.ppm ruby

is based on one-site Orbach processes at metastable Cr^{3+} [4] experiencing diagonal exchange with nearby Cr^{3+} in the $^4A_2$ ground state, and further relates, as required by experiment, the mean free path with the typical dimension of the excited zone. (iii) Information is retrieved from the complex development of T_{eff} with increasing magnetic field, providing a critical test of the way the phonon packets resonant with the various Zeeman components of $2\bar{A}(^2E)$-$\bar{E}(^2E)$ communicate. As it appears, the model is consistent with these results. Exactly the same information is further derived from the development of the R_2 intensity with field under the conditions of continuous optical pumping.

As concerns the model, the basic assumptions are in more detail the following: (i) A spin-nonflip transition $\bar{E}(^2E)$ to $2\bar{A}(^2E)$ followed by a spin-flip transition back to $\bar{E}(^2E)$ invokes an energy shift equal to the diagonal exchange parameter J (Fig.2), where it is noted that the spin-nonflip transitions virtually stay in resonance with one another irrespective of the magnitude of J. (ii) The shifted phonons are wiped out, i.e., do not interact with metastable Cr^{3+} in the wings of the inhomogeneously broadened line, if their mean free path Λ against reabsorption by another Cr^{3+} with a suitable splitting surpasses the typical dimension of the excited zone. The latter is identified with R. In working out the model, we take J to decrease with distance according to $J = J_0 \exp(-ar)$, in which $J_0 = 330$ cm^{-1} and a = 0.1 nm. On the assumption of a random distribution of Cr^{3+} ions, we then derive for the distribution of the exchange parameters

$$D_J = \frac{4\pi N_0[\ln(J/J_0)]^2}{a^3 J} \exp\left(\frac{4\pi N_0[\ln(J/J_0)]^3}{3a^3}\right) . \quad (1)$$

For a typical ground-state concentration N_0 = 2500 at.ppm, D_J is depicted in Fig.3. For reasons of tractability, the metastable Cr^{3+} are divided in those with a $J > J_{crit}$ sufficient to provide phonon wipeout (type II), and those with an exchange splitting too small to do so (type I). Here, J_{crit} is determined by Λ in relation to R. We have

$$\Lambda = 2\rho T_d^{(f)} v / D_J N^* , \quad (2)$$

in which ρ is the density of phonon states, $T_d^{(f)}$ is the spontaneous spin-flip decay time, and v is the velocity of sound. For $N^* = 10^{18}$ cm^{-3} and L = 200 μm, for example, J_{crit} = 0.05 cm^{-1}, which is already 3 times the width of the transition, and corresponds to a Cr^{3+}-Cr^{3+} separation of 0.9 nm. The concentration of optically excited Cr^{3+} active in a wipeout N^{II} is now found by integration over all J above J_{crit}. In our example, $N^{II}/N^* = 0.27$. Solving the rate equations for N^I,N^{II}, and the phonon populations, we finally have under the relevant prevailing conditions

$$T_{eff} = T_d^{(f)}(N^* + \rho\Delta\nu)/\alpha N^{II} , \qquad (3)$$

in which $\Delta\nu$ is the resonant width and α expresses the coupling of the resonant phonons with type-II Cr^{3+} with respect to type-I Cr^{3+}. The solid curve in Fig.1 essentially represents Eq.(3), where all parameters entering are known, except for α, which was set to 6 in consistency with evidence from the magnetic-field data. At low N*, boundary-limited spatial diffusion is effective. When its effects are added to Eq.(3), resulting in the dashed curve in Fig.1, excellent overall agreement is achieved.

The dependence of T_{eff} on a magnetic field parallel to the trigonal axis is quite complicated. For fields large enough to separate the Zeeman transitions, but smaller than J_{crit}, there are four distinct phonon packets, each of which communicates with as many type-II Cr^{3+} as in the zero-field case. The number of participating modes has, however, quadrupled, and accordingly T_{eff} drops to one fourth of its zero-field value. Upon further increase of the field, the spin-flip-phonon packets become out of resonance with the active Cr^{3+} ions, and consequently no longer decay. The result is a recovery of T_{eff} by a factor of two. At very strong fields, a spin-nonflip-phonon packet can only be reabsorbed by a type-II ion in a single spin-nonflip transition, yielding another factor of two of recovery. The half-value of the initial drop vs field thus is a measure of the resonant-phonon packet width, while the typical field at which T_{eff} is restoring is a gauge for the transition width of the active ions. These effects are observed in the experiments [3].

In summary, a model based one-site Orbach processes in metastable Cr^{3+} coupled by the diagonal part of the exchange to nearby Cr^{3+} is found to provide an adequate description of the $2\bar{A}(^2E)$ relaxation, in particular the flattening out at high N*. This, of course, does not preclude other mechanisms [2] to be in effect to some extent. A full account of this work can be found in [3].

References

1. K.F. Renk, J. Deisenhofer: Phys. Rev. Lett. 26, 764 (1971)
2. R.S. Meltzer, J.E. Rives, W.C. Egbert: Phys. Rev. B 25, 3026 (1982)
3. R.J.G. Goossens, J.I. Dijkhuis, H.W. de Wijn: Phys. Rev. B 32, 7065 (1985)
4. J.I. Dijkhuis, A. van der Pol, H.W. de Wijn: Phys. Rev. Lett. 37, 1554 (1976)

Monte Carlo Study of Effective Decay Time of High-Frequency Phonons

*S. Tamura**

Department of Physics, University of Illinois at Urbana-Champaign, Urbana, IL 61801, USA

Anharmonic interaction of acoustic phonons satisfying $h\nu \gg k_BT$ (T being the ambient temperature) is highly frequency dependent. The decay probability of phonons proportional to ν^5T^0 has been predicted by nonlinear elasticity theory [1]. This ν^5-law is characteristic to the spontaneous two-phonon decays and applicable to both the LA and TA phonons in anisotropic solids [2]. With the use of an optical technique of tunable phonon detection BAUMGARTNER et al. [3] have verified experimentally this strong frequency dependence of the down-conversion rate of the phonons. The observed phonon lifetime is, however, not the one being calculated directly from the anharmonic Hamiltonian. This is because there exists strong mass-defect scattering of the phonons in $CaF_2:Eu^{2+}$ sample used in the experiment [3]. In the experimental frequency range (1-3 THz) this elastic scattering time is shorter than the inelastic scattering time and the population of the TA phonons should dominate the LA-phonon population. The lifetime of the TA phonons is generally considered to be very long by the kinematical reason and usually assumed to be infinite. The purpose of this paper is to study by the Monte Carlo method the effective down-conversion time of high-frequency phonons. The presence of strong mass-defect scattering and TA phonon decay will be explicitly taken into account.

Anharmonic decay rate of a phonon $\lambda=(\vec{q},j)$ ($\vec{q}$ being the wave-vector and j the phonon mode) is given by

$$\tau_A^{-1}(\lambda)= \alpha(\lambda)\, \nu^5(\lambda). \tag{1}$$

The coefficients α's for three phonon modes in CaF_2 are calculated by using the nonlinear elasticity theory with published values of the second- and third-order elastic constants [4]. The theoretical decay rates of 1-THz phonons with $\vec{q}$ being oriented in the $(1\bar{1}0)$ plane are shown in Fig. 1. The decay rates are very anisotropic especially in TA modes. (T1 and T2 denote the lower and upper TA branches, respectively.) It should be noted that along certain directions the spontaneous decay of the T2 phonons which is prohibited by the isotropic approximation becomes the same order of magnitude as that of the LA phonons, though the decay of the T1 phonons can be neglected practically. The averaged decay rates $\bar{\tau}_{A,LA}^{-1}=6.9\times10^5$ sec^{-1} and $\bar{\tau}_{A,T2}^{-1}=1.7\times10^5$ sec^{-1} of the LA and T2 phonons at 1-THz frequency are also indicated in this figure together with the decay rate $\tau_{A,iso}^{-1}=5.8\times10^5$ sec^{-1} of the LA phonons in the isotropic approximation at the same frequency.

The mass-defect scattering of a phonon λ into another phonon λ' is described by the following differential scattering rate

* Permanent address: Department of Engineering Science, Hokkaido University, Sapporo 060, Japan

$$\Delta\tau_D^{-1}(\lambda,\lambda') = \pi^2 V_0 g \nu^4(\lambda) \frac{|\vec{\varepsilon}(\lambda)\cdot\vec{\varepsilon}(\lambda')|^2}{c^3(\lambda')} \Delta\Omega(\lambda')|_{\nu(\lambda')=\nu(\lambda)}, \tag{2}$$

where V_0 is the volume of a unit cell, $\vec{\varepsilon}$ is the unit-polarization vector of the phonons, c is the phase velocity, $\Delta\Omega$ is the small solid angle, and g is the constant representing the magnitude of mass fluctuations. In a CaF_2 sample used in the experiment by BAUMGARTNER et al., 0.003 mole % of Eu^{2+} ions are doped. The important sources of mass-defect scattering in this sample are isotopic Ca atoms and doped Eu^{2+} ions. If we assume that the latters occupy the lattice sites of Ca^{2+} ions it can be deduced that $g=1.39 \times 10^{-4}$. Integrating Eq. (2) over solid angles subtended by the surfaces of constant frequency $\nu(\lambda')= \nu(\lambda)$ in $\vec{q}'$ space and summing over j', we obtain the total mass-defect scattering rate of the phonon λ

$$\tau_D^{-1}(\lambda)=2.65\times 10^{-42}\ \nu^4(\lambda)\ \ \text{sec}^{-1}. \tag{3}$$

The explicit procedures for the Monte Carlo calculations of the effective decay time are as follows: At each initial frequency ν, a large number of starting phonons (e.g., $n_0=10^4$) are assumed to be excited in the bulk of CaF_2 sample. The directions of the wave vectors of initial phonons are then given randomly. For relative populations of the initial phonons of each mode we have considered two cases, that is, the case in which purely one mode out of three acoustic branches is excited and the other case in which the mixing of three modes is realized. In the latter case the relative phonon populations are assumed to be given by the densities of states proportional to c^{-3}. For each phonon provided in this manner the total probability of interaction per unit time is given by $\tau_{tot}^{-1} = \tau_A^{-1} + \tau_D^{-1}$ and this is used to generate the time for the first collision. The branching ratio τ_A^{-1}: τ_D^{-1} is then applied to determine whether the collision process is the mass-defect scattering or the anharmonic decay. If the process is the mass-defect scattering the mode and direction of the scattered

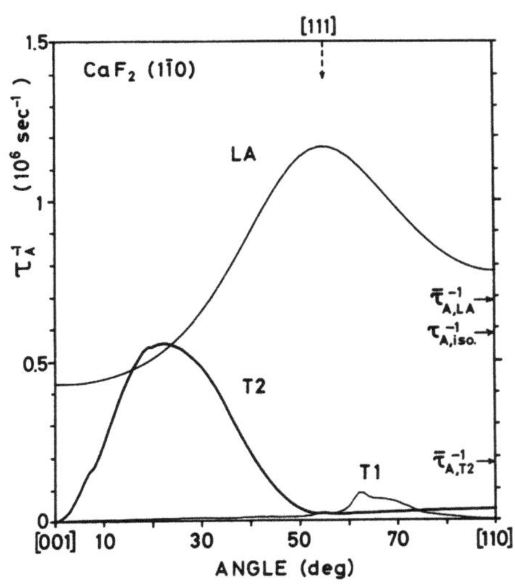

Fig. 1 Calculated anharmonic decay rates in the (1$\bar{1}$0) plane of CaF_2

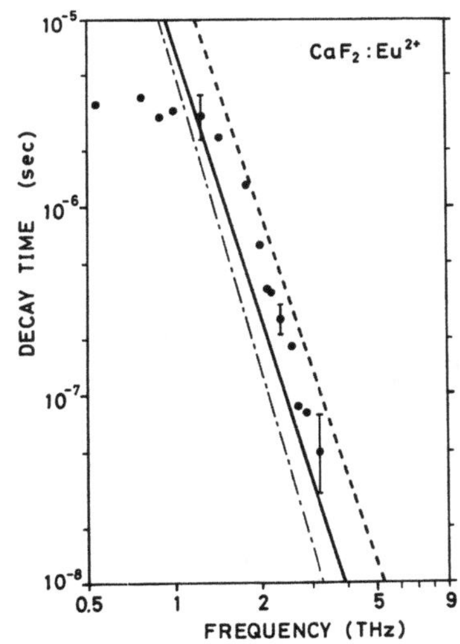

Fig. 2 Frequency dependence of effective decay time. (See text.) Experimental data are from ref. 3

phonon are calculated by Eq.(2) and the process is repeated. If the down-conversion occurs, we record the elapsed time and then start the calculation for the next initial phonon. After all phonons being tracked are down-converted, we sort the times needed for the first down-conversion and identify the effective decay time of the phonons by the time at which the number of nondecaying phonons n is reduced to $n/n_0=e^{-1}$.

The calculated frequency dependences of the effective decay times are displayed in Fig. 2 together with the experimental data [3]. Solid line shows the result derived by assuming the complete mode mixing for the initial phonons. Both the LA and TA phonon decays are taken into account. The result obtained by assuming that only the LA phonons decay anharmonically is plotted by dashed line. Dot-dash line represents the decay rate obtained in the case that only the LA phonons are populated initially. (Both the LA and TA phonon decays are assumed.) The corresponding plots of the decay times obtained in the cases that only the T1 or T2 phonons are populated initially are very close to the solid line (they lie slightly upwards and downwards of the solid line, respectively) and then being omitted from this figure. The agreement of the solid line with the experimental data is good at frequencies for which the spontaneous, anharmonic decay governs the origin of the frequency down-conversion.

Here, some further comments should be added. In order to consider the possible effects of the foreign impurity scattering other than that due to Eu^{2+} ions, we have repeated the Monte Carlo calculations by varying artificially the strength of the mass-defect scattering. It is found that with increasing the scattering rate the effective decay times are shortened only slightly and approach values about 20% smaller than those obtained here. Taking these considerations together with the results of Fig. 2 into account, it may be concluded that the T2-phonon-decay rate (Fig.1) has been overestimated to some degree because the calculated decay time (solid line) is shorter than the experiment. It would be due to overestimation of the T2 phonon decay by the inclusion of collinear process $T2\rightarrow T2+T2$, which has been automatically counted in our numerical calculations based on the zone-integration method [5]. The collinear decays are, however, prohibited if the dispersive effects are more important than the energy broadening by collisions. This is indeed the case being held for high-frequency phonons we have considered in the present paper.

References

1. R. Orbach and L. A. Vredevoe: Physics 1, 91 (1964).
2. S. Tamura and H. J. Maris: Phys. Rev. B31, 2595 (1985).
3. R. Baumgartner, M. Engelhardt, and K. F. Renk: Phys. Rev. Lett. 47, 1403 (1981).
4. S. Alterovitz and D. Gerlich: Phys. Rev. 184, 999 (1969).
5. G. Gilat: in Methods in Computational Physics, edited by G. Gilat (Academic, New York, 1976) 15, 317.

Vibrational Dynamics of Cadmium Ions in β-Alumina Crystals

G. Mariotto and E. Cazzanelli

Dipartimento di Fisica and Unita' CISM, Universita' di Trento, I-38050 Povo, Italy

Recently we have reported the results of a room temperature Raman study of the vibrational dynamics of mixed sodium-cadmium β-alumina crystals [1,2]. Mixed compositions Na_{1-y}-Cd_{y-2} , where y is the fraction of replaced sodium, were obtained by ionic exchange technique from sodium β-alumina crystals, containing ~22% of soda excess [3]. Raman spectra show remarkable effects related to cation substitution both as concerns bandwidths and intensities of the spinel block modes. For compositions with y>0.4, the overall spectral shape approaches the one typical of a partial density of states, due to the presence of charge-compensating vacancies in the mirror plane . On the other hand, the widths of spinel block bands show a more or less pronounced minimum at about 30% of exchanged sodium. These results have been interpreted as the evidence that the maximum degree of order in the mixed system occurs for the pseudostoichiometric composition, i.e. y~0.36 [1].

In this work the vibrational dynamics of the mirror plane sublattice will be more carefully considered. Our analysis will also concern the temperature dependence, in the range 30 - 300 K, of low-frequency spectra of the crystal with about 88% of sodium replaced by cadmium.

Figures 1 and 2 show experimental Raman spectra of some mixed Na-Cd β-alumina crystals, carried out at room temperature in a'(a,a+a')c and a'(c,c)a polarizations, respectively, where a' and a are a couple of unoriented, perpendicular axes of the mirror plane. Raman bands connected with the cadmium ions vibrations occur in the region below 50 cm^{-1} [2]. For compositions with y<0.4 they consist of two well separated peaks, α and β, centered at ~24 cm^{-1} (α) and ~34 cm^{-1} (β). With more than 40% of exchanged sodium, a third feature (β'), peaked at ~40 cm^{-1}, is present with the same polarization character as β. On the contrary, no evidence of a similar feature, arising near the band α, can be gained from our data, even if the maximum of this band shifts to 27 cm^{-1} with increasing sodium replacement. The intensity evolution of the β' band seems to suggest that, in sub-stoichiometric compositions (y>0.4), cadmium ions feel slightly different potentials. They must locate in low-symmetry sites, as revealed by the polarization properties of their Raman bands. These bands show predominant xx+yy and zz Raman activities, i.e. A_{1g} symmetry, as espected for ions in low-symmetry sites of D^4_{6h} space group. More specifically, the peak α shows a relatively stronger activity in zz polarization than in the xx+yy one. On the contrary, the features β and β' seem to be more connected with the in-plane vibrations. In addition, the crystals with high cadmium content show a broad γ feature, centered at ~85 cm^{-1}, which is the sum of two components, as is clearly observed at low cadmium concentrations [see Fig. 2]. The shape of this γ band recalls a vibrational density of states, while its intensity is strongly dependent on crystal composition. Moreover, γ band is totally zz polarized [see Fig. 3]. Finally, its temperature behaviour is, at first sight, quite surprising. In Fig. 4 we report the experimental Raman spectra obtained from nearly pure cadmium β-alumina

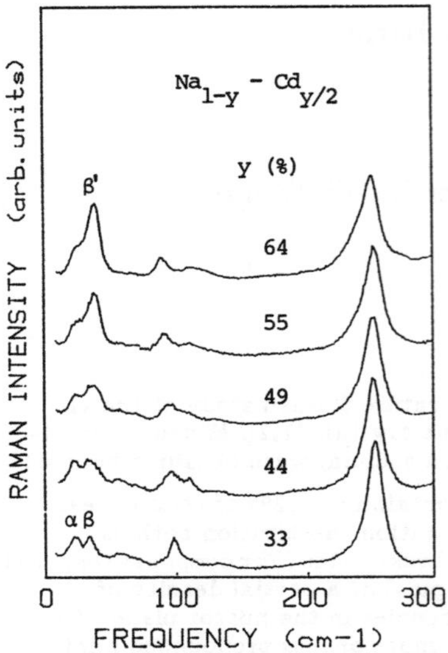

Fig. 1. Low-frequency Raman spectra of some mixed Na_{1-y}-$Cd_{y/2}$ β-alumina crystals, obtained at room temperature in a'(a,a+a')c polarization

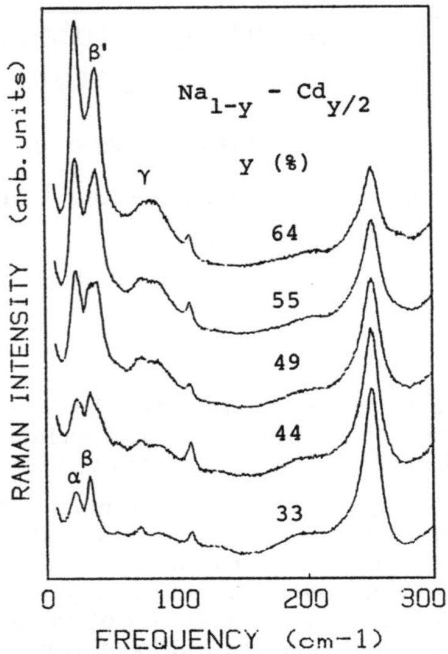

Fig. 2. Low-frequency Raman spectra of some mixed Na_{1-y}-$Cd_{y/2}$ β-alumina crystals, obtained at room temperature in a'(c,c)a polarization

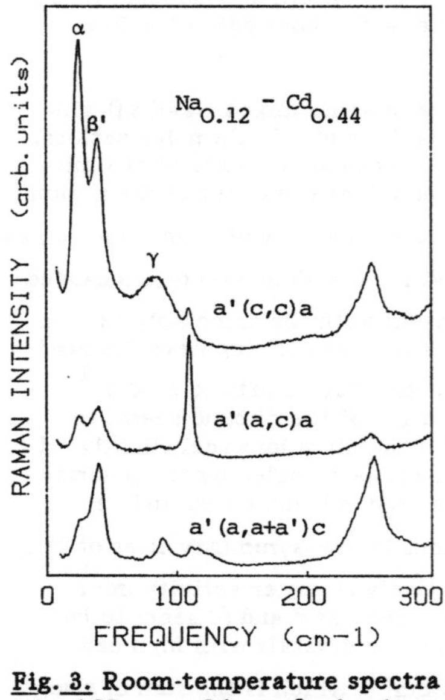

Fig. 3. Room-temperature spectra of mixed $Na_{0.12}$-$Cd_{0.44}$ β-alumina crystal for some polarization settings

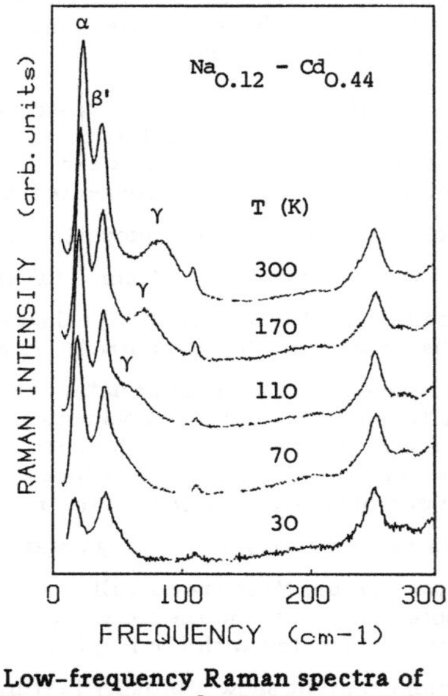

Fig. 4. Low-frequency Raman spectra of mixed $Na_{0.12}$-$Cd_{0.44}$ β-alumina crystal at different temperatures

crystal (y≃0.88) in a'(c,c)a polarization at different temperatures, down to liquid helium. The γ feature shows a sharp decrease both in the peak frequency and, apparently, in the intensity, as the temperature decreases, until it merges down in the β+β' structure at 30 K. In the meantime the peak α undergoes an analogous shift, passing from ~27 cm^{-1} at 300 K, to ~19 cm^{-1} at 30 K.

In order to account for these experimental findings we should remember that we are dealing with a strongly disordered system from both the chemical and structural point of view. Therefore, it seems reasonable to connect the temperature evolution of the γ band, both in frequency and in intensity, with the continuous activation of phonon branches at non-zero momenta. This would be due to the breaking of wavevector selection rules operated by the diffuse disorder of the conduction plane. Optical branches with dispersion behaviours consistent with this hypothesis have been found at room temperature in sodium β-alumina by SHAPIRO and REIDINGER [4], with an elastic neutron scattering. These branches approach the energy value of ~5 meV near the Γ-point (k≃0), while for increasing wavevectors they behave like acoustical ones. The observed shift of the γ band maximum can be accounted for by supposing an activation of these branches up to some temperature-dependent cut-off value of the wavevector. No definitive picture of the disorder responsible for such an activation can be given on the basis of only our data, even if it seems reasonable to postulate a displacement of the cadmium ion equilibrium position away from the mirror plane, along the c-axis direction. In this way the observed intensity could be related to the mean displacement, which increases with the temperature. Instead, the shift of the γ band could be connected with the temperature decrease in the correlation length of the displacement fluctuations. Within this model we can justify the frequency shift of the α band, which reflects a hardening of the potentials felt by cadmium ions along the c-axis direction. Finally, in this framework the vibrational frequency of cadmium ions for in-plane modes would not be affected, as it results from the temperature independence of β and β' peak position.

REFERENCES:

1. G. Mariotto, G.C. Farrington and E. Cazzanelli: in Transport-Structure Relations in Fast Ion and Mixed Conductors, ed. by F.W. Poulsen, N. Hessel Andersen, K. Clausen, S. Skaarup and O. Toft Sorensen (RISO National Laboratory, Roskilde 1985) p. 455
2. G. Mariotto and G.C. Farrington: Solid State Ionics 18&19, 619 (1986)
3. P.H. Sutter, L. Cratty, M. Saltzberg and G.C. Farrington: Solid State Ionics 9&10, 295 (1983)
4. S.M. Shapiro and F. Reidinger: in Physics of Superionic Conductors, ed. by M.B. Salomon (Springer, Berlin Heidelberg 1979) p.45

Terahertz-Phonons in Diamond

H. Schwartz[1], *K.F. Renk*[1], *A. Berke*[2], *A.P. Mayer*[2], and *R.K. Wehner*[2]

[1]Institut für Angewandte Physik, Universität Regensburg, D-8400 Regensburg, Fed. Rep. of Germany

[2]Institut für Theoretische Physik II, Westfälische Wilhelms-Universität, D-4400 Münster, Fed. Rep. of Germany

We have performed measurements and calculations of spontaneous lifetimes of acoustic phonons in diamond.

The phonons were generated [1] by defect-induced one-phonon absorption of CO_2 laser radiation in a natural diamond crystal of type Ia; i.e. the crystal contained nitrogen impurity centers, among others the N3 center that gives rise to blue fluorescence. The laser radiation frequency, 28 THz (wavelength 10.6 µm), lies within the acoustic band of the phonon dispersion curves of diamond. We observed decay products of the originally generated phonons by vibronic sideband spectroscopy [2]. By means of a cw dye laser, we excited resonantly N3 centers at a wavelength of 415.2 nm and investigated phonon-induced fluorescence. The anti-Stokes fluorescence intensity at frequency separation ν from the zero-phonon line (415.2 nm) is a measure of the phonon occupation number $n(\nu)$.

A diamond crystal, size $(2.5\ \mathrm{mm})^3$, with an absorption coefficient of $5\ \mathrm{cm}^{-1}$ at 10.6 µm, was immersed in superfluid He at 2 K. A dye laser beam (power 180 mW) was focused to the center of the crystal for excitation of N3 centers. Phonons were generated by irradiating the crystal almost homogeneously with the pulses of a Q-switched CO_2 laser (peak power 2 kW; pulse duration 250 ns, repetition rate 80 Hz and, improved, 100 ns, 3.5 kHz). Fluorescence radiation was analysed with a double-grating monochromator (resolution 250 GHz) and a photon counting system (time resolution 20 ns). We detected phonons in the linear part of the dispersion curves from 1 to 7 THz. The laser pulse energy was kept low, $n(\nu) \ll 1$, to avoid phonon recombination.

Phonon-induced fluorescence curves are shown in Fig. 1 for two detection frequencies. At $\nu = 5$ THz, the signal increases within a time comparable to the rise time of the CO_2 laser pulse and decays with a time constant of about 400 ns. The signal at 1 THz reaches its maximum delayed and decreases with a larger time constant. The signal decay time, corresponding to the phonon lifetime, is plotted in Fig. 2 for different frequencies. The lifetime is almost constant up to a frequency of 3 THz and decreases rapidly above 4 THz. The decrease is consistent with the ν^{-5} dependence that is characteristic for spontaneous, anharmonic decay [3]; for recent theoretical studies see [4,5].

In a theoretical study, we have calculated the phonon damping constant $\Gamma_{\hat{q}j}(\nu) := \Gamma_{\vec{q}j}$ for different normalized wavevector directions $\hat{q}$ and polarizations j, taking account of cubic anharmonicity using the anisotropic nonlinear elastic continuum model as described in [5]. The calculations have been carried out for 36 directions in the irreducible part of the solid angle. For the third order elastic constants, only room temperature data are available. Because of the compensations between linear and nonlinear elastic constants, which usually occur in the Kun Huang tensor entering the third order coupling constants, we used the following set of consistent data [6,7]; unit $10^{11}\ \mathrm{N/m^2}$:

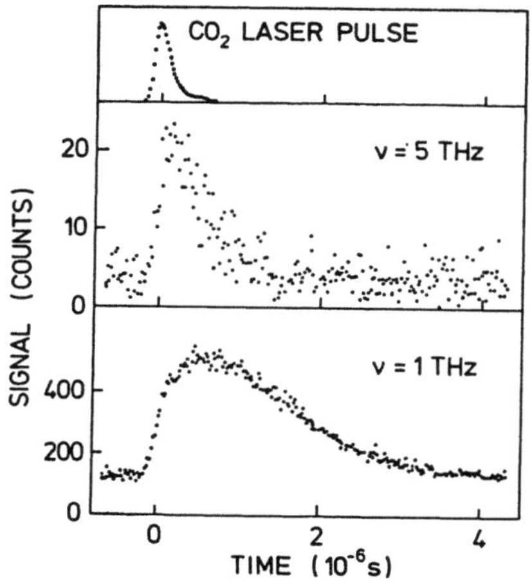

Fig. 1. Phonon-induced fluorescence signals and CO_2 laser pulse

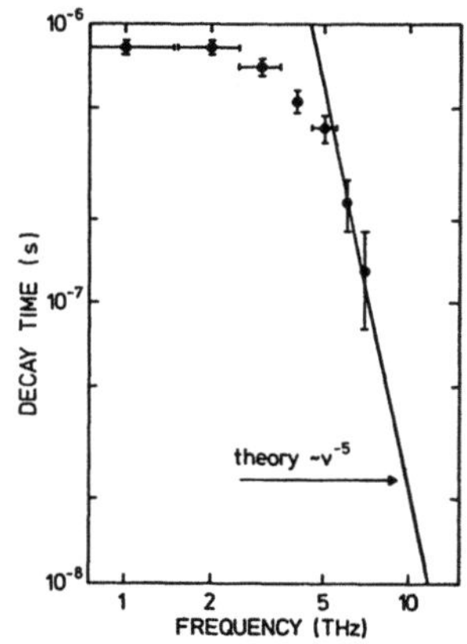

Fig. 2. Phonon lifetime; experiment (points) and theory (solid line)

$c_{11} = 10.76$, $c_{12} = 1.25$, $c_{44} = 5.75$; $c_{111} = -62.60$, $c_{112} = -22.60$, $c_{123} = 1.12$, $c_{144} = -6.74$, $c_{166} = -28.60$, $c_{456} = -8.23$. In order to achieve an accurate calculation of the density of final states, being of particular importance for the decay of transverse acoustic phonons, we have used the low-temperature elastic constants [8] $c_{11} = 10.808$, $c_{12} = 1.250$, $c_{44} = 5.789$ for determination of the sound velocities $v_{\hat{q}j}$ and eigenvectors; for the mass density we used ρ = 3 593 kg/m^3. For wave vectors in the symmetry planes the results referring to the decay of longitudinal (L) and fast transverse phonons (FT) are represented in Fig. 3 by smoothed curves.

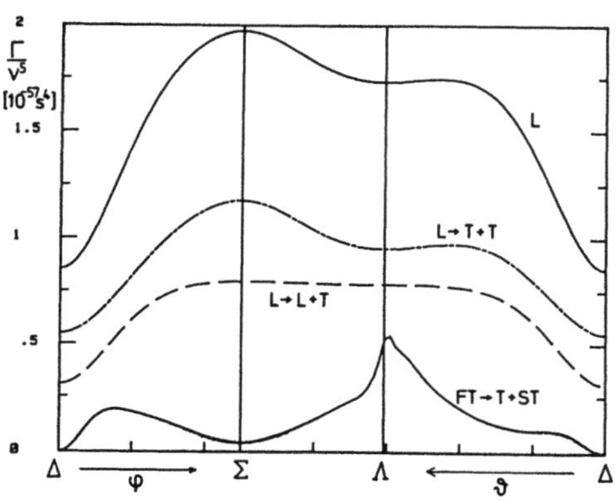

Fig. 3. Theoretical reduced damping constants for longitudinal (L) and fast transverse (FT) phonons

Concerning the experimental results, we suggest that a fast mode conversion of the phonons, due to elastic scattering at impurities, with scattering times short compared to the anharmonic lifetimes, leads to a balanced mode mixture of phonons at fixed frequency. A comparison of the experimental and theoretical data is therefore done by calculating mean damping constants by averaging over all $\hat{q}$-directions at constant frequency with the density of states, $D_{\hat{q}j} \sim \nu^2 v_{\hat{q}j}^{-3}$, as weight factor,

$$\bar{\Gamma}_j(\nu) = \int d\Omega v_{\hat{q}j}^{-3} \Gamma_{\hat{q}j}(\nu) \;/\; \int d\Omega v_{\hat{q}j}^{-3} \;.$$

Neglecting the decay of slow transverse (ST) phonons, we obtain for the total damping constant

$$\bar{\Gamma}_{tot}(\nu) = [D_L \bar{\Gamma}_L(\nu) + D_{FT} \bar{\Gamma}_{FT}(\nu)] \; (D_L + D_{FT} + D_{ST})^{-1}.$$

The values of the average damping constants are $\bar{\Gamma}_L(\nu)/\nu^5 \simeq 1.7\cdot10^{-57} s^4$, $\bar{\Gamma}_{FT}(\nu)/\nu^5 \simeq 1.4\cdot10^{-58}\ s^4$, $\bar{\Gamma}_{tot}(\nu)/\nu^5 \simeq 2.8\cdot10^{-58}\ s^4$; accordingly the ratio of the average rates for longitudinal and fast transverse phonon decay is about 4. From the value of $\bar{\Gamma}_{tot}$, we derive a decay time $\tau = \frac{1}{2}\,\bar{\Gamma}_{tot}^{-1} \simeq 1.8\cdot10^{57}\ s^{-4}\cdot\nu^{-5}$ shown in Fig.2.

The slow transverse phonons have been found to decay into two transverse phonons only in a very narrow region of the solid angle (predominantly near the Λ-direction), where their damping constant can however become considerably high. The almost singular angular distribution of the decay products suggests in this case, that the damping constant of the ST phonons will be strongly influenced by the effects of dispersion and anharmonic processes of higher order than calculated here.

The deviation from the ν^{-5} dependence at lower frequencies is attributed to spatial phonon escape from the crystal. The experimental decay rates may then be described by the sum $\tau^{-1} + \tau_d^{-1}$ where τ_d^{-1} is a diffusive-escape rate. It follows from the experiment that $\tau_d \simeq 750$ ns at 4 THz and that the phonon lifetime against elastic scattering is 5 ns. This indicates that our assumption of fast mode-mixing is justified.

Our analysis suggests that both longitudinal-phonon decay and fast transverse-phonon decay contribute in the same order of magnitude to the decay of a mixture of acoustic phonons in diamond.

We would like to thank J. Giordmaine and the de Beers Industrial Diamond Division for lending us diamond crystals. Our work was supported by the Deutsche Forschungsgemeinschaft.

References

1. H. Schwartz and K.F. Renk: Solid State Commun. 54, 925 (1985).
2. "Nonequilibrium Phonon Dynamics", ed. W.E.Bron (Plenum Press, N.Y. 1985), NATO ASI Series B: Physics, Vol. 124.
3. R. Baumgartner, M. Engelhardt, and K.F. Renk: Phys. Rev. Lett. 47, 1403 (1981).
4. S. Tamura and H.J. Maris: Phys. Rev. B 30, 2595 (1985).
5. A. Berke, A.P. Mayer, and R.K. Wehner, Solid State Commun. 54, 395 (1985).
6. P.N. Keating: Phys. Rev. 145, 637 (1966).
7. M.H. Grimsditch, E. Anastassakis, and M. Cardona: Phys. Rev. B18, 901 (1978).
8. H.J. McSkimin and P. Andreatch: J. Appl. Phys. 43, 2944 (1972).

Part IX

New Methods and Phenomena

New Developments in Tunnel-Junction Phonon Spectroscopy

W. Dietsche

Physik Department E10, Technische Universität München, D-8046 Garching, Fed. Rep. of Germany

1. Introduction

The physics of high-frequency ballistic phonons has profited immensely from the use of superconducting tunnel junctions. Since the pioneering experiments of Eisenmenger and Dayem [1] the phonon work with tunnel junctions has developed in different directions. Some of them will be reviewed here.

The generation of monochromatic phonons by using the bremsstrahlung of quasiparticles in tunnel junctions was first demonstrated by Kinder in 1972 [2]. Frequencies as high as 2000 GHz [3] and more can now be reached with this technique. Thus, solid-state spectroscopy in a large number of materials became possible [4]. The frequency resolution was greatly improved to better than 40 MHz with the phonons emitted during ac Josephson tunneling [5]. This allowed the direct observation of the zero-field hyperfine splitting in $Al_2O_3 : V^{3+}$ [4].

2. Narrow-band detector

On the detector side too, the use of tunnel junctions has made progress. It is well known that an important property of these detectors is the detection-threshold frequency at $2\Delta/h$ above which the breaking of Cooper pairs becomes possible. In the case of Pb the threshold frequency is 650 GHz. With many sample materials, e.g. Ge or Si, this property is complemented by an upper frequency limit imposed by elastic scattering in the bulk. Isotope scattering reduces the intensity of the ballistic phonons with increasing frequency f by a factor exp $(-Af^4l)$ where A is a material constant and l is the phonon-path length.

These two properties together form a narrow-band detector extending from 650 to about 800 GHz. In this frequency range the phonon-dispersion relation starts to deviate from the linear behavior in many materials. Although the deviation is only a few percent it leads to dramatic effects in the phonon focusing which could be studied by the phonon-imaging technique developed by Northrop and Wolfe [6].

In order to study the nonlinear dispersion, a broad distribution of phonon frequencies was generated by optical excitation of a Ge (110) surface [7]. A Pb tunnel junction was placed on the opposite surface. The excitation spot was raster-scanned across the sample allowing the probing of the phonon intensities in all propagation directions. The results were displayed as brightness on a TV monitor (Fig. 1 (a)). The phonon-flux distribution seen in this measurement is an example of the phonon focusing. The result of Fig. 1(a) is, however, different from the one observed with phonons of lower frequencies. In Fig. 1(b) an image recorded with the same

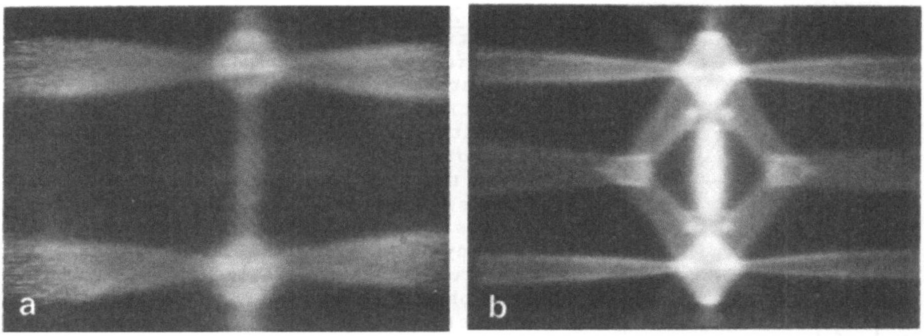

Fig. 1. Focusing images taken with a Pb detector (a) and a bolometer (b)

sample but with a bolometer detector is shown. A comparison shows that the ridges extending from the two <100> diamonds are wider in the high-frequency band. This as well as other differences could be traced back to the phonon-dispersion relation in Ge [7,8].

More recently, very precise data were obtained by G.A. Northrop et.al. in GaAs with a similar technique [9]. They compared the shape of the focusing patterns with the predictions of several competing lattice - dynamics models. Only a shell model predicted the phonon-focusing behavior correctly. This is an impressive demonstration that focusing can be a very sensitive probe for subtle properties of the lattice dynamics.

The narrow-band detection behavior of Pb junctions can also be used to study the frequency spectra of phonon sources. A very interesting example is the two-dimensional electron gas (2DEG) which can be created in MOS structures. At low temperatures and moderate electron densities the electrons occupy the lowest subband only which is a circle in the k plane parallel to the surface. The electron density and hence the Fermivector k_F can be varied with gate voltage. If the 2DEG is heated, e.g. by passing a current through it, then phonons are emitted. The largest value of the phonon-wavevector component within the 2DEG plane is of the order of $2k_F$. Thus the phonon frequency spectrum extend up to a maximum which depends on the emission angle and on the phonon polarization [10].

The existence of the cut-off was demonstrated with Pb junction detectors (Fig.2). These were placed on the opposite side of the (100) Si crystal onto which the MOS structure had been prepared (inset of Fig. 2). By variation of the electron density the 2 k_F-cut-off was tuned over the sensitivity band. This led to a sharp signal increase (see Fig.2).

Not only the phonon-frequency spectrum but also the angular distribution of phonon emission of a 2DEG show interesting effects. This will be demonstrated in the following section. In order to get a global picture of the angular distribution a spatially resolving phonon detector is necessary. The phonon-imaging technique described earlier in this article is not suitable for this task since it requires the scanning of the phonon source.

3. Spatially-resolving detector

It was demonstrated that a standard Al tunnel-junction detector can be operated in spatially resolving mode by inducing a nonequilibrium state

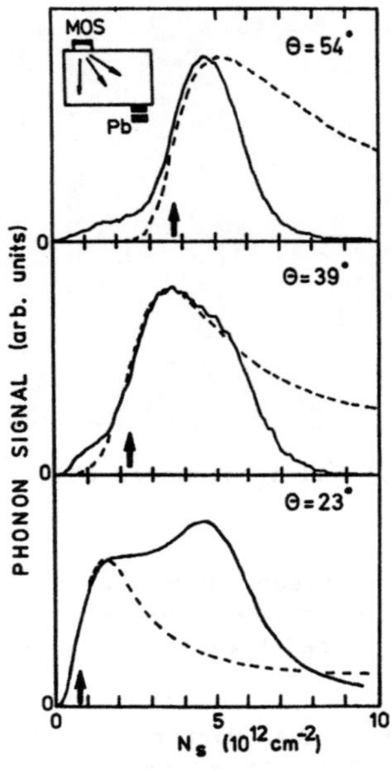

Fig. 2. Phonon signal as function of electron density (transverse phonon polarization). The three panels correspond to three emission angles, respectively. All traces show a steep increase at the density (arrows) where the high-frequency cut-off of the phonon spectra is swept over the detector window. This occurs at the smaller densities the smaller the emission angle because the cut-off frequency is higher. The dashed lines are the results of a detailed calculation of the phonon emission [10]. The agreement is very good at low carrier densities. The deviations at higher densities are probably due to the population of higher electron subbands [11].

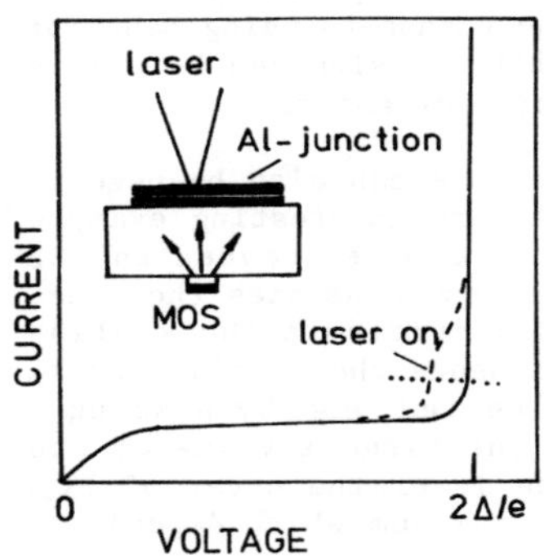

Fig. 3. Current-voltage characteristic of an Al tunnel junction used as spatially resolving phonon detector. In the experiment described in the text the junction was a rectangle measuring 6 x 4 mm^2. The laser power was typically 100 μW focused onto 10^{-2} mm^2, i.e. only a fraction of 5 x 10^{-4} of the junction area was illuminated. Inset: Experimental set-up. The Si crystal was (100) oriented. The MOS structure was prepared on the opposite side of the detector. It was a square with 0.4 x 0.4 mm^2.

with a laser [12]. The current-voltage characteristic of such a junction is shown in Fig.3 (solid line). To get the junction into the spatially resolving mode it has to be illuminated with a focused laser beam. The optical excitation leads to pair breaking and, consequently, to a reduction of the superconducting energy gap in the focal area. Thus, the characteristic in Fig. 3 gets a precursor (dashed line). Along the dashed portion of the characteristic the single particle tunnel current flows preferentially through the illuminated spot. By raster-scanning this spot can be moved over the junction area. A phonon flux impinging on the detector from the opposite side of the substrate leads also to a gap reduction. The phonons, however, reduce the voltage at the precursor only if the phonon flux and the laser hit the same spot on the junction.

Operation as a spatially resolved detector is now straightforward. As bias point the intersection with the dotted load line is chosen. The pho-

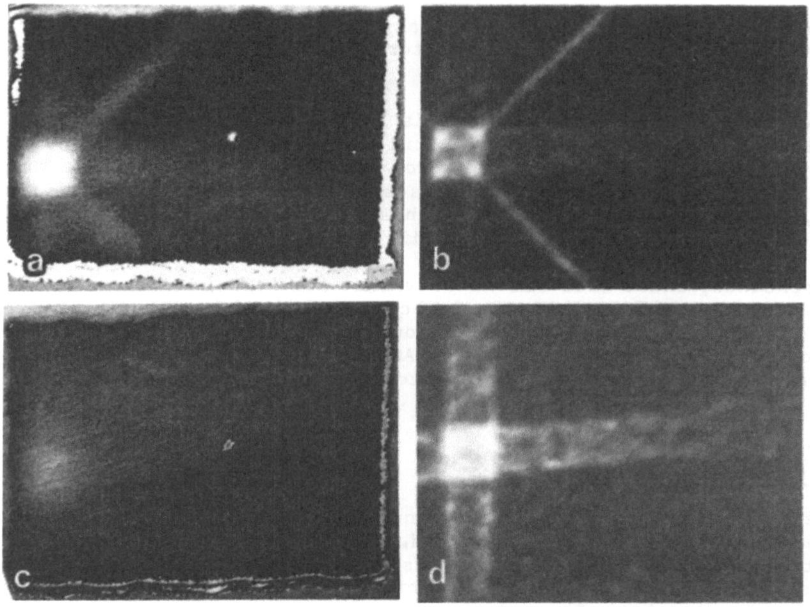

Fig. 4. (a) "Phonograph" of a metal-film heater on a (100) Si crystal; (b) computer simulated focusing pattern; (c) phonon emission of a 2DEG; (d) same as (c) but with the phonon intensities weighed with the deformation potential

nons are either pulse or ac modulated. The modulation of the junction voltage along a laser scan is displayed as brightness on a suitable video system. So far, this detector has been used to observe the phonon focusing in Si [12] and to obtain micro "phonographs" with less than 10 μ m spatial resolution of buried doping structures in Si wafers [13]. In this article, results on the angular distribution of the phonon emission of MOS structures will be presented.

The data were obtained with MOS phonon sources measuring 0.4 x 0.4 mm^2 which were prepared on (100) oriented Si crystals of 3 mm thickness. First, a "phonograph" of a metal-film heater placed atop of the MOS structure and having the same dimensions was taken (Fig. 4(a)). It showed a large intensity along the [100] direction (the bright square) and sharp ridges extending from it towards the <110> directions. The latter ones consisted of FTA phonons. In Fig 4(b) a computer simulation is shown. The experimental resolution was limited because of the relatively large size of the phonon source.

In Fig. 4(c) a phonograph of the 2DEG emission is shown (electron density: 10^{12} cm^{-2}, E_{SD} = 100 V/cm). It differs from the one taken with the heater in several respect. First, there is considerable less contrast and, second, the FTA ridges have disappeared. These effects can be understood if the anisotropy of the deformation potential is taken into account. In the computer-simulated focusing pattern of Fig. 4(d) each randomly chosen phonon direction is weighed with the respective deformation potential. As a consequence the FTA ridges vanished because the potential is zero for them. Furthermore, the contrast in the area around the [100] direction is reduced. Thus, the main features of the MOS phonograph of Fig. 4(c) can be qualitatively explained.

In addition to these features there is a faint but sharp structure at a point along the original FTA ridge. Strikingly, it is better resolved spatially than it were to be expected from the dimension of the 2DEG. Thus it is likely an edge effect. It was observed at all electron densities studied, but never with the metalfilm heater. Probably, it is connected with interface-phonon modes but the details of the process are not known.

Nevertheless, these data demonstrate that new information can be gathered with a spatially resolving detector. It would be interesting to combine it with monochromatic phonon sources and to measure angular and frequency dependence at the same time. Another point to be checked is if the narrow band Pb detector can be operated in a spatially resolving mode.

4. High sensitivity detector

The high sensitivity of tunnel-junction detectors has always been acknowledged by the workers in the field. Clearly, such a property should also put to use in other research fields. One interesting application is the measurement of the optical absorption spectra of submonolayers of adsorbed molecules [14]. The underlying idea is that the optical energy, after absorption by the molecule, is partly converted to phonons during a nonradiative relaxation process. The phonons can then be detected by the tunnel junctions. The technique is obviously related to the much older photoacoustical and photothermal techniques [15].

The experimental set-up is shown in Fig. 5 (a). A Si substrate was part of a vacuum chamber immersed in liquid Helium. Two Al tunnel-junction detectors were placed onto the outer side of the Si crystal. The inner side could be illuminated with quasi-monochromatic light coming from a 50 W halogen lamp in conjunction with a 1/4 m Czerny-Turner monochromator. The optical absorption of the substrate was minimized by covering it with an Ag film of 2000 Å thickness. Molecules could be deposited by means of a mini effusion cell. It consisted of a Ni tube, 1 mm in diameter and 10 mm in length, which was indirectly heated by resistive wire. The tube was filled

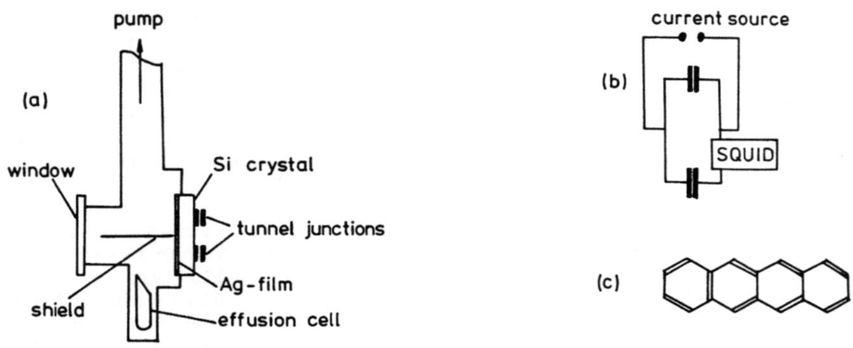

Fig. 5 (a) Set-up of the optical absorption experiment; (b) SQUID circuit; (c) tetracene molecule

with the substance to be deposited. The vapor pressure or, equivalently, the deposition rate could be controlled with a thermocouple soldered to the mouth of the tube. This design allowed the preparation of submonolayers with a substrate at a temperature of 1 K as well as at room temperature.

The optical absorption is determined from the phonons emanating from the illuminated surface. After thermalization in the Ag film they will travel ballistically through the substrate crystal. In order to get the optical absorption spectra of the molecules it is necessary to subtract out the background phonons caused by the light absorption in the Ag film. To this end, only the lower half of the crystal was covered with molecules and the difference between the two phonon detectors was measured. It turned out that by utilizing a SQUID in a suitable circuit this difference could be determined with a ten to a hundred times higher sensitivity as compared with our standard techniques.

The circuit is shown in Fig. 5 (b). The two tunnel junctions were connected in parallel using superconducting metal. The SQUID was inserted into one of the two arms. With equal phonon flux the junction current divided equally between the two branches. With differing flux, however, the junction impedances differed and the current distribution became asymmetric. This caused a signal in the SQUID.

The first molecule studied with this technique has been tetracene. Its structure formula is shown in Fig. 5 (c). Tetracene is an aromatic hydrocarbon molecule having an electronic S_0-S_1 transition at 471 nm if the molecule is in solution. The substance can form single crystals with two molecules per unit cell. The excited states become excitons which are split in two branches (Davydov-splitting [16]). The observation of the splitting is generally viewed as evidence of an ordered solid. Evaporated films with thicknesses of the order of 1000 Å are known to consist of crystallites if they are prepared on room-temperature substrates [17]. On cold (<100 K) substrates an amorphous film is observed [18].

In Fig. 6 the results of the phonon-emission measurements are shown. The signal, or more precisely, the difference between the detectors, was first normalized by the intensity of the Ag background signal. This way the spectral distribution of the light source was compensated. The different parts of Fig. 6 correspond to average film thicknesses ranging from 120 to 4 Å or 9 to 0.3 monolayers. A number of sharp absorption lines is visible which are at the same wavelengths as the lines observed in bulk single crystals (dashed vertical lines). These data were taken with films deposited on a room-temperature substrate. Apparently, at this temperature, the molecules are sufficiently mobile so that they cluster to crystallites.

This is in contrast to the results shown in Fig. 7 (a). In this case, the molecules were deposited at a substrate temperature of 1 K. Strikingly, the wavelengths of the absorption lines agree with the ones observed with dissolved molecules (dashed vertical lines). Thus, one can conclude that, on the cold substrate, the molecules remain isolated from each other after deposition. Surprisingly, the substrate effect on the electronic transition is the same as that of a solvent. The reason for this is not yet known.

Annealing of the sample of Fig. 7 (a) at room temperature and subsequent cooling led to the rearrangement of the molecules which is evident from Fig. 7 (b). The spectrum changed to the one observed with single crystals.

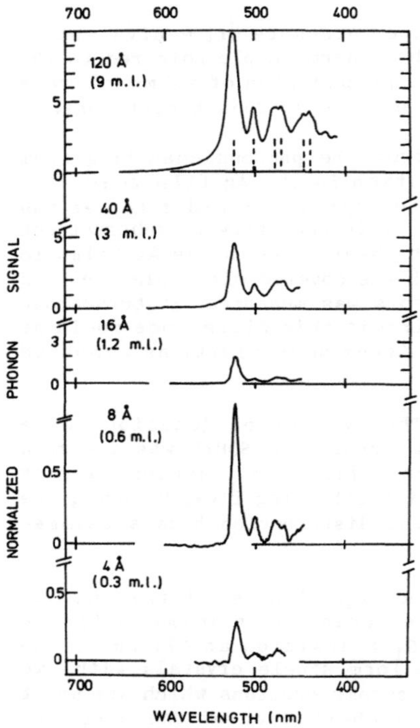

Fig. 6. Optical spectra of several tetracene coverages. Substrate temperature at deposition was 300 K.

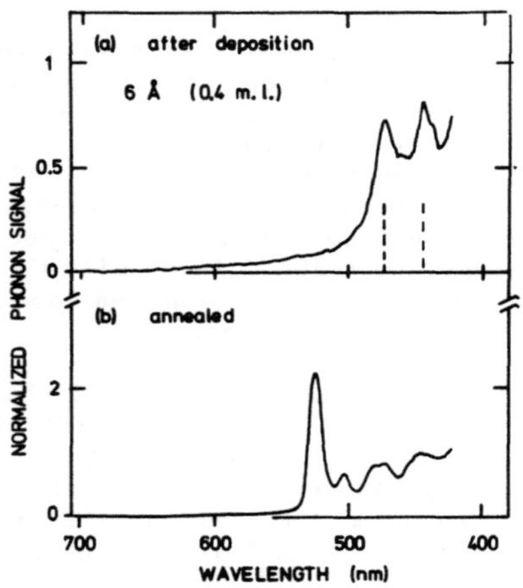

Fig. 7. (a) Optical spectrum measured with the molecules deposited on a substrate at 1K; (b) the same sample after annealing at room temperature

In conclusion, these data demonstrate that optical spectroscopy is possible by means of the phonon emission. With this technique it was possible to obtain information on the arrangement of adsorbed molecules in the submonolayer regime. The optical power incident on the sample was between 1 to 20 μW of which less than 1 % was absorbed. Presently the sensitivity is still limited by external noise which is picked up by the SQUID from the current leads.

This technique holds promise to be particularly useful in the infrared where surface studies are very often hindered by the limited sensitivity of the conventional techniques. In the visible, it should be possible to measure extremely low molecules coverages by exciting them with the high optical power of dye lasers.

5. Acknowledgements

It is a pleasure to acknowledge the contributions to this work of H.C. Basso, H. Kinder, L. Koester, G.A. Northrop, Th. Rapp, M. Rothenfusser, H. Schreyer, and J.P. Wolfe.

References

1. W. Eisenmenger and A.H. Dayem: Phys. Rev. Lett. 18, 125 (1967)
2. H. Kinder: Phys. Rev. Lett. 28, 1564 (1972)

3. W.Forkel, M. Welte, and W. Eisenmenger: Phys. Rev. Lett. 31, 215 (1973); see also K. Lassmann: this volume
4. For a review, see: H. Kinder: Nonequilibrium phonon dynamics, edited by W.E. Bron, Nato Advanced Study Institutes Series, 124 (Plenum, 1985) pg. 129
5. P. Berberich, R. Buemann, and H. Kinder: Phys. Rev. Lett. 49, 1500 (1982)
6. G.A. Northrop and J.P. Wolfe: Phys. Rev. Lett. 43,1424 (1979)
7. W. Dietsche, G.A. Northrop, and J.P. Wolfe: Phys. Rev. Lett. 47, 660 (1981)
8. S. Tamura: Phys. Rev. B 25, 1415 (1982)
9. G.A. Northrop, S.E. Hebboul, and J.P. Wolfe: Phys. Rev. Lett. 55, 95 (1985)
10. M. Rothenfusser, L. Koester, and W. Dietsche: submitted to Phys. Rev. B
11. M. Rothenfusser, L. Koester, and W. Dietsche: paper C 1, this volume.
12. H. Schreyer, W. Dietsche, and H. Kinder: Proceedings of the LT 17-Contributed-papers, edited by U.Eckern et al. (North-Holland, 1984) pg. 665
13. H. Schreyer, W. Dietsche, and H. Kinder: paper P 21, this volume
14. W. Dietsche, Th. Rapp, and H.C. Basso: J. Appl. Phys. 59, 1431 (1986)
15. C.K.N. Patel and A.C. Tam: Rev. Mod. Phys. 53, 517 (1981)
16. A.S. Davydov: Theory of Molecular Excitons (Plenum, N.Y., 1971)
17. W. Hofberger: Phys.Status Sol. A 30, 271 (1975)
18. R. Hesse, W. Hofberger, and H. Bässler: Chem. Phys. 49, 201 (1980)

Studies of High-Frequency Acoustic Phonons Using Picosecond Optical Techniques

H.J. Maris, C. Thomsen, and J. Tauc

Department of Physics and Division of Engineering, Brown University, Providence, RI 02912, USA

1. Introduction

In the last fifteen years much work has been devoted to the development of new techniques for the generation and detection of high frequency phonons. [1] Nevertheless, the ideal phonon generation-detection scheme has not yet been invented. One would like to be able to:

(1) Generate and detect phonons over a wide range of frequency from the upper limit of the ultrasonic range (~ 10 GHz or 1/2 K) to the highest frequencies of a lattice (~ 6000 GHz or 300K).

(2) Make measurements over a broad range of temperature.

(3) Apply the method to a wide variety of materials without the requirement that the materials be modified. For example, techniques in which it is necessary to introduce ions into a crystal to generate or detect phonons have the disadvantage that the ions also scatter phonons.

To make measurements at high frequencies and high temperatures one has to be prepared to study phonons with very short lifetimes τ and short mean free paths Λ. At room temperature a typical thermal phonon has a τ of 10 psecs. This short lifetime means that if one wants to do a phonon experiment in "real time" (as distinct from a resonance experiment, for example,) time-resolution in the picosecond range is required. This is impossible with conventional electronics. However, experiments in the picosecond time-domain are possible with optical techniques, and so one is naturally led to consider this method.

2. Phonon Generation and Detection by Picosecond Light Pulses

The first experiment [2] of this type that we performed is shown schematically in Fig. 1. A thin film of amorphous As_2Te_3 was deposited onto a sapphire substrate. A "pump" light pulse (duration 1 psec, energy 1nJ) was focussed onto a 50μ diameter spot on the surface. The light was absorbed within a distance ζ (~ 300 A) of the surface. The heating of this layer set up a thermal stress, which in turn launched a strain wave (coherent phonon pulse) into the film. This pulse bounced back and forth inside the film, with diminishing amplitude because of the combined effects of attenuation and partial transmission into the sapphire substrate. To detect the existence of the strain wave we measured the optical transmission through the film as a function of time t after the application of the pump pulse. We used a "probe pulse" (also of duration 1 psec) which was produced by splitting off a fraction of the energy from the pump pulse and then delaying it by means of a variable optical path length. The results in Fig. 2 clearly show oscillations in optical transmission due to the strain pulse bouncing back and forth in the film. The optical transmission is sensitive to the motion of the acoustic pulse because the optical constants of the film (particularly the absorption coefficient α) depend on strain.

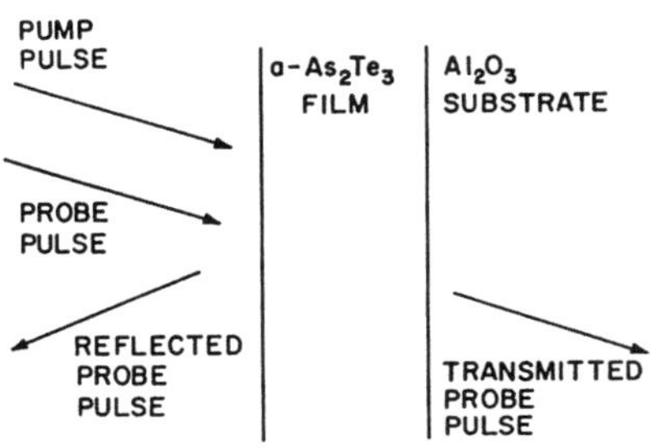

Fig. 1. Schematic diagram of experiment to generate and detect phonons in a semiconducting film

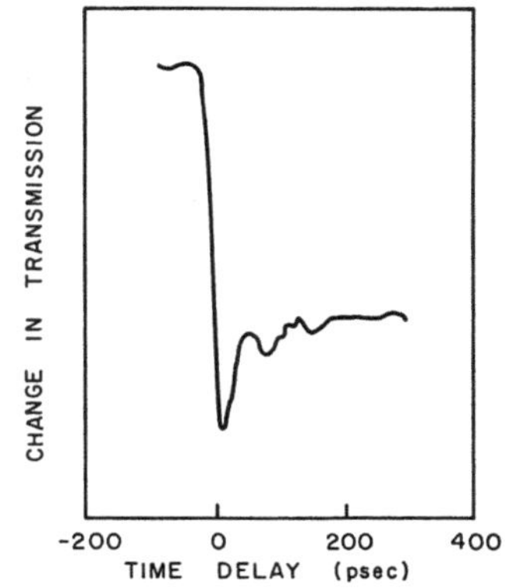

Fig. 2. Results obtained from the experiment shown in Fig. 1; the oscillations are due to an elastic wave bouncing back and forth in an a–As_2Te_3 film 470 A thick

This experiment has the disadvantage that it is hard to work backwards from the observed change ΔT in transmission to obtain details about the strain in the film. This is because ΔT is actually the sum of effects due to the change in reflectivity at the two surfaces of the film, and due to changes in optical absorption in the film interior. A better approach is to study the change ΔR in reflectivity induced by the propagating strain. This has the advantage that, at least for a film through which the light transmission is small, ΔR is only influenced by the strain near to the surface of the film. Thus, one can write [3]

$$\Delta R(t) = \int f(z)\eta(z,t)dz \tag{1}$$

where $\eta(z,t)$ is the strain at z at time t, and

$$f(z) = B\cos(4\pi nz/\lambda_0 - \psi)\, e^{-z/\zeta} \tag{2}$$

λ_0 is the wavelength of light in free-space, and B and ψ are constants which can be expressed [3] in terms of the optical constants n and α, and their strain derivatives $\partial n/\partial\eta$ and $\partial\alpha/\partial\eta$. In Fig. 3 we show this "sensitivity function" f for amorphous As_2Te_3. One can see that for this material the change in reflectivity is only influenced by the strain in the region within about 300 A of the surface. Results of an experiment using reflectivity detection are shown in Fig. 4. These data are for an a–As_2Te_3 film which is 2400 A thick and were taken with pulses of duration 0.2 psec energy 0.1 nJ, and repetition rate 110 MHz. Three echoes are visible, and Fig. 5 shows an expanded view of the first echo. To understand the shape of this echo we have to calculate the spatial form of the propagating strain pulse. The stress set up by the absorbed light pulse varies in space as $e^{-z/\zeta}$. This stress is set up essentially instantaneously since the elastic wave only moves 4 A in the 0.2 psec duration of the light pulse. The stress produces a strain pulse ($\propto e^{-z/\zeta}$) propagating into the film, and another pulse of the same shape which propagates towards the free surface. The second part of the strain is reflected with a sign change, and consequently the total strain propagating into the film has a shape

$$\eta(z,t) \propto -\operatorname{sgn}(z - vt)e^{-|z-vt|/\zeta} \tag{3}$$

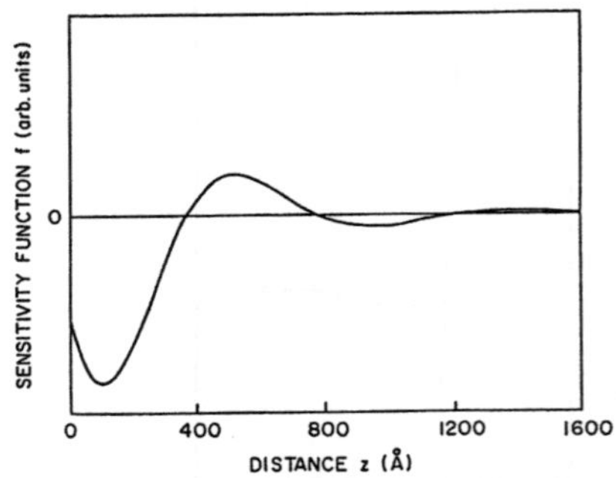

Fig. 3. The sensitivity function f(z)

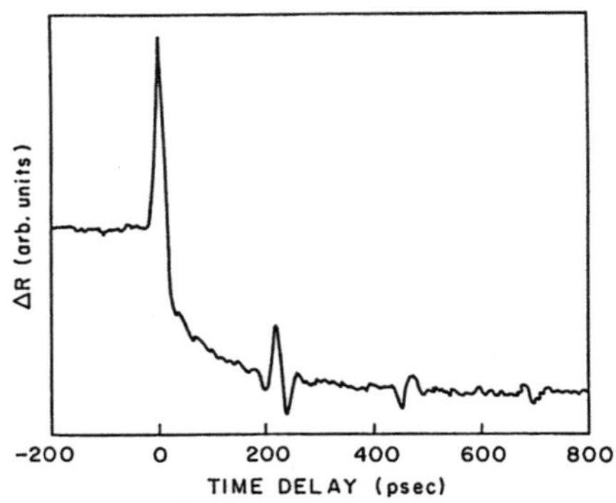

Fig. 4. A strain pulse bouncing back and forth in a 2400 A thick film of a–As_2Te_3; the pulse is detected by a measurement of the optical reflectivity as shown in Fig. 1

This propagates across the film and is reflected at the substrate. When the pulse returns to the free surface the strain near the surface is the sum of the returning pulse, and its reflection at the free surface (inverted). Thus,

$$\eta(z,t) \propto - \operatorname{sgn}(-z-vt)e^{-|-z-vt|/\zeta} + \operatorname{sgn}(z-vt)e^{-|z-vt|/\zeta} \tag{4}$$

where now time t is measured from the arrival at the free surface of the midpoint of the returning pulse. The shape of η is shown for a series of times in Fig. 6. For every value of z the strain is an <u>even</u> function of time. Thus, one can see from (1) that the reflectivity change $\Delta R(t)$ should also be an even function of time, as is observed. The dashed line in Fig. 5 is the calculated shape of the echo using the sensitivity function of Fig. 3.

The generation process is probably more complicated than we have just described. The effects which must be considered include:

(1) Diffusion of heat. The heat will spread out of the thin layer in which it is absorbed, and thus the thermal stress in the film will vary with time. It can be shown [3] that this effect causes a change in the shape of the strain pulse which depends on the parameter $D/v\zeta$, where D is the thermal diffusivity. For semiconducting glasses $D/v\zeta \ll 1$ and this effect is unimportant. For metals $D/v\zeta$ is of the order of 1 and there is a large change in pulse shape [3].

(2) Non-equilibrium effects. At the macroscopic level the stress is considered a consequence of the temperature rise of the film. This assumes implicitly that the thermal phonons in the film are describable by a thermal distribution. However, for times after the absorption of the light pulse but before one thermal phonon lifetime has elapsed this will not be the case. Then one must use a microscopic expression for the stress [4], i.e.

$$\sigma_{\alpha\beta} = \sum_k \hbar\omega_k \, \delta n_k \, \gamma_{\alpha\beta}(k) \tag{5}$$

where δn_k is the change in the number of phonons of wave vector k, and $\gamma_{\alpha\beta}$ is the Gruneisen tensor. As the phonons relax the stress will change, thus modifying the shape of the generated strain pulse.

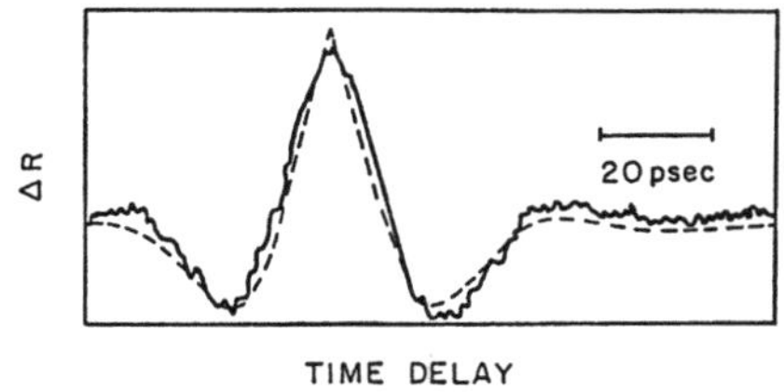

Fig. 5. The shape of the first returning echo in Fig. 4, compared with a theoretical estimate based on the sensitivity function shown in Fig. 3

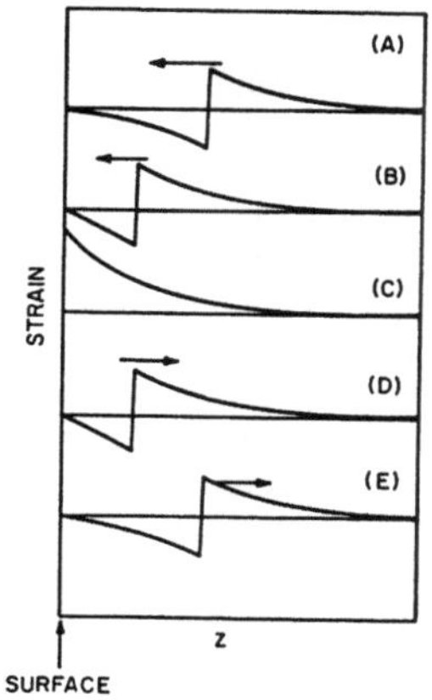

Fig. 6. The strain near the surface due to the returning pulse; the time increases as one goes from (A) to (E), and (C) corresponds to the time when the midpoint of the pulse reaches the surface

(3) Electrons. In a semiconductor the absorption of light of photon energy greater than the bandgap produces electrons and holes. These give an electronic contribution to the stress, similar to (5). When the electrons and holes have relaxed to the band-edge, their contribution to the stress is

$$\sigma^{el}_{\alpha\beta} = \delta n_{el}\, \partial E_g/\partial\eta_{\alpha\beta} \tag{6}$$

where E_g is the band gap. This contribution to the stress persists for a long time since electron-hole recombination is a relatively slow process. In metals the energy given by the light pulse to the electrons will be transferred very rapidly to the phonons (in $\tau_{el-ph} \sim 0.01$ psec at room temperature), and after this the change in the electron distribution function, and $\sigma^{el}_{\alpha\beta}$, will be very small. The electrons may change the stress distribution, however, because in the time τ_{el-ph} they are able to diffuse a significant distance.

In the case of a–As_2Te_3 the signs of the electron and phonon terms are opposite, and the electron contribution is larger. Diffusion is not important. We have also observed strain pulses in a-Ge, a–As_2Se_3, Ni and Cu. In Ni and Cu diffusion is presumably important, but we have not made a quantitative comparison of the observed pulse shapes with theory.

In these experiments the strain pulse contains a broad distribution of phonon frequencies. One can study phonon scattering at a definite frequency by the following technique. [5] We measure the first and second echoes, and find their Fourier transforms $\Delta R^{(1)}(\Omega)$ and $\Delta R^{(2)}(\Omega)$. Then, the attenuation $\gamma(\Omega)$ per unit distance for a phonon of frequency Ω is

$$\gamma(\Omega) = \frac{1}{2d} \log\left[\frac{r_{FS}\,|\Delta R^{(1)}(\Omega)|}{|\Delta R^{(2)}(\Omega)|}\right] \tag{7}$$

where d is the thickness of the film, and r_{FS} is the reflection coefficient of the pulse at the interface between the film and the substrate (assumed to be frequency independent and given by acoustic-mismatch theory). In this way one can measure the attenuation of phonons throughout the frequency range where the pulse has appreciable Fourier components. For the measurements on a–As_2Te_3 this range is about 10 to 40 GHz. In general, the peak in the spectrum is at a frequency of about

$$\nu_{max} \sim \frac{v}{2\zeta} \quad (8)$$

3. Generation and Detection Using Transducers

The method just described can only be used to generate phonons in materials which strongly absorb light. In addition, the frequency range of the phonons which are generated is determined entirely by the optical properties of the material. (There is the possibility of changing the phonon frequency by making a change in the photon energy in the pump pulse.) For these reasons there are advantages in using a second thin film as a transducer (Fig. 7). To consider how the transducer works we assume that it is very thin; i.e. the film thickness d is 100 A or less. In films of this thickness the pump light pulse produces a uniform stress throughout the film. For a semiconducting film ζ is much larger than 100 A, and hence the light absorption is uniform. For a metal film ζ may be comparable to or less than 100 A, but the diffusion will be rapid enough to spread the heat throughout the thickness of the film. The shape of the generated strain wave is then as shown in Fig. 8. For $Z_T \geq Z_F$ (Z_T and Z_F are the acoustic impedances of the transducer and the sample film) the dominant frequency generated is

$$\nu = \frac{v}{2d}. \quad (9)$$

When $Z_T = Z_F$ just one cycle is produced. If $Z_T \ll Z_F$ the transducer is a $\lambda/4$-resonator and has frequency v/4d.

To work well as a detector the film material should have a large value of $\partial R/\partial\eta$. This is most likely to occur in a material for which R changes rapidly [6] with photon energy E at the probe photon energy E_P. This happens when E_P is close to a singularity in the interband transition spectrum, or for a metal in the region of the plasma edge. Thus, for a given material the sensitivity of the detector is expected to vary strongly with the energy of the photons in the probe pulse. In the experiments we have performed so far this has been fixed at 2.0 eV, so we have been forced to search for a sensitive detector by looking at different

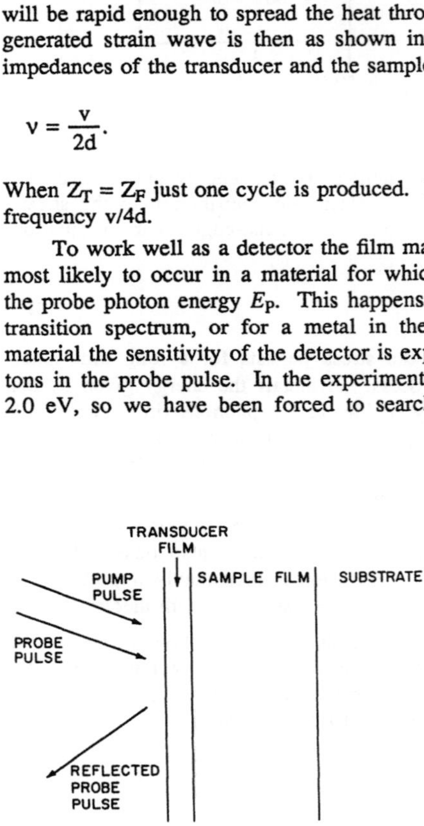

Fig. 7. Experiment to study phonon attenuation using a transducer film

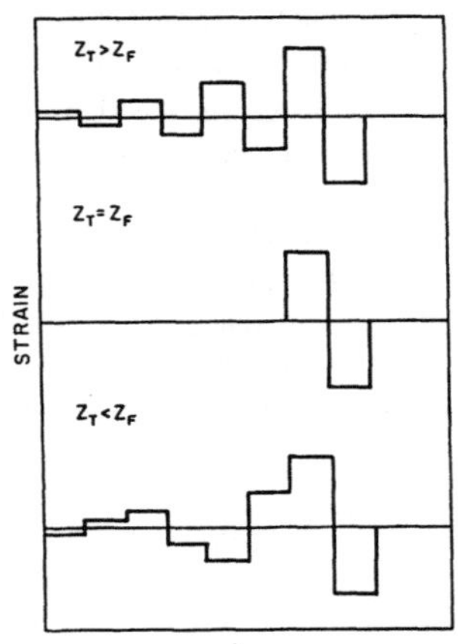

Fig. 8. Strain pulses generated in the sample film by the transducer for different ratios of the acoustic impedances of the transducer and sample

materials, rather than by adjusting the energy. We have found strong responses in a–As_2Te_3, a–As_2Se_3, In_2Te_3, and InSb and Al. Amorphous As_2Te_3 and As_2Se_3 have the disadvantage of low sound velocities (~$2\times10^5\,cm\,sec^{-1}$), and hence generate low frequencies. In addition, As_2Se_3 has a low absorption coefficient for 2.0 eV photons which reduces its efficiency as a phonon generator. We have used a–As_2Te_3 to produce phonons up to ~ 50 GHz. In_2Te_3 has a sound velocity of $2.6\times10^5\,cm\,sec^{-1}$, and with very thin films we have generated and detected phonons of frequency up to 230 GHz. In_2Te_3 and other chalcogenides have the disadvantage that they are somewhat sensitive to light exposure, and undergo structural changes when illuminated for extended periods of time. InSb has the advantage that it is stable, and from preliminary measurements appears to be at least as sensitive as In_2Te_3. We have to date used this as a transducer up to 130 GHz.

So far we have obtained the highest frequencies with aluminum transducers, but we have not yet used these for phonon attenuation measurements. Figure 9 shows the response of the transducers when deposited onto a thick pyrex substrate. Figure 9(a) and (b) are for transducers which resonate at 290 and 400 GHz respectively. The transducers and substrate are fairly well-matched acoustically, but it is clear that the bonding is not perfect, especially in 9(a). We have tried to fit these data by postulating that between the Al and the pyrex there is a layer of "organic dirt" ($\rho = 1\,g\,cm^{-3}$, $v = 2\times10^5\,cm\,sec^{-1}$). To explain the results of 9(a) one needs only a dirt thickness of 3-4 A to explain the poor bonding.

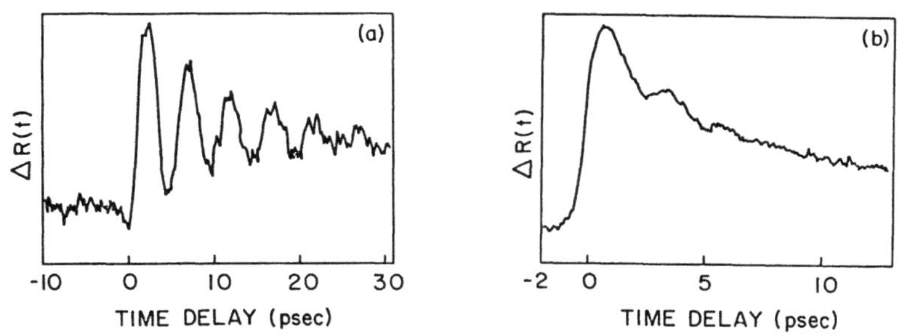

Fig. 9. Responses of two Al film transducers

4. Measurements of Phonon Attenuation in SiO_2

We have used transducers to make room-temperature measurements of the attenuation of longitudinal phonons in fused quartz. We used an InSb transducer on a 2400 A SiO_2 film for measurements between 60 and 130 GHz, and covered the range from 140 to 230 GHz with an In_2Te_3 transducer on a 600 A SiO_2 film. The SiO_2 was electron-beam evaporated onto a sapphire substrate. Figure 10 shows the 1st and 2nd echo received in the 2400 A film. The attenuation as a function of frequency was obtained by comparing the Fourier transforms of the 1st and 2nd echoes, as already described. The Fourier transforms of the 1st echo are shown in Fig. 11.

Preliminary results for the attenuation at room temperature are shown in Fig. 12, together with the result of Vacher and Pelous [7] at 33 GHz. These data are consistent with an ω^2 frequency-dependence to within the experimental uncertainty. At 230 GHz the wavelength of the phonons studied is ~ 200 A, the attenuation is $7\times10^5\,dB\,cm^{-1}$, and the mean-free-path is ~ 600 A. We are currently extending these measurements to higher frequencies and to low temperatures. The maximum frequency that can be reached by this method is probably between 500 and 1000 GHz. In this range very thin transducers and samples will be required, and the bonding of the transducer to the sample will be of critical importance.

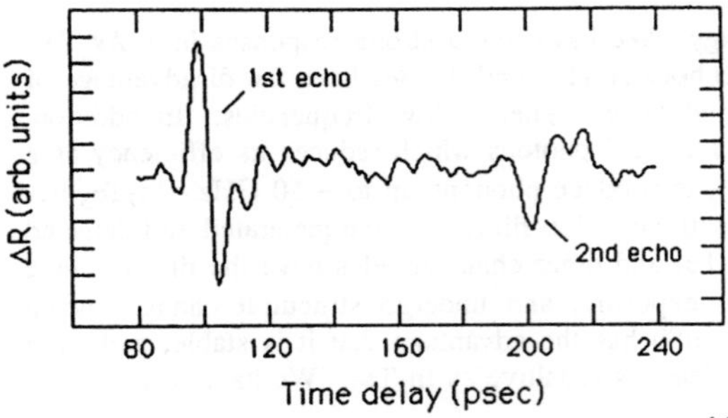

Fig. 10. Echoes in a 2400 A SiO_2 film with an InSb transducer

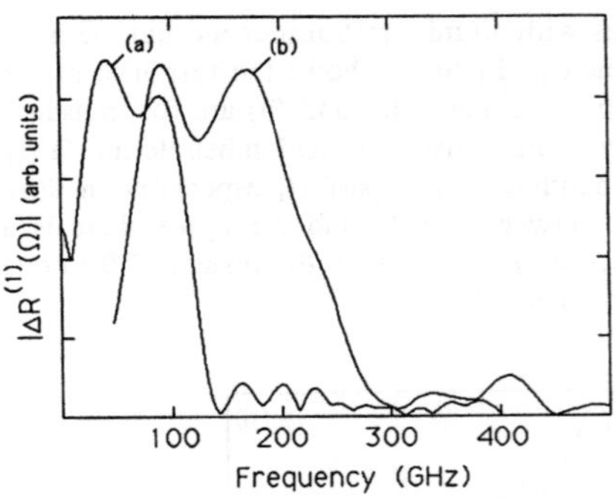

Fig. 11. Fourier transforms of the 1st echo in (a) a 2400 A SiO_2 film with an InSb transducer and (b) a 600 A film with an In_2Te_3 transducer

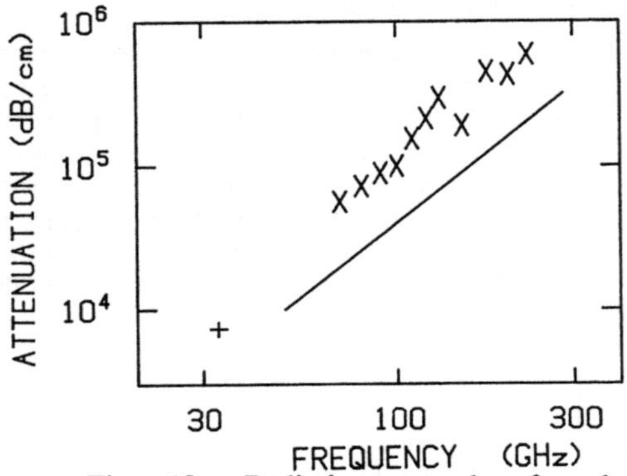

Fig. 12. Preliminary results for the attenuation of longitudinal phonons in SiO_2 at room temperature; the solid line shows an ω^2 frequency dependence

We thank H. T. Grahn, D. A. Young, and T. R. Kirst for help with the experiments. This work was supported in part by the National Science Foundation through the Materials Research Laboratory at Brown University.

1. Phonon Scattering in Condensed Matter, ed. by W. Eisenmenger, K. Lassmann, and S. Dottinger (Springer, Berlin, 1984)
2. C. Thomsen, J. Strait, Z. Vardeny, H. J. Maris, J. Tauc, and J. J. Hauser: Phys. Rev. Lett. 53, 989 (1984)
3. C. Thomsen, H. T. Grahn, H. J. Maris, and J. Tauc: submitted to Physical Review.
4. H. J. Maris: In Physical Acoustics, ed. by W. P. Mason and R. N. Thurston (Academic, New York, 1971), Vol. 7, p. 279
5. C. Thomsen, H. T. Grahn, H. J. Maris, and J. Tauc: J. Phys., Colloq. 46, C10-765 (1985)
6. See, for example, M. Cardona: Modulation Spectroscopy (Academic, New York, 1969).
7. R. Vacher and J. Pelous: In Phonon Scattering in Solids, ed. by L. J. Challis, V. W. Rampton, and A. F. G. Wyatt (Plenum, New York, 1975), p. 147

materials, rather than by adjusting the energy. We have found strong responses in a–As_2Te_3, a–As_2Se_3, In_2Te_3, and InSb and Al. Amorphous As_2Te_3 and As_2Se_3 have the disadvantage of low sound velocities (~2×10^5 cm sec^{-1}), and hence generate low frequencies. In addition, As_2Se_3 has a low absorption coefficient for 2.0 eV photons which reduces its efficiency as a phonon generator. We have used a–As_2Te_3 to produce phonons up to ~ 50 GHz. In_2Te_3 has a sound velocity of 2.6×10^5 cm sec^{-1}, and with very thin films we have generated and detected phonons of frequency up to 230 GHz. In_2Te_3 and other chalcogenides have the disadvantage that they are somewhat sensitive to light exposure, and undergo structural changes when illuminated for extended periods of time. InSb has the advantage that it is stable, and from preliminary measurements appears to be at least as sensitive as In_2Te_3. We have to date used this as a transducer up to 130 GHz.

So far we have obtained the highest frequencies with aluminum transducers, but we have not yet used these for phonon attenuation measurements. Figure 9 shows the response of the transducers when deposited onto a thick pyrex substrate. Figure 9(a) and (b) are for transducers which resonate at 290 and 400 GHz respectively. The transducers and substrate are fairly well-matched acoustically, but it is clear that the bonding is not perfect, especially in 9(a). We have tried to fit these data by postulating that between the Al and the pyrex there is a layer of "organic dirt" ($\rho = 1$ g cm^{-3}, $v = 2\times10^5$ cm sec^{-1}). To explain the results of 9(a) one needs only a dirt thickness of 3-4 A to explain the poor bonding.

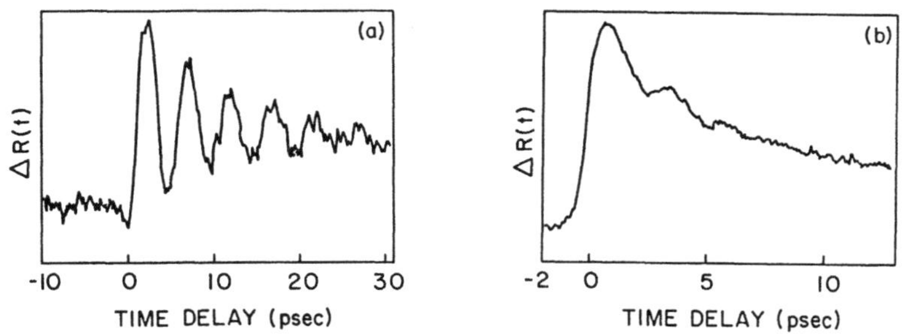

Fig. 9. Responses of two Al film transducers

4. Measurements of Phonon Attenuation in SiO_2

We have used transducers to make room-temperature measurements of the attenuation of longitudinal phonons in fused quartz. We used an InSb transducer on a 2400 A SiO_2 film for measurements between 60 and 130 GHz, and covered the range from 140 to 230 GHz with an In_2Te_3 transducer on a 600 A SiO_2 film. The SiO_2 was electron-beam evaporated onto a sapphire substrate. Figure 10 shows the 1st and 2nd echo received in the 2400 A film. The attenuation as a function of frequency was obtained by comparing the Fourier transforms of the 1st and 2nd echoes, as already described. The Fourier transforms of the 1st echo are shown in Fig. 11.

Preliminary results for the attenuation at room temperature are shown in Fig. 12, together with the result of Vacher and Pelous [7] at 33 GHz. These data are consistent with an ω^2 frequency-dependence to within the experimental uncertainty. At 230 GHz the wavelength of the phonons studied is ~ 200 A, the attenuation is 7×10^5 dB cm^{-1}, and the mean-free-path is ~ 600 A. We are currently extending these measurements to higher frequencies and to low temperatures. The maximum frequency that can be reached by this method is probably between 500 and 1000 GHz. In this range very thin transducers and samples will be required, and the bonding of the transducer to the sample will be of critical importance.

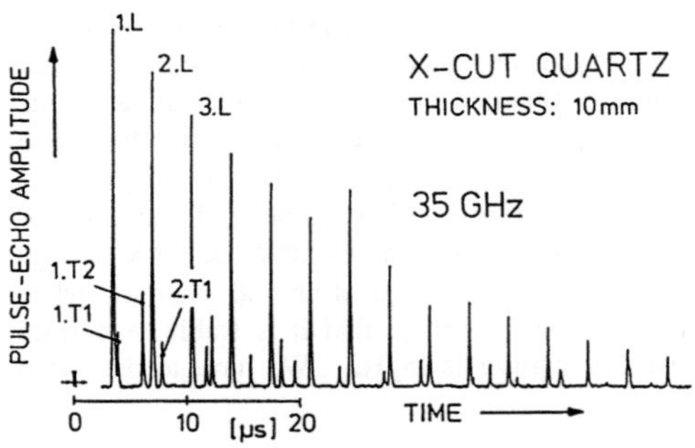

Fig. 2: Typical pulse-echo pattern for a movable sound beam in a 10 mm thick X-cut quartz at 35 GHz and 2 K

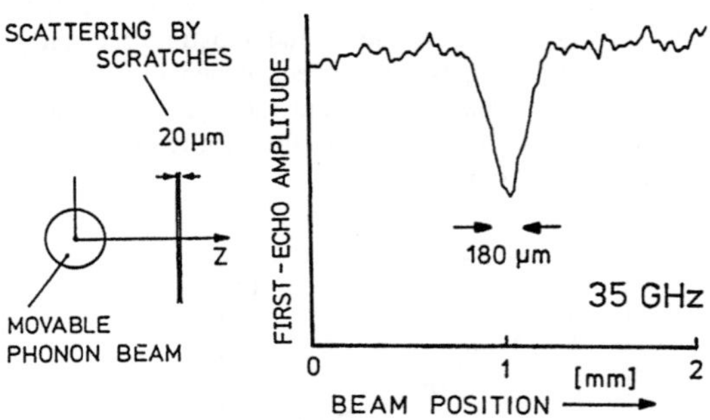

Fig. 3: Determination of the width and profile of a longitudinal sound beam at 35 GHz using the scattering at a straight scratch which was made with a diamond tip. The coupling hole had a diameter of 0.6 mm and the tapered post of 0.3 mm. The distance between orifice and crystal was about 40 µm

We observed for the L-pulse a halfwidth of 0.18 mm at 35 GHz. The resonator at 24 GHz produced a beam halfwidth of 0.3 mm for the L-pulse and 0.45 mm for the T2-pulse.

3. Multiple-Beam Interferometry in Thin Films

The attenuation and phase velocity of hypersound can be determined in thin films by multiple-beam interferometry /6/. In this method, the reflection is measured at normal incidence as a function of frequency or film thickness. From such a 'phonon interferogram' the attenuation and phase velocity can be deduced. Frequency tuning with film thickness kept constant was introduced by SPLITT /7/ with coherent phonons at 9 GHz and, recently, by ROTHENFUSSER et al. /8/ with incoherent phonons above 100 GHz. Alternatively, the frequency can be kept constant and the film thickness changed. This method was first used with monochromatic incoherent phonons between 18 and 58 GHz in superfluid films by ANDERSON and SABISKY /9/. Later on, this method was applied with hypersound of 24 GHz to growing solid films by ZUR NIEDEN and WEIS /6/. Fig. 4 reproduces such a typical multiple-beam interferogram obtained with growing solid oxygen. The main characteristics are: neighbouring minima are half a wavelength apart and hence determine the phase velocity if the sound frequency is known and the film thickness is measured. The two envelopes and especially the zero value of the lower envelope give information about the attenuation per wavelength. At large film thickness the reflection between halfspace is reached asymptotically. Using movable sound beams, the growing cryogenic films can be replaced by wedge-shaped films which may be prepared at room temperature under most fa-

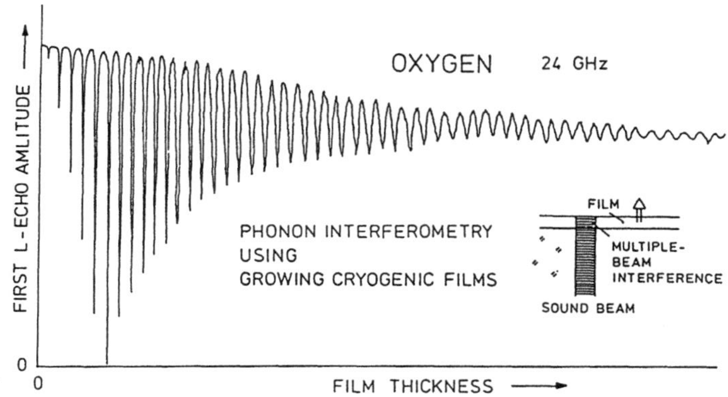

Fig. 4: Experimental phonon interferogram taken with longitudinal 24 GHz hypersound pulses in a growing oxygen film on an X-cut quartz /6/

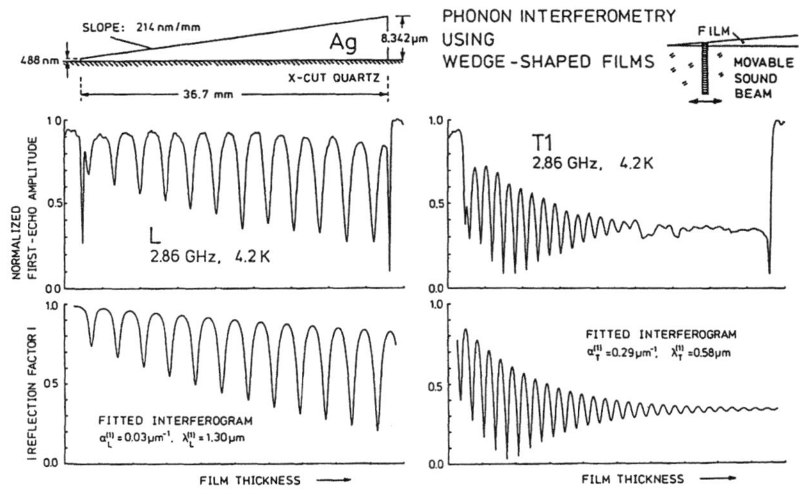

Fig. 5: Experimental and fitted theoretical phonon interferogram for longitudinal and fast transverse movable hypersound pulses in wedge-shaped silver films at a frequency of 2.86 GHz /4/

vourable experimental conditions /4/. The wedge-shaped films are produced in an evaporator chamber by means of a special, fast rotating shutter placed in the vapour beam. The outer contour of the shutter has a radius which increases proportional to the angle of rotation (Archimedian spiral). Polycrystalline wedge-shaped films of silver, copper and constantan have been investigated in a frequency range from 1 GHz to 24 GHz by LEHR et al. /10/, Silver films were also measured at 35 GHz by PIETZSCHMANN and BIALAS /5/. Experimental results for silver and copper are plotted in Fig. 6 in a dimensionless form. The mean crystallite diameter $\langle a \rangle$ was about 60 nm in these evaporated films. The amplitude-attenuation coefficient for polarization σ is denoted by α_σ, the wavelength λ_σ and the wavevector q_σ. The ob-

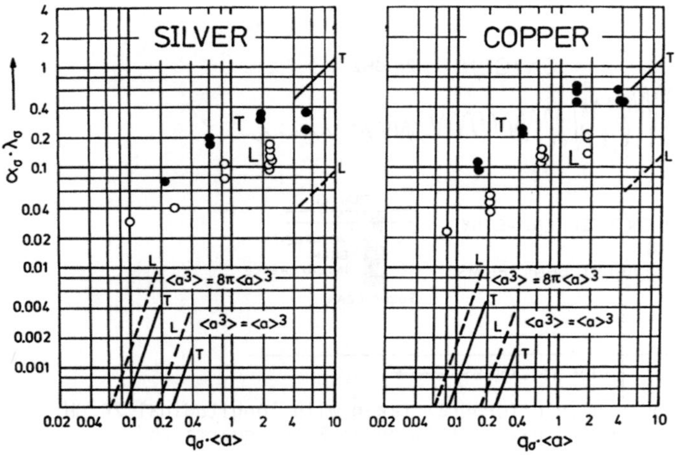

Fig. 6: Comparison between measured hypersound attenuation /5,10/ and calculated asymptotic behaviour due to the theory of LIFSHITS and PARKHOMOVSKII /11/ for longitudinal and transverse polarized sound in polycrystalline silver and copper.

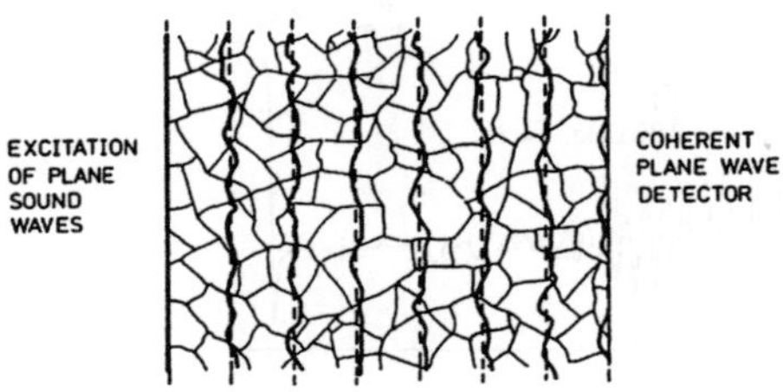

Fig. 7: Illustration of plane wavefront distortion during sound propagation in a lossless but polycrystalline material with ideal grain boundaries. The observed attenuation is thus only due to this distortion which includes small- as well as large-angle scattering

served attenuation is too large to be explained only by electron-phonon interaction. Furthermore, measurements at 1 GHz showed no significant change of attenuation over the temperature range up to room temperature. Hence, we concluded that scattering at grain boundaries dominates. LIFSHITS and PARKHOMOVSKII /11/ developed an appropriate theory of sound propagation in polycrystals assuming 'ideal' grain boundaries with abrupt change of elastic constants (see Fig. 7). From this theory the asymptotic behaviour for $q_\sigma \cdot \langle a \rangle$ larger, as well as smaller than unity is easily deduced and plotted in Fig. 6 for comparison. Dependent on the choice of correlation for two points at a distance r to lie in the same crystallite, different expressions result for the mean volume $\langle a^3 \rangle$. An exponential correlation function yields $8 \cdot \pi \langle a \rangle^3$ for the mean volume whereas a Gaussian correlation function yields $\langle a \rangle^3$. Independent of this choice, the observed attenuation in the Rayleigh-scattering regime $q_\sigma \cdot \langle a \rangle \ll 1$ is about one order of magnitude higher than predicted. This can be explained by absorption and scattering at real grain boundaries which are more important with small grain diameters than in ultrasonic work where the grains in cast metals are about three orders of magnitude larger.

In addition to the amplitude-reflection interferogram there exists the possibility (see Fig. 8) to record the change of absorption heat as a function of thickness using wedge-shaped films and movable beams. This can be done by detecting the second-sound pulses which are emitted from the pulse

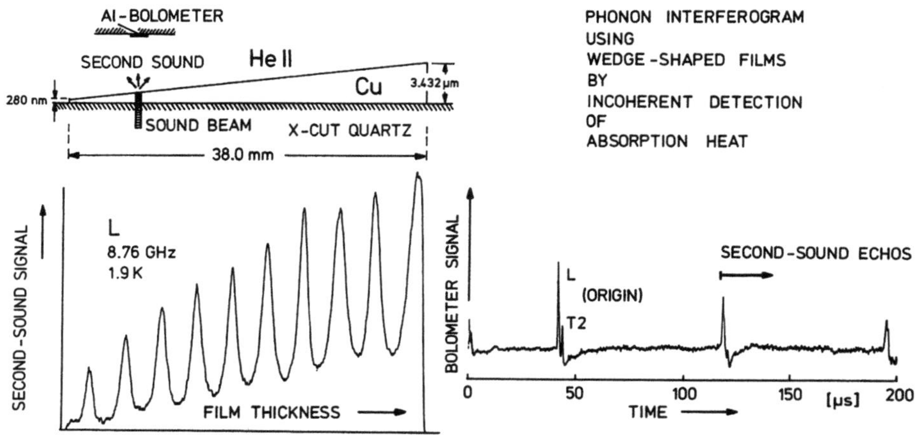

Fig. 8: Change of hypersound-absorption heat in wedge-shaped metal films by recording the magnitude of emitted second-sound pulses /12/

heated parts of the film if placed in superfluid helium. The aluminium bolometer is less than 1 mm above the film and is moved together with the hypersound beam. This method was developed by KLEIN /2/. The second-sound interferogram of Fig. 8 shows clearly the large change of absorption with film thickness in accordance to the thickness-dependent sound energy stored in the multiple-beam field of the film. In the limiting case of local transformation of sound power to heat, the absorption-heat interferogram is expected to be proportional to the following phonon absorption probability

$$a(\ell/\lambda_\sigma) = 1 - |R_\sigma(\ell/\lambda_\sigma)|^2$$

where ℓ denotes the film thickness. $|R_\sigma|$ represents the amplitude-reflection factor at normal incidence for sound of polarization σ as measured by an amplitude-reflection interferogram already discussed. The interferogram of transfer from the sound field to heat will occur in a nonlocal way at low temperatures. Further experiments are necessary for better understanding of the absorption-heat interferograms.

4. Sound-Beam Topography

Interferometric records with movable sound beams can be disturbed by crystal defects. This was already demonstrated by LEHR et al. /4/ who measured the position dependence of echo amplitudes without any evaporated film. As can be seen from the records of Fig. 9, the first-echo amplitude of longitudinal sound beams reveals with rising frequency a higher sensitivity to defects. This is partly due to the shorter wavelength and partly due to the narrowing of the sound beam. It is also demonstrated that there exists a polarization dependence. We may have different causes for a reduction of the echo amplitudes. They are compiled in Fig. 10. In addition to absorption of beam power and to scattering out of the beam, there may occur a distortion of the phase fronts due to small-angle scattering which reduces the coherent detection signal, too. Furthermore, a microscopic damage layer may exist near the surface as a consequence of mechanical crystal preparation and may also influence the piezoelectric conversion efficiency. This was shown by LAMB and SEGUIN /13/.

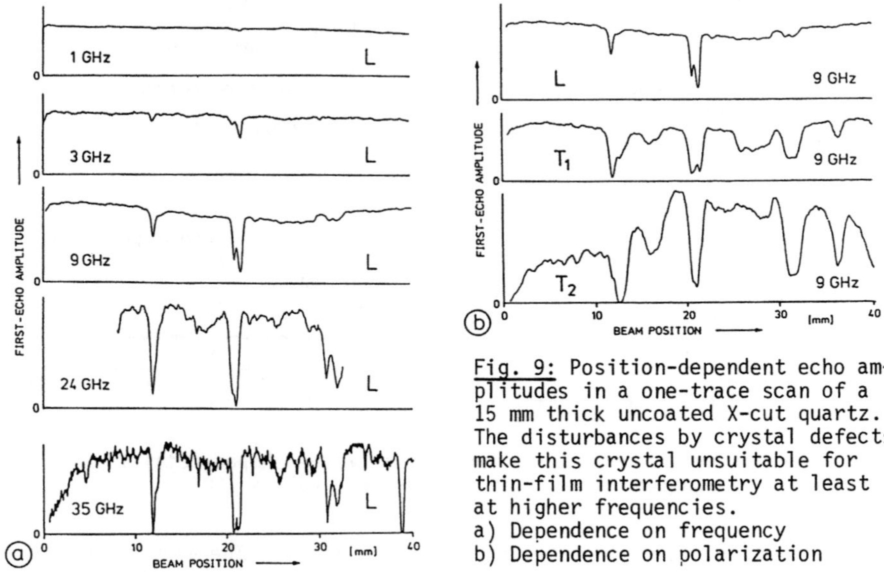

Fig. 9: Position-dependent echo amplitudes in a one-trace scan of a 15 mm thick uncoated X-cut quartz. The disturbances by crystal defects make this crystal unsuitable for thin-film interferometry at least at higher frequencies.
a) Dependence on frequency
b) Dependence on polarization

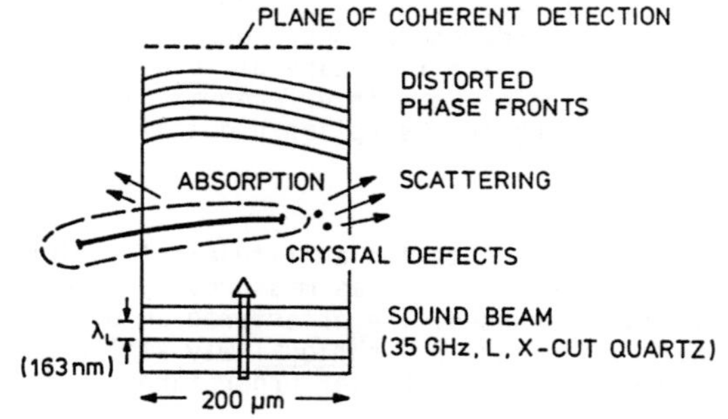

Fig. 10: This schematic drawing illustrates the influence of crystal defects on the propagation of sound beams. The beam width and indicated wavelength corresponds to L-waves in X-cut quartz at a frequency of 35 GHz

In view of the urgent demand for characterizing crystals concerning their quality for highfrequency phonon-propagation experiments, we started a systematic investigation of natural quartz crystals and developed the new method of sound-beam topography /14,15/. In this method a two-dimensional scan with movable sound beams is performed and the position-dependent echo amplitudes are presented as a brightness modulated image. Examples for an X-cut quartz are given in Fig. 11 and for Y-cut quartz in Fig. 12. The gain on resolution

→

Fig. 11: Sound-beam topographs of an X-cut Quartz (10 x 10 x 50 mm^3) made at 24 and 35 GHz with longitudinal phonons /15/. The crystal is dextrorotary and its orientation corresponds to the drawing of the crystal habit on the left at the bottom. The photograph is taken from a CRT

Fig. 12: Sound-beam topographs of Y-cut quartz produced by means of a matrix plotter. The large dark areas are a consequence of beam reflection at the side wall. The line pattern can be identified with rz-edge defects which are produced during crystal growth /15/

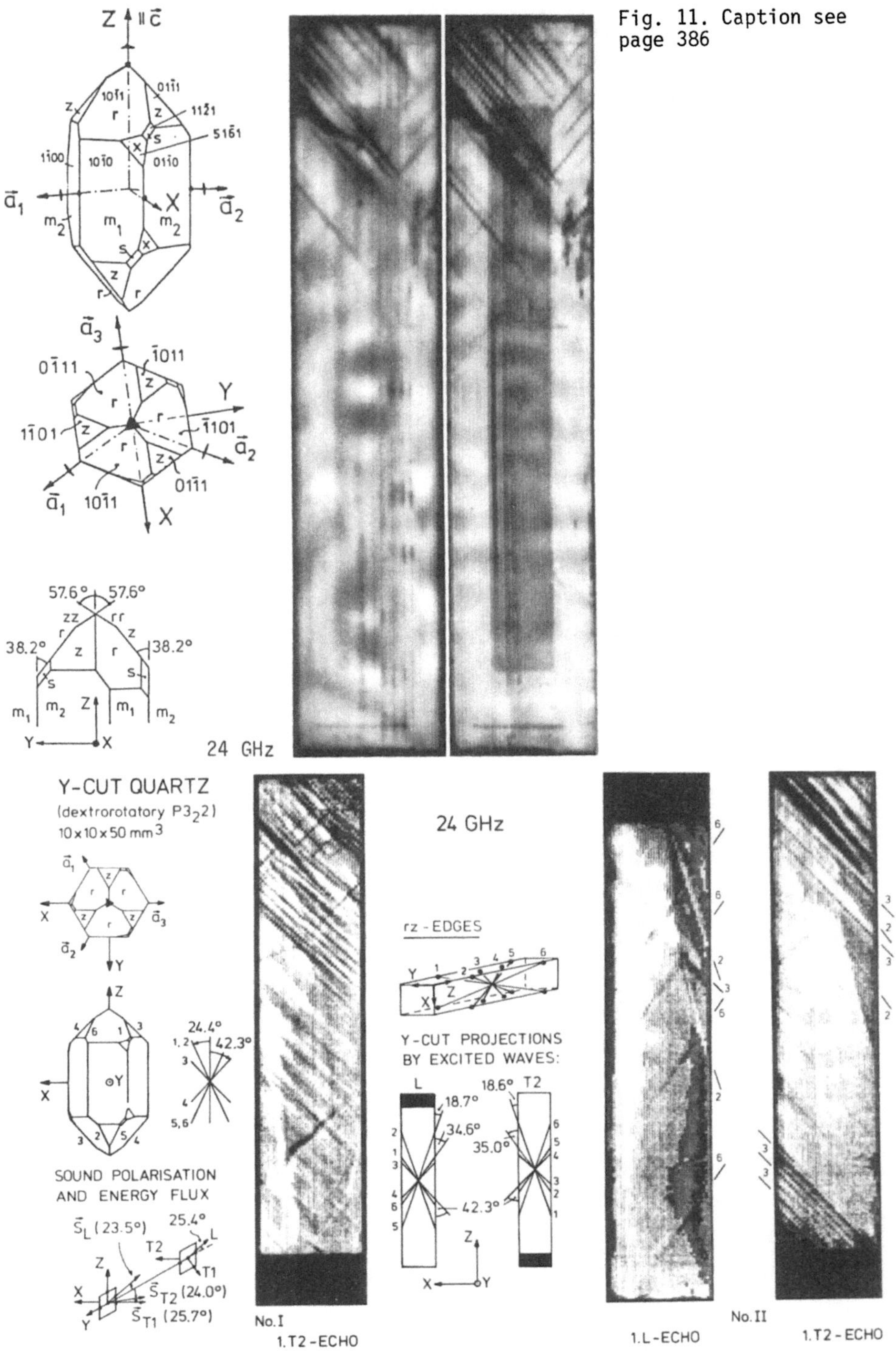

Fig. 11. Caption see page 386

at 35 GHz in comparison to 24 GHz is clearly visible in Fig. 11. At 35 GHz the position of the removed wedge-shaped metal films is visible as a dark area. The diamond scratch at the bottom and a line pattern at the top due to growing defects at the edges between r- and z-faces are visible in both topographs. Such defects at rz-edges appear also in the Y-cut topographs of Fig. 12 which are taken with two other crystals. The energy flux of the sound beams is oblique in this cut. Therefore, the six denoted kinds of edges are projected in a different way and can be identified in the topographs.

References

1. H. Bömmel and K. Dransfeld: Phys. Rev. Letters 1, 234 (1958) and 2, 298 (1959)
2. E.H. Jacobsen: J. Acoust. Soc. Am. 32, 949 (1960)
3. H. Ulrich and O. Weis: Z. Phys. B-Condensed Matter 29, 185 (1978)
4. B. Lehr, H. Ulrich and O. Weis: Z. Phys. B-Condensed Matter 44, 167 (1981)
5. U. Pietzschmann and H. Bialas: J. Phys. F: Met. Phys. 16, 277 (1986)
6. Th. zur Nieden and O. Weis: Z. Phys. B 21, 11 (1975)
7. G. Splitt: Z. Physik 225, 60 (1969)
8. M. Rothenfusser, W. Dietsche and H. Kinder: Phys. Rev. B 27, 5196 (1983) and earlier work cited therein
9. C.H. Anderson and E.S. Sabisky: Phys. Rev. Lett. 24, 1049 (1970)
10. B. Lehr, H. Ulrich and O. Weis: Z. Phys. B-Condensed Matter 48, 23 (1982)
11. I.M. Lifshits and G.D. Parkhomovskii: Karkovskiy Gosudartsvenny Universitet. Uchenye Zapiski 27, 25 (1948), Zh. Eksp. Teor. Fiz. 20, 175 (1950)
12. H. Klein: present work, unpublished
13. J. Lamb and H. Seguin: J. Acoust. Soc. Am. 39, 752 (1966)
14. R. Stowasser and O. Weis: In Fortschritte der Akustik-DAGA '85, p. 875, DPG-GmbH, Bad Honnef 1985
15. H. Edel, H. Bialas and O. Weis: submitted for publication in Z. Phys. B-Condensed Matter

The Use of Phonon Echoes to Study Defects in Dielectric Crystals

J.K. Wigmore, D.J. Meredith, A. Newton, and I. Terry

Physics Department, University of Lancaster, Lancaster, UK

It is increasingly recognised that non-linear properties of crystal lattices may be modified drastically by the presence of structural and electronic defects. Phonon echoes, because they arise directly through the presence of non-linear interactions between strain and electric field, can provide valuable additional information for the characterisation of such defects. In this paper, we describe results obtained on two systems of interest, irradiation defects in quartz and shallow traps in bismuth germanate, and discuss possible mechanisms for the effects we observed. All the experiments were carried out at a temperature of 1.5 K. Echoes were excited by a sequence of two or three sub-microsecond pulses at a microwave frequency of 9.3 GHz and at power levels up to 1 kW, in a single cavity. Only parametric, backward-wave (two-pulse) echoes were observed from the irradiated quartz, but in $Bi_{12}GeO_{20}$ the echoes were all of the storage type.

The quartz samples had received either a $10^{19}cm^{-2}$ dose of 1 MeV neutrons (NIRQ), or 10^6 rads of 2 MeV X-rays (XIRQ). The echoes observed were of approximately the same amplitude, up to 10 dB above noise, in all crystals, regardless of their treatment, whereas no echoes were visible from unirradiated samples. However, the maximum amplitude of the echo in the NIRQ was not obtained on the first run at 1.5 K, but was reached asymptotically as the sample was cycled successively between room and helium temperatures. The echo disappeared again if the sample remained at room temperature for any length of time, but could be restored by repeated thermal cycling. The XIRQ echoes did not show this effect, but in other ways the behaviour of the different samples was similar. Figure 1 shows the power dependence of the echoes on the pump electric field, produced by the second exciting pulse, and Figure 2 the effect of a "pre-pulse", that is, a third microwave pulse applied some time before the two-pulse sequence which stimulated the echo. The result was to decrease the amplitude of the subsequent parametric echo; the size of the decrease depended on the timing of the pre-pulse.

These parametric echoes observed at 9.3 GHz and 1.5 K were quite different in kind from the pseudo-spin echoes observed by Golding and Graebner |1| in NIRQ at 0.995 GHz and 0.015 K, and from the electric dipolar echoes seen by Bernard et al |2| in γ-irradiated quartz at 0.360 GHz and .006 K. Both these experiments were interpreted in terms of spin-type echoes of defect two-level systems (TLS), the former associated with amorphous tunnelling states of unknown microscopic structure, and the latter with a hole tunnelling between O^- sites surrounding an Al^{3+} impurity. We believe that TLS were also responsible for the echoes we observed, through a rather different mechanism that deserves wider appreciation. The Bloch equations that describe the dynamical behaviour of TLS are inherently non-linear.

Shiren et al |3| showed that even when the transverse relaxation was too short for pseudo-spin echoes to be developed, the system might still be sufficiently non-linear to generate phonon echoes by parametric interaction. The size of the non-linearity depends on the extent to which the TLS is saturated. Melcher |4| gave an expression for the resulting non-linear stress. Although correlation is only qualitative so far, the form of the power dependence is characteristic, and is totally different from the $(\text{power})^2$ given by a defect-enhanced intrinsic non-linearity |5|. In addition, the echo depends on the amplitude of the exciting pulse and not on its area, as with spin echoes. Finally the fact that the effect of a pre-pulse has a decay time much longer than the T_2 of the echoes themselves also suggests a relation with dielectric loss, which is determined in irradiated quartz by Al related defects. If its origins can be confirmed quantitatively, the parametric echo may be more useful in characterising defects than the true spin echo since the conditions for its appearance are so much less rigorous.

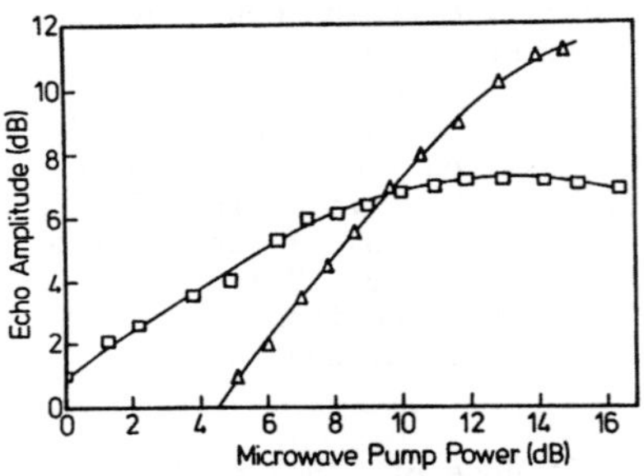

Fig. 1: Dependence of parametric echo on pump power, for NIRQ (Δ) and XIRQ (□) samples.

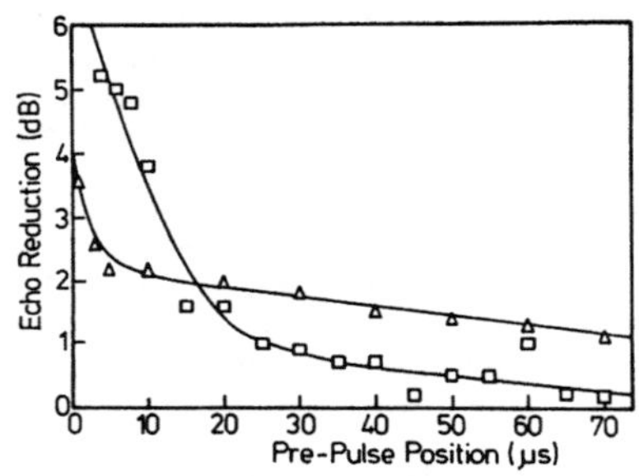

Fig. 2: Variation of echo reduction as a function of pre-pulse timing for NIRQ (Δ) and XIRQ (□).

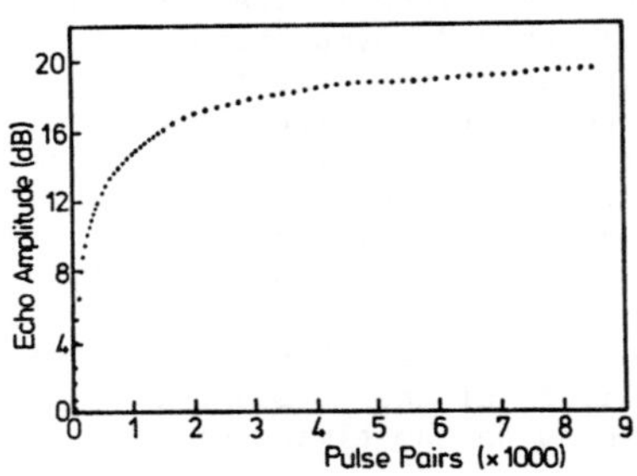

Fig. 3: Growth of Storage echo in BGO:Zn with number of exciting pulse pairs.

The $Bi_{12}GeO_{20}$ that we studied contained approximately 1% of zinc. Figure 3 shows the build-up of a storage echo as the number of excitation pulse pairs was increased. The amplitudes depended strongly on the presence of light, and after cooling, no echoes were observed until the sample had been illuminated for a few seconds by white light, or by a He-Ne laser. On removal of the light, the echo remained until the sample was warmed up again to room temperature. The non-linearity in this system arises, we believe, through the exponential transition probability for electric field-enhanced ionisation of shallow Coulomb-type traps | 6 | which mixes the piezoelectric field E_1 of the acoustic pulse and the pump field E_2, both at frequency ω.

In addition to the 2ω and higher harmonics generated by the exponential non-linearity, there exists a zero frequency component of the ionisation probability which retains the cos kx spatial variation of the acoustic wave. The net effect is to ionise selectively traps in the regions of high E and increase the relative population of filled traps in the low-field regions, thus producing a space charge hologram of the initial acoustic signal. A third electric field pulse interacts electrostrictively with the hologram to produce a storage echo.

Echoes of this type had previously been seen in the II-VI semiconductors CdS, CdTe and CdSe, and whilst $Bi_{12}GeO_{20}$ often exhibits large parametric echoes, the Zn-doped sample is the only one we have found to show storage effects. We believe the explanation to be as follows. Although many shallow electron traps are present in undoped $Bi_{12}GeO_{20}$ due to lack of stoichiometry, if they are all populated no redistribution can take place by the mechanism described above. Zinc impurities replace Ge to form double acceptors and hence to compensate the donor traps. The effect of light is to repopulate these, but the Coulomb barrier around an ionised double acceptor prevents the electrons from being recaptured at helium temperatures. We were able to estimate the depth of the shallow traps from the dependence of the initial slope of the curve in Figure 3 on pump field, obtaining a value of a few meV. The reasoning contains one or two questionable steps, but there are very few methods available for probing these traps. It is probable that the model is incomplete, since thermal stimulated conductivity measurements reveal several other states in the energy range below 1 eV.

We are grateful to Dr. C. Laermans for the NIRQ sample.

1. B. Golding and J.E. Graebner: "Phonon Scattering in Condensed Matter" ed. Maris (Plenum, New York) 11 (1980).

2. L. Bernard and M. Saint-Paul: J. de Physique L-40 593 (1979).

3. N.S. Shiren, W. Arnold and T.G. Kazyaka: Phys. Rev. Letters 39 239 (1977).

4. R.L. Melcher, Phys. Rev. Letters 43 939 (1979).

5. D.J. Meredith, T. Miyasato and J.K. Wigmore, Phys. Rev. Letters 52 843 (1984).

6. N.S. Shiren and R.L. Melcher, IEEE Ultrasonics Symposium Proceedings, 588 (1974).

Picosecond Acoustic Interferometry

C. Thomsen, H.T. Grahn, D. Young, H.J. Maris, and J. Tauc

Department of Physics and Division of Engineering, Brown University, Providence, RI 02912, USA

Spontaneous, stimulated [1-4] and impulsive [5] Brillouin scattering are well established optical techniques to study phonon velocity and attenuation in solids and liquids. The Brillouin frequency ν is

$$\nu = (2nv/\lambda)\sin\psi/2 \tag{1}$$

where ψ is the scattering angle, n is the index of refraction, v is the sound velocity and λ the wavelength of the incident light. The sound velocity in a material can be determined from (1); the sound attenuation in transparent materials is related to the linewidth of the scattered light. Due to instrumental linewidth limitations weak attenuation cannot be easily measured by Brillouin scattering. In this paper we present a new technique to study phonons in transparent materials in the Brillouin frequency range using picosecond optical pulses. The technique is first order in strain, can measure weak attenuation, and is particularly well adapted to study thin films and regions close to interfaces where acoustical properties may vary.

The geometry of the experiment is shown schematically in Fig. 1. A film, opaque to the light wavelength used, is deposited on the surface of a transparent material. A pump light pulse passes through the material and is absorbed near the interface. The absorption of light produces a thermal stress in the film, which leads to a strain pulse propagating in the direction normal to the interface. The strain pulse is detected by a second optical pulse. The probe is incident also through the transparent material and is partly reflected at the stationary interface and partly at the propagating acoustic strain. The reflection coming off of the moving strain pulse undergoes a Doppler shift with respect to the incident light. When the total change in reflectivity due to the acoustic pulse is measured, a beating between the two reflections is expected; the beating frequency is proportional to the sound velocity v.

The Doppler frequency shift in the light reflected off of the moving strain pulse is $\Delta\nu = \mathbf{k}\cdot\mathbf{v}/\pi$ where $\mathbf{v}$ is the velocity of the strain pulse and $\mathbf{k}$ is the wavevector of the light in the transparent material. The reflection coming from the stationary interface is unshifted. If we overlap the two reflections in our detector we find the beating period to be

$$\tau = \lambda/2nv\cos\theta \tag{2}$$

where θ is the angle between the incident probe beam and the normal of the surface. For normal incidence, (2) is seen to yield the same frequency as Brillouin backscattering. The frequency range covered by our method hence lies in the same range as Brillouin scattering.

More specifically, if we denote the strain pulse propagating in the $\mathbf{z}$ direction by $\eta(z,t)$ we can write the change in reflectivity $\Delta R(t)$ as [6]

$$\Delta R(t) = r_0^* a \int_0^\infty \eta(z,t)\exp(-2i\mathbf{k}\cdot\mathbf{z})\,dz + \text{c.c.} \tag{3}$$

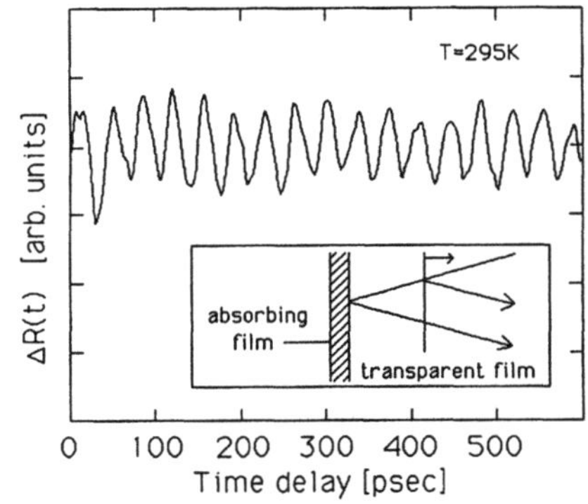

Fig. 1. Picosecond interferometric signal in pyrex due to a propagating phonon pulse

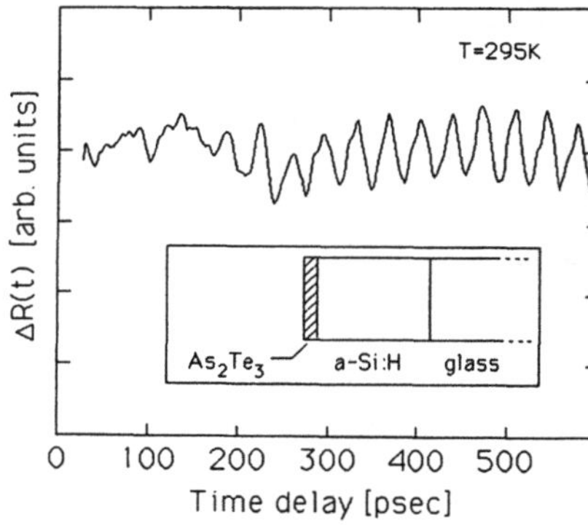

Fig. 2. Interferometric signal in a glass; the oscillations are delayed by the transit time of sound through the a-Si:H

where a is a constant proportional to the change in index of refraction with strain, and r_0 is the reflection coefficient of the light at the opaque film. We express $\eta(z,t)$ in terms of its Fourier components η_K of frequency ω_K and damping γ_K. Then we find

$$\Delta R(t) = 2\pi r_0^* a \eta_K \exp(-i\omega_K t)\exp(-\gamma_K t) + c.c. \qquad (4)$$

where $K = 2k\cos\theta$. The change in reflectivity is, in general, a damped oscillatory function, sensitive to strain components with wave vector $2k\cos\theta$. The oscillation period $\tau = (\omega_K/2\pi)^{-1}$ of $\Delta R(t)$ is thus directly related to the sound velocity through (2). The sound attenuation is found from the damping coefficient γ_K of the oscillations.

We used a colliding-pulse modelocked ring dye laser with a pulse duration of 0.2psec, a photon energy of 2eV(λ = 6150Å) and a repetition rate of 108MHz. The output of the laser was split into pump and probe beams, the probe was passed through a variable optical delay line. In this experiment, we monitored the change in intensity of the reflected probe light pulse due to the propagating strain. Results of a measurement of $\Delta R(t)$ for a sample of pyrex are shown in Fig. 1. The absorbing film was aluminum, dc-sputtered onto one surface of the pyrex. We can clearly see the oscillations caused by the beating of the light Doppler shifted with respect to reflection from the stationary interface. From the oscillation period τ we find the longitudinal sound velocity in pyrex $v = (5.8 \pm 0.05)\times10^5$cm/sec at ν = 28GHz. This value is in reasonable agreement with the low frequency result of $v = 5.64\times10^5$cm/sec [7]. We find no time dependence of the oscillation amplitude in Fig. 1, indicating that there is only negligible sound attenuation on our timescale.

The amplitude $\Delta R(t)/R$ (R = unperturbed reflectivity) of the oscillations in Fig. 1 is of the order of 10^{-5}. We can estimate the amplitude by considering thermal expansion as a generation mechanism [8]. The volume of film heated is $A\zeta$ where A is the area of the pump beam, and ζ is the absorption length of the opaque film. The temperature increase corresponding to one absorbed light pulse is approximated by $\Delta T \approx Q/A\zeta C$ where Q and C are the energy absorbed in the film and the heat capacity per unit volume of the film respectively. The order of magnitude of the amplitude of the strain pulse is therefore clearly $\beta\Delta T$, where β is the thermal expansion coefficient. Materials with a strong absorption for the laser wavelength produce a large temperature increase and hence a large strain amplitude. As is seen from (4), however, the change in reflectivity is most sensitive to strains with Fourier components near $2k\cos\theta$. The dominant Fourier component of the generated strain pulse has a wavelength of approximately 2ζ maximizing $\Delta R(t)$ if a material with $\zeta \sim \lambda/4n$ is used.

A number of general conditions must be met in order to observe a beating signal as in Fig. 1. (i) The spatial extent of the acoustic strain must be smaller than the wavelength of light in the medium; (ii) there must be spatial and temporal overlap of the shifted and unshifted reflections; (iii) the material parameters must be such that the magnitude of $\Delta R(t)$ is above the detection limit. Whether condition (iii) is fulfilled or not, is given by the material studied. We have seen these oscillations also in a–As_2Se_3 and in a Corning glass [6].

In Fig. 2 we show an example where condition (i) is not met. The sample is amorphous hydrogenated silicon (a-Si:H) deposited on a fused quartz substrate (thickness ~ 1μm; see inset of Fig. 2). We dc-sputtered a thin absorbing film of amorphous As_2Te_3 on top of the a-Si:H. At the wavelength of our laser the film of a-Si:H can be considered transparent so that most of the pump light energy is absorbed in the thin layer of a–As_2Te_3. The measurement of $\Delta R(t)$ shown in Fig. 2 has oscillations which, however, start only at t=180psec. These oscillations are due to the strain pulse propagating in the substrate and do not come from the a-Si:H. The starting time of the oscillations in Fig. 2 is consistent with the transit time of the acoustic strain pulse through the film of a-Si:H and the period of oscillations leads to a substrate sound velocity of 5.7×10^5cm/sec at ν = 28GHz.

The probable reason that we do not see oscillations in a-Si:H is that the spatial extent of the acoustic pulse is larger than that of the wavelength of light in a-Si:H. Taking the absorption length in a–As_2Te_3 to be 300Å at 2eV [9] the acoustic pulse length in the a-Si:H is ~ 2400Å, where we have taken into account the ratio of the velocities in the two materials. The light wavelength in a-Si:H is 1760Å [10] so that condition (i) is not fulfilled.

We thank T. R. Kirst for technical assistance. This work was supported by the National Science Foundation through the Materials Research Laboratory at Brown University.

1. M. Cardona: In Light Scattering in Solids II, ed. by M. Cardona and G. Güntherodt, Topics Appl. Phys., Vol. 50 (Springer-Verlag, New York 1975) p. 69
2. J. R. Sandercock: In Light Scattering in Solids III, ed. by M. Cardona and G. Güntherodt, Topics Appl. Phys., Vol. 51 (Springer-Verlag, New York 1982) Chap. 5
3. R. Vacher and J. Pelous, Phys. Rev. B 14 (1976) 823
4. W. Heinicke, G. Winterling and K. Dransfeld, J. Acoust. Soc. Am. 49 (1971) 954
5. K. A. Nelson, R. J. D. Miller and M. D. Fayer, J. Appl. Phys. 53 (1982) 1144
6. C. Thomsen, H. T. Grahn, H. J. Maris, and J. Tauc, submitted to Optics Commun. (1986)
7. W. P. Mason: In The American Institute of Physics Handbook: Acoustics, edited by R. K. Cook (Mc-Graw Hill, New York) p. 3-88
8. For a detailed description of generation mechanisms see C. Thomsen, H. T. Grahn, H. J. Maris, and J. Tauc, submitted to Phys. Rev. B (1986)
9. H. K. Rockstad, J. Non-Cryst. Sol. 2, 192 (1970)
10. We have taken an index of n=4.2 from G. D. Cody, C. R. Wronski, B. Abeles, R. B. Stephens, and B. Brooks, Solar Cells 2, 227 (1980)

Fast Response GeAu Heat Pulse Bolometers for Use in Ballistic Phonon, High Magnetic Field Experiments

M.A. Chin, K.W. Baldwin, V. Narayanamurti, and H.L. Stormer

AT & T Bell Laboratories, Murray Hill, NJ 07974, USA

Introduction: GeAu films operating near the metal insulator transition are ideally suited for high magnetic field thermometry. Until now, difficulties in fabrication, unpredictable temperature responsivity and their inherently high impedances(R up to 100M ohms) have prevented their uses as fast heat pulse detectors. It has been found [1] that within a narrow window of Au concentration that these films have excellent temperature responses. However, these concentrations were determined by Rutherford backscattering after fabrication. Variations on either side of this window resulted in films that are metallic(small R(T)) or insulating (exceedingly large resistances). In addition the responsivity was strongly dependent on whether or not these films were amorphous or polycrystalline.

We have developed empirical procedures which repeatedly and predictably generate films with the desired responsivity for the temperature range of interest. The problems of high impedances was solved by using either interdigitated Au electrodes to reduce the resistances to 50 to 200 ohms or subsequent low temperature FETS and PMODFETS[2] in source follower configurations. PMODFETS were especially suited for high magnetic field operation and at temperatures below 1K. In this configuration we observed for the first time ballistic phonon propagation in magnetic fields to 10 T.

Experimental Details: Attempts using known techniques such as evaporating prefabricated alloys of GeAu resulted in films with fluctuating Au concentrations and hence inconsistent temperature responses. We have developed methods by which a desired temperature characteristic can be reproducibly achieved. The fabrication procedures were performed at room temperature in a vacuum chamber. A tungsten evaporation basket was loaded with 0.5g Au stopped by 1.5g Ge. With the shutter closed, this load was slowly heated to a melt, to prevent splashing, and an evaporation of 1500 Å was performed at 40 Å /sec. Five subsequent and similar depositions onto different glass substrates followed. Resistivity measurements after each evaporation, checked that the resistance increased predictably. A test slide and the sample was then evaporated on during the sixth or seventh cycle depending on the preselected operational temperature range of the bolometer. Sixth for temperatures above 1.5K and the seventh below 1.5 to 0.3 K. The resistances of the bolometer were reduced by interdigitated Au electrodes fabricated by photolithography onto our samples prior to the above procedures. A 50 Å Al film under the electrodes improved adherence. Heat treating the samples at 80 ° C for about 30 to 45 minutes, was done when necessary.

Results: Plots of the resistance per square as a function of evaporation cycle for four different runs are shown in Fig. 1a. It is evident that the resistivity for all runs

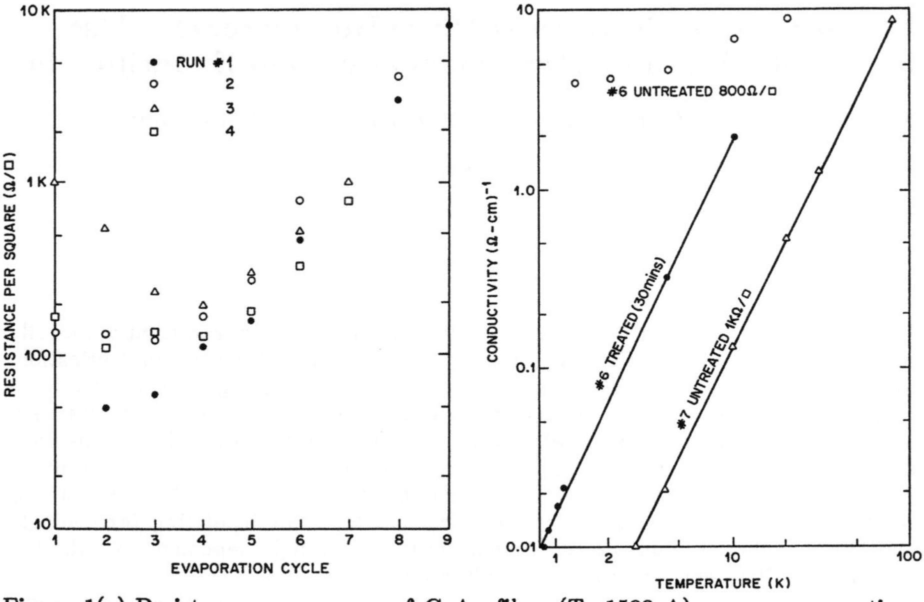

Figure 1(a) Resistance per square, of GeAu films (T=1500 A), versus evaporation cycle for 4 different fabrication runs. Figure 1(b) Conductivity versus Temperature for samples #6 and #7 untreated, and for sample #6 after annealing for 30 minutes @ 80 ° C.

converges by the fourth cycle and its magnitude predictably increases on further successive evaporations. The initial deviation on the third run was a result of the load falling out of the boat but the above observations were still valid.

Conductivity versus temperature plots are shown in Fig. 1b for films #6 and #7. No low temperature sensitivity was observed for the untreated sample #6 or earlier films. The exact sample, annealed for 30 minutes at 80 ° C, is shown to have a similar dR/dT as that of sample #7 which had a resistance of 1 k-ohm per square at 300 K. This ideal resistivity resulted in films having the largest operational temperature range, the least magnetic field susceptibility and tolerable final resistance. These inherently high impedances were further reduced by factors of 10^5 by predeposited photolithographic 2.5μ Au interdigitated electrodes. With these procedures, we have fabricated films having typical temperature sensitivities of 6 ohms/mK for a reduced final resistance of 66 ohms.

Incomplete photolithograpy sometimes yielded bolometers with R >500 ohms. For fast time response in these cases, we have utilized commercial 3n158 FETS and PMODFETS in source follower configurations at low temperatures to impedance match into 50 ohm coaxial lines. The FETS were used for temperatures above 1K and in zero magnetic field due to typical dissipation of 30mW and a magnetic casing. For T < 1K and in magnetic fields, PMODFETS were used as typical dissipation was a mere 5μW and an absence of a casing allows proper orientation for minimum magnetic field susceptibility.

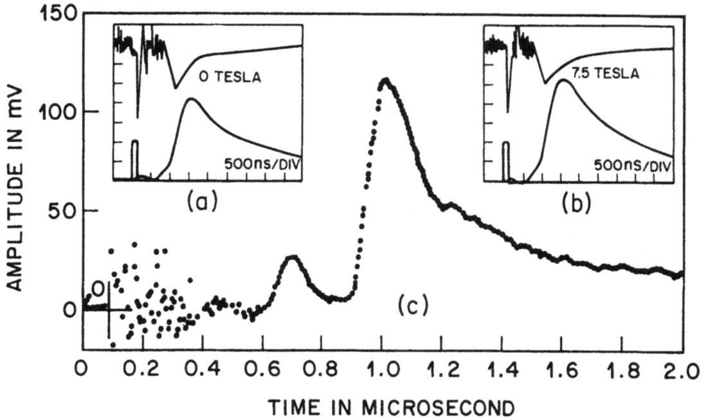

Figure 2. Phonon pulses detected with GeAu bolometers. Fig. 2 a and b show signals using the PMODFET in magnetic fields of zero or 10 T respectively. Fig. 2c shows an optimized signal of the same bolometer.

A fast ballistic phonon high magnetic field bolometer, fabricated as above on a 3mm InP substrate, was used at 1.2K. Shown in Fig. 2 a,b are phonon pulses obtained in conjunction with a PMODFET in magnetic fields of zero and 10T respectively. The input to the PMODFET (lower trace) was monitored by a 1M ohm differential amplifier with a 1Mhz bandwidth. The output is shown in the upper traces. These results demonstrate that (1)GeAu bolometers are sensitive phonon detectors even in high magnetic fields, and (2)the use of impedance matching PMODFETS for time resolved high impedance experiments-especially in high magnetic fields and at very low temperatures. In this case the PMODFET was a necessity for signal detection. The time response was limited by the excess capacitance of the probes used for characterization. This same bolometer, coupled to the FET, without the excess probes, resulted in the phonons shown in Fig. 2c. Clearly two modes are resolvable and the risetime of 100ns was limited by the experiment. An input energy of 8 nanojoules was detectable even without signal averaging.

Conclusion: We have clearly demonstrated, for the first time, that GeAu films can be used as fast sensitive heat pulse bolometers especially at very high magnetic fields. The empirical fabrication methods for these GeAu films allows us to generate reproducibly bolometers and in addition, fabrication is done on room temperature substrates. Swept pulsed phonon-high magnetic field measurements can now be performed directly using these bolometers or in conjunction with PMODFETS. In addition these FETS can be utilized for 2DEG, 2DHG experiments in which the impedances are similar to our GeAu bolometers.

REFERENCES

1. B. W. Dodson, W. L. McMillan, J. M. Mochel, R. C. Dynes, Phys. Rev. Lett. *46,* 46 (1980).
2. H. L. Stormer, K. Baldwin, A. C. Gossard, W. Wiegmann, Appl. Phys. Lett. *44,* 1062 (1984).

New Superconducting Tunneling Spectroscopy and Its Applications to Studies of Nonequilibrium Phonons from Superconductors and Semiconductors

I. Iguchi and Y. Kasai

Institute of Materials Science, University of Tsukuba,
Sakura-mura, Ibaraki 305, Japan

There has been a great deal written about detecting ultrahigh-frequency phonons utilizing a superconducting tunnel junction, as originated from the pioneering work of Eisenmenger and Dayem [1]. In this discussion, we report two new phonon spectrocopies utilizing a tunnel junction, one is related to energy spectroscopy and the other to photoexcited spectroscopy. Since these two spectroscopies are qualitatively different, we describe them separately in the following way.

1. Quasiparticle-Injected Gap-Tunable Spectroscopy

The energy distribution of nonequilibrium phonons has been so far investigated by a few methods such as represented by the stress-tunable spectroscopy [2] or Bremsstrahlung spectroscopy [3]. Our new method, in contrast to the above techniques, makes use of the effect of gap reduction in detector films by quasiparticle injection using a double-tunnel-junction consisting of a detector junction and an injector junction. By changing the bias current of the injector junction, the detector gap could be continuously tunable from its equilibrium value $2\Delta_{eq}$ down to about 40% of $2\Delta_{eq}$. Below this value, the gap structure was smeared out, resulting in loss of sensitivity.

Fig. 1 shows an example of the I-V characteristics of a detector junction for various injection currents. The spectroscopic measurements were performed by recording the external phonon signal for different injection currents with the detector junction biased at gap voltage. The gap reduction $\delta\Delta_s^*$ due to external phonon injection is approximately given by, in case of ballistic phonon propagation,

$$\delta\Delta_s^* \propto S^* \int_{2\Delta_s^*}^{\infty} [n(\omega) - n(\omega, T)]\, \omega^2 d\omega$$

where $n(\omega)$ and $n(\omega, T)$ are the nonequilibrium and thermal equilibrium phonon distribution respectively [4]. S^* is the sensor sensitivity given by $\tau_{es}\langle\tau_R\rangle/N(0)\langle\tau_B\rangle$, where $\langle\tau_R\rangle$ is the averaged quasiparticle recombination time, $\langle\tau_B\rangle$ is the averaged phonon pair-breaking time and τ_{es} is the phonon escape time [5]. Therefore the differential characteristic of the observed data as a function of the nonequilibrium gap $2\Delta_s^*$ directly gives the information on $n(\omega)$ in the stationary state.

Fig. 2 shows an example of the calculated results for gap reduction for various phonon distributions such as realized in the modified T^* model [6] in a superconductor when the generator gap $2\Delta_g$ is equal to $1.22\Delta_{eq}$. The rapid decrease just below $2\Delta_{eq}$ was due to the effect of quasiparticle recombination. A slope discontinuity at $\Delta_s^* = \Delta_g$ is expected for a $2\Delta_g$-peaked phonon distribution.

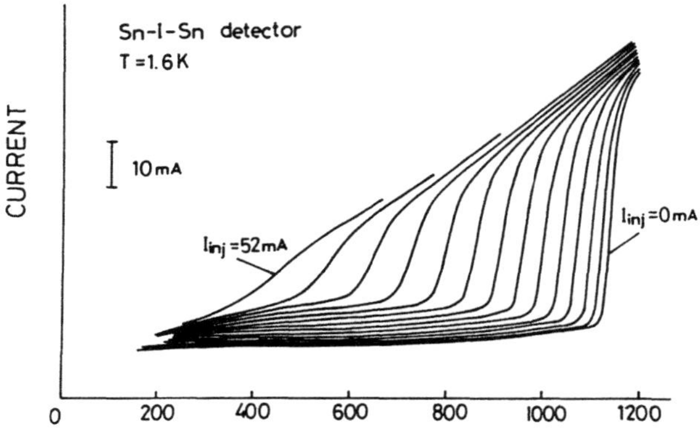

Fig. 1. Current-voltage characteristics of a detector junction for different injection currents ranging from 0mA to 52mA with 4mA interval for a Sn-I-Sn double-tunnel-junction configuration.

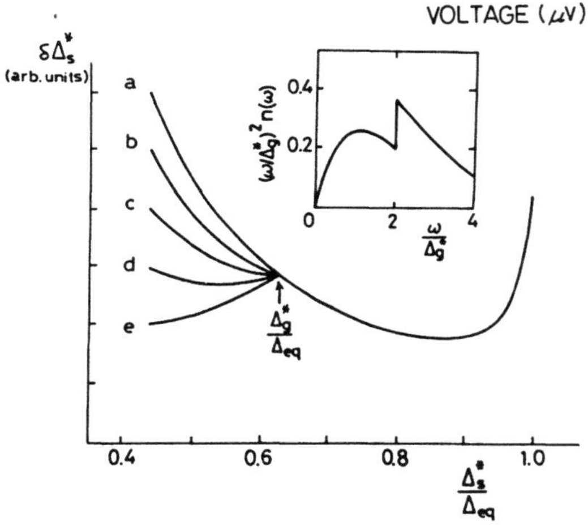

Fig. 2. Calculated results for gap reduction as a function of nonequilibrium gap parameter for various phonon distributions. Curve a represents a Planck distribution. The inset shows the phonon distribution corresponding to curve c.

In principle, this technique might be applicable to all phonon detection systems, including the conventional generator - single crystal - detector geometry and the generator - helium layer or thin film - detector geometry provided that the phonons escaping from the injector junction do not perturb the sample system significantly.

2. Superconducting Photo-relaxation Spectroscopy

This spectroscopy provides a new technique for investigating the gap structures of the condensed matter, particularly, those of semiconductors including deep levels by detecting the phonons emitted through nonradiative transition of optically illuminated materials utilizing a superconducting tunnel junction. As for phonon detection, the so-called "photoacoustic spectroscopy (PAS)" has been developed so far [7]. This method is, however, not so powerful because the detection sensitivity is not high enough and it detects phonons rather indirectly by a piezoelectric transducer or a microphone. In fact, sufficient information on deep levels of semiconductors such as GaAs has not been obtained by this technique. It is proved that our method provides not only direct detection of phonons but also is highly sensitive.

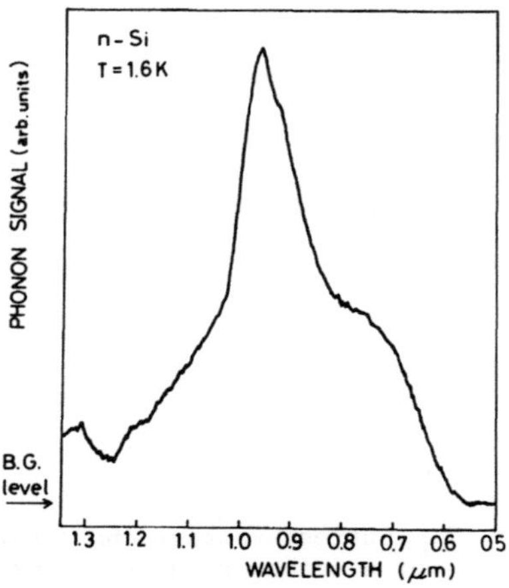

Fig. 3. Phonon signal as a function of wavelength of light for n-Si. The chopper frequency was 300Hz and two different filters were used below and above 0.7μm. The presence of the bandgap near 1μm is clear. The dip around =1.25μm was due to absorption of light by the fiber. Note that the light intensity from a monochromator was not normalized.

The sample was prepared first by evaporating an aluminum film of 300nm thick onto one side of a polished wafer, then depositing a silicon nitride film of ≃400nm thick, finally fabricating a Sn-I-Sn tunnel junction on it. Light from W lamp passed through a monochromator and a chopper and then was focused by a concave mirror into an optical fiber. The fiber entered a cryostat and illuminated the sample surface. The phonons produced by non-radiative processes propagate and contribute to pair-breaking in a superconducting film, being detected by a tunnel junction biased at a certain voltage using a lock-in amplifier. The observed phonon signal ranged from of the order of 10nV to 100nV, mainly due to the mismatching of light at the fiber face and use of a junction of not highest quality. Noise level of the system was, however, only a few hundred picovolts owing to operation in superfluid helium.

Fig. 3 shows the detected phonon signal as a function of wavelength of light for a n-Si wafer. The phonon signal did not change with the chopper frequency (100Hz- 1kHz), in contrast to PAS technique. It is concluded that almost no structure in the deep level region was seen for n-Si. On the contrary, the results for GaAs exhibited significant structures below the gap energy for which the analysis is going on in light of the deep levels observed previously.

References

1. W. Eisenmenger and A.H. Dayem: Phys. Rev. Lett. 18, 125 (1967).
2. R.C. Dynes, V. Narayanamurti and M.A. Chin: Phys. Rev. Lett. 4, 181 (1971).
3. H. Kinder: Phys. Rev. Lett. 28, 1564 (1972).
4. I. Iguchi, Y. Kasai and Y. Suzuki: Phys. Rev. B33, 4574 (1986).
5. I. Iguchi: J. Appl. Phys. 59, 533 (1986).
6. J.-J. Chang, W.Y. Lai and D.J. Scalapino: Phys. Rev. B20, 2739 (1979).
7. See for example, K. Wasa, K. Tsubouchi and N. Mikoshiba: Jpn. J. Appl. Phys. 19, L475 (1980); 19, L653 (1980).

Quantum Hall Effect Experiments Using Ballistic Phonons

*J.P. Eisenstein, V. Narayanamurti, H.L. Stormer, A.Y. Cho, and J.C.M. Hwang**

AT & T Bell Laboratories, Murray Hill, NJ 07974, USA

The intense interest over the last few years in the quantum Hall effect[1], occuring in the 2-D electron gas (2DEG), has been sustained largely by electronic transport measurements. For example, the prediction of a gap in the spectrum of elementary excitations of the fractionally quantized 2DEG has been tested only via observation of activated conductivity. No spectroscopic measurement of the gap has been made. We outline here a new phonon spectroscopy technique for the study of the 2-D electron gas and present data on the integrally quantized 2DEG.

Our technique involves the generation of short (~100ns) phonon pulses with a thin-film gold heater, their propagation through 1.31mm of superfluid helium at T<0.2K, followed by transmission across the surface of a MBE-grown GaAs/AlGaAs heterostructure containing a 2DEG. A magnetic field, typically 1-10T perpendicular to the 2DEG plane, biases the system into the quantum Hall effect regime. The phonons are detected by their effect on the conductivity of the 2DEG; see Fig. 1. By changing the hydrostatic pressure of the helium a spectral sweep through the heat pulse distribution occurs and the resulting changes in the detected phonon signal are recorded.

It is now well established[2] that the phonon dispersion in liquid ^4He at small wavevector curves slightly upward before finally bending down to form the roton minimum. Upward dispersion allows 3-phonon events whereas at higher energy the onset of normal dispersion cuts this process off. For low energy phonons the mean free path is determined by these 3-phonon processes. The quasi-blackbody pulse of phonons generated by the heater will rapidly split into two separate groups as it propagates. Phonons with energy in excess of the cutoff E_c will travel ballistically while those below will rapidly down-convert, leaving a gap in the distribution. The spectrum of down-converted phonons is difficult to assess but it is unlikely that

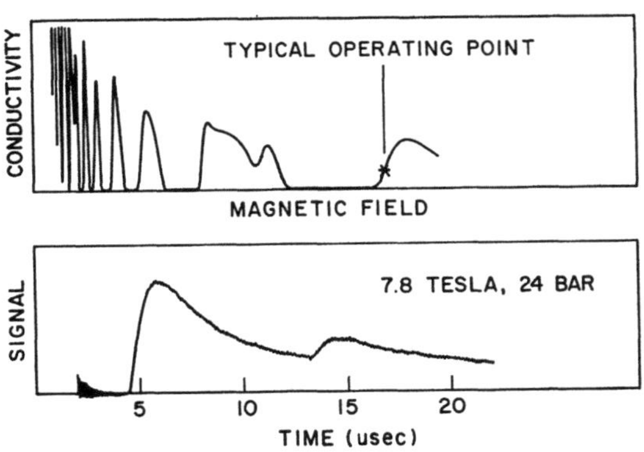

Fig. 1. Upper panel, Corbino conductivity vs. field for 2DEG. Lower panel, typical detected phonon signal, main pulse and echo.

phonons with energy above about 1K reach the detector. The cutoff energy drops from 9.85K at zero pressure to zero at about 20 bar[3]. Varying the pressure shifts spectral weight between these two groups of phonons allowing study of the spectral response of the detector.

We must assume that the phonon distribution arriving at the heterostructure is not radically altered by transmission across the interface or by propagation through the layered structure itself. To check this an aluminum tunnel junction detector was evaporated onto the backside of the heterostructure. Two nearly identical heaters were used; one facing the front surface of the heterostructure and the other the back. Thus phonon pulses could be made to impinge directly upon the tunnel junction or only after crossing the GaAs-helium interface and travelling 0.25mm through the sample. In this experiment the 2DEG, about 1000Å below the front surface, was not used and only a small 20 gauss field was applied. In Fig. 2 the detected pulse amplitude versus pressure for both paths are shown. The tunnel junction only detects phonons with energy greater than the 2 Δ energy gap, 4.1K for this device. Those which have down-converted via 3-phonon decays are not seen. The observed signal represents those phonons with energy above both Ec and the 4.1K gap. We estimate the heat pulse temperature to be about 1K indicating that at zero pressure where E_c is 9.85K there should be very little signal. As the pressure is increased and E_c falls, the signal should begin to rise. Around 14 bar E_c becomes equal to 4.1K and the signal should stop rising. This is essentially what we see, for the phonons directly incident upon the tunnel junction as well as those which cross the GaAs-^4He interface. Aside from showing the filtering provided by the helium, we also see that a significant number of phonons with energy greater than 4K are generated by the heaters and that they are not noticeably degraded on crossing the interface or substrate.

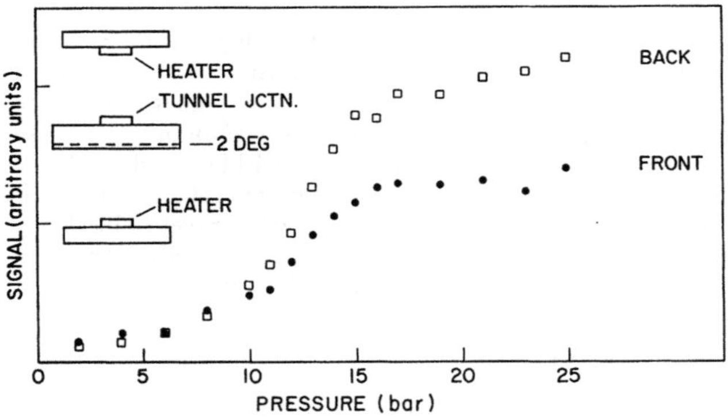

Fig. 2. Pulse amplitude vs. pressure using tunnel junction. Inset depicts experimental arrangement.

Fig. 3 shows the pressure dependence of the phonon pulse amplitude as observed using the 2DEG detector. This sample has a 2D carrier density of $1.5x10^{11}$ cm^{-2} and a mobility of $40m^2/V/s$. Since we are using a Corbino geometry it is the diagonal conductivity σ_{xx} that is measured. The magnetic fields shown bias the 2DEG onto steeply sloping flanks of the deep conductivity minima characteristic of the quantum Hall effect. At these fields the fermi level E_f should

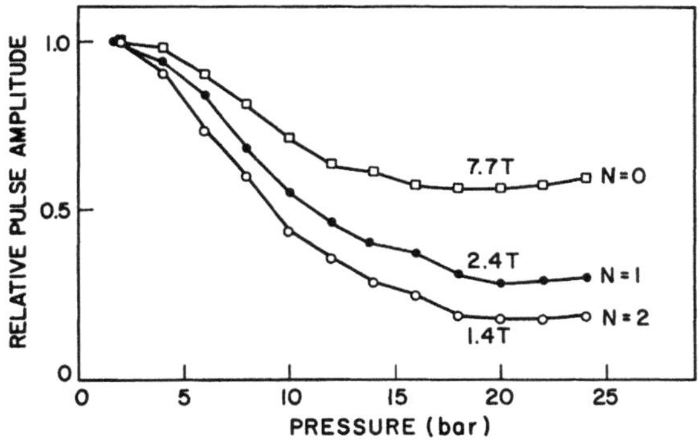

Fig. 3. Pulse amplitude vs. pressure using 2DEG detector at various magnetic fields. Data is normalized by value at 2 bar.

be near the center of a given Landau level spin subband. The low phonon energies generated excite intra-level transitions only. At all fields studied we observe strong signals at low pressures. This implies that the 2DEG, unlike the tunnel junction, is a good absorber of the low energy (<1K) phonons that dominate the pulses at these pressures. At higher pressures the detected pulse amplitudes depend strongly on magnetic field. At "low" magnetic fields, when E_f is above the ground (N=0, spin-split) Landau level, the observed signals smoothly decrease with increasing pressure. The magnitude of the drop becomes greater as the magnetic field is lowered and E_f passes to higher Landau levels. As shown in Fig. 3 at 1.4T, with E_f in the N=2 Landau level, the signals can drop as much as a factor of five. This drop is consistent with a reduced absorption cross-section for the higher energy phonons that are present in the heat pulses at high pressure. Matrix element arguments predict such an energy dependence; phonons with wavevector much larger than the inverse magnetic length ($l^{-1} = \sqrt{eB/\hbar}$) can not be absorbed. Estimates show this effect is important at the phonon energies involved here and can reproduce qualitatively the observed dependence on field and Landau level index.

At the highest fields (5-10T) with E_f in the ground N=0 Landau level the data is not reproducible from one cool-down to the next. Aside from strong signals at low pressure they are also significant at high pressure. While a gentle drop, similar to but less steep than at lower magnetic fields, is shown in Fig. 3, the signals have also been seen to rise above about 12 bar and reach amplitudes comparable to low pressure. Apparently there is greater sensitivity to high energy phonons at high fields in the ground Landau level. The irreproducibility of the pressure dependence is as yet unexplained.

In summary, we have described a novel technique for phonon spectroscopy of the 2DEG. The technique has been tested using a tunnel junction quantum detector. Our studies of the 2DEG have revealed a significant absorptivity for low energy phonons which is dependent on magnetic field and Landau level index.

* Present address: Gain Electronics Corp., P.O. Box 921, Somerville, NJ 08876

1. For a recent review see H.L. Stormer, Festkorperprobleme, Adv. in Sol. State Phys. XXIV, 25 (1983) ed. P. Grosse, Vieweg, Braunschweig
2. H.J. Maris, Rev. Mod. Phys. *49,* 341 (1977)
3. R. C. Dynes and V. Narayanamurti, Phys. Rev. B, *12,* 1720 (1975)

Design of Phonon Detectors for Neutrinos

H.J. Maris

Department of Physics, Brown University, Providence, RI 02912, USA

The development of detectors for very low energy neutrinos is one of the more challenging problems in experimental high-energy physics. Important topics in neutrino physics include the neutrino spectrum of the sun (the solar neutrino problem), and the related question of neutrino mass and oscillations [1]. For neutrinos of reactor energies or greater, conventional neutrino detectors rely principally on nuclear transmutations induced by the neutrinos. At low energies the cross-sections for these processes are extremely small. In the chlorine experiments of Davis et al [2], for example, the detector has a mass of 600 tons, but the solar neutrino flux (6×10^{10} cm^{-2} sec^{-1}) produces only of the order of one event per day.

Alternative detection schemes have recently been proposed in which the neutrino scatters elastically off a nucleus N or an electron in a crystal. The recoiling N or e^- produces ionization and some displacement of atoms, and phonons are generated. Drukier and Stodolsky [3] proposed a detector made of a large number of small grains of a superconductor. If a ν was scattered in a grain, the grain would be heated sufficiently to bring it into the normal state. This transition would then be detected through the Meissner effect.

Another scheme has been proposed by Cabrera et al [4]. Each element of the detector is a 1 kg block of Si, immersed in the mixing chamber of a dilution refrigerator at a few mK. If it is assumed that the specific heat of Si in this temperature range is solely due to phonons, the energy deposited by the recoiling e^- is sufficient to raise the temperature by a few mK. This would therefore be easily measurable. To have a reasonable detection rate from solar neutrinos one would need a large number of Si blocks, e.g., 1000. There are several difficulties with this experiment and these are connected with problems of current interest in phonon physics. After a ν is elastically scattered by an e^- (valence or core) in Si, the recoiling e^- of energy a few hundred keV makes a track of length of the order of 1 mm. Along this track electrons and holes relax rapidly to the band edge (energy-loss rate of order eV.psec^{-1}) and emit high-energy phonons. These phonons undergo rapid anharmonic decays, initially with a lifetime of the order of 1 psec. Once they have decayed to the point that their energy E is much less than θ_D, their anharmonic decay time τ_A can be calculated from the known 3rd order elastic constants. The result is [5]

$$\tau_A = a\,E^{-5}\ \text{secs} \tag{1}$$

where E is in K, and a is between 1000 and 2000 depending on direction. This rate is for L-phonons splitting into L+T or T+T. The fast transverse phonons T1 have a similar lifetime, but the slow transverse phonons T2 are stable against anharmonic decay. T2 phonons do not get trapped at high energy because isotope scattering also occurs, and this can change their polarization into L or T1. It is straightforward to show that the isotope scattering lifetime is

$$\tau_I = 2.2\ E^{-4} \tag{2}$$

In each anharmonic decay two phonons of energy roughly half the energy of the decaying phonons are produced. Thus, for these new phonons τ_A and τ_I are larger by factors of 2^5 and 2^4, respectively.

The difficulties with the Si calorimeter experiment include:
1) It may be impossible to prepare Si crystals of sufficiently high quality that the heat capacity at a few mK will still be dominated by phonons. There may be contributions from impurities, TLS, and the surface.
2) A 1 kg block of Si is a cube of side 7.5 cm. The ν scatters at a random point inside the block. Thus, the phonons travel, on the average, ~ 4 cm before they reach the block surface. From (1) and (2) one can estimate that the phonon energy is typically 20 to 40K after travelling this distance. Phonons of this energy have a high probability [6] (0.2-0.7) of transmission into the liquid helium, and will therefore not contribute to the heating of the Si. One could try to reduce the phonon transmission by laser annealing [7], or the Si could perhaps be cooled without helium contact. But in any event, anharmonic processes clearly cannot lead to a Planck phonon distribution in the mK range. For example, Eq. (1) gives $\tau_A \cong 10^{13}$ secs for a 10 mK phonon. One might be able to decrease the thermalization time by the addition of a metal film on the Si surface, or by suitable impurities in the bulk, but this would most likely increase the overall specific heat.

These difficulties suggest that a ballistic-phonon experiment might be more promising. In such an experiment the amplitude received at a bolometer will depend on the distance it is from the ν event, and also on the direction because of phonon focusing. To determine the energy deposited in the crystal it is therefore necessary to have several bolometers so that the event position can be located from the differences in arrival times. Note that the time the event occurs is not known <u>a priori</u>. We discuss the design of this experiment assuming that the phonons are detected by conventional superconducting bolometers. (This leaves open the possibility that some other device, such as a tunnel junction, may be more sensitive). At 0.1K a metal film bolometer on Si has a thermal time constant $\tau_B \gtrsim 1$ μsecs, even if the film is only 500 Å thick and acts as a perfect phonon radiator. Thus, $\tau_B \gtrsim \tau_E$, where τ_E is the time over which the ballistic phonons arrive. The bolometer then acts as an <u>integrator</u> of phonon flux and the incoming phonons produce $\delta T \propto T^{-1}$, because of the linear specific heat of the bolometer. For a given resistance R and area A the maximum bias current I is limited by self-heating of the bolometer. Since the power radiated goes as T^4 (assuming perfect phonon emissivity [8], $I \propto T^2$. Thus, the voltage output δV is

$$\delta V = I \frac{dR}{dT}\, \delta T \propto \beta T^{o} \tag{3}$$

where $\beta \equiv d\,\ell n\, R/d\,\ell n\, T$. In general, it is harder to make bolometers with large β at low T, so clearly one does not gain in sensitivity by going below 0.1K. Thus, an important problem is the development of a superconducting bolometer of high sensitivity with transition around 0.1K. The bolometer needs to be thick enough to absorb efficiently the incident phonons of energy $\sim$ 30K, and should be acoustically well-matched to Si. For a 1 cm^2 bolometer of resistance 50 Ω and $\beta = 100$, one can estimate that 500 keV energy deposited in the crystal gives a voltage output of the order to 50 μV, thus indicating that the experiment is feasible. It is clear that the development of low-temperature bolometers with large β, or alternative high-sensitivity detectors [9], is crucial for the success of the experiment.

If neutrinos can be detected in this way, the next step would be to find the direction of the ν flux. Information about this can be obtained from a study of the distribution of direction of the tracks of the recoiling electrons or nuclei. To find the track direction one may be able to use a large number of detectors distributed over the surface of the crystal, and compare the arrival times at these detectors. To tell that the phonons come from an extended source (the track is of length ∿ 1 mm) rather than a point will require measurement of arrival times to ∿ 0.01 μsec. (For nuclei the recoil distance is so short that the method is impractical). An alternative method has been proposed by Cabrera, Martoff, and Neuhauser [10]. They propose to use a very large number of detectors. A point source should give the usual phonon focusing pattern on each face of the Si cube. (We assume the cube has faces perpendicular to the cubic axes). The center of the pattern is the projection of the event location onto the face. The fact that the source is a track, rather than a point, should cause an analysable change in the phonon intensity pattern.

These ideas have a problem in common. The phonons which reach the detectors will have travelled a distance of about L/2 (L = side of cube). Thus, their mean-free-path Λ_1 is at least this distance. The previous generation phonons had energy twice as large, and had a Λ_2 smaller by 2^4 or 2^5, i.e., roughly $L/2^4$ or $L/2^5$. Thus, these phonons had travelled at least this distance in a random direction [11] from the track before being scattered into the direction of the detector. The effect is to smear out the effective source over a distance of $\sim \Lambda_2$, and effectively make it impossible to determine the direction of the track, whether by time-of-flight or focusing pattern methods. A quantitative estimate of this effect requires an estimate of Λ_2 averaged over all phonons which reach the detector (some of which have $\Lambda_1 \gg L/2$), and averaged over scattering angles, etc. This same effect should also cause a rounding of cusp edges in focusing patterns observed in conventional experiments. This rounding should occur when the excitation method generates high-energy phonons (e.g., by direct laser absorption in a semiconductor), but not when the source is a metal film heated to just a few K. High-resolution studies of phonon focusing would thus be of great value. They would determine the amount of the rounding, and would thus indicate the feasibility of experiments to determine the direction of the electron track.

I should like to thank R. E. Lanou, G. M. Seidel, B. Cabrera, and J. P. Wolfe for helpful discussions. This work was supported in part by NSF Grant No. DMR-8304224.

1. Solar Neutrinos and Neutrino Astronomy, ed. by M. L. Cherry, K. Lande, and W. A. Fowler: (Am. Inst. Phys., New York, 1985)
2. See ref. 1, p. 1 and p. 22
3. A. Drukier and L. Stodolsky: Phys. Rev. D30, 2295 (1984)
4. B. Cabrera, L. M. Krauss, and F. Wilczek: Phys. Rev. Lett. 55, 25 (1985)
5. A. Berke, A. P. Mayer, and R. K. Wehner: Sol. St. Comm. 54, 395 (1985)
6. E. S. Sabisky and C. H. Anderson, Sol. St. Comm. 17, 1095 (1975); C.-J. Guo and H. J. Maris: Phys. Rev. Lett. 29, 855 (1972)
7. H. C. Basso, W. Dietsche, H. Kinder, and P. Leiderer: in Phonon Scattering in Condensed Matter, edited by W. Eismenger, K. Lassman, and S. Dottinger (Springer, Berlin, 1984), p. 212.
8. The phonon emissivity should decrease as T goes down, causing I to vary more rapidly than T^2.
9. S. H. Moseley, J. C. Mather, and D. McCammon: J. Appl. Phys. 56, 1257 (1984); D. McCammon, S. H. Moseley, J. C. Mather, and R. F. Mushotzky: J. Appl. Phys. 56, 1263 (1984)
10. B. Cabrera, J. Martoff, and B. Neuhauser: preprint
11. For a related discussion, see R. S. Markiewicz: Phys. Rev. B21, 4674 (1980)

Index of Contributors

[illegible] Other topical areas of work [illegible] include the stability and chaotic behaviour [illegible], the quantum theory of [illegible] tests of general relativity, and the [illegible] governing the [illegible]

[illegible] and Fabrication of Micro-Structures and Microdevices

[illegible] Les Houches, France [illegible] 1985

[illegible] Kelly

[illegible] figures. Approx. 450 pages

ISBN [illegible]

[illegible] Proceedings of the 1985 Les Houches [illegible] and Fabrication of Microstructures and Microdevices [illegible] a wide range [illegible] physics [illegible] of microstructures [illegible] to the fabrication of micro-devices.

[illegible]

Biophysical Effects of Steady Magnetic Fields

[illegible] Les Houches, France [illegible] 1986

[illegible]

[illegible] ISBN [illegible]

[illegible]

Magnetic Properties of [illegible]

[illegible] Dimensional Systems

[illegible] Magnetic Properties of Low-Dimensional Systems, [illegible], Mexico, [illegible]

[illegible]

[illegible] ISBN [illegible]

Springer-Verlag
Berlin Heidelberg New York
London Paris Tokyo

Springer Series in Solid-State Sciences

Editors: M. Cardona P. Fulde H.-J. Queisser

Springer Series in Solid-State Sciences

Editors: M. Cardona P. Fulde H.-J. Queisser

1 Principles of Magnetic Resonance 2nd Edition 2nd Printing By C. P. Slichter
2 Introduction to Solid-State Theory By O. Madelung
3 Dynamical Scattering of X-Rays in Crystals By Z. G. Pinsker
4 Inelastic Electron Tunneling Spectroscopy Editor: T. Wolfram
5 Fundamentals of Crystal Growth I Macroscopic Equilibrium and Transport Concepts 2nd Printing By F. Rosenberger
6 Magnetic Flux Structures in Superconductors By R. P. Huebener
7 Green's Functions in Quantum Physics By E. N. Economou
8 Solitons and Condensed Matter Physics Editors: A. R. Bishop and T. Schneider
9 Photoferroelectrics By V. M. Fridkin
10 Phonon Dispersion Relations in Insulators By H. Bilz and W. Kress
11 Electron Transport in Compound Semiconductors By B. R. Nag
12 The Physics of Elementary Excitations By S. Nakajima, Y. Toyozawa, and R. Abe
13 The Physics of Selenium and Tellurium Editors: E. Gerlach and P. Grosse
14 Magnetic Bubble Technology 2nd Edition By A. H. Eschenfelder
15 Modern Crystallography I Symmetry of Crystals, Methods of Structural Crystallography By B. K. Vainshtein
16 Organic Molecular Crystals Their Electronic States By E. A. Silinsh
17 The Theory of Magnetism I Statics and Dynamics By D. C. Mattis
18 Relaxation of Elementary Excitations Editors: R. Kubo and E. Hanamura
19 Solitons Mathematical Methods for Physicists By G. Eilenberger
20 Theory of Nonlinear Lattices By M. Toda
21 Modern Crystallography II Structure of Crystals By B. K. Vainshtein, V. M. Fridkin, and V. L. Indenbom
22 Point Defects in Semiconductors I Theoretical Aspects By M. Lannoo and J. Bourgoin
23 Physics in One Dimension Editors: J. Bernasconi, T. Schneider
24 Physics in High Magnetic Fields Editors: S. Chikazumi and N. Miura
25 Fundamental Physics of Amorphous Semiconductors Editor: F. Yonezawa
26 Elastic Media with Microstructure I One-Dimensional Models By I. A. Kunin
27 Superconductivity of Transition Metals Their Alloys and Compounds By S. V. Vonsovsky, Yu. A. Izyumov, and E. Z. Kurmaev
28 The Structure and Properties of Matter Editor: T. Matsubara
29 Electron Correlation and Magnetism in Narrow-Band Systems Editor: T. Moriya
30 Statistical Physics I By M. Toda, R. Kubo, N. Saito
31 Statistical Physics II By R. Kubo, M. Toda, N. Hashitsume
32 Quantum Theory of Magnetism By R. M. White
33 Mixed Crystals By A. I. Kitaigorodsky
34 Phonons: Theory and Experiments I Lattice Dynamics and Models of Interatomic Forces By P. Brüesch
35 Point Defects in Semiconductors II Experimental Aspects By J. Bourgoin and M. Lannoo
36 Modern Crystallography III Crystal Growth By A. A. Chernov
37 Modern Crystallography IV Physical Properties of Crystals By L. A. Shuvalov
38 Physics of Intercalation Compounds Editors: L. Pietronero and E. Tosatti
39 Anderson Localization Editors: Y. Nagaoka and H. Fukuyama

MIX
Papier aus verantwortungsvollen Quellen
Paper from responsible sources
FSC® C105338

If you have any concerns about our products,
you can contact us on
ProductSafety@springernature.com

In case Publisher is established outside the EU,
the EU authorized representative is:
Springer Nature Customer Service Center GmbH
Europaplatz 3, 69115 Heidelberg, Germany

Printed by Libri Plureos GmbH
in Hamburg, Germany